Geometric Integrators for Differential Equations with Highly Oscillatory Solutions

Xinyuan Wu • Bin Wang

Geometric Integrators for Differential Equations with Highly Oscillatory Solutions

Science Press
Beijing

Springer

Xinyuan Wu
Department of Mathematics
Nanjing University
Nanjing, Jiangsu, China

Bin Wang
School of Mathematics and Statistics
Xi'an Jiaotong University
Xi'an, Shaanxi, China

ISBN 978-981-16-0149-1   ISBN 978-981-16-0147-7   (eBook)
https://doi.org/10.1007/978-981-16-0147-7

Jointly published with Science Press
The print edition is not for sale in China (Mainland). Customers from China (Mainland) please order the print book from: Science Press.

This Springer imprint is published by the registered company Springer Nature Singapore Pte Ltd.
The registered company address is: 152 Beach Road, #21-01/04 Gateway East, Singapore 189721, Singapore

Dedicated to the memory of Professor Feng Kang on the centenary of his birth.

The profound and seminal contributions of Professor Feng Kang, on symplectic geometric algorithms for Hamiltonian systems, have opened up a rich and new field of numerical mathematics research in China and throughout the world.

Foreword

The numerical integration of ordinary differential equations has a long and distinguished history inaugurated by Euler in the eighteenth century. The use of numerical methods was of course extremely limited before computers became available and it is only around 1955 that the subject took off. Important theoretical developments, which used very beautiful mathematics, were carried out by Lax (the celebrated equivalence theorem), Dahlquist (linear multistep methods) and Butcher (Runge–Kutta methods). The theory made it possible, starting in the 1960s, to write powerful general-purpose software that may be routinely used by practitioners to solve initial or boundary value problems. For the first time in history, it was possible for engineers and scientists facing a possibly very complicated, nonlinear differential equation to find its solutions in next to no time by plugging it into one of the general-purpose codes that became widely available. This was a revolution whose importance is easily overlooked by those that have grown up surrounded by computers and software. One of the strengths of the numerical differential equations solvers in software libraries is that, as I have mentioned, they are general purpose: they are carefully crafted black boxes that will deal with all initial value problems (to be more precise, one needs one black box for stiff problems and a second black box for nonstiff problems). Without questioning the strengths and validity of general-purpose numerical methods and software (that we may call "classical"), a very different new paradigm appeared in the 1980s. The new paradigm brought many innovative changes. In the classical approach what was demanded of the numerical method was to find accurate numerical value of the solution at any desired time. The new paradigm wanted to identify the long-time behaviour of the solutions or perhaps the existence of conservation laws or some other qualitative feature of the dynamics. The success of the classical software was based on a one-size-fits-all approach; in the new paradigm, the numerical method and the software were tailored to the application at hand, trying to capture as much as possible the structure of the problem being dealt with. In addition, the new paradigm often used mathematical techniques, mainly from differential geometry, that had not hitherto been connected with numerical integration. Finally, the new paradigm introduced or popularized new ways of analysing errors, notably the modified equation approach.

In 1996, a State of the Art in Numerical Analysis conference was held in York (England) with the purpose of analysing the main developments that had taken place in the field of numerical computation in the previous decade. I was asked to prepare a survey presentation on the area of ordinary differential equations, and for that talk and the subsequent publication, I chose the title "Geometric Integration", because I thought at the time that geometry was the unifying theme (or at least one of the unifying themes) in the new paradigm. The terminology caught on and it is now widely used in the literature. As discussed above, one of the salient features of geometric integration is that specific classes of problems now take centre stage. The first class to be considered was that of Hamiltonian problems, whose study was pioneered by Professor Feng Kang; we commemorate his birth centenary the year I wrote this foreword. Professor Feng Kang, who had made other important contributions to mathematics, was the most important figure in convincing the international community that, first, it is important for the applications to create numerical algorithms specifically designed to solve Hamiltonian problems and that, second, geometric notion of symplecticness is essential to build the new integrators—traditional ideas of stability and consistency were not sufficient. Professor Feng Kang's work was followed and is still being followed by many hundreds or perhaps thousands of publications in China and throughout the world. Another area that has kept growing in importance within geometric integration is the study of highly oscillatory problems: problems where the solutions are periodic or quasiperiodic and have to be studied in time intervals that include an extremely large number of periods. Examples abound in manifold applications: for instance, we may wish to ascertain the future evolution of the solar system in intervals of time where the planets have performed billions of revolutions. The authors of the present book, Professors Xinyuan Wu and Bin Wang, are among the scientists who, in the international scene, have contributed most to the development and analysis of geometric integrators for highly oscillatory differential equations, ordinary or partial. This book will no doubt be a valuable addition to a long list of publications that started with the seminal papers of Professor Feng Kang and his students.

President, Royal Academy of Sciences of Spain   J. M. Sanz-Serna
Excellence Chair in Applied Mathematics, Universidad
Carlos III de Madrid
SIAM Fellow
Fellow of the American Mathematical Society
Fellow of the Institute of Mathematics and its Applications

Preface

Differential equations that have highly oscillatory solutions arise in a variety of fields in science and engineering such as astrophysics, classical and quantum physics, and molecular dynamics, and their computation presents numerous challenges. As is known, these equations cannot be solved efficiently using conventional methods. Although notable progress has been made in numerical integrators for highly oscillatory differential equations, it is not obvious for these integrators what effects on their long-time behaviour are produced by preserving certain geometric properties. A further study of novel geometric integrators has become increasingly important in recent years. The objective of this book is to explore further geometric integrators for highly oscillatory problems that can be formulated as systems of ordinary and partial differential equations.

This book has grown out of recent research work published in professional journals by the research group of the authors. This book is divided into two parts. The first part deals with highly oscillatory systems of ordinary differential equations (ODEs), and the second part is concerned with time-integration of partial differential equations (PDEs) having oscillatory solutions.

The first part includes six chapters, dealing with highly oscillatory ODEs, and the second part consists of eight chapters, providing some novel insights into geometric integrators for PDEs. Chapter 1 is a review of oscillation-preserving integrators for systems of second-order ODEs with highly oscillatory solutions. As is known, continuous-stage Runge–Kutta–Nyström (RKN) methods have been developed for this class of problems. Chapter 2 proposes and derives continuous-stage extended Runge–Kutta–Nyström (ERKN) integrators for second-order ODEs with highly oscillatory solutions. Since stability and convergence are essential aspects of numerical analysis, we provide nonlinear stability and convergence analysis of ERKN integrators for second-order ODEs with highly oscillatory solutions in Chap. 3. Poisson systems occur very frequently in physics, so in Chap. 4, we investigate functionally fitted energy-preserving integrators for Poisson systems. We then consider exponential collocation methods for conservative or dissipative systems in Chap. 5. It is known that various dynamical systems including all Hamiltonian systems preserve volume in phase space, and hence we discuss volume-preserving exponential integrators for first-order ODEs in Chap. 6. Chapter 7 analyses global error bounds of one-stage ERKN integrators for semilinear wave equations. In Chap. 8, we derive linearly fitted conservative (dissipative) schemes for efficiently solving conservative (dissipative) nonlinear wave PDEs. Chapter 9 focuses on the formulation and analysis of energy-preserving schemes for solving high-dimensional nonlinear Klein–Gordon equations. In Chap. 10, we introduce symmetric and arbitrarily high-order Hermite–Birkhoff time integrators for solving nonlinear Klein–Gordon equations. Chapter 11 describes a symplectic approximation with nonlinear stability and convergence analysis for efficiently solving semilinear Klein–Gordon equations. Chapter 12 proposes and analyses a continuous-stage modified leap-frog scheme for high-dimensional semilinear Hamiltonian wave equations. Chapter 13 is concerned with semi-analytical exponential RKN integrators for efficiently solving high-dimensional nonlinear wave equations based on fast Fourier transform (FFT) techniques. Chapter 14 considers long-time momentum and actions behaviour of energy-preserving methods for wave equations.

The presentation in this book provides some new perspectives of the subject which is based on theoretical derivations and mathematical analysis, facing challenging scientific computational problems, and providing high-performance numerical simulations. In order to show the long-time numerical behaviour of the simulation, all the integrators presented in this book have been tested and verified on highly oscillatory systems from a wide range of applications in the field of science and engineering. They are more efficient than existing schemes in the literature for differential equations that have highly oscillatory solutions.

We take this opportunity to thank all colleagues and friends for their selfless help during the preparation of this book. Among them, we particularly express our heartfelt thanks to John Butcher of the University of Auckland, Christian Lubich of Universität Tübingen, Arieh Iserles of the University of Cambridge, J. M. Sanz-Serna of Universidad Carlos III de Madrid, and Reinout Quispel of La Trobe University for their encouragement.

The authors are also grateful to many colleagues and friends for reading the manuscript and for their valuable comments. Special thanks go to Robert Peng Kong Chan of the University of Auckland, Qin Sheng of Baylor University, Jichun Li of the University of Nevada Las Vegas, David McLaren of La Trobe University, Adrian Hill of the University of Bath, Xiaowen Chang of McGill University, Jianlin Xia of Purdue University and Marcus David Webb of the University of Manchester.

Thanks go as well to the following colleagues and friends for their help and support in various forms: Zuhe Shen, Jinxi Zhao, Yiqian Wang and Jiansheng Geng of Nanjing University; Fanwei Meng of Qufu Normal University; Yaolin Jiang and Jing Gao of Xi'an Jiaotong University; Yongzhong Song, Yushun Wang and Qikui Du of Nanjing Normal University; Chunwu Wang of Nanjing University of Aeronautics and Astronautics; Xinru Wang of Nanjing Medical University; Qiying Wang of the University of Sydney; Shixiao Wang of the University of Auckland; Robert Mclachlan of Massey University; Tianhai Tian of Monash University; Choi-Hong Lai of University of Greenwich; Jialin Hong, Zaijiu Shang, Yifa Tang and

Yajuan Sun of the Chinese Academy of Sciences; Yuhao Cong of Shanghai Customs College; Guangda Hu of Shanghai University; Zhizhong Sun and Hongwei Wu of Southeast University; Shoufo Li, Aiguo Xiao and Liping Wen of Xiangtan University; Chuanmiao Chen of Hunan Normal University; Siqing Gan and Xiaojie Wang of Central South University; Chengjian Zhang, Chengming Huang and Dongfang Li of Huazhong University of Science & Technology; Hongjiong Tian and Wansheng Wang of Shanghai Normal University; Yongkui Zou of Jilin University; Jingjun Zhao of Harbin Institute of Technology; Xiaofei Zhao and Jiwei Zhang of Wuhan University; Xiong You of Nanjing Agricultural University; Wei Shi of Nanjing Tech University, Qinghe Ming and Yonglei Fang of Zaozhuang University; Qinghong Li of Chuzhou University, Fan Yang, Xianyang Zeng and Hongli Yang of Nanjing Institute of Technology; Kai Liu of Nanjing University of Finance and Economics; Jiyong Li of Hebei Normal University; and Fazhan Geng of Changshu Institute of Technology.

We would like to thank Kai Hu and Ji Luo for their help with the editing, and the production team of Science Press and Springer-Verlag.

We are grateful to our family members for their love and support throughout all these years.

The work on this book was supported in part by the National Natural Science Foundation of China under Grant No. 11671200.

Nanjing, China   Xinyuan Wu
Xi'an, China   Bin Wang

Contents

Chapter 1
Oscillation-Preserving Integrators for Highly Oscillatory Systems of Second-Order ODEs

In this chapter, from the point of view of Geometric Integration, i.e. the numerical solution of differential equations using integrators that preserve as many as possible the geometric/physical properties of them, we first introduce the concept of oscillation preservation for Runge–Kutta–Nyström (RKN)-type methods and then analyse the oscillation-preserving behaviour of RKN-type methods in detail. This chapter is also accompanied by numerical experiments which show the importance of the oscillation-preserving property for a numerical method, and the remarkable superiority of oscillation-preserving integrators for solving nonlinear multi-frequency highly oscillatory systems.

1.1 Introduction

This chapter focuses on oscillation-preserving integrators for ordinary differential equations and time-integration of partial differential equations with highly oscillatory solutions. As is known, one of the most difficult problems in the numerical simulation of evolutionary problems is to deal with highly oscillatory problems, and here we refer to two important review articles on this subject by Petzold et al. [1] and Cohen et al. [2]. These type of problems occur in a variety of fields in science and engineering such as quantum physics, fluid dynamics, acoustics, celestial mechanics and molecular dynamics, including the semidiscretisation of nonlinear wave equations and Klein–Gordon (KG) equations. The computation of highly oscillatory problems contains numerous enduring challenges (see, e.g. [1–8]). It is important to note that standard methods need a very small stepsize and hence a long runtime to reach an acceptable accuracy for highly oscillatory differential equations.

In this chapter, we focus on the following initial value problem of nonlinear multi-frequency highly oscillatory second-order ordinary differential equations

$$
\begin{cases}
y'' + My = f(y, y'), & t \in [0, T], \\
y(0) = y_0, \quad y'(0) = y_0',
\end{cases}
\tag{1.1}
$$

where $y \in \mathbb{R}^d$, and $M \in \mathbb{R}^{d \times d}$ is a positive semi-definite matrix that implicitly contains the dominant frequencies of the highly oscillatory problem and $\|M\| \gg \max\left\{1, \left\|\dfrac{\partial f}{\partial y}\right\|\right\}$. In some applications, the dimension d of the matrix M refers to the number of degrees of freedom in the space semidiscretisation such as semilinear wave equations, and then $\|M\|$ will tend to infinity as finer space semi-discretisations are carried out. Among typical examples of this type are semi-discretised KG equations (see, e.g. [9–11]).

In the case where $M = 0$, (1.1) reduces to the conventional initial value problem of second-order differential equations

$$
\begin{cases}
y'' = f(y, y'), & t \in [0, T], \\
y(0) = y_0, \quad y'(0) = y_0'.
\end{cases}
\tag{1.2}
$$

As is known, the standard RKN methods (see [12]) are very popular for solving (1.2). However, it may be believed that the standard RKN methods were not initially designed for the nonlinear multi-frequency highly oscillatory system (1.1). The standard RKN methods, including symplectic and symmetric RKN methods may result in unfavorable numerical behaviour when applied to highly oscillatory systems (see, e.g. [10, 13]). As a result, various RKN-type methods for solving highly oscillatory differential equations have received a lot of attention (see, e.g. [1, 2, 6, 8, 11, 13–28]).

In designing numerical integrators for efficiently solving (1.1), the so-called matrix-variation-of-constants formula plays an important role, which is summarised as follows:

Theorem 1.1 (Wu et al. [25]) *If $M \in \mathbb{R}^{d \times d}$ is a positive semi-definite matrix and $f : \mathbb{R}^d \times \mathbb{R}^d \to \mathbb{R}^d$ in (1.1) is continuous, then the exact solution of (1.1) and its derivative satisfy the following formula*

$$
\begin{cases}
y(t) = \phi_0(t^2 M) y_0 + t\phi_1(t^2 M) y_0' + \displaystyle\int_0^t (t - \tau)\phi_1((t - \tau)^2 M) \hat{f}(\tau) d\tau, \\
y'(t) = -tM\phi_1(t^2 M) y_0 + \phi_0(t^2 M) y_0' + \displaystyle\int_0^t \phi_0((t - \tau)^2 M) \hat{f}(\tau) d\tau,
\end{cases}
\tag{1.3}
$$

for $t \in [0, T]$, where

$$\hat{f}(\tau) = f(y(\tau), y'(\tau))$$

and the matrix-valued functions $\phi_0(V)$ and $\phi_1(V)$ of $V \in \mathbb{R}^{d \times d}$ are defined by

$$\phi_i(V) = \sum_{k=0}^{\infty} \frac{(-1)^k V^k}{(2k+i)!}, \tag{1.4}$$

for $i = 0, 1, 2, \cdots$.

Remark 1.1 Actually, the matrix-variation-of-constants formula (1.3) provides an implicit expression of the solution of the nonlinear multi-frequency highly oscillatory system (1.1), which gives a valuable insight into the underlying highly oscillatory solution. The formula (1.3) also makes it possible to gain a new insight into the standard RKN methods for (1.2) (see Sect. 1.2 for details).

If $f(y, y') = 0$, (1.3) yields

$$\begin{cases} y(t) = \phi_0(t^2 M) y_0 + t\phi_1(t^2 M) y_0', \\ y'(t) = -tM\phi_1(t^2 M) y_0 + \phi_0(t^2 M) y_0', \end{cases} \tag{1.5}$$

which exactly solves the system of multi-frequency highly oscillatory linear homogeneous equations

$$\begin{cases} y'' + My = 0, \\ y(0) = y_0, \quad y'(0) = y_0', \end{cases} \tag{1.6}$$

associated with the nonlinear highly oscillatory system (1.1).

Assume that both $y(t_n)$ and $y'(t_n)$ at $t = t_n \in [0, T]$ are prescribed, it follows from the formula (1.3) that

$$\begin{cases} y(t_n + \mu h) = \phi_0(\mu^2 V) y(t_n) \\ \qquad + \mu h \phi_1(\mu^2 V) y'(t_n) + h^2 \int_0^{\mu} (\mu - \zeta) \phi_1((\mu - \zeta)^2 V) \hat{f}(t_n + \zeta h) \, d\zeta, \\ y'(t_n + \mu h) = -\mu h M \phi_1(\mu^2 V) y(t_n) \\ \qquad + \phi_0(\mu^2 V) y'(t_n) + h \int_0^{\mu} \phi_0((\mu - \zeta)^2 V) \hat{f}(t_n + \zeta h) \, d\zeta, \end{cases} \tag{1.7}$$

where $V = h^2 M$ and $0 < \mu \leqslant 1$. The special case where $\mu = 1$ in (1.7) gives

$$
\begin{cases}
y(t_n + h) = \phi_0(V)y(t_n) + h\phi_1(V)y'(t_n) + h^2 \int_0^1 (1 - z)\phi_1\big((1 - z)^2 V\big)\hat{f}(t_n + hz)dz, \\
\\
y'(t_n + h) = -hM\phi_1(V)y(t_n) + \phi_0(V)y'(t_n) + h \int_0^1 \phi_0\big((1 - z)^2 V\big)\hat{f}(t_n + hz)dz.
\end{cases}
$$
(1.8)

Remark 1.2 We here remark that since the formula (1.3) is an implicit expression of the solution of the nonlinear multi-frequency highly oscillatory system (1.1), the formula (1.7) with $0 < \mu < 1$ exposes the structure of the internal stages, and (1.8) expresses the structure of the updates in the design of an RKN-type integrator specially for solving the nonlinear multi-frequency highly oscillatory system (1.1).

In applications, an important special case of (1.1) is that the right-hand side function f does not depend on y', i.e.,

$$
\begin{cases}
y'' + My = f(y), & t \in [0, T], \\
y(0) = y_0, & y'(0) = y'_0.
\end{cases}
$$
(1.9)

The case where $M = 0$ in (1.9) gives

$$
\begin{cases}
y'' = f(y), & t \in [0, T], \\
y(0) = y_0, & y'(0) = y'_0.
\end{cases}
$$
(1.10)

Remark 1.3 Here it is important to realise that the matrix-variation-of-constants formula (1.3) is also valid for the nonlinear multi-frequency highly oscillatory system (1.9), and so are the formulae (1.5), (1.7) and (1.8), provided we replace $\hat{f}(\tau) = f\big(y(\tau), y'(\tau)\big)$ appearing in (1.3) with $\hat{f}(\tau) = \hat{f}\big(y(\tau)\big)$.

Obviously, the formula (1.5) implies that if $y_n = y(t_n)$ and $y'_n = y'(t_n)$, then we have

$$
\begin{cases}
y(t_n + c_i h) = \phi_0(c_i^2 V)y_n + c_i h\phi_1(c_i^2 V)y'_n, \\
y'(t_n + c_i h) = -c_i hM\phi_1(c_i^2 V)y_n + \phi_0(c_i^2 V)y'_n,
\end{cases}
$$
(1.11)

for any t_n, $t = t_n + c_i h \in [0, T]$, where $h > 0$ and $0 < c_i \leqslant 1$ for $i = 1, \cdots, s$.

In what follows, it is convenient to introduce the block vector which will be used in the analysis of oscillation preservation in Sect. 1.4:

$$
\hat{Y} = (\hat{Y}_1^\mathsf{T}, \cdots, \hat{Y}_s^\mathsf{T})^\mathsf{T},
$$
(1.12)

where

$$\hat{Y}_i = y(t_n + c_i h),$$

express the exact solutions to the multi-frequency highly oscillatory linear homogeneous equation (1.6) at $t = t_n + c_i h$ for $i = 1, \cdots, s$. It is clear from (1.11) and (1.12) that $\hat{Y}$ is a block vector, which can be expressed in the block-matrix notation with Kronecker products as

$$\hat{Y} = \phi_0(C^2 \otimes V)(e \otimes y_n) + h(C \otimes I_d)\phi_1(C^2 \otimes V)(e \otimes y_n'), \qquad (1.13)$$

where $e = (1, 1, \cdots, 1)^\mathsf{T}$ is an $s \times 1$ vector,

$$C = \mathrm{diag}(c_1, \cdots, c_s)$$

is an $s \times s$ diagonal matrix, and the block diagonal matrices are given by

$$\phi_0(C^2 \otimes V) = \mathrm{diag}(\phi_0(c_1^2 V), \cdots, \phi_0(c_s^2 V)),$$
$$(C \otimes I_d)\phi_1(C^2 \otimes V) = \mathrm{diag}(c_1\phi_1(c_1^2 V), \cdots, c_s\phi_1(c_s^2 V)).$$

If $t = t_n + h$, namely, $c_i = 1$, the formula (1.11) is identical to

$$\begin{cases} y(t_n + h) = \phi_0(V)y_n + h\phi_1(V)y_n', \\ y'(t_n + h) = -hM\phi_1(V)y_n + \phi_0(V)y_n'. \end{cases} \qquad (1.14)$$

Historically, the ARKN methods and ERKN integrators were successively proposed and investigated in order to solve the highly oscillatory system (1.1) and (1.9), respectively. Although both ARKN methods and ERKN integrators were proposed and developed from single frequency to multi-frequency oscillatory problems in chronological order, *throughout this chapter we are only interested in nonlinear multi-frequency highly oscillatory systems.*

1.2 Standard Runge–Kutta–Nyström Schemes from the Matrix-Variation-of-Constants Formula

It is interesting to point out that the formula (1.7) provides an enlightening approach to standard RKN methods for solving second-order initial value problems (1.2) numerically, although Nyström established them in 1925 (see Nyström [12]). To clarify this, using the matrix-variation-of-constants formula (1.7) with $M = 0$, we are easily led to the following formulae of integral equations for second-order initial value problems (1.2):

$$
\begin{cases}
y(t_n + \mu h) = y(t_n) + \mu h y'(t_n) + h^2 \displaystyle\int_0^\mu (\mu - z)\hat{f}(t_n + hz)\,\mathrm{d}z, \\[2ex]
y'(t_n + \mu h) = y'(t_n) + h \displaystyle\int_0^\mu \hat{f}(t_n + hz)\,\mathrm{d}z,
\end{cases}
\tag{1.15}
$$

for $0 < \mu < 1$, and

$$
\begin{cases}
y(t_n + h) = y(t_n) + h y'(t_n) + h^2 \displaystyle\int_0^1 (1 - z)\hat{f}(t_n + hz)\,\mathrm{d}z, \\[2ex]
y'(t_n + h) = y'(t_n) + h \displaystyle\int_0^1 \hat{f}(t_n + hz)\,\mathrm{d}z,
\end{cases}
\tag{1.16}
$$

for $\mu = 1$, where $\hat{f}(v) := f\big(y(v), y'(v)\big)$.

Clearly, the formulae (1.15) and (1.16) contain and generate the structure of the internal stages and updates of a Runge–Kutta-type integrator for solving (1.2), respectively. This indicates the standard RKN scheme in a quite simple and natural way compared with the original idea (with the block vector $(y^\mathsf{T}, y'^\mathsf{T})^\mathsf{T}$ regarded as the new variable, (1.2) can be transformed into a system of first-order differential equations of doubled dimension, and then we apply Runge–Kutta methods to the system of first-order differential equations, accompanying some simplifications). Approximating the integrals in (1.15) and (1.16) by using a suitable quadrature formula with nodes $c_1, \cdots, c_s$, we straightforwardly obtain the standard RKN methods (see Nyström [12]) as follows.

Definition 1.1 An s-stage RKN method for the initial value problem (1.2) is defined by

$$
\begin{cases}
Y_i = y_n + c_i h y'_n + h^2 \displaystyle\sum_{j=1}^s \bar{a}_{ij} f(Y_j, Y'_j), \quad i = 1, \cdots, s, \\[2ex]
Y'_i = y'_n + h \displaystyle\sum_{j=1}^s a_{ij} f(Y_j, Y'_j), \qquad\qquad i = 1, \cdots, s, \\[2ex]
y_{n+1} = y_n + h y'_n + h^2 \displaystyle\sum_{i=1}^s \bar{b}_i f(Y_i, Y'_i), \\[2ex]
y'_{n+1} = y'_n + h \displaystyle\sum_{i=1}^s b_i f(Y_i, Y'_i),
\end{cases}
\tag{1.17}
$$

where $\bar{a}_{ij},\, a_{ij},\, \bar{b}_i,\, b_i,\, c_i$ for $i, j = 1, \cdots, s$ are real constants.

The standard RKN method (1.17) also can be expressed in the partitioned Butcher tableau as follows:

$$
\frac{c \mid \bar{A} \mid A}{\bar{b}^{\mathsf{T}} \mid b^{\mathsf{T}}} =
\begin{array}{c|ccc|ccc}
c_1 & \bar{a}_{11} & \cdots & \bar{a}_{1s} & a_{11} & \cdots & a_{1s} \\
\vdots & \vdots & \ddots & \vdots & \vdots & \ddots & \vdots \\
c_s & \bar{a}_{s1} & \cdots & \bar{a}_{ss} & a_{s1} & \cdots & a_{ss} \\
\hline
 & \bar{b}_1 & \cdots & \bar{b}_s & b_1 & \cdots & b_s
\end{array} ,
$$

where $\bar{b} = (\bar{b}_1, \cdots, \bar{b}_s)^{\mathsf{T}}$, $b = (b_1, \cdots, b_s)^{\mathsf{T}}$ and $c = (c_1, \cdots, c_s)^{\mathsf{T}}$ are s-dimensional vectors, and $\bar{A} = (\bar{a}_{ij})$ and $A = (a_{ij})$ are $s \times s$ constant matrices.

1.3 ERKN Integrators and ARKN Methods Based on the Matrix-Variation-of-Constants Formula

The integration of highly oscillatory differential equations has been a challenge for numerical computation for a long time. Much effort has been focused on preserving important high-frequency oscillations. The adapted RKN (ARKN) methods and extended RKN (ERKN) integrators were proposed one after another.

1.3.1 ARKN Integrators

What is the difference between a standard RKN method and an ARKN method for (1.1)? Inheriting the internal stages of standard RKN methods (ignoring the matrix-variation-of-constants formula (1.7)) and approximating the integrals appearing in (1.8) by a suitable quadrature formula with nodes $c_1, \cdots, c_s$ to *modify only the updates* of standard RKN methods yields the ARKN methods for the nonlinear multi-frequency highly oscillatory system (1.1).

Definition 1.2 (Wu et al. [29]) An s-stage ARKN method with stepsize $h > 0$ for solving the multi-frequency highly oscillatory system (1.1) is defined by

$$
\begin{cases}
Y_i = y_n + hc_i y_n' + h^2 \sum_{j=1}^{s} \bar{a}_{ij}\big(f(Y_j, Y_j') - MY_j\big), & i = 1, \cdots, s, \\[2ex]
Y_i' = y_n' + h \sum_{j=1}^{s} a_{ij}\big(f(Y_j, Y_j') - MY_j\big), & i = 1, \cdots, s, \\[2ex]
y_{n+1} = \phi_0(V)y_n + h\phi_1(V)y_n' + h^2 \sum_{i=1}^{s} \bar{b}_i(V)f(Y_i, Y_i'), & \\[2ex]
y_{n+1}' = -hM\phi_1(V)y_n + \phi_0(V)y_n' + h \sum_{i=1}^{s} b_i(V)f(Y_i, Y_i'), &
\end{cases}
$$

$$(1.18)$$

where $\bar{a}_{ij}, a_{ij}, c_i$ for $i, j = 1, \cdots, s$ are real constants, and $\bar{b}_i(V), b_i(V)$ for $i = 1, \cdots, s$ in the updates are matrix-valued functions of $V = h^2 M$. The ARKN method (1.18) can also be denoted by the partitioned Butcher tableau

$$
\begin{array}{c|c|c}
c & \bar{A} & A \\ \hline
 & \bar{b}^{\mathsf T}(V) & b^{\mathsf T}(V)
\end{array}
=
\begin{array}{c|ccc|ccc}
c_1 & \bar{a}_{11} & \cdots & \bar{a}_{1s} & a_{11} & \cdots & a_{1s} \\
\vdots & \vdots & \ddots & \vdots & \vdots & \ddots & \vdots \\
c_s & \bar{a}_{s1} & \cdots & \bar{a}_{ss} & a_{s1} & \cdots & a_{ss} \\ \hline
 & \bar{b}_1(V) & \cdots & \bar{b}_s(V) & b_1(V) & \cdots & b_s(V)
\end{array}.
$$

In the block-matrix notation with Kronecker products, (1.18) can be expressed as

$$
\begin{cases}
Y = e \otimes y_n + hc \otimes y_n' + h^2(\bar{A} \otimes I_d)\big(f(Y, Y') - (I_s \otimes M)Y\big), \\
Y' = e \otimes y_n' + h(A \otimes I_d)\big(f(Y, Y') - (I_s \otimes M)Y\big), \\
y_{n+1} = \phi_0(V)y_n + h\phi_1(V)y_n' + h^2\bar{b}^{\mathsf T}(V)f(Y, Y'), \\
y_{n+1}' = -hM\phi_1(V)y_n + \phi_0(V)y_n' + hb^{\mathsf T}(V)f(Y, Y'),
\end{cases}
$$

$$(1.19)$$

where e is an $s \times 1$ vector of units, and the block vectors involved are defined by

$$
Y = \big(Y_1^{\mathsf T}, \cdots, Y_s^{\mathsf T}\big)^{\mathsf T}, \quad Y' = \big(Y_1'^{\mathsf T}, \cdots, Y_s'^{\mathsf T}\big)^{\mathsf T},
$$

$$
f(Y, Y') = \big(f(Y_1, Y_1')^{\mathsf T}, \cdots, f(Y_s, Y_s')^{\mathsf T}\big)^{\mathsf T}.
$$

It is noted again that the internal stages of an ARKN method are the same as those of standard RKN methods, and only its updates have been modified. Concerning single-frequency ARKN methods, readers are referred to [14, 30], and the research on symplectic ARKN methods can be found in [31, 32]. Besides, Franco was the first to attempt to extend his single-frequency ARKN methods in [14] to multi-frequency systems (1.9), but his order conditions are based on single-frequency theory (see [30, 33]).

It is also important to emphasise that the internal stages and the updates for an RKN-type method when applied to (1.1) should play the same role in the approximation based on its matrix-variation-of-constants formula (1.7), and the well-known fact that

$$Y_i \approx y(t_n + c_i h), \quad Y_i' \approx y'(t_n + c_i h),$$

for $i = 1, \cdots, s$ and

$$y_{n+1} \approx y(t_{n+1}) = y(t_n + h), \quad y_{n+1}' \approx y'(t_{n+1}) = y'(t_n + h).$$

Unfortunately, from this point of view, it can be observed from (1.18) that the internal stages of an ARKN method are not put on an equal footing in the light of the matrix-variation-of-constants formula (1.7). This means that the revision or modification of an ARKN method for the multi-frequency highly oscillatory system does not go far enough and is still far from being satisfactory from both a theoretical and practical perspective. This key observation motivates ERKN integrators for the nonlinear multi-frequency highly oscillatory system (1.9), which can also be thought of as improved ARKN methods.

1.3.2 ERKN Integrators

Since we have mentioned that the ARKN method is still not satisfactory due to its internal stages, it is natural to improve both the internal stages and updates of an RKN method in the light of the matrix-variation-of-constants formulae (1.7) and (1.8) with $\hat{f}(\zeta) = f(y(\zeta))$. To this end, approximating the integrals appearing in the formulae by using a suitable quadrature formula with nodes $c_1, \cdots, c_s$ leads to the following ERKN integrator for the nonlinear multi-frequency highly oscillatory system (1.9).

Definition 1.3 (Wu et al. [25]) An s-stage ERKN integrator for the numerical integration of the nonlinear multi-frequency highly oscillatory system (1.9) with stepsize $h > 0$ is defined by

$$
\begin{cases}
Y_i = \phi_0(c_i^2 V) y_n + c_i h \phi_1(c_i^2 V) y_n' + h^2 \sum_{j=1}^{s} \bar{a}_{ij}(V) f(Y_j), \quad i = 1, \cdots, s, \\[2mm]
y_{n+1} = \phi_0(V) y_n + h \phi_1(V) y_n' + h^2 \sum_{i=1}^{s} \bar{b}_i(V) f(Y_i), \\[2mm]
y_{n+1}' = -h M \phi_1(V) y_n + \phi_0(V) y_n' + h \sum_{i=1}^{s} b_i(V) f(Y_i),
\end{cases}
$$

$$\tag{1.20}$$

where c_i for $i = 1, \cdots, s$ are real constants, $b_i(V)$, $\bar{b}_i(V)$ for $i = 1, \cdots, s$, and $\bar{a}_{ij}(V)$ for $i, j = 1, \cdots, s$ are matrix-valued functions of $V = h^2 M$.

The scheme (1.20) can also be denoted by the following partitioned Butcher tableau

$$
\begin{array}{c|c}
c & \bar{A}(V) \\
\hline
\bar{b}^{\mathsf{T}}(V) \\
\hline
b^{\mathsf{T}}(V)
\end{array}
=
\begin{array}{c|ccc}
c_1 & \bar{a}_{11}(V) & \cdots & \bar{a}_{1s}(V) \\
\vdots & \vdots & \ddots & \vdots \\
c_s & \bar{a}_{s1}(V) & \cdots & \bar{a}_{ss}(V) \\
\hline
& \bar{b}_1(V) & \cdots & \bar{b}_s(V) \\
\hline
& b_1(V) & \cdots & b_s(V)
\end{array}.
$$

It will be convenient to express the equations of (1.20) in block-matrix notation in terms of Kronecker products

$$
\begin{cases}
Y = \phi_0(C^2 \otimes V)(e \otimes y_n) + h(C \otimes I_d) \phi_1(C^2 \otimes V)(e \otimes y_n') + h^2 \bar{A}(V) f(Y), \\
y_{n+1} = \phi_0(V) y_n + h \phi_1(V) y_n' + h^2 \bar{b}^{\mathsf{T}}(V) f(Y), \\
y_{n+1}' = -h M \phi_1(V) y_n + \phi_0(V) y_n' + h b^{\mathsf{T}}(V) f(Y),
\end{cases}
$$

$$\tag{1.21}$$

where $e = (1, 1, \cdots, 1)^\mathsf{T}$ is an $s \times 1$ vector of units, $c = (c_1, \cdots, c_s)^\mathsf{T}$ is an $s \times 1$ vector of nodes, $C = \mathrm{diag}(c_1, \cdots, c_s)$ is an $s \times s$ diagonal matrix, and the block vectors and block diagonal matrices are given by

$$
Y = \begin{bmatrix} Y_1 \\ \vdots \\ Y_s \end{bmatrix}, \qquad f(Y) = \begin{bmatrix} f(Y_1) \\ \vdots \\ f(Y_s) \end{bmatrix},
$$

$$
\phi_0(C^2 \otimes V) = \mathrm{diag}(\phi_0(c_1^2 V), \cdots, \phi_0(c_s^2 V)),
$$

$$
(C \otimes I_d)\phi_1(C^2 \otimes V) = \mathrm{diag}(c_1\phi_1(c_1^2 V), \cdots, c_s\phi_1(c_s^2 V)).
$$

Here, it should be remarked that both internal stages and updates of an ERKN integrator have been revised and improved in terms of the matrix-variation-of-constants formulae (1.7) and (1.8) with $\hat{f}(\zeta) = f(y(\zeta))$. This class of ERKN integrators has been well developed, and we will further present their stability and convergence analysis in Chap. 3. Moreover, we will also make an attempt to discuss ERKN integrators combined with Fourier pseudospectral discretisation for solving semilinear wave equations in Chap. 3.

If $f(y) = 0$ in (1.9), then accordingly (1.20) reduces to

$$
\begin{cases}
Y_i = \phi_0(c_i^2 V)y_n + c_i h\phi_1(c_i^2 V)y_n', \quad i = 1, \cdots, s, \\
y_{n+1} = \phi_0(V)y_n + h\phi_1(V)y_n', \\
y_{n+1}' = -hM\phi_1(V)y_n + \phi_0(V)y_n'.
\end{cases} \tag{1.22}
$$

In terms of Kronecker products with block-matrix notation, (1.22) can be expressed by

$$
\begin{cases}
Y = \phi_0(C^2 \otimes V)(e \otimes y_n) + h(C \otimes I_d)\phi_1(C^2 \otimes V)(e \otimes y_n'), \\
y_{n+1} = \phi_0(V)y_n + h\phi_1(V)y_n', \\
y_{n+1}' = -hM\phi_1(V)y_n + \phi_0(V)y_n'.
\end{cases} \tag{1.23}
$$

It follows from (1.13) and (1.14) that both the internal stages and updates of an ERKN integrator exactly solve the multi-frequency highly oscillatory linear homogeneous equation (1.6) on noticing the fact that

$$
Y = \hat{Y}, \quad y_{n+1} = y(t_{n+1}), \quad y_{n+1}' = y'(t_{n+1}). \tag{1.24}
$$

It is worth pointing out that (1.24) is an essential feature of ERKN integrators, especially for the effective treatment of nonlinear multi-frequency highly oscillatory systems and this property inherits and develops the idea of the Filon-type method for highly oscillatory integrals (see, e.g. [34, 35]), since the dominant oscillation source introduced by the linear term My has been calculated explicitly.

It is known that energy-preserving methods can be expressed as so-called continuous stage Runge–Kutta methods. Here, from the perspective of the continuous-stage Runge–Kutta methods (see, e.g. [36–41]), it is also worth noting that continuous-stage ERKN integrators for (1.9) have not received enough attention. We will next introduce the definition of continuous-stage ERKN integrators.

Definition 1.4 A continuous-stage ERKN integrator for solving the nonlinear Hamiltonian system (1.9) is given by

$$
\begin{cases}
Y_\tau = \phi_0(\tau^2 V)y_n + \tau h \phi_1(\tau^2 V)y'_n + h^2 \int_0^1 \bar{A}_{\tau,\sigma}(V)f(Y_\sigma)\mathrm{d}\sigma, \quad 0 \leqslant \tau \leqslant 1, \\[2mm]
y_{n+1} = \phi_0(V)y_n + h\phi_1(V)y'_n + h^2 \int_0^1 \bar{b}_\tau(V)f(Y_\tau)\mathrm{d}\tau, \\[2mm]
y'_{n+1} = -hM\phi_1(V)y_n + \phi_0(V)y'_n + h \int_0^1 b_\tau(V)f(Y_\tau)\mathrm{d}\tau,
\end{cases}
$$

$$(1.25)$$

where $\bar{b}_\tau(V), b_\tau(V)$ are matrix-valued functions of τ and V, and $\bar{A}_{\tau,\sigma}(V)$ is a matrix-valued function depending on τ, σ and V.

We will further discuss continuous-stage extended Runge–Kutta–Nyström methods for highly oscillatory Hamiltonian systems in Chap. 2. Continuous-stage Leapfrog schemes for semilinear Hamiltonian wave equations will be investigated in detail in Chap. 12.

1.4 Oscillation-Preserving Integrators

It is well known that efficiency is often an important consideration for solving multiple high-frequency oscillatory ordinary differential equations over long-time intervals, although standard RKN methods are popular and effective for second-order ordinary differential equations in many applications. One needs to select an appropriate mathematical or numerical approach to track the high-frequency oscillation in order to use larger stepsizes over long-time intervals.

In the last few decades, geometric numerical integration for differential equations has received more and more attention in order to respect their structural invariants and geometry. The geometric numerical integration for nonlinear differential equations has led to the development of numerical schemes which systematically incorporate qualitative features of the underlying problem into their structures. Accordingly, first of all, a numerical algorithm should respect the highly oscillatory structure of the underlying continuous system (1.1) or (1.9), in the sense of Geometric Integration. On noticing that the Filon-type method (see, e.g. [34, 35]) for highly oscillatory integrals is very successful, and the idea behind this method is that the oscillatory part involved in these integrals must be calculated explicitly, this point is also essential for efficiently solving the nonlinear multi-frequency highly oscillatory system (1.1) or (1.9). This idea for exponential or trigonometric integrators, in fact, has been used for decades by many authors (see, e.g. [4, 15, 16, 42–46]).

It is noted that a comprehensive review of exponential integrators can be found in Hochbruck and Ostermann [46], in which Gautschi-type methods, impulse and mollified impulse methods (see, e.g. Grubmüller et al. [47]), multiple time-stepping methods (see Hairer et al. [15], Chapter VIII. 4), and adiabatic integrators (see Lorenz et al. [48]) were reviewed in detail for the highly oscillatory second-order differential equation, and for the singularly perturbed second-order differential equation, respectively. Hence, in this chapter, we won't cover them again.

Here, it is clear that high oscillations are brought by the linear part My of (1.1) or (1.9) which should be solved explicitly and exactly for an efficient numerical integrator. Therefore, it will be convenient to introduce the concept of oscillation-preserving numerical methods for solving the nonlinear highly oscillatory system (1.1) or (1.9). Taking into account the significant fact that the internal stages Y_i for $i = 1, \cdots, s$, must be nonlinearly involved in the updates y_{n+1} and y'_{n+1} at each time step for an RKN-type method when applied to (1.1) or (1.9), we present the following definition of oscillation-preserving numerical methods for efficiently solving the nonlinear multi-frequency highly oscillatory system (1.1) or (1.9).

Definition 1.5 An RKN-type method for solving the nonlinear multi-frequency highly oscillatory system (1.1) or (1.9) is oscillation preserving, if its internal stages Y_i for $i = 1, \cdots, s$, together with its updates y_{n+1} and y'_{n+1} at each time step explicitly and exactly solve the highly oscillatory homogeneous linear equation (1.6) associated with (1.1) or (1.9). Apart from this, if only the updates of an RKN-type method can exactly solve the highly oscillatory homogeneous linear equation (1.6), then the RKN-type method is called to be partly oscillation preserving.

Theorem 1.2 *An ERKN integrator is oscillation preserving, but an ARKN method is partly oscillation preserving, and a standard RKN method is neither oscillation preserving, nor partly oscillation preserving.*

Proof In the light of Definition 1.5, it is very clear from (1.22) or (1.23) that an ERKN integrator is oscillation preserving.

Unfortunately, *an ARKN method is not oscillation preserving* due to its internal stages. In fact, applying the internal stages of the ARKN method (1.18) to (1.6) gives

$$Y_i = y_n + c_i h y'_n - h^2 M \sum_{j=1}^{s} \bar{a}_{ij} Y_j,$$

for $i = 1, \cdots, s$, which leads to

$$Y = e \otimes y_n + h(C \otimes I_d)(e \otimes y'_n) - h^2(\bar{A} \otimes I_d)(I_s \otimes M)Y, \tag{1.26}$$

where e is an $s \times 1$ vector of units,

$$h(C \otimes I_d) = \text{diag}(hc_1 I_d, \cdots, hc_s I_d),$$

and $\otimes$ represents Kronecker products. We then obtain

$$\begin{aligned}
Y &= \left(I_s \otimes I_d + h^2(\bar{A} \otimes I_d)(I_s \otimes M)\right)^{-1}\left(e \otimes y_n + h(C \otimes I_d)(e \otimes y'_n)\right) \\
&= \left(I_s \otimes I_d + h^2(\bar{A} \otimes I_d)(I_s \otimes M)\right)^{-1}\left(e \otimes y_n\right) \\
&\quad + \left(I_s \otimes I_d + h^2(\bar{A} \otimes I_d)(I_s \otimes M)\right)^{-1}h(C \otimes I_d)\left(e \otimes y'_n\right), \tag{1.27}
\end{aligned}$$

provided $\det\left(I_s \otimes I_d + h^2(\bar{A} \otimes I_d)(I_s \otimes M)\right) \neq 0$. In comparison with $\hat{Y}$ defined in (1.13), this implies that

$$Y \neq \hat{Y} = \phi_0(C^2 \otimes V)(e \otimes y_n) + h(C \otimes I_d)\phi_1(C^2 \otimes V)(e \otimes y'_n),$$

i.e.,

$$Y_i \neq \hat{Y}_i = \phi_0\left(c_i^2 V\right)y_n + c_i h\phi_1\left(c_i^2 V\right)y'_n,$$

for $i = 1, \cdots, s$, on noticing the fact that

$$(C \otimes I_d)\phi_1(C^2 \otimes V) = \phi_1(C^2 \otimes V)(C \otimes I_d)$$

and

$$\phi_0\left(c_i^2 V\right) \neq \phi_1\left(c_i^2 V\right),$$

for $i = 1, \cdots, s$. Therefore, it follows from Definition 1.5 that an ARKN method cannot be oscillation preserving, although it is partly oscillation preserving due to its updates. Since the internal stages of standard RKN methods are the same as those of ARKN methods, a standard RKN method is not oscillation preserving. Moreover, in a similar way, it can be shown that the updates of a standard RKN method cannot exactly solve (1.6). This implies that a standard RKN method is neither oscillation preserving nor partly oscillation preserving, because both its internal stages and updates fail to exactly solve the highly oscillatory homogeneous linear equation (1.6) associated with (1.1) or (1.9).

The proof is complete. □

Clearly, it follows from (1.25) that a continuous-stage ERKN method for (1.9) is also oscillation preserving.

Theorem 1.2 presents and confirms a fact that an ERKN integrator possesses excellent oscillation-preserving behaviour for solving the nonlinear highly oscillatory system (1.9) in comparison with RKN and ARKN methods.

With regard to the construction of arbitrary order ERKN integrators for (1.9), readers are referred to a recent paper (see [18]). Concerning the order conditions of ERKN integrators for (1.9), readers are referred to [26, 49].

1.5 Towards Highly Oscillatory Nonlinear Hamiltonian Systems

As is known, Hamiltonian systems have very important applications. Nonlinear Hamiltonian systems with highly oscillatory solutions frequently occur in areas of physics and engineering such as molecular dynamics, classical and quantum mechanics. Numerical methods used to treat them also depend on the knowledge of certain other characteristics of the solution besides high-frequency oscillation.

We now consider the initial value problem of the nonlinear multi-frequency highly oscillatory Hamiltonian system

$$
\begin{cases} \ddot{q} + Mq = f(q), & t \in [0, T], \\ q(0) = q_0, \quad \dot{q}(0) = \dot{q}_0, \end{cases} \tag{1.28}
$$

where M is a $d \times d$ symmetric positive semi-definite matrix and $f : \mathbb{R}^d \to \mathbb{R}^d$ is a continuous nonlinear function of q with $f(q) = -\nabla U(q)$ for a real-valued function $U(q)$. Then, the highly oscillatory Hamiltonian system (1.28) can be rewritten as the standard format

$$
\begin{cases} \dot{p} = -\nabla_q H(p, q), \\ \dot{q} = \nabla_p H(p, q), \end{cases} \tag{1.29}
$$

with the initial values $q(0) = q_0$, $p(0) = p_0 = \dot{q}_0$ and the Hamiltonian

$$
H(p, q) = \frac{1}{2} p^\mathsf{T} p + \frac{1}{2} q^\mathsf{T} M q + U(q). \tag{1.30}
$$

It is well known that two remarkable features of a Hamiltonian system are the symplecticity of its flow and the conservation of the Hamiltonian. Consequently, for a numerical integrator for (1.29), in addition to oscillation preservation, these two features should be respected as much as possible in the spirit of geometric numerical integration. In the development of symplectic integration, the earliest significant contributions to this field were due to Feng Kang, who was a pioneer in stressing the importance of using symplectic integrators when the equations to be solved are Hamiltonian systems (see [50–52]). It is also worth noting the earlier important work on symplectic integration by J. M. Sanz-Serna, who first found and analysed symplectic Runge–Kutta schemes for Hamiltonian systems (see Sanz-Serna [53]). For the survey papers and monographs on numerical approaches to dealing with nonlinear Hamiltonian differential equations with highly oscillatory solutions, readers are referred to [1, 2, 13, 15, 54, 55].

1.5.1   SSMERKN Integrators

Symplecticity is an important characteristic property of Hamiltonian systems and symplectic methods have been well developed (see, e.g. [15, 52, 53, 56–60]). Symplectic ERKN methods for highly oscillatory Hamiltonian systems have been analysed (see Wu et al. [24]). Symplectic and symmetric multi-frequency ERKN integrators (SSMERKN integrators) have been proposed and analysed for the nonlinear multi-frequency highly oscillatory Hamiltonian system (1.29) in Wu et al. [49].

We now state the coupled conditions of explicit SSMERKN integrators for (1.29).

Theorem 1.3 *An s-stage explicit multi-frequency ERKN integrator for integrating* (1.29) *is symplectic and symmetric if its coefficients are given by*

$$
\begin{cases}
c_i = 1 - c_{s+1-i}, \quad d_i = d_{s+1-i} \neq 0, & i = 1, 2, \cdots, s, \\
b_i(V) = d_i \phi_0(c_{s+1-i}^2 V), & i = 1, 2, \cdots, s, \\
\bar{b}_i(V) = d_i c_{s+1-i} \phi_1(c_{s+1-i}^2 V), & i = 1, 2, \cdots, s, \\
\bar{a}_{ij}(V) = \dfrac{1}{d_i} \big(b_i(V) \bar{b}_j(V) - \bar{b}_i(V) b_j(V) \big), & i > j, \ i, \ j = 1, 2, \cdots, s.
\end{cases}
$$

$$(1.31)$$

The detailed proof of this theorem can be found in Wu et al. [49]. It is noted that when $V \to \mathbf{0}_{d \times d}$, the ERKN methods reduce to standard RKN methods for solving Hamiltonian systems with the Hamiltonian $H(p, q) = \dfrac{1}{2} p^\mathsf{T} p + U(q)$. The following result can be deduced from Theorem 1.3.

Theorem 1.4 *An s-stage explicit RKN method with the coefficients*

$$
\begin{cases}
c_i = 1 - c_{s+1-i}, \quad d_i = d_{s+1-i} \neq 0, & i = 1, 2, \cdots, s, \\
b_i = d_i, \quad \bar{b}_i = d_i c_{s+1-i}, & i = 1, 2, \cdots, s, \\
\bar{a}_{ij} = \dfrac{1}{d_i} \big(b_i \bar{b}_j - \bar{b}_i b_j \big), & i > j, \ i, \ j = 1, 2, \cdots, s,
\end{cases}
$$

$$(1.32)$$

is symplectic and symmetric. In (1.32), d_i *for* $i = 1, 2, \cdots, \left\lfloor \dfrac{s+1}{2} \right\rfloor$, *are real numbers and can be chosen based on the order conditions of RKN methods or other requirements, where* $\left\lfloor \dfrac{s+1}{2} \right\rfloor$ *denotes the integer part of* $\dfrac{s+1}{2}$.

The proof of Theorem 1.4 can be found in [49].

Theorem 1.5 *An SSMERKN integrator is oscillation preserving. However, a symplectic and symmetric RKN method is neither oscillation preserving, nor partly oscillation preserving.*

Proof It follows directly from the definition of oscillation preservation (Definition 1.5). □

Hence, we conclude from Theorem 1.5 that a symplectic and symmetric RKN method may not be a good choice for efficiently solving the nonlinear multi-frequency and highly oscillatory Hamiltonian system (1.29) due to its lack of oscillation preservation, whereas an SSMERKN integrator is preferred. This point will also be observed from the results of numerical experiments in Sect. 1.7.

With regard to energy-preserving continuous-stage extended Runge–Kutta–Nyström methods for nonlinear Hamiltonian systems with highly oscillatory solutions, see Chap. 2 for details.

1.5.2 Trigonometric Fourier Collocation Methods

Geometric numerical integration is still a very active subject area and much work has yet to be done. Accordingly, the exponential/trigonometric integrator has become increasingly important (see, e.g. [8, 13, 15, 20, 61–64]). The original attempts at exploring exponential/trigonometric algorithms for the oscillatory system (1.28) with the special structure brought by the linear term Mq were motivated by many fields of research such as mechanics, astronomy, quantum physics, theoretical physics, molecular dynamics, semidiscrete wave equations approximated by the method of lines or spectral discretisation. The exponential/trigonometric methods take advantage of the special structure to achieve an improved qualitative behaviour, and produce a more accurate long-time integration than standard methods.

We next consider the highly oscillatory system (1.28) which is restricted to the interval $[0, h]$:

$$\ddot{q}(t) + Mq(t) = f(q(t)), \qquad q(0) = q_0, \quad \dot{q}(0) = \dot{q}_0, \qquad t \in [0, h]. \qquad (1.33)$$

It follows from the matrix-variation-of-constants formula that the exact solution of the system (1.33) and its derivative satisfy

$$
\begin{cases}
q(h) = \phi_0(V)q_0 + h\phi_1(V)p_0 + h^2 \int_0^1 (1-z)\phi_1\big((1-z)^2 V\big) f(q(hz))\mathrm{d}z, \\[2mm]
p(h) = -hM\phi_1(V)q_0 + \phi_0(V)p_0 + h \int_0^1 \phi_0\big((1-z)^2 V\big) f(q(hz))\mathrm{d}z,
\end{cases}
$$

$$(1.34)$$

for stepsize $h > 0$, where $V = h^2 M$.

Choose an orthogonal polynomial basis $\{\widetilde{P}_j\}_{j=0}^\infty$ on the interval $[0, 1]$: e.g., the shifted Legendre polynomials over the interval $[0, 1]$, scaled in order to be orthonormal. Hence, we have

$$
\int_0^1 \widetilde{P}_i(x)\widetilde{P}_j(x)\mathrm{d}x = \delta_{ij}, \qquad \deg\left(\widetilde{P}_j\right) = j, \qquad i, j \geqslant 0,
$$

where δ_{ij} is the Kronecker symbol. The right-hand side of (1.33) can be rewritten as

$$
f(q(\xi h)) = \sum_{j=0}^\infty \widetilde{P}_j(\xi)\gamma_j(q), \quad \xi \in [0, 1]; \qquad \gamma_j(q) := \int_0^1 \widetilde{P}_j(\tau) f(q(\tau h))\mathrm{d}\tau.
$$

$$(1.35)$$

For the sake of simplicity we now use $\gamma_j(q)$ to denote the coefficients $\gamma_j(h, f(q))$ involved in the Fourier expansion.

We now state a result which follows from (1.34) and (1.35), and the proof can be found in Wang et al. [8].

Theorem 1.6 *The solution of* (1.33) *and its derivative satisfy*

$$
\begin{cases}
q(h) = \phi_0(V)q_0 + h\phi_1(V)p_0 + h^2 \sum_{j=0}^\infty I_{1,j}\gamma_j(q), \\[2mm]
p(h) = -hM\phi_1(V)q_0 + \phi_0(V)p_0 + h \sum_{j=0}^\infty I_{2,j}\gamma_j(q),
\end{cases}
$$

$$(1.36)$$

where

$$
I_{1,j} := \int_0^1 \widetilde{P}_j(z)(1-z)\phi_1\big((1-z)^2 V\big)\mathrm{d}z, \quad I_{2,j} := \int_0^1 \widetilde{P}_j(z)\phi_0\big((1-z)^2 V\big)\mathrm{d}z.
$$

$$(1.37)$$

Naturally, a practical scheme to solve (1.28) needs to truncate the series (1.35) after r ($r \geqslant 2$) terms and this means replacing the initial value problem (1.28) with the following approximate problem

$$
\begin{cases}
\tilde{q}'(\xi h) = \tilde{p}(\xi h), \quad \tilde{q}(0) = q_0, \\
\tilde{p}'(\xi h) = -M\tilde{q}(\xi h) + \displaystyle\sum_{j=0}^{r-1} \tilde{P}_j(\xi)\gamma_j(\tilde{q}), \quad \tilde{p}(0) = p_0.
\end{cases}
\tag{1.38}
$$

We then obtain the implicit solution of (1.38) as follows:

$$
\begin{cases}
\tilde{q}(h) = \phi_0(V)q_0 + h\phi_1(V)p_0 + h^2 \displaystyle\sum_{j=0}^{r-1} I_{1,j}\gamma_j(\tilde{q}), \\
\tilde{p}(h) = -hM\phi_1(V)q_0 + \phi_0(V)p_0 + h \displaystyle\sum_{j=0}^{r-1} I_{2,j}\gamma_j(\tilde{q}).
\end{cases}
\tag{1.39}
$$

The analysis stated above leads to the following definition of the trigonometric Fourier collocation methods.

Definition 1.6 (Wang et al. [8]) A trigonometric Fourier collocation (TFC) method for integrating the oscillatory system (1.28) or (1.29) is defined by

$$
\begin{cases}
v_i = \phi_0(c_i^2 V)q_0 + c_i h\phi_1(c_i^2 V)p_0 + (c_i h)^2 \displaystyle\sum_{j=0}^{r-1} I_{1,j,c_i} \sum_{l=1}^{k} b_l \tilde{P}_j(c_l)f(v_l), \\
\quad i = 1, 2, \cdots, k, \\
v(h) = \phi_0(V)q_0 + h\phi_1(V)p_0 + h^2 \displaystyle\sum_{j=0}^{r-1} I_{1,j} \sum_{l=1}^{k} b_l \tilde{P}_j(c_l)f(v_l), \\
u(h) = -hM\phi_1(V)q_0 + \phi_0(V)p_0 + h \displaystyle\sum_{j=0}^{r-1} I_{2,j} \sum_{l=1}^{k} b_l \tilde{P}_j(c_l)f(v_l),
\end{cases}
\tag{1.40}
$$

where h is the stepsize, r is an integer satisfying $2 \leqslant r \leqslant k$, $\tilde{P}_j$ are defined by

$$
\tilde{P}_j(x) = (-1)^j \sqrt{2j+1} \sum_{k=0}^{j} \binom{j}{k}\binom{j+k}{k}(-x)^k, \qquad j = 0, 1, \cdots, \qquad x \in [0, 1],
\tag{1.41}
$$

and c_l, b_l for $l = 1, 2, \cdots, k$, are the nodes, and the quadrature weights of a quadrature formula, respectively. $I_{1,j}$, $I_{2,j}$, and I_{1,j,c_i} are well determined by the generalised hypergeometric functions (see Wang et al. [8] for details):

$$
{}_mF_n\begin{bmatrix} \alpha_1, \alpha_2, \cdots, \alpha_m; \\ \beta_1, \beta_2, \cdots, \beta_n; \end{bmatrix} = \sum_{l=0}^{\infty} \frac{\prod_{i=1}^{m}(\alpha_i)_l}{\prod_{i=1}^{n}(\beta_i)_l} \frac{x^l}{l!}. \tag{1.42}
$$

In (1.42), the *Pochhammer symbol* $(z)_l$ is recursively defined by $(z)_0 = 1$ and $(z)_l = z(z + 1) \cdots (z + l - 1)$, $l \in \mathbb{N}$, and the parameters α_i and β_i are arbitrary complex numbers, except that β_i can be neither zero nor a negative integer.

Remark 1.4   We remark that $\phi_0(V)$ and $\phi_1(V)$ defined by (1.4) can also be expressed by the generalised hypergeometric function $_0F_1$:

$$
\phi_0(V) = {}_0F_1\begin{bmatrix} -; \\ \frac{1}{2}; \end{bmatrix} -\frac{V}{4}, \quad \phi_1(V) = {}_0F_1\begin{bmatrix} -; \\ \frac{3}{2}; \end{bmatrix} -\frac{V}{4}. \tag{1.43}
$$

The other $\phi_j(V)$ for $j \geq 2$ can be recursively obtained from $\phi_0(V)$ and $\phi_1(V)$ (see, e.g. [54]). This hypergeometric representation is useful, and most modern software, e.g., Maple, Mathematica, and Matlab, is well equipped for the calculation of generalised hypergeometric functions.

Remark 1.5 Although the TFC method (1.40) approximates the solution $q(t)$, $p(t)$ of the system (1.28) or (1.29) only in the time interval $[0, h]$, the values $v(h)$, $u(h)$ can be considered as the initial values for a new initial value problem approximating $q(t)$, $p(t)$ in the next time interval $[h, 2h]$. In such a time-stepping routine manner, we can extend the TFC methods to the interval $[(i - 1)h, ih]$ for any $i \geq 2$ and finally obtain a TFC method for $q(t)$, $p(t)$ in an arbitrary interval $[0, Nh]$. For more details, readers are referred to Wang et al. [8].

Concerning the order of TCF methods, we assume that the quadrature formula for $\gamma_j(q)$ in (1.35) is of order $m - 1$. Then the order of TFC methods is of order $n = \min\{m, 2r\}$. The details can be found in Wang et al. [8].

Theorem 1.7 *The TFC method* (1.40) *is oscillation preserving.*

Proof Clearly, it follows from the definition of TFC method (1.40) that the TFC method (1.40) is a kind of k-stage RKN-type method, and both its internal stages and updates exactly solve the system of multi-frequency highly oscillatory linear homogeneous equations (1.6). Consequently, the TFC method (1.40) is oscillation preserving.   □

It is worth mentioning that the TFC method (1.40) is based on the variation-of-constants formula and a local Fourier expansion of the underlying problem, via the approximation of orthogonal polynomial basis. The approximation of orthogonal trigonometric basis is another possible strategy in the effort to solve (1.33).

1.5.3   The AAVF Method and AVF Formula

It is known that one of the important characteristic properties of a Hamiltonian system is energy conservation. The study of numerical energy conservation for oscillatory systems has appeared in the literature (see, e.g. Hairer et al. [65, 66], Li et al. [7]). In particular, the average-vector-field (AVF) formula (see, e.g. [67, 68]) for (1.10) is of great importance, once (1.10) is a Hamiltonian system.

It follows from the matrix-variation-of-constants formula (1.3) that the solution of (1.28) and its derivative satisfy the following equations:

$$
\begin{cases}
q(t) = \phi_0(t^2 M)q_0 + t\phi_1(t^2 M)p_0 + \displaystyle\int_0^t (t - \zeta)\phi_1((t - \zeta)^2 M)\hat{f}(\zeta)\mathrm{d}\zeta, \\
p(t) = -tM\phi_1(t^2 M)q_0 + \phi_0(t^2 M)p_0 + \displaystyle\int_0^t \phi_0((t - \zeta)^2 M)\hat{f}(\zeta)\mathrm{d}\zeta,
\end{cases}
$$
$$(1.44)$$

where t is any real numbers and $\hat{f}(\zeta) = f(q(\zeta))$.

The formula (1.44) motivates the following integrator with stepsize h of the form:

$$
\begin{cases}
q_{n+1} = \phi_0(V)q_n + h\phi_1(V)p_n + h^2 I Q^1, \\
p_{n+1} = -hM\phi_1(V)q_n + \phi_0(V)p_n + hI Q^2,
\end{cases}
$$
$$(1.45)$$

where $V = h^2 M$, and $I Q^1$, $I Q^2$ can be determined by the energy-preserving condition at each time step:

$$
H(p_{n+1}, q_{n+1}) = H(p_n, q_n).
$$

We now state a sufficient condition (see, e.g. Wang and Wu [21]) for the scheme (1.45) to yield energy preservation.

Theorem 1.8 *If*

$$
I Q^1 = \phi_2(V) \int_0^1 f((1 - \tau)q_n + \tau q_{n+1})\mathrm{d}\tau, \quad I Q^2 = \phi_1(V) \int_0^1 f((1 - \tau)q_n + \tau q_{n+1})\mathrm{d}\tau,
$$
$$(1.46)$$

then the scheme (1.45) *exactly preserves the Hamiltonian* (1.30), *i.e.,*

$$H(p_{n+1}, q_{n+1}) = H(p_n, q_n), \quad n = 0, 1, \cdots. \tag{1.47}$$

Thus, we state the adapted average-vector-field (AAVF) method (see [21, 23]) as follows:

Definition 1.7 An adapted average-vector-field (AAVF) method with stepsize h for the multi-frequency highly oscillatory Hamiltonian system (1.28) is defined by

$$\begin{cases} q_{n+1} = \phi_0(V)q_n + h\phi_1(V)p_n + h^2\phi_2(V) \displaystyle\int_0^1 f\big((1-\tau)q_n + \tau q_{n+1}\big)d\tau, \\ p_{n+1} = -hM\phi_1(V)q_n + \phi_0(V)p_n + h\phi_1(V) \displaystyle\int_0^1 f\big((1-\tau)q_n + \tau q_{n+1}\big)d\tau, \end{cases} \tag{1.48}$$

where $\phi_0(V)$, $\phi_1(V)$ and $\phi_2(V)$ are determined by (1.4).

It follows from Theorem 1.8 that the AAVF method (1.48) is energy preserving.

It can be observed that when $M = 0$ in (1.48), the AAVF method reduces to the well-known AVF formula for (1.10) with $y = q$ and $f(q) = -\nabla U(q)$ (see, e.g. [67, 68]):

$$\begin{cases} q_{n+1} = q_n + hp_n + \dfrac{h^2}{2} \displaystyle\int_0^1 f\big((1-\tau)q_n + \tau q_{n+1}\big)d\tau, \\ p_{n+1} = p_n + h \displaystyle\int_0^1 f\big((1-\tau)q_n + \tau q_{n+1}\big)d\tau. \end{cases} \tag{1.49}$$

Remark 1.5.1 This class of discrete gradient methods is very important in Geometric Integrators, and the first actual appearance of the integrator that came to be known as the AVF method was in [68]. On the basis of this idea, we will analyse linearly-fitted conservative (dissipative) schemes for nonlinear wave equations in Chap. 8. We also consider the volume-preserving exponential integrators for different vector fields in Chap. 6. Furthermore, we will present energy-preserving integrators for Poisson systems in Chap. 4 and energy-preserving schemes for high-dimensional nonlinear KG equations in Chap. 9.

Many physical problems have time reversibility and this structure of the original continuous system can be preserved by symmetric integrators (readers are referred to Chapter V of Hairer et al. [15] for a rigorous definition of reversibility). The AAVF methods were also proved to be symmetric (see, e.g. [21]). However, it follows from the definition of oscillation preservation (Definition 1.5) that an AAVF method is neither oscillation preserving nor is the AVF method. Fortunately, an AAVF method is partly oscillation preserving due to its updates, and the result is stated as follows.

Theorem 1.9 *The AAVF method for* (1.28) *is partly oscillation preserving.*

Proof Similarly to the AVF method, the AAVF method defined by (1.48) is also dependent on the integral and, in practice, the integral usually must be approximated by a numerical integral formula (see, e.g. [23]), a weighted summation of the evaluations of function f at s different values $Y_i = (1 - \tau_i)q_n + \tau_i q_{n+1}$ for $i = 1, \cdots, s$, which can be regarded as the internal stages of the AAVF method. Obviously, the updates of an AAVF method can exactly solve the highly oscillatory homogeneous linear equation (1.6), but the internal stages cannot. The proof is complete.   □

Remark 1.6 It is worth mentioning that in a recent paper, the AAVF methods have been extended to the computation of high-dimensional semilinear KG equations. Readers are referred to Chap. 9 for details. Moreover, long-time momentum and actions behaviour of the AAVF methods for Hamiltonian wave equations are presented in Chap. 14. Furthermore, the global error bounds of one-stage ERKN integrators for semilinear wave equations are analysed in Chap. 7. We also discussed the resonance instability for AAVF methods (see [54]).

1.6   Other Concerns Relating to Highly Oscillatory Problems

Gautschi-type methods have been intensively studied in the literature, and general ERKN methods for highly oscillatory problems have been proposed. Here, it is also important to recognise that the numerical solution of semilinear Hamiltonian wave equations is closely related to oscillation-preserving integrators.

1.6.1   Gautschi-Type Methods

This section starts from the Gautschi-type methods which have been well investigated in the literature (see, e.g. [4, 16, 69]). Gautschi-type methods for the nonlinear highly oscillatory Hamiltonian system (1.29) can be traced back to a profound paper of Gautschi [43]. Gautschi-type methods are special explicit ERKN methods of order two (see [10]). An error and stability analysis of the Gautschi-type methods can be found in [16]. Thus, Gautschi-type methods are oscillation preserving in the light of Definition 1.5. However, it is noted that ERKN methods for the highly oscillatory Hamiltonian system (1.29) can be of an arbitrarily high order which can be thought of as generalised Gautschi-type methods.

1.6.2   General ERKN Methods for (1.1)

We next turn to the general ERKN methods for solving nonlinear multi-frequency highly oscillatory second-order ordinary differential equations (1.1).

Definition 1.8 (You et al. [27]) An s-stage general extended Runge–Kutta–Nyström (ERKN) method for the numerical integration of the IVP (1.1) is defined by

$$
\left\{
\begin{aligned}
Y_i &= \phi_0(c_i^2 V)y_n + c_i\phi_1(c_i^2 V)hy_n' + h^2 \sum_{j=1}^{s} \bar{a}_{ij}(V)f(Y_j, Y_j'), && i = 1, \cdots, s, \\[2mm]
hY_i' &= -c_i V\phi_1(c_i^2 V)y_n + \phi_0(c_i^2 V)hy_n' + h^2 \sum_{j=1}^{s} a_{ij}(V)f(Y_j, Y_j'), && i = 1, \cdots, s, \\[2mm]
y_{n+1} &= \phi_0(V)y_n + \phi_1(V)hy_n' + h^2 \sum_{i=1}^{s} \bar{b}_i(V)f(Y_i, Y_i'), \\[2mm]
hy_{n+1}' &= -V\phi_1(V)y_n + \phi_0(V)hy_n' + h^2 \sum_{i=1}^{s} b_i(V)f(Y_i, Y_i'),
\end{aligned}
\right.
$$

$$(1.50)$$

where $\phi_0(V)$, $\phi_1(V)$, $\bar{a}_{ij}(V)$, $a_{ij}(V)$, $\bar{b}_i(V)$ and $b_i(V)$ for $i, j = 1, \cdots, s$, are matrix-valued functions of $V = h^2 M$.

The general ERKN method (1.50) in Definition 1.8 can also be represented briefly in a partitioned Butcher tableau of the coefficients:

$$
\begin{array}{c|c|c}
c & A(V) & \bar{A}(V) \\
\hline
 & b^{\mathsf{T}}(V) & \bar{b}^{\mathsf{T}}(V)
\end{array}
$$

$$
=
\begin{array}{c|cccc|cccc}
c_1 & \bar{a}_{11}(V) & \bar{a}_{12}(V) & \cdots & \bar{a}_{1s}(V) & a_{11}(V) & a_{12}(V) & \cdots & a_{1s}(V) \\
c_2 & \bar{a}_{21}(V) & \bar{a}_{22}(V) & \cdots & \bar{a}_{2s}(V) & a_{21}(V) & a_{22}(V) & \cdots & a_{2s}(V) \\
\vdots & \vdots & \vdots & \ddots & \vdots & \vdots & \vdots & \ddots & \vdots \\
c_s & a_{s1}(V) & a_{s2}(V) & \cdots & a_{ss}(V) & a_{s1}(V) & a_{s2}(V) & \cdots & a_{ss}(V) \\
\hline
 & \bar{b}_1(V) & \bar{b}_2(V) & \cdots & \bar{b}_s(V) & b_1(V) & b_2(V) & \cdots & b_s(V)
\end{array}
. \qquad (1.51)
$$

Obviously, the general ERKN method (1.50) for the nonlinear multi-frequency highly oscillatory system (1.1) is oscillation preserving in the light of Definition 1.5. The general ERKN method (1.50) can be of an arbitrarily high order and the analysis of order conditions for the general ERKN method (1.50) can be found in [13, 28].

1.6.3   Towards the Application to Semilinear KG Equations

We note a fact that one of the major applications of oscillation-preserving integrators is to solve semilinear Hamiltonian wave equations such as semilinear KG equations:

$$\begin{cases} u_{tt} - a^2 \Delta u = f(u), & t_0 < t \leqslant T, \quad x \in \Omega, \\ u(x, t_0) = \varphi_1(x), \quad u_t(x, t_0) = \varphi_2(x), \quad x \in \bar{\Omega}, \end{cases} \tag{1.52}$$

where $u(x, t)$ represents the wave displacement at position x and time t, and the nonlinear function $f(u)$ is the negative derivative of a potential energy $V(u) \geqslant 0$:

$$V(u) = -\int_0^u f(\sigma) d\sigma.$$

Here, suppose that the initial value problem (1.52) is subject to the periodic boundary condition on the domain $\Omega = (-\pi, \pi)$,

$$u(x, t) = u(x + 2\pi, t), \qquad x \in (-\pi, \pi], \tag{1.53}$$

where 2π is the fundamental period with respect to x. It is known that, as a relativistic counterpart of the Schrödinger equation, the *KG equation* is used to model diverse nonlinear phenomena, such as the propagation of dislocations in crystals and the behaviour of elementary particles and of Josephson junctions (see [70] Chap. 2). Its efficient computation, without a doubt, induces numerous enduring challenges (see, e.g. [9, 11, 71]).

In practice, a suitable space semidiscretisation for semilinear KG equations can lead to (1.9), where the matrix M is derived from the space semidiscretisation. If we denote the total number of spatial mesh grids by N, then the larger N is, the larger $\|M\|$ becomes. This means that the semidiscrete wave equation is a highly oscillatory system. In our recent work (see Mei et al. [10]), it has been proved under the so-called finite-energy condition that the error bound of ERKN integrators when applied to semilinear wave equations is independent of $\|M\|$. This point is crucial to the numerical solution of the underlying semilinear KG equation.

Another approach to the numerical solution of KG equations is that we try to gain an abstract formulation for the problem (1.52)–(1.53), and then deal with it numerically. To this end, we first consider the following differential operator $\mathscr{A}$ defined by

$$(\mathscr{A}v)(x) = -a^2 v_{xx}(x). \tag{1.54}$$

In (1.54), $\mathscr{A}$ is a linear, unbounded positive semi-definite operator, whose domain is

$$D(\mathscr{A}) := \left\{ v \in H^1(\Omega) : v(x) = v(x + 2\pi) \right\}.$$

Fortunately, however, the operator $\mathscr{A}$ has a complete system of orthogonal eigen-functions $\{e^{ikx} : k \in \mathbb{Z}\}$ and the linear span of all these eigenfunctions

$$X := \lin\{e^{i\ell x} : \ell \in \mathbb{Z}\} \tag{1.55}$$

is dense in the Hilbert space $L^2(\Omega)$. We then obtain the orthonormal basis of eigenvectors of the operator $\mathscr{A}$ with the corresponding eigenvalues $a^2\ell^2$ for $\ell \in \mathbb{Z}$.

Define the bounded functions through the following series (see [72])

$$\phi_k(x) := \sum_{j=0}^{\infty} \frac{(-1)^j x^j}{(2j+k)!}, \qquad k \in \mathbb{N} \quad \text{for} \quad \forall x \geqslant 0. \tag{1.56}$$

Accordingly, these functions (1.56) can induce the bounded operators

$$\phi_k(t\mathscr{A}) : \ L^2(\Omega) \rightarrow L^2(\Omega)$$

for $k \in \mathbb{N}$ and $t_0 \leqslant t \leqslant T$:

$$\phi_k(t\mathscr{A})v(x) = \sum_{\ell=-\infty}^{\infty} \hat{v}_\ell \phi_k(ta^2\ell^2)e^{i\ell x} \quad \text{for} \quad v(x) = \sum_{\ell=-\infty}^{\infty} \hat{v}_\ell e^{i\ell x}, \tag{1.57}$$

and the boundedness follows from the definition of the operator norm that

$$\|\phi_k(t\mathscr{A})\|_*^2 = \sup_{\|v\| \neq 0} \frac{\|\phi_k(t\mathscr{A})v\|^2}{\|v\|^2} \leqslant \sup_{t_0 \leqslant t \leqslant T} |\phi_k(ta^2\ell^2)|^2 \leqslant \gamma_k^2, \tag{1.58}$$

where $\| \cdot \|_*$ is the Sobolev norm $\| \cdot \|_{L^2(\Omega) \leftarrow L^2(\Omega)}$, and γ_k for $k \in \mathbb{N}$ are the uniform bounds of the functions $|\phi_k(x)|$ for $k \in \mathbb{N}$ and $x \geqslant 0$. With regard to the analysis for the boundedness, readers are referred to Liu and Wu [72].

We are now in a position to define $u(t)$ as the function that maps x to $u(x, t)$, $u(t) := [x \mapsto u(x, t)]$, and in this way the system (1.52)–(1.53) can be formulated as an abstract second-order ordinary differential equation

$$\begin{cases} u''(t) + \mathscr{A}u(t) = f(u(t)), & t_0 < t \leqslant T, \\ u(t_0) = \varphi_1(x), \quad u'(t_0) = \varphi_2(x). \end{cases} \tag{1.59}$$

on the closed subspace

$$\mathscr{X} := \{u(x, \cdot) \in X \mid u(x, \cdot) \text{ satisfies the corresponding boundary conditions}\} \subseteq L^2(\Omega). \tag{1.60}$$

With this premise, the solution of the abstract second-order ordinary differential equations (1.59) can be expressed by the following operator-variation-of-constants formula (see, e.g. [73–75]).

Theorem 1.10 *The solution of (1.59) and its derivative satisfy the following operator-variation-of-constants formula*

$$
\begin{cases}
u(t) = \phi_0\big((t - t_0)^2 \mathscr{A}\big)u(t_0) + (t - t_0)\phi_1\big((t - t_0)^2 \mathscr{A}\big)u'(t_0) \\
\qquad + \displaystyle\int_{t_0}^{t} (t - \zeta)\phi_1\big((t - \zeta)^2 \mathscr{A}\big) f\big(u(\zeta)\big)\mathrm{d}\zeta, \\
u'(t) = -(t - t_0)\mathscr{A}\phi_1\big((t - t_0)^2 \mathscr{A}\big)u(t_0) + \phi_0\big((t - t_0)^2 \mathscr{A}\big)u'(t_0) \\
\qquad + \displaystyle\int_{t_0}^{t} \phi_0\big((t - \zeta)^2 \mathscr{A}\big) f\big(u(\zeta)\big)\mathrm{d}\zeta,
\end{cases}
\tag{1.61}
$$

for $t_0 \leqslant t \leqslant T$, where both $\phi_0\big((t - t_0)^2 \mathscr{A}\big)$ and $\phi_1\big((t - t_0)^2 \mathscr{A}\big)$ are bounded operators.

Remark 1.6.1 We here remark that the special case where $f(u) = 0$, the operator-variation-of-constants formula (1.61) yields the closed-form solution to (1.59). Moreover, the idea of the operator-variation-of-constants formula (1.61) also provides a useful approach to the development of the so-called semi-analytical ERKN integrators for solving high-dimensional nonlinear wave equations. See Chap. 13 for details.

According to Theorem 1.10, the solution of (1.59) and its derivative at a time point $t_{n+1} = t_n + \Delta t$, $n \in \mathbb{N}$ are given by

$$
\begin{cases}
u(t_{n+1}) = \phi_0\big(\mathscr{V}\big)u(t_n) + \Delta t\phi_1\big(\mathscr{V}\big)u'(t_n) + \Delta t^2 \displaystyle\int_0^1 (1 - z)\phi_1\big((1 - z)^2 \mathscr{V}\big)\tilde{f}(z)\mathrm{d}z, \\
u'(t_{n+1}) = -\Delta t\mathscr{A}\phi_1\big(\mathscr{V}\big)u(t_n) + \phi_0\big(\mathscr{V}\big)u'(t_n) + \Delta t \displaystyle\int_0^1 \phi_0\big((1 - z)^2 \mathscr{V}\big)\tilde{f}(z)\mathrm{d}z,
\end{cases}
\tag{1.62}
$$

where $\mathscr{V} = \Delta t^2 \mathscr{A}$ and $\tilde{f}(z) = f\big(u(t_n + z\Delta t)\big)$.

If the nonlinear integrals

$$
I_1 := \int_0^1 (1 - z)\phi_1\big((1 - z)^2 \mathscr{V}\big)\tilde{f}(z)\mathrm{d}z \quad \text{and} \quad I_2 := \int_0^1 \phi_0\big((1 - z)^2 \mathscr{V}\big)\tilde{f}(z)\mathrm{d}z
\tag{1.63}
$$

are efficiently approximated, then we are hopeful of obtaining some new integrators based on (1.62). For example, using the operator-variation-of-constants formula (1.62) and the two-point Hermite interpolation, we developed a class of arbitrarily high-order and symmetric time integration formulae (see Chap. 10). The preservation of symmetry by a numerical scheme is also very important because the KG equation (1.52) is time reversible. Hairer et al. [15] have emphasised that symmetric methods have excellent long-time behaviour when solving reversible differential systems. Therefore, the preservation of time symmetry for a numerical scheme is also one of the favourable features.

Here it is worth emphasising that since $\mathscr{A}$ is a linear, unbounded positive semi-definite operator, it is a wise choice to approximate the operator $\mathscr{A}$ by a symmetric and positive semi-definite differentiation matrix M on a d-dimensional space when spatial discretisations of the underlying KG equation are carried out, and this will assist in structure preservation.

It is noted that a symmetric and arbitrarily high order time integration formula can be designed in operatorial terms in an infinite-dimensional function space $\mathscr{X}$ (see, e.g. [73, 74]). Using this approach, we also consider symplectic approximations for semilinear KG equations in Chap. 11. In practice, the differential operator $\mathscr{A}$ must be replaced with a suitable differentiation matrix M so that we may obtain a proper full discrete numerical scheme. Fortunately, there exist many research publications discussing the replacement of spatial derivatives of the semilinear KG equation (1.52) with periodic boundary conditions (1.53) in the literature. Thus, it is not a main point in this chapter. Here, however, again it is notable that the operator $\mathscr{A}$ should be approximated by a symmetric and positive semi-definite differentiation matrix M and the norm of M will change with the accuracy requirement of spatial discretisations. The higher the accuracy of spatial discretisations is required, the larger $\|M\|$ will be. This implies that the spatial structure preservation is required for the full discretisation of KG equation (1.52). A full discretisation for the KG equation (1.52) with periodic boundary conditions (1.53) is spatially structure-preserving if the operator $\mathscr{A}$ is approximated by a $d \times d$ symmetric and positive semi-definite differentiation matrix M, and the norm of M, $\|M\|$, tends to infinity as d tends to infinity, where d is the number of degrees of freedom in the space discretisation.

Obviously, the global error of a fully discrete scheme for the KG equation (1.52) depends on the accuracy of both time integrators and space discretisations. As the mesh partition in the space discretisation increases for (1.59), $\|M\|$ will increase, and the larger $\|M\|$ is, the higher the accuracy will be increased in space approximations.

Remark 1.7 Actually, the family of matrices $\{M_{d \times d}\}$ approximates the infinite-dimensional, unbounded, operator $\mathscr{A}$. $\|M_{d \times d}\|$ tends to infinity with d, where d is the dimension of the matrix $M_{d \times d}$; i.e., the number of degrees of freedom in the spatial discretisation. Consequently, the family of matrices $\{M_{d \times d}\}$ inherits the unbounded property of $\mathscr{A}$. This objectively reflects an important fact that the norm of the differentiation matrix $M_{d \times d}$ could be arbitrarily large, depending on the requirement of computational accuracy, and the corresponding system of second-order differential equations must be a multi-frequency highly oscillatory system once high global accuracy is required. In this case, an oscillation-preserving time integrator is needed for the numerical simulation of semilinear wave equations, including the KG equation (1.52), during a long-time computation. Moreover, the accuracy of the oscillation-preserving time integrator will be required to match that of the space discretisation. Hence, oscillation-preserving ERKN integrators of arbitrarily high order are favourable in applications, especially when applied to the semidiscrete KG equation, and high-accuracy time integrators will be required for the underlying PDEs in practice.

1.7 Numerical Experiments

This section concerns numerical experiments, and we will consider four problems which are closely related to (1.1) or (1.9). Since explicit methods are cheaper (use less CPU time in general) than implicit methods, we use three explicit RKN-type methods and two implicit methods. These methods are chosen as follows:

- ERKN3s4: the explicit three-stage ERKN method of order four presented in [18] (with its Butcher tableau given by Table 1.1);
- ARKN3s4: the explicit three-stage ARKN method of order four proposed in [76] (with its Butcher tableau given by Table 1.2);
- ERKN7s6: the explicit seven-stage ERKN method of order six derived in [77] with the coefficients

$$c_5 = 1 - c_3 = 0.06520862987680341024,$$
$$c_6 = 1 - c_2 = 0.65373769483744778901,$$
$$c_7 = 1 - c_1 = 0.05586607811787376572,$$
$$c_4 = 0.5,$$
$$d_4 = 0.26987577187133640373,$$
$$d_5 = d_3 = 0.92161977504885189358,$$

$$d_6 = d_2 = 0.13118241020105280626,$$
$$d_7 = d_1 = -0.68774007118557290171,$$
$$\bar{b}_i(V) = d_i c_{8-i} \phi_1(c_{8-i}^2 V), \quad b_i(V) = d_i \phi_0(c_{8-i}^2 V),$$
for $i = 1, 2, \cdots, 7,$
$$\bar{a}_{ij}(V) = d_j(c_i - c_j)\phi_1((c_i - c_j)^2 V),$$
for $i = 2, 3, \cdots, 7, \quad j = 1, 2, \cdots, i - 1;$

Table 1.1 Butcher tableau of ERKN3s4

c			
$\dfrac{1}{2(2-\sqrt[3]{2})}$	0	0	0
$\dfrac{1}{2}$	$\dfrac{-1+\sqrt[3]{2}}{2(2-\sqrt[3]{2})^2}\phi_1\left(\dfrac{2-\sqrt[3]{2}}{24}v\right)$	0	0
$1-\dfrac{1}{2(2-\sqrt[3]{2})}$	$\dfrac{1-\sqrt[3]{2}}{(2-\sqrt[3]{2})^2}\phi_1\left(\left(1-\dfrac{1}{2-\sqrt[3]{2}}\right)^2 v\right)$	$\dfrac{2+\sqrt[3]{4}}{12}\phi_1\left(\dfrac{2-\sqrt[3]{2}}{24}v\right)$	0
	$\dfrac{4+\sqrt[3]{2}}{12}\phi_1\left(\left(1-\dfrac{1}{2(2-\sqrt[3]{2})}\right)^2 v\right)$	$\dfrac{1}{\sqrt[3]{2^2}(-2+\sqrt[3]{2})}\phi_1\left(\dfrac{1}{4}v\right)$	$\dfrac{1}{2(-2+\sqrt[3]{2})^2}\phi_1\left(\dfrac{1}{4(2-\sqrt[3]{2})^2}v\right)$
	$\dfrac{1}{2(2-\sqrt[3]{2})}\phi_0\left(\left(1-\dfrac{1}{2(2-\sqrt[3]{2})}\right)^2 v\right)$	$\dfrac{1}{2}\phi_0\left(\dfrac{1}{4}v\right)$	$\left(1-\dfrac{1}{2(2-\sqrt[3]{2})}\right)\phi_0\left(\dfrac{1}{4(2-\sqrt[3]{2})^2}v\right)$

Table 1.2 Butcher tableau of ARKN3s4

$\dfrac{1}{8}$	0		0
$\dfrac{23}{42}$	$\dfrac{71}{441}$		0
$\dfrac{11}{12}$	$\dfrac{2641}{14058}$		$\dfrac{4123}{18744}$
	$\dfrac{2024\phi_2(V) - 5904\phi_3(V) + 8064\phi_4(V)}{1349}$		$\dfrac{-1617\phi_2(V) + 14700\phi_3(V) - 28224\phi_4(V)}{2201}$
	$\dfrac{2024\phi_1(V) - 5904\phi_2(V) + 8064\phi_3(V)}{1349}$		$\dfrac{-1617\phi_1(V) + 14700\phi_2(V) - 28224\phi_3(V)}{2201}$
$\dfrac{1}{8}$	0		
$\dfrac{23}{42}$	0		
$\dfrac{11}{12}$	0		
	$\dfrac{138\phi_2(V) - 1356\phi_3(V) + 4032\phi_4(V)}{589}$		
	$\dfrac{138\phi_1(V) - 1356\phi_2(V) + 4032\phi_3(V)}{589}$		

Table 1.3 Butcher tableau of TFCr2

c_1	$c_1^2 b_1 \displaystyle\sum_{j=0}^{r-1} I_{1,j,c_1} \widehat{P}_j(c_1)$	$c_1^2 b_2 \displaystyle\sum_{j=0}^{r-1} I_{1,j,c_1} \widehat{P}_j(c_2)$
c_2	$c_2^2 b_1 \displaystyle\sum_{j=0}^{r-1} I_{1,j,c_2} \widehat{P}_j(c_1)$	$c_2^2 b_2 \displaystyle\sum_{j=0}^{r-1} I_{1,j,c_2} \widehat{P}_j(c_2)$
	$\displaystyle\sum_{j=0}^{r-1} I_{1,j} b_1 \widehat{P}_j(c_1)$	$\displaystyle\sum_{j=0}^{r-1} I_{1,j} b_2 \widehat{P}_j(c_2)$
	$\displaystyle\sum_{j=0}^{r-1} I_{2,j} b_1 \widehat{P}_j(c_1)$	$\displaystyle\sum_{j=0}^{r-1} I_{2,j} b_2 \widehat{P}_j(c_2)$

- TFCr2: the TFC method (1.40) of order four described in [8] (with its Butcher tableau given by Table 1.3 with the coefficients $c_1 = \dfrac{3 - \sqrt{3}}{6}, c_2 = \dfrac{3 + \sqrt{3}}{6}$, $b_1 = b_2 = \dfrac{1}{2}$, and $r = 2$);
- TFCr3: the TFC method (1.40) of order six described in [8] (with its Butcher tableau given by Table 1.4 with the coefficients $c_1 = \dfrac{5 - \sqrt{15}}{10}, c_2 = \dfrac{1}{2}, c_3 = \dfrac{5 + \sqrt{15}}{10}, b_1 = \dfrac{5}{18}, b_2 = \dfrac{4}{9}, b_3 = \dfrac{5}{18}$, and $r = 3$).

We remark that an ERKN method reduces to an RKN method when $M \to \mathbf{0}$. Hence, the reduced method ERKN3s4 is assigned as the corresponding RKN method, which is denoted by RKN3s4. In the numerical experiments, we use fixed-

Table 1.4 Butcher tableau of TFCr3

c_1	$c_1^2 b_1 \sum\limits_{j=0}^{r-1} I_{1,j,c_1} \widehat{P}_j(c_1)$	$c_1^2 b_2 \sum\limits_{j=0}^{r-1} I_{1,j,c_1} \widehat{P}_j(c_2)$	$c_1^2 b_3 \sum\limits_{j=0}^{r-1} I_{1,j,c_1} \widehat{P}_j(c_3)$
c_2	$c_2^2 b_1 \sum\limits_{j=0}^{r-1} I_{1,j,c_2} \widehat{P}_j(c_1)$	$c_2^2 b_2 \sum\limits_{j=0}^{r-1} I_{1,j,c_2} \widehat{P}_j(c_2)$	$c_2^2 b_3 \sum\limits_{j=0}^{r-1} I_{1,j,c_2} \widehat{P}_j(c_3)$
c_3	$c_3^2 b_1 \sum\limits_{j=0}^{r-1} I_{1,j,c_3} \widehat{P}_j(c_1)$	$c_3^2 b_2 \sum\limits_{j=0}^{r-1} I_{1,j,c_3} \widehat{P}_j(c_2)$	$c_3^2 b_3 \sum\limits_{j=0}^{r-1} I_{1,j,c_3} \widehat{P}_j(c_3)$
	$\sum\limits_{j=0}^{r-1} I_{1,j} b_1 \widehat{P}_j(c_1)$	$\sum\limits_{j=0}^{r-1} I_{1,j} b_2 \widehat{P}_j(c_2)$	$\sum\limits_{j=0}^{r-1} I_{1,j} b_3 \widehat{P}_j(c_3)$
	$\sum\limits_{j=0}^{r-1} I_{2,j} b_1 \widehat{P}_j(c_1)$	$\sum\limits_{j=0}^{r-1} I_{2,j} b_2 \widehat{P}_j(c_2)$	$\sum\limits_{j=0}^{r-1} I_{2,j} b_3 \widehat{P}_j(c_3)$

point iteration for the implicit TFC methods. We set 10^{-16} as the error tolerance and 10 as the maximum number of each iteration. It will be observed from the numerical experiments that the numerical behaviour of the ERKN methods and TFC methods is much better than that of the ARKN and RKN methods.

In these methods, the matrix-valued functions $\phi_i(V)$, for $i = 0, 1, \cdots, 4$, are defined by (1.4).

Problem 1.1 Consider the Duffing equation (see, e.g. [10, 13, 78, 79])

$$\begin{cases} \ddot{q} + \omega^2 q = k^2(2q^3 - q), \\ q(0) = 0, \quad \dot{q}(0) = \omega, \end{cases}$$

where $0 \leqslant k < \omega$. As is known, this is a Hamiltonian system with the Hamiltonian

$$H(p, q) = \frac{1}{2}p^2 + \frac{1}{2}\omega^2 q^2 + \frac{k^2}{2}(q^2 - q^4).$$

The analytic solution is given by

$$q(t) = sn(\omega t, k/\omega),$$

where sn denotes the Jacobian elliptic function (see, e.g. [80]).

Problem 1.1 is solved on the interval $[0, 10000]$ with $k = 0.03$ and $\omega = 50$. Figure 1.1a presents the global errors results (in logarithmic scale) with the stepsizes $h = \dfrac{0.1}{2^j}$ for $j = 1, \cdots, 4$. We here remark that some global errors for RKN3s4 are too large to be plotted in Fig. 1.1a due to its instability and nonconvergence with the stepsize $h = 0.05$. In the next problems, similar situations are encountered and the corresponding points are not plotted either. We also show the global errors against the CPU time in Fig. 1.1b. It can be observed from these figures, ERKN3s4, ERKN7s6 and TFC methods are much more accurate than ARKN3s4 and RKN3s4, and RKN3s4 gives disappointing accuracy in comparison with the other methods, although it is a symplectic and symmetric method. *This observation implies that the property of oscillation preservation for numerical methods is also of great importance in Geometric Integration.* What can we learn from this observation? This experiment demonstrates that for a nonlinear highly oscillatory differential equation, the most important consideration should be the oscillation preservation when concerning numerical solutions.

Meanwhile, we also show the curves of the Hamiltonian error growth with $h = \dfrac{1}{40}$ as the integration interval is extended in Fig. 1.1c for all the methods, where the ERKN methods and TFC methods show better numerical energy preservation than the reduced RKN method: RKN3s4, and ARKN3s4 method. It follows from Fig. 1.1c that both the ERKN and RKN methods can preserve the energy approximately, whereas the ARKN3s4 method cannot. In fact, it is clear from Fig. 1.1c that the energy of the ARKN3s4 method grows as the integration interval is extended. This is because both the ERKN and RKN methods are symplectic and symmetric methods, whereas the ARKN3s4 method is not a symplectic method. Another important aspect is that, just as its algebraic accuracy, the accuracy of energy preservation of RKN3s4 method is also disappointing, even though RKN3s4 method possesses both favourable properties of symplecticity preservation and symmetry preservation. It is worth noting that, although the TFC methods are not symplectic, they are oscillation preserving and preserve the energy approximately. Hence, we should take full account of the oscillation-preserving structure in the design of numerical methods for efficiently solving a highly oscillatory nonlinear Hamiltonian system, although we cannot ignore the other structures.

Problem 1.2 Consider the sine-Gordon equation (see, e.g. [81])

$$\frac{\partial^2 u}{\partial t^2} = \frac{\partial^2 u}{\partial x^2} - \sin u,$$

on the region $-10 \leqslant x \leqslant 10$ and $t_0 \leqslant t \leqslant T$ with the initial conditions

$$u(x, t_0) = -4\arctan\left(c^{-1}\operatorname{sech}(\kappa x)\sin(-t_0 c\kappa)\right),$$

$$u_t(x, t_0) = \frac{4\kappa \cos(-t_0 c\kappa)\operatorname{sech}(\kappa x)}{1 + c^{-2}\operatorname{sech}^2(\kappa x)\sin^2(-t_0 c\kappa)},$$

and the boundary conditions

$$u(-10, t) = u(10, t) = -4\arctan\left(c^{-1}\operatorname{sech}(10\kappa)\sin(c\kappa t)\right),$$

where $\kappa = 1/\sqrt{1 + c^2}$. The exact solution is

$$u(x, t) = -4\arctan\left(c^{-1}\operatorname{sech}(\kappa x)\sin(c\kappa t)\right).$$

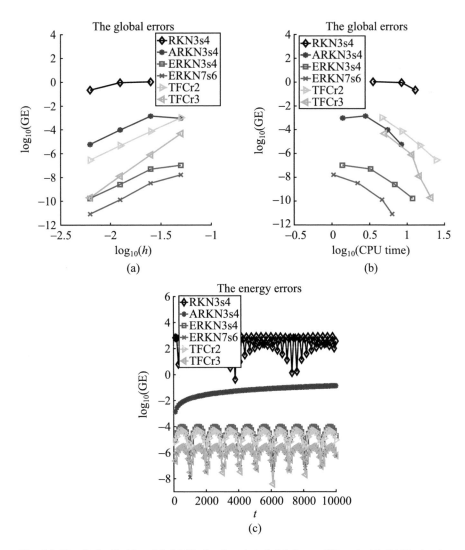

Fig. 1.1 Results for Problem 1.1. (**a**) The log-log plot of global error GE against h. (**b**) The log-log plot of global error GE against CPU time. (**c**) The logarithm of the global energy error GE against t

For this problem, we use the Chebyshev pseudospectral discretisation with 240 spatial mesh grids and select the parameter $c = 0.5$, which leads to a discretisation of the type (1.9). This equation is solved on the interval [0, 100]. Figure 1.2a, b show the global errors results (in logarithmic scale) with the stepsizes $h = \dfrac{1}{2^k}$ for $k = 1, \cdots, 4$. We then integrate this equation with the stepsize $h = \dfrac{1}{10}$ on the interval [0, 10000] and the numerical energy conservation is presented in Fig. 1.2c.

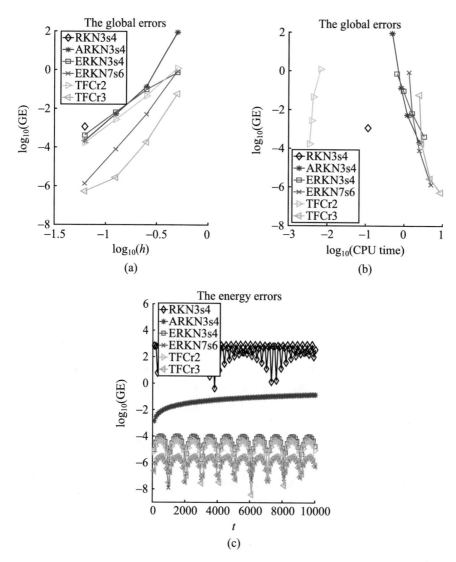

Fig. 1.2 Results for Problem 1.2. (**a**) The log-log plot of global error GE against h. (**b**) The log-log plot of global error GE against CPU time. (**c**) The logarithm of the global energy error GE against t

Again it can be observed from the numerical results that the numerical behaviour of ERKN methods and TFC methods is much better than that of ARKN3s4 and RKN3s4. In summary, an oscillation-preserving numerical method gives much better results than those methods which are not oscillation preserving. In particular, the symplectic and symmetric RKN3s4 performs badly and leads to completely disappointing numerical results in this numerical experiment.

Problem 1.3 Consider the semilinear wave equation

$$
\begin{cases}
\dfrac{\partial^2 u}{\partial t^2} - a(x)\dfrac{\partial^2 u}{\partial x^2} + 92u = f(t, x, u), \quad 0 < x < 1, \quad t > 0, \\[2mm]
u(x, 0) = a(x), \quad u_t(x, 0) = 0, \\[2mm]
u(0, t) = 0, \quad u(1, t) = 0,
\end{cases}
$$

where

$$
a(x) = 4x(1 - x)
$$

and

$$
f(t, x, u) = u^5 - a^2(x)u^3 + \frac{a^5(x)}{4} \sin^2(20t) \cos(10t).
$$

The exact solution of this problem is

$$
u(x, t) = a(x) \cos(10t),
$$

which represents a vibrating string.

Differently from Problem 1.2, we now consider semidiscretisation of the spatial variable with second-order symmetric differences, and this results in

$$
\frac{d^2 U}{dt^2} + MU = F(t, U), \quad U(0) = \big(a(x_1), \cdots, a(x_{N-1})\big)^\mathsf{T}, \quad U'(0) = \mathbf{0}, \quad (1.64)
$$

where $U(t) = \big(u_1(t), \cdots, u_{N-1}(t)\big)^\mathsf{T}$ with $u_i(t) \approx u(x_i, t)$, $x_i = i\Delta x$ for $i = 1, \cdots, N - 1$, and $\Delta x = 1/N$.

$$
M = 92 I_{N-1} + \frac{1}{\Delta x^2}
\begin{pmatrix}
2a(x_1) & -a(x_1) \\
-a(x_2) & 2a(x_2) & -a(x_2) \\
& \ddots & \ddots & \ddots \\
& & -a(x_{N-2}) & 2a(x_{N-2}) & -a(x_{N-2}) \\
& & & -a(x_{N-1}) & 2a(x_{N-1})
\end{pmatrix},
$$

and

$$
F(t, U) = \big(f(t, x_1, u_1), \cdots, f(t, x_{N-1}, u_{N-1})\big)^\mathsf{T}.
$$

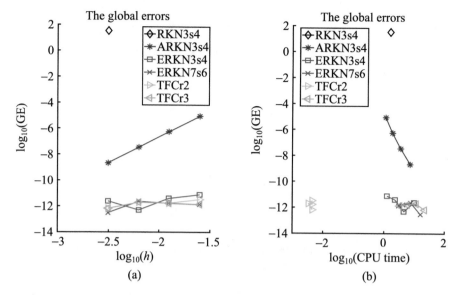

Fig. 1.3 Results for Problem 1.3. (**a**) The log-log plot of global error GE against h. (**b**) The log-log plot of global error GE against CPU time

(1.64) is a highly oscillatory system, but not a Hamiltonian system. We solve this problem on the interval $[0, 10]$ with $N = 256$ and $h = \dfrac{0.1}{2^j}$ for $j = 1, \cdots, 4$. The global errors are shown in Fig. 1.3. Once again, it can be observed from Fig. 1.3 that the numerical behaviour of the oscillation-preserving ERKN methods and TFC methods is much better than the others. In Fig. 1.3, the global errors figures of both ERKN methods and TFC methods almost coincide with each other.

We next consider a damped wave equation. For this problem, we choose time integrators as follows:

- ERKN3s3: the explicit three-stage ERKN method of order three proposed in [27] and denoted by the Butcher tableau

0	0	0	0	0	0	0
$\dfrac{1}{3}$	$\dfrac{1}{3}$	0	0	0	0	0
$\dfrac{2}{3}$	0	$\dfrac{2}{3}$	0	$\dfrac{2}{9}$	0	0
	$\bar{b}_1(V)$	$\bar{b}_2(V)$	$\bar{b}_3(V)$	$b_1(V)$	$b_2(V)$	$b_3(V)$

where

$$b_1(V) = \phi_1(V) - \frac{9}{2}\phi_2(V) + 9\phi_3(V),$$

$$b_2(V) = 6\phi_2(V) - 18\phi_3(V),$$

$$b_3(V) = -\frac{3}{2}\phi_2(V) + 9\phi_3(V),$$

$$\bar{b}_1(V) = \phi_2(V) - \frac{9}{2}\phi_3(V) + 9\phi_4(V),$$

$$\bar{b}_2(V) = 6\phi_3(V) - 18\phi_4(V),$$

$$\bar{b}_3(V) = -\frac{3}{2}\phi_3(V) + 9\phi_4(V).$$

- ARKN3s3: the explicit three-stage ARKN method of order three given in [82] and denoted by the Butcher tableau

0	0	0	0	0	0	0
$\dfrac{1}{2}$	$\dfrac{1}{2}$	0	0	$\dfrac{1}{8}$	0	0
1	-1	2	0	$\dfrac{1}{2}$	0	0
	$\bar{b}_1(V)$	$\bar{b}_2(V)$	$\bar{b}_3(V)$	$b_1(V)$	$b_2(V)$	$b_3(V)$

where

$$b_1(V) = \phi_1(V) - 3\phi_2(V) + 4\phi_3(V),$$

$$b_2(V) = 4\phi_2(V) - 8\phi_3(V),$$

$$b_3(V) = -\phi_2(V) + 4\phi_3(V),$$

$$\bar{b}_1(V) = \phi_2(V) - \frac{3}{2}\phi_3(V),$$

$$\bar{b}_2(V) = \phi_3(V),$$

$$\bar{b}_3(V) = \frac{1}{2}\phi_3(V).$$

- RKN3s3: the explicit three-stage RKN method of order three with the Butcher tableau

$$
\begin{array}{c|ccc|ccc}
0 & 0 & 0 & 0 & 0 & 0 & 0 \\
\dfrac{1}{2} & \dfrac{1}{2} & 0 & 0 & \dfrac{1}{8} & 0 & 0 \\
1 & -1 & 2 & 0 & \dfrac{1}{2} & 0 & 0 \\
\hline
 & \dfrac{1}{6} & \dfrac{2}{3} & \dfrac{1}{6} & \dfrac{1}{4} & \dfrac{1}{6} & \dfrac{1}{12}
\end{array}
$$

Problem 1.4 Consider the damped wave equation (see, e.g. [27, 82])

$$
\begin{cases}
\dfrac{\partial^2 u}{\partial t^2} + \dfrac{\partial u}{\partial t} - \dfrac{\partial^2 u}{\partial x^2} = f(u), & -1 < x < 1, \ t > 0, \\[2mm]
u(-1, t) = u(1, t).
\end{cases}
$$

A semidiscretisation in the spatial variable by using second-order symmetric differences yields the type of (1.1)

$$
\ddot{U} + MU = F(U, \dot{U}), \tag{1.65}
$$

where $U(t) = \big(u_1(t), \cdots, u_N(t)\big)^{\mathsf{T}}$ with $u_i(t) \approx u(x_i, t)$, $x_i = -1 + i\Delta x$ for $i = 1, \cdots, N$, $\Delta x = 2/N$,

$$
M = \frac{1}{\Delta x^2}
\begin{pmatrix}
2 & -1 & & & -1 \\
-1 & 2 & -1 & & \\
 & \ddots & \ddots & \ddots & \\
 & & -1 & 2 & -1 \\
-1 & & & -1 & 2
\end{pmatrix},
$$

and

$$
F(U, \dot{U}) = \big(f(u_1) - \dot{u}_1, \cdots, f(u_N) - \dot{u}_N\big)^{\mathsf{T}}.
$$

In this experiment, we consider the damped sine-Gordon equation with $f(u) = -\sin u$ and with the initial conditions

$$
U(0) = (\pi)_{i=1}^{N}, \quad \dot{U}(0) = \sqrt{N}\Big(0.01 + \sin(\frac{2\pi i}{N})\Big)_{i=1}^{N}.
$$

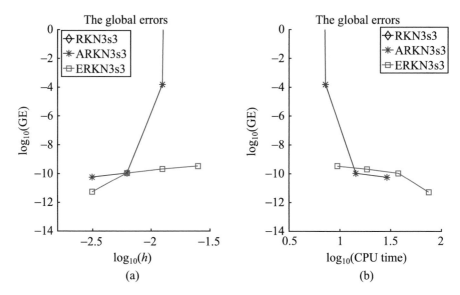

Fig. 1.4 Results for Problem 1.4. (**a**) The log-log plot of global error GE against h. (**b**) The log-log plot of global error GE against CPU time

This equation is integrated on $[0, 100]$ with $N = 256$ and $h = \dfrac{0.1}{2^i}$ for $i = 2, 3, 4, 5$. The global errors against the stepsizes and the CPU time are shown in Fig. 1.4. It can be observed again from the results that the oscillation-preserving integrator ERKN3s3 performs much better than the other methods. It is easy to see that the ERKN3s3 integrator provides a considerably more accurate numerical solution than other methods.

1.8   Conclusions and Discussion

In practice, nonlinear second-order differential equations with highly oscillatory solution behaviour are ubiquitous in science and engineering applications. The overarching question now is how to preserve high frequency oscillations in the numerical treatment of nonlinear multi-frequency highly oscillatory second-order ordinary differential equations (1.1) or (1.9). This chapter presented systematic oscillation-preserving analysis, which began with the concept of oscillation preservation for RKN-type methods, and then analysed oscillation-preserving behaviour for RKN-type methods, including ERKN integrators, TFC methods, AVF methods, AAVF methods, ARKN methods, symplectic and symmetric RKN methods, and standard RKN methods, designed to solve the initial value problem of nonlinear multi-frequency highly oscillatory second-order ordinary differential equations (1.1) or (1.9). It was found that the ERKN integrators and TFC methods are oscillation preserving, whereas neither the ARKN methods nor the standard RKN methods, including symplectic and symmetric RKN methods, and AVF methods, are oscillation preserving. However, ARKN and AAVF methods are partly oscillation preserving. An oscillation-preserving integrator shows much better numerical behaviour than those methods which are not oscillation preserving when applied to nonlinear multi-frequency highly oscillatory second-order ordinary differential equations. The least favourable results are for the RKN method, in comparison with the ARKN method and ERKN method when solving nonlinear multi-frequency highly oscillatory problems. Here *the most interesting conclusion is that the oscillation-preserving property depends essentially on the internal stages rather than the updates of an RKN-type method when applied to highly oscillatory second-order systems.* This chapter also mentioned the potential developments of ERKN integrators and TCF methods for the nonlinear multi-frequency highly oscillatory second-order ordinary differential equation (1.9).

An important concern relating to oscillation-preserving integrators is to efficiently solve a semidiscrete nonlinear wave equation, which usually is approximated by a system of nonlinear highly oscillatory second-order ordinary differential equations derived from a suitable space discretisation of semilinear wave equations such as KG equations, i.e., the operator $\mathscr{A}$ appearing in (1.59) is approximated by a $d \times d$ symmetric and positive semi-definite differentiation matrix M. Therefore, the analysis of oscillation-preserving behaviour for RKN-type methods in this chapter is also significant for numerical PDEs. In this case, the standard RKN method, in comparison with an oscillation-preserving integrator such as the ERKN method, may not be a satisfactory choice for efficiently dealing with such highly oscillatory problems.

This chapter focuses on highly oscillatory second-order differential equations (1.1) or (1.9). Other highly oscillatory systems will also be discussed in this monograph. For instance, in Chap. 5, using exponential collocation methods, we will deal with the following highly oscillatory system:

$$q''(t) - Nq'(t) + \Upsilon q(t) = -\nabla U(q(t)), \qquad q(0) = q_0, \ q'(0) = q_0', \qquad t \in [0, T],$$

where N is a symmetric negative semi-definite matrix, Υ is a symmetric positive semi-definite matrix, and $U : \mathbb{R}^d \to \mathbb{R}$ is a differentiable function.

Last, but not least, we believe that the oscillation-preserving concept introduced and analysed in this chapter for numerical methods for solving nonlinear multi-frequency highly oscillatory differential equations is significant and interesting within the broader framework of the subject of Geometric Integration. The results of numerical experiments in this chapter have strengthened the impression that an oscillation-preserving integrator is required when efficiently solving a nonlinear multi-frequency highly oscillatory system, or a semidiscrete nonlinear wave equation.

The material in this chapter is based on the work by Wu et al. [83].

References

1. Petzold, L.R., Jay, L.O., Yen, J.: Numerical solution of highly oscillatory ordinary differential equations. Acta Numer. **7**, 437–483 (1997)
2. Cohen, D., Jahnke, T., Lorenz, K., et al.: Numerical integrators for highly oscillatory Hamiltonian systems: a review mielke A. In: Analysis, Modeling and Simulation of Multiscale Problems, pp. 553–576. Springer, Berlin (2006)
3. Brugnano, L.J., Montijano, I., Rández, L.: On the effectiveness of spectral methods for the numerical solution of multi-frequency highly oscillatory Hamiltonian problems. Numer. Algor. **81**, 345–376 (2019)
4. García-Archillay, B., Sanz-Serna, J.M., Skeel, R.D.: Long-time-step methods for oscillatory differential equations. SIAM J. Sci. Comput. **20**, 930–963 (1998)
5. González, A.B., Martín, P., Farto, J.M.: A new family of Runge–Kutta type methods for the numerical integration of perturbed oscillators. Numer. Math. **82**, 635–646 (1999)
6. van der Houwen, P.J., Sommeijer, B.P.: Explicit Runge–Kutta (–Nyström) methods with reduced phase errors for computing oscillating solution. SIAM J. Numer. Anal. **24**, 595–617 (1987)
7. Li, Y.W., Wu, X.: Functionally fitted energy-preserving methods for solving oscillatory nonlinear Hamiltonian systems. SIAM J. Numer. Anal. **54**, 2036–2059 (2016)
8. Wang, B., Iserles, A., Wu, X.: Arbitrary-order trigonometric Fourier collocation methods for multi-frequency oscillatory systems. Found. Comput. Math. **16**, 151–181 (2016)
9. Bao, W.Z., Dong, X.C.: Analysis and comparison of numerical methods for the Klein-Gordon equation in the nonrelativistic limit regime. Numer. Math. **120**, 189–229 (2012)
10. Mei, L., Liu, C., Wu, X.: An essential extension of the finite-energy condition for extended Runge–Kutta–Nyström integrators when applied to nonlinear wave equations. Commun. Comput. Phys. **22**, 742–764 (2017)
11. Wang, B., Wu, X.: The formulation and analysis of energy-preserving schemes for solving high-dimensional nonlinear Klein-Gordon equations. IMA J. Numer. Anal. **39**, 2016–2044 (2019)
12. Nyström, E.J.: Uber die numerische Integration von differentialgleichungen. Acta. Soc. Sci. Fennicae **50**(13), 1–54 (1925)
13. Wu, X., Wang, B.: Recent developments. In: Structure-Preserving Algorithms for Oscillatory Differential Equations. Springer Nature Singapore Pte Ltd., Singapore (2018)
14. Franco, J.M.: Runge–Kutta–Nyström methods adapted to the numerical integration of perturbed oscillators. Comput. Phys. Commun. **147**, 770–787 (2002)
15. Hairer, E., Lubich, C., Wanner, G.: Geometric Numerical Integration: Structure-Preserving Algorithms for Ordinary Differential Equations, 2nd edn. Springer, Berlin (2006)
16. Hochbruck, M., Lubich, C.h.: A Gautschi-type method for oscillatory second-order differential equations. Numer. Math. **83**, 403–426 (1999)
17. Li, Y.W., Wu, X.: Exponential integrators preserving first integrals or Lyapunov functions for conservative or dissipative systems. SIAM J. Sci. Comput. **38**, 1876–1895 (2016)
18. Mei, L., Wu, X.: The construction of arbitrary order ERKN methods based on group theory for solving oscillatory Hamiltonian systems with applications. J. Comput. Phys. **323**, 171–190 (2016)
19. Tocino, A., Vigo-Aguiar, J.: Symplectic conditions for exponential fitting Runge–Kutta–Nyström methods. Math. Comput. Model. **42**, 873–876 (2005)
20. Wang, B., Meng, F., Fang, Y.: Efficient implementation of RKN-type Fourier collocation methods for second-order differential equations. Appl. Numer. Math. **119**, 164–178 (2017)
21. Wang, B., Wu, X.: A new high precision energy-preserving integrator for system of oscillatory second-order differential equations. Phys. Lett. A **376**, 1185–1190 (2013)
22. Wang, B., Wu, X.: Global error bounds of one-stage extended RKN integrators for semilinear wave equations. Numer. Algor. **81**, 1203–1218 (2019)

23. Wu, X., Wang, B., Shi, W.: Efficient energy-preserving integrators for oscillatory Hamiltonian systems. J. Comput. Phys. **235**, 587–605 (2013)
24. Wu, X., Wang, B., Xia, J.: Explicit symplectic multidimensional exponential fitting modified Runge–Kutta–Nyström methods. BIT Numer. Math. **52**, 773–795 (2012)
25. Wu, X., You, X., Shi, W., et al.: ERKN integrators for systems of oscillatory second-order differential equations. Comput. Phys. Commun. **181**, 1873–1887 (2010)
26. Yang, H., Zeng, X., Wu, X., et al.: A simplified Nyström-tree theory for extended Runge–Kutta–Nyström integrators solving multi-frequency oscillatory systems. Comput. Phys. Commun. **185**, 2841–2850 (2014)
27. You, X., Zhao, J., Yang, H., et al.: Order conditions for RKN methods solving general second-order oscillatory systems. Numer. Algor. **66**, 147–176 (2014)
28. Zeng, X., Yang, H., Wu, X.: An improved tri-colored rooted-tree theory and order conditions for ERKN methods for general multi-frequency oscillatory systems. Numer. Algor. **75**, 909–935 (2017)
29. Wu, X., You, X., Xia, J.: Order conditions for ARKN methods solving oscillatory systems. Comput. Phys. Commun. **180**, 2250–2257 (2009)
30. Wu, X., You, X., Li, J.: Note on derivation of order conditions for ARKN methods for perturbed oscillators. Comput. Phys. Commun. **180**, 1545–1549 (2009)
31. Li, J., Shi, W., Wu, X.: The existence of explicit symplectic ARKN methods with several stages and algebraic order greater than two. J. Comput. Appl. Math. **353**, 204–209 (2019)
32. Shi, W., Wu, X.: A note on symplectic and symmetric ARKN methods. Comput. Phys. Commun. **184**, 2408–2411 (2013)
33. Franco, J.M.: New methods for oscillatory systems based on ARKN methods. Appl. Numer. Math. **56**, 1040–1053 (2006)
34. Filon, L.N.G.: On a quadrature formula for trigonometric integrals. Proc. Royal Soc. Edin. **49**, 38–47 (1928)
35. Iserles, A., Levin, D.: Asymptotic expansion and quadrature of composite highly oscillatory integrals. Math. Comput. **80**, 279–296 (2011)
36. Baker, T.S., Dormand, J.R., Gilmore, J.P., et al.: Continuous approximation with embedded Runge–Kutta methods. Appl. Numer. Math. **22**, 51–62 (1996)
37. Li, J., Wu, X.: Energy-preserving continuous stage extended Runge–Kutta–Nyström methods or oscillatory Hamiltonian systems. Appl. Numer. Math. **145**, 469–487 (2019)
38. Owren, B., Zennaro, M.: Order barriers for continuous explicit Runge–Kutta methods. Math. Comput. **56**, 645–661 (1991)
39. Owren, B., Zennaro, M.: Derivation of efficient, continuous, explicit Runge–Kutta methods. SIAM J. Sci. Stat. Comput. **13**, 1488–1501 (1992)
40. Papakostas, S.N., Tsitouras, C.: Highly continuous interpolants for one-step ODE solvers and their application to Runge–Kutta methods. SIAM J. Numer. Anal. **34**, 22–47 (1997)
41. Verner, J.H., Zennaro, M.: The orders of embedded continuous explicit Runge–Kutta methods. BIT Numer. Math. **35**, 406–416 (1995)
42. Deuflhard, P.: A study of extrapolation methods based on multistep schemes without parasitic solutions. Z. Angew. Math. Phys. **30**, 177–189 (1979)
43. Gautschi, W.: Numerical integration of ordinary differential equations based on trigonometric polynomials. Numer. Math. **3**, 381–397 (1961)
44. Grimm, V., Hochbruck, M.: Error analysis of exponential integrators for oscillatory second order differential equations. J. Phys. A **39**, 5495 (2006)
45. Hersch, J.: Contribution à la méthode des équations aux differences. ZAMP **9**, 129–180 (1958)
46. Hochbruck, M., Ostermann, A.: Exponential integrators. Acta Numer. **19**, 209–286 (2010)
47. Grubmüller, H., Heller, H., Windemuth, A., et al.: Generalized Verlet algorithm for efficient molecular dynamics simulations with long-range interactions. Molecul. Simul. **6**, 121–142 (1991)

48. Lorenz, K., Jahnke, T., Lubich, C.: Adiabatic integrators for highly oscillatory second-order linear differential equations with time-varying eigen decomposition. BIT Numer. Math. **45**, 91–115 (2005)
49. Wu, X., Liu, K., Shi, W.: Structure-Preserving Algorithms for Oscillatory Differential Equations II. Springer, Heidelberg (2015)
50. Feng, K.: On difference schemes and symplectic geometry. In: Proceedings of the 5th International Symposium on Differential Geometry & Differential Equations, pp. 42–58. Science Press, Beijing (1985)
51. Feng, K.: Difference schemes for Hamiltonian formalism and symplectic geometry. J. Comp. Math. **4**, 279–289 (1986)
52. Feng, K., Qin, M.: Symplectic Geometric Algorithms for Hamiltonian Systems. Springer, Berlin (2010)
53. Sanz-Serna, J.M.: Runge–Kutta schemes for Hamiltonian systems. BIT Numer. Math. **28**, 877–883 (1988)
54. Wu, X., You, X., Wang, B.: Structure-Preserving Algorithms for Oscillatory Differential Equations. Springer, Berlin (2013)
55. Leimkuhler, B., Reich, S.: Simulating Hamiltonian Dynamics. Cambridge University Press, Cambridge (2004)
56. Blanes, S., Casas, F.: A Concise Introduction to Geometric Numerical Integration. CRC Press, Taylor & Francis Group, Florida (2016)
57. McLachlan, R.I., Quispel, G.R.W.: Geometric integrators for ODEs. J. Phys. A **39**, 5251 (2006)
58. McLachlan, R.I., Quispel, G.R.W., Tse, P.S.P.: Linearization-preserving self-adjoint and symplectic integrators. BIT Numer. Math. **49**, 177–197 (2009)
59. Sanz-Serna, J.M.: Symplectic integrators for Hamiltonian problems: an overview. Acta Numer. **1**, 243–286 (1992)
60. Sanz-Serna, J.M., Calvo, M.P.: Numerical Hamiltonian Problems. Chapman & Hall, London (1994)
61. Hairer, E.: Energy-preserving variant of collocation methods. Am. J. Numer. Anal. Ind. Appl. Math. **5**, 73–84 (2010)
62. Hairer, E., Nörsett, S.P., Wanner, G.: Solving Ordinary Differential Equations I: Nonstiff Problems. Springer, Berlin (1993)
63. Iserles, A.: A First Course in the Numerical Analysis of Differential Equations, 2nd edn. Cambridge University Press, Cambridge (2008)
64. Wright, K.: Some relationships between implicit Runge–Kutta, collocation and Lanczost methods, and their stability properties. BIT Numer. Math. **10**, 217–227 (1970)
65. Celledoni, E., McLachlan, R.I., Owren, B., et al.: Energy-preserving integrators and the structure of B-series. Found. Comput. Math. **10**, 673–693 (2010)
66. Hairer, E., Lubich, C.: Long-time energy conservation of numerical methods for oscillatory differential equations. SIAM J. Numer. Anal. **38**, 414–441 (2000)
67. McLachlan, R.I., Quispel, G.R.W., Robidoux, N.: Geometric integration using discrete gradients. Philos. Trans. R. Soc. A **357**, 1021–1046 (1999)
68. Quispel, G.R.W., McLaren, D.I.: A new class of energy-preserving numerical integration methods. J. Phys. A Math. Theor. **41**, 045206 (2008)
69. Grimm, V.: On error bounds for the Gautschi-type exponential integrator applied to oscillatory second-order differential equations. Numer. Math. **100**, 71–89 (2005)
70. Drazin, P.J., Johnson, R.S.: Solitons: An Introduction. Cambridge University Press, Cambridge (1989)
71. Bratsos, A.G.: On the numerical solution of the Klein-Gordon equation. Numer. Methods Partial Differ. Equ. **25**, 939–951 (2009)
72. Liu, C., Wu, X.: The boundness of the operator-valued functions for multidimensional nonlinear wave equations with applications. Appl. Math. Lett. **74**, 60–67 (2017)

73. Liu, C., Iserles, A., Wu, X.: Symmetric and arbitrarily high-order Birkhoff-Hermite time integrators and their long-time behavior for solving nonlinear Klein-Gordon equations. J. Comput. Phys. **356**, 1–30 (2018)
74. Liu, C., Wu, X.: Arbitrarily high-order time-stepping schemes based on the operator spectrum theory for high-dimensional nonlinear Klein–Gordon equations. J. Comput. Phys. **340**, 243–275 (2017)
75. Wu, X., Liu, C.: An integral formula adapted to different boundary conditions for arbitrarily high-dimensional nonlinear Klein–Gordon equations with its applications. J. Math. Phys. **57**, 021504 (2016)
76. Wu, X., Wang, B.: Multidimensional adapted Runge–Kutta–Nyström methods for oscillatory systems. Comput. Phys. Commun. **181**, 1955–1962 (2010)
77. Wang, B., Yang, H., Meng, F.: Sixth order symplectic and symmetric explicit ERKN schemes for solving multi-frequency oscillatory nonlinear Hamiltonian equations. Calcolo **54**, 117–140 (2017)
78. Kovacic, I., Brennan, M.J.: The Duffing Equation: Nonlinear Oscillators. Wiley, Hoboken (2011)
79. Liu, K., Shi, W., Wu, X.: An extended discrete gradient formula for oscillatory Hamiltonian systems. J. Phys. A Math. Theor. **46**, 165203 (2013)
80. Abramowitz, M., Stegun, I.A.: Handbook of Mathematcal Functions with Formulas, Graphs, and Mathematical Tables. National Bureau of Standards, Washington (1964)
81. Schiesser, W.E., Griffiths, G.W.: A Compendium of Partial Differential Equation Models: Method of Lines Analysis with Matlab. Cambridge University Press, Cambridge (2009)
82. Liu, K., Wu, X., Shi, W.: Extended phase properties and stability analysis of RKN-type integrators for solving general oscillatory second-order initial value problems. Numer. Algor. **77**, 37–56 (2018)
83. Wu, X., Wang, B., Mei, L.: Oscillation-preserving algorithms for efficiently solving highly oscillatory second-order ODEs. Numer. Algor. **86**, 693–727 (2021)

Chapter 2
Continuous-Stage ERKN Integrators for Second-Order ODEs with Highly Oscillatory Solutions

In this chapter, continuous-stage extended Runge–Kutta–Nyström (CSERKN) integrators for solving highly oscillatory systems of second-order ODEs are derived and analysed. These integrators are incorporated into the special structure of highly oscillatory systems so that their internal stages and updates can integrate the associated highly oscillatory homogeneous systems exactly. When the underlying highly oscillatory systems are Hamiltonian systems, sufficient conditions for energy preservation are shown for CSERKN methods. The symmetry and stability of CSERKN integrators are also analysed in detail. Preliminary numerical results highlight the effectiveness of CSERKN methods.

2.1 Introduction

We consider the following system of second-order ordinary differential equations with oscillatory solutions

$$\begin{cases} q''(t) + Mq(t) = f(q(t)), & t \in [t_0, T], \\ q(t_0) = q_0, \quad q'(t_0) = q'_0, \end{cases} \tag{2.1}$$

where $M \in \mathbb{R}^{d \times d}$ is a symmetric positive semi-definite matrix that implicitly contains the dominant frequencies of the system, $q \in \mathbb{R}^d$ and $f(q) : \mathbb{R}^d \to \mathbb{R}^d$ is a nonlinear function which is independent of q'. If $\|M\| \gg \max\left\{1, \left\|\frac{\partial f}{\partial q}\right\|\right\}$ then (2.1) is a highly oscillatory problem. This kind of problem frequently occurs in science and engineering fields such as quantum mechanics, astrophysics, quantum chemistry and electronics. It is particularly interesting when this highly oscillatory problem is obtained from a spatial semidiscretisation of a semilinear wave equation within the framework of the method of lines [1]. In practice, the system (2.1) can

X. Wu, B. Wang, *Geometric Integrators for Differential Equations with Highly Oscillatory Solutions*, https://doi.org/10.1007/978-981-16-0147-7_2

be integrated with general purpose methods [2, 3] or other codes adapted to its special structure. However, it is worth noting that adaptive methods will be more efficient than general purpose methods since adaptive methods make good use of the information transmitted from the special structure of (2.1) introduced by the linear term $Mq(t)$.

For the particular case where $M = \omega^2 I_d$ with a single frequency $\omega > 0$ and the $d \times d$ identity matrix I_d, methods with frequency-dependent coefficients using techniques like trigonometrical/exponential fitting can be traced back to the 1960s (see, e.g. [4]). Here, we refer the reader to the reviews of the literature (see, [5, 6]) and the relevant papers (see, e.g. [7–18]). If M is a symmetric positive semi-definite matrix, exponential integrators (see, e.g. [19, 20]), adapted Runge–Kutta–Nyström (ARKN) methods (see [21, 22]) and other adaptive methods (see, e.g. [23–28]) have been developed. Wu et al. proposed and analysed extended Runge–Kutta–Nyström (ERKN) methods (see, e.g. [29–31]), whose internal stages and updates exactly integrate the following highly oscillatory homogeneous linear system

$$q''(t) + Mq(t) = 0 \qquad (2.2)$$

associated with (2.1). This property plays an important role in oscillation-preserving integrators as stated in Chap. 1. The global error analysis of ERKN methods was presented and collocation techniques were also studied in [32–34].

If $f(q) = -\nabla U(q)$ for some smooth function $U(q)$, the system (2.1) is identical to a separable Hamiltonian system of the following form

$$\begin{cases} p'(t) = -\nabla_q H(p, q), \\ q'(t) = \nabla_p H(p, q), \end{cases} \qquad (2.3)$$

with the initial values $q(t_0) = q_0$, $p(t_0) = p_0 = q_0'$, and the Hamiltonian

$$H(p, q) = \frac{1}{2} p^{\mathsf{T}} p + \frac{1}{2} q^{\mathsf{T}} Mq + U(q), \qquad (2.4)$$

where $q : \mathbb{R} \to \mathbb{R}^d$ and $p : \mathbb{R} \to \mathbb{R}^d$ are known as generalised position and generalised momenta, respectively. It is clear that (2.3) is a highly oscillatory Hamiltonian system once $\|M\| \gg \max \left\{ 1, \left\| \frac{\partial f}{\partial q} \right\| \right\}$. As is known, for Hamiltonian system (2.3), the corresponding map is symplectic and the true solution preserves the energy $H(p, q)$ for all $t \in [t_0, T]$ (see, e.g. [35]). In the spirit of geometric numerical integration, an integrator that inherits such geometric properties as much as possible would be preferable. Unfortunately, however, it is often difficult to design numerical integrators which inherit both symplecticity and energy preservation. A numerical method which is energy preserving at each step and defined by a symplectic map has been discussed in [36, 37]. Since there is no symplectic B-series method that conserves arbitrary Hamiltonians [38, 39], methods satisfying one of these properties have been developed in the past few decades [20]. Research work has shown that symplectic methods perform very well in approximately preserving the energy of Hamiltonian systems and we refer the reader to [20], for instance. However, it is worth noting that symplectic methods just approximately, rather than exactly, preserve the energy (2.4). In practical applications, apart from the accuracy of approximate solutions, high-precision energy-preserving integrators are also required. Moreover, in comparison with symplectic methods, energy-preserving integrators have better nonlinear stability characteristics, are easier to adapt the time step for, and are more suitable for the integration of chaotic systems (see, e.g. [40–43]). Therefore, energy-preserving algorithms are becoming more popular.

As is known, for first-order ordinary differential equations of the form

$$y'(t) = G(y(t)), \quad y(t_0) = y_0, \quad t \in [t_0, T], \tag{2.5}$$

continuous-stage Runge–Kutta (CSRK) methods were firstly researched in [44, 45]. Then, some relevant papers appeared (see, e.g. [46, 47]). Hairer proposed a family of CSRK methods and studied the corresponding energy conservation (see [48]). Miyatake and Butcher proved a sufficient and necessary energy-preserving condition of CSRK methods (see [49]). Recently, some developments in this field have been made (see, e.g. [50–52]). The exponentially and functionally-fitted version of the CSRK method appeared in [53, 54]. The conservation of energy has been approached by means of the definition of the discrete line integral [55–57].

More recently, for second-order ordinary differential equations of the form

$$q''(t) = F(q(t)), \quad q(t_0) = q_0, \quad q'(t_0) = q_0', \quad t \in [t_0, T], \tag{2.6}$$

Tang et al. [58] discussed continuous-stage Runge–Kutta–Nyström (CSRKN) methods and studied symplecticity-preserving algorithms. Energy-preserving CSRKN methods were studied in [59]. The corresponding result in [58] has been extended to high-order symplectic CSRKN methods [60].

2.2 Extended Runge–Kutta–Nyström Methods

Suppose that M is a positive semi-definite matrix. We begin with the following matrix-valued ϕ-functions

$$\phi_j(M) = \sum_{k=0}^{\infty} \frac{(-1)^k M^k}{(2k+j)!}, \quad j \geqslant 0, \quad M \in \mathbb{R}^{d \times d}. \tag{2.7}$$

which originally appeared in [22]. It can be observed from (2.7) that $j! \phi_j(M) \to I_d$ when $M \to 0$, where I_d is the $d \times d$ identity matrix. The following proposition establishes the properties of matrix-valued ϕ-functions which will be used in Sect. 2.4 of this chapter.

Proposition 2.1 *The matrix-valued ϕ-functions defined by (2.7) satisfy:*

- (i) $M \in \mathbb{R}^{d \times d}$,

$$\phi_{j+2}(M) = \int_0^1 \frac{(1-z)\phi_1\big(M(1-z)^2\big)z^j}{j!} \, dz, \quad j = 0, 1, \cdots,$$

$$\phi_{j+1}(M) = \int_0^1 \frac{\phi_0\big(M(1-z)^2\big)z^j}{j!} \, dz, \quad j = 0, 1, \cdots; \tag{2.8}$$

- (ii) *If M is invertible, then*

$$\phi_{j+2}(M) = M^{-1}\left(\frac{1}{j!}I_d - \phi_j(M)\right), \quad j = 0, 1, \cdots; \tag{2.9}$$

- (iii) $\phi_0^2(M) + M\phi_1^2(M) = I_d$.

The proofs of Proposition 2.1 and further details about the matrix-valued ϕ-functions can be found in [22, 31]. As shown in Chap. 1 (see also [31]), an s-stage ERKN method for the numerical integration of the system (2.1) is defined as

$$\begin{cases} Q_i = \phi_0(c_i^2 V)q_n + hc_i\phi_1(c_i^2 V)q_n' + h^2\sum_{j=1}^s a_{ij}(V)f(Q_j), \quad i = 1, \cdots, s, \\[2mm] q_{n+1} = \phi_0(V)q_n + h\phi_1(V)q_n' + h^2\sum_{i=1}^s \bar{b}_i(V)f(Q_i), \\[2mm] q_{n+1}' = -hM\phi_1(V)q_n + \phi_0(V)q_n' + h\sum_{i=1}^s b_i(V)f(Q_i), \end{cases}$$

$$\tag{2.10}$$

where c_i are real numbers, and $a_{ij}(V)$, $\bar{b}_i(V)$ and $b_i(V)$ for $i, j = 1, \cdots, s$ are matrix-valued functions of $V = h^2 M$. The method (2.10) can be represented briefly in Butcher's notation by the following block tableau of coefficients:

c	$\phi_0(c^2 V)$	$c\phi_1(c^2 V)$	$A(V)$
	$\phi_0(V)$	$\phi_1(V)$	$\bar{b}^{\mathsf{T}}(V)$
	$-hM\phi_1(V)$	$\phi_0(V)$	$b^{\mathsf{T}}(V)$

$$
= \begin{array}{c|cc|ccc}
c_1 & \phi_0(c_1^2 V) & c_1\phi_1(c_1^2 V) & a_{11}(V) & \cdots & a_{1s}(V) \\
\vdots & \vdots & \vdots & \vdots & \ddots & \vdots \\
c_s & \phi_0(c_s^2 V) & c_s\phi_1(c_s^2 V) & a_{s1}(V) & \cdots & a_{ss}(V) \\
\hline
& \phi_0(V) & \phi_1(V) & \bar{b}_1(V) & \cdots & \bar{b}_s(V) \\
& -hM\phi_1(V) & \phi_0(V) & b_1(V) & \cdots & b_s(V)
\end{array}
\tag{2.11}
$$

The order conditions for an ERKN method (2.10) have been investigated in [31] by using the B-series theory associated with the set of extended special Nyström trees (see [61]). To learn more about this point the reader is referred to the relevant references for all the definitions and notations.

Let

$$
a_{ij}(V) = \sum_{k=0}^{\infty} a_{ij}^{(2k)} V^k,
\tag{2.12}
$$

where the coefficients $a_{ij}^{(2k)}$ define the expansion of $a_{ij}(V)$. Then the local truncation errors of q_{n+1} and q'_{n+1} can be expanded in the form

$$
\begin{aligned}
e_{n+1} &= q_{n+1} - q(t_{n+1}) \\
&= \sum_{\beta\tau \in \text{ESNT}} h^{\rho(\beta\tau)+1} \left(\frac{\gamma(\beta\tau)}{\rho(\beta\tau)!} \sum_{i=1}^{s} \bar{b}_i(V)\Phi_i(\beta\tau) - \phi_{\rho(\beta\tau)+1}(V) \right) \alpha(\beta\tau) F(\beta\tau)(q_n, q'_n),
\end{aligned}
$$

$$
\begin{aligned}
e'_{n+1} &= q'_{n+1} - q'(t_{n+1}) \\
&= \sum_{\beta\tau \in \text{ESNT}} h^{\rho(\beta\tau)} \left(\frac{\gamma(\beta\tau)}{\rho(\beta\tau)!} \sum_{i=1}^{s} b_i(V)\Phi_i(\beta\tau) - \phi_{\rho(\beta\tau)}(V) \right) \alpha(\beta\tau) F(\beta\tau)(q_n, q'_n),
\end{aligned}
$$

where the set ESNT of extended special Nyström trees $\beta\tau$, functions $\rho(\beta\tau)$, $\alpha(\beta\tau)$ and elementary differential $F(\beta\tau)(q, q')$ are defined in [31, 61]. The following theorem states the order conditions for ERKN methods.

Theorem 2.1 *The ERKN method* (2.10) *is convergent of order p if and only if*

$$
\sum_{i=1}^{s} \bar{b}_i(V)\Phi_i(\beta\tau) = \frac{\rho(\beta\tau)!}{\gamma(\beta\tau)} \phi_{\rho(\beta\tau)+1}(V) + \mathcal{O}(h^{p-\rho(\beta\tau)}), \quad \rho(\beta\tau) \leqslant p-1,
$$

$$
\sum_{i=1}^{s} b_i(V)\Phi_i(\beta\tau) = \frac{\rho(\beta\tau)!}{\gamma(\beta\tau)} \phi_{\rho(\beta\tau)}(V) + \mathcal{O}(h^{p+1-\rho(\beta\tau)}), \quad \rho(\beta\tau) \leqslant p,
$$

$$
\tag{2.13}
$$

where $\beta\tau \in$ ESNT.

With regard to the follow-up work of ERKN methods, we refer the reader to [32], in which trigonometric Fourier collocation methods were studied. The symplectic conditions for ERKN methods were derived and analysed in [29, 30].

2.3   Continuous-Stage ERKN Methods and Order Conditions

Similarly to the CSRK method, Tang et al. considered the continuous-stage Runge–Kutta–Nyström (CSRKN) method for (2.6) as follows (see [58]).

Definition 2.1 Let $A_{\tau\sigma}$ be a function of variables τ, $\sigma \in [0, 1]$ and $\bar{B}_\tau$, B_τ and C_τ be functions of $\tau \in [0, 1]$. A continuous-stage Runge–Kutta–Nyström (CSRKN) method for solving (2.6) is defined by

$$
\begin{cases}
Q_\tau = q_n + h C_\tau q'_n + h^2 \displaystyle\int_0^1 A_{\tau\sigma} F(Q_\sigma)\, \mathrm{d}\sigma, \quad \tau \in [0, 1], \\[2mm]
q_{n+1} = q_n + h q'_n + h^2 \displaystyle\int_0^1 \bar{B}_\tau F(Q_\tau)\, \mathrm{d}\tau, \\[2mm]
q'_{n+1} = q'_n + h \displaystyle\int_0^1 B_\tau F(Q_\tau)\, \mathrm{d}\tau.
\end{cases}
\tag{2.14}
$$

The order conditions for CSRKN methods (2.14) have been given as those for classical RKN methods with $\sum$, c_i, a_{ij}, $\bar{b}_i$, and b_i, replaced by $\displaystyle\int_0^1$, C_τ, $A_{\tau\sigma}$, $\bar{B}_\tau$, and B_τ, respectively. For a more detailed description of the order conditions of the CSRKN methods, we refer the reader to [58].

On the basis of the matrix-variation-of-constants formula of (2.1), applying the continuous-stage idea to the ERKN methods leads to continuous-stage extended Runge–Kutta–Nyström methods as follows.

Definition 2.2 An s-degree continuous-stage extended Runge–Kutta–Nyström (CSERKN) method for the numerical integration of the system (2.1) is defined by

$$
\begin{cases}
Q_\tau = C_\tau(V) q_n + h D_\tau(V) q'_n + h^2 \displaystyle\int_0^1 \bar{A}_{\tau\sigma}(V) f(Q_\sigma)\, \mathrm{d}\sigma, \quad \tau \in [0, 1], \\[2mm]
q_{n+1} = \phi_0(V) q_n + h \phi_1(V) q'_n + h^2 \displaystyle\int_0^1 \bar{b}_\tau(V) f(Q_\tau)\, \mathrm{d}\tau, \\[2mm]
q'_{n+1} = -h M \phi_1(V) q_n + \phi_0(V) q'_n + h \displaystyle\int_0^1 b_\tau(V) f(Q_\tau)\, \mathrm{d}\tau,
\end{cases}
$$

$$
\tag{2.15}
$$

where Q_τ is a polynomial of degree s with respect to τ satisfying $Q_0 = q_n$ and $Q_1 = q_{n+1}$, $C_\tau(V)$, $D_\tau(V)$, $\bar{b}_\tau(V)$, and $b_\tau(V)$ are polynomials of degree s and depend on V, $\bar{A}_{\tau\sigma}(V)$ is a polynomial of degree s for τ, and $s-1$ for σ and depend on V, where $\tau, \sigma \in [0, 1]$ and $V = h^2M$. In addition, the relations $\bar{A}_{0\sigma}(V) = \mathbf{0}$ and $\bar{A}_{1\sigma}(V) = \bar{b}_\sigma(V)$ hold. The polynomials $C_\tau(V)$ and $D_\tau(V)$ satisfy

$$C_{c_i}(V) = \phi_0(c_i^2 V), \quad D_{c_i}(V) = c_i\phi_1(c_i^2 V), \tag{2.16}$$

where c_i for $i = 0, \cdots, s$ are the fitting nodes, and one of them should be 1. We take $c_0 = 0$ and $c_s = 1$ in general. $C_\tau(V)$ and $D_\tau(V)$ can be expressed as

$$C_\tau(V) = \sum_{i=0}^{s} L_i(\tau)\phi_0(c_i^2 V), \quad D_\tau(V) = \sum_{i=0}^{s} L_i(\tau)c_i\phi_1(c_i^2 V), \tag{2.17}$$

where $L_i(\tau)$ for $i = 0, \cdots, s$ are the following Lagrange interpolations functions

$$L_i(\tau) = \prod_{j=0, j\neq i}^{s} \frac{\tau - c_j}{c_i - c_j}.$$

The CSERKN method can be expressed by the following block Butcher tableau

$$\begin{array}{c|c|c|c} C_\tau & C_\tau(V) & D_\tau(V) & \bar{A}_{\tau\sigma}(V) \\ \hline & \phi_0(V) & \phi_1(V) & \bar{b}_\tau(V) \\ \hline & -hM\phi_1(V) & \phi_0(V) & b_\tau(V) \end{array} \cdot \tag{2.18}$$

A CSERKN method (2.15) is of order p, if for sufficiently smooth problem (2.1), the local truncation errors satisfy

$$e_{n+1} = q_{n+1} - q(t_{n+1}) = \mathcal{O}(h^{p+1}), \quad e'_{n+1} = q'_{n+1} - q'(t_{n+1}) = \mathcal{O}(h^{p+1}),$$

under the so-called local assumptions. In order to obtain the order conditions for CSERKN methods, it is assumed that

$$\bar{A}_{\tau\sigma}(V) = \sum_{k=0}^{\infty} \bar{A}_{\tau\sigma}^{(2k)} V^k, \tag{2.19}$$

where the coefficients $\bar{A}_{\tau\sigma}^{(2k)}$ define the expansion of $\bar{A}_{\tau\sigma}(V)$. Similarly to the analysis of the paper [31], we have

$$e_{n+1} = q_{n+1} - q(t_{n+1})$$

$$= \sum_{\beta\tau\in\mathrm{ESNT}} h^{\rho(\beta\tau)+1} \left(\frac{\gamma(\beta\tau)}{\rho(\beta\tau)!} \int_0^1 \bar{b}_\tau(V)\Phi_\tau(\beta\tau)\mathrm{d}\tau - \phi_{\rho(\beta\tau)+1}(V) \right) \alpha(\beta\tau)F(\beta\tau)(q_n, q_n'),$$

$$e_{n+1}' = q_{n+1}' - q'(t_{n+1})$$

$$= \sum_{\beta\tau\in\mathrm{ESNT}} h^{\rho(\beta\tau)} \left(\frac{\gamma(\beta\tau)}{\rho(\beta\tau)!} \int_0^1 b_\tau(V)\Phi_\tau(\beta\tau)\mathrm{d}\tau - \phi_{\rho(\beta\tau)}(V) \right) \alpha(\beta\tau)F(\beta\tau)(q_n, q_n').$$

The weights $\Phi_\tau(\beta\tau)$ can be given similarly to the ones for ERKN methods with $\sum$, c_i^k, $a_{ij}^{(2k)}$, $\bar{b}_i(V)$, and $b_i(V)$ replaced by $\int_0^1$, $\sum_{i=1}^s L_i(\tau)c_i^k$, $\bar{A}_{\tau\sigma}^{(2k)}$, $\bar{b}_\tau(V)$, and $b_\tau(V)$, respectively. Therefore, we obtain the order conditions for a CSERKN method as follows.

Theorem 2.2 *The CSERKN method* (2.15) *is convergent of order p if and only if*

$$\int_0^1 \bar{b}_\tau(V)\Phi_\tau(\beta\tau)\mathrm{d}\tau = \frac{\rho(\beta\tau)!}{\gamma(\beta\tau)}\phi_{\rho(\beta\tau)+1}(V) + \mathcal{O}(h^{p-\rho(\beta\tau)}), \quad \rho(\beta\tau) \leqslant p - 1,$$

$$\int_0^1 b_\tau(V)\Phi_\tau(\beta\tau)\mathrm{d}\tau = \frac{\rho(\beta\tau)!}{\gamma(\beta\tau)}\phi_{\rho(\beta\tau)}(V) + \mathcal{O}(h^{p+1-\rho(\beta\tau)}), \quad \rho(\beta\tau) \leqslant p,$$

$$(2.20)$$

where $\beta\tau$ is the extended special Nyström-tree.

In what follows, we provide a list of the p-th order conditions (2.20) for the CSERKN method (2.15) up to the extended special Nyström trees with $\rho(\beta\tau) \leqslant 4$.

- For the tree $\beta\tau$ with $\rho(\beta\tau) = 1$, (2.20) gives
$$\int_0^1 \bar{b}_\tau(V)\mathrm{d}\tau = \phi_2(V) + \mathcal{O}(h^{p-1}), \quad \int_0^1 b_\tau(V)\mathrm{d}\tau = \phi_1(V) + \mathcal{O}(h^p).$$

- For the tree $\beta\tau$ with $\rho(\beta\tau) = 2$, it follows from (2.20) that
$$\int_0^1 \bar{b}_\tau(V) \sum_{i=0}^s L_i(\tau)c_i\mathrm{d}\tau = \phi_3(V) + \mathcal{O}(h^{p-2}),$$

$$\int_0^1 b_\tau(V) \sum_{i=0}^s L_i(\tau)c_i\mathrm{d}\tau = \phi_2(V) + \mathcal{O}(h^{p-1}).$$

- For the trees $\beta\tau$ with $\rho(\beta\tau) = 3$, the order conditions are

$$\int_0^1 \bar{b}_\tau(V) \sum_{i=0}^s L_i(\tau)c_i^2 d\tau = 2\phi_4(V) + \mathcal{O}(h^{p-3}),$$

$$\int_0^1 \int_0^1 \bar{b}_\tau(V)\bar{A}_{\tau\sigma}^{(0)} d\tau d\sigma = \phi_4(V) + \mathcal{O}(h^{p-3}),$$

$$\int_0^1 b_\tau(V) \sum_{i=0}^s L_i(\tau)c_i^2 d\tau = 2\phi_3(V) + \mathcal{O}(h^{p-2}),$$

$$\int_0^1 \int_0^1 b_\tau(V)\bar{A}_{\tau\sigma}^{(0)} d\tau d\sigma = \phi_3(V) + \mathcal{O}(h^{p-2}).$$

- For the trees $\beta\tau$ with $\rho(\beta\tau) = 4$, we have

$$\int_0^1 \bar{b}_\tau(V) \sum_{i=0}^s L_i(\tau)c_i^3 d\tau = 6\phi_5(V) + \mathcal{O}(h^{p-4}),$$

$$\int_0^1 \int_0^1 \bar{b}_\tau(V) \sum_{i=0}^s L_i(\tau)c_i \bar{A}_{\tau\sigma}^{(0)} d\tau d\sigma = 3\phi_5(V) + \mathcal{O}(h^{p-4}),$$

$$\int_0^1 \int_0^1 \bar{b}_\tau(V)\bar{A}_{\tau\sigma}^{(0)} \sum_{i=0}^s L_i(\sigma)c_i d\tau d\sigma = \phi_5(V) + \mathcal{O}(h^{p-4}),$$

$$\int_0^1 b_\tau(V) \sum_{i=0}^s L_i(\tau)c_i^3 d\tau = 6\phi_4(V) + \mathcal{O}(h^{p-3}),$$

$$\int_0^1 \int_0^1 b_\tau(V) \sum_{i=0}^s L_i(\tau)c_i \bar{A}_{\tau\sigma}^{(0)} d\tau d\sigma = 3\phi_4(V) + \mathcal{O}(h^{p-3}),$$

$$\int_0^1 \int_0^1 b_\tau(V)\bar{A}_{\tau\sigma}^{(0)} \sum_{i=0}^s L_i(\sigma)c_i d\tau d\sigma = \phi_4(V) + \mathcal{O}(h^{p-3}).$$

Likewise, we can list more order conditions for trees with $\rho(\beta\tau) \geqslant 5$. It should be pointed out that, when $s \geqslant p$ and the abscissae $c_1, c_2, \cdots, c_s$ are distinct, we have $\sum_{i=0}^s L_i(\tau)c_i^p = \tau^p$.

2.4 Energy-Preserving Conditions and Symmetric Conditions

In what follows, we show sufficient conditions for energy preservation for a CSERKN method (2.15) when applied to the highly oscillatory Hamiltonian system (2.1).

Theorem 2.3 *A CSERKN method* (2.15) *solving highly oscillatory Hamiltonian systems* (2.1) *is energy preserving if the coefficients satisfy*

$$V\phi_0(V)\bar{b}_\tau(V) - V\phi_1(V)b_\tau(V) = C'_\tau(V),$$

$$\phi_0(V)b_\tau(V) + V\phi_1(V)\bar{b}_\tau(V) = D'_\tau(V), \tag{2.21}$$

$$b_\tau(V)b_\sigma(V) + V\bar{b}_\tau(V)\bar{b}_\sigma(V) = \bar{A}'_{\tau\sigma}(V) + \bar{A}'_{\sigma\tau}(V),$$

where $\bar{A}'_{\tau\sigma}(V) = \dfrac{\partial}{\partial\tau}\bar{A}_{\tau\sigma}(V)$, $C'_\tau(V) = \dfrac{\mathrm{d}}{\mathrm{d}\tau}C_\tau(V)$ *and* $D'_\tau(V) = \dfrac{\mathrm{d}}{\mathrm{d}\tau}D_\tau(V)$.

Proof For a CSERKN method (2.15) and Hamiltonian $H(p, q)$ determined by (2.4) with $p = q'$, we have

$$H(p_{n+1}, q_{n+1}) - H(p_n, q_n)$$

$$= \frac{1}{2}p_{n+1}^\mathsf{T}p_{n+1} + \frac{1}{2}q_{n+1}^\mathsf{T}Mq_{n+1} + U(q_{n+1}) - \frac{1}{2}p_n^\mathsf{T}p_n - \frac{1}{2}q_n^\mathsf{T}Mq_n - U(q_n)$$

$$= \frac{1}{2}\left(-hM\phi_1(V)q_n + \phi_0(V)p_n + h\int_0^1 b_\tau(V)f(Q_\tau)\,\mathrm{d}\tau\right)^\mathsf{T}$$

$$\cdot\left(-hM\phi_1(V)q_n + \phi_0(V)p_n + h\int_0^1 b_\tau(V)f(Q_\tau)\,\mathrm{d}\tau\right)$$

$$+ \frac{1}{2}\left(\phi_0(V)q_n + h\phi_1(V)p_n + h^2\int_0^1 \bar{b}_\tau(V)f(Q_\tau)\,\mathrm{d}\tau\right)^\mathsf{T}$$

$$\cdot M\left(\phi_0(V)q_n + h\phi_1(V)p_n + h^2\int_0^1 \bar{b}_\tau(V)f(Q_\tau)\,\mathrm{d}\tau\right)$$

$$+ \int_0^1 \left[\nabla U(Q_\tau)\right]^\mathsf{T}\mathrm{d}Q_\tau - \frac{1}{2}p_n^\mathsf{T}p_n - \frac{1}{2}q_n^\mathsf{T}Mq_n. \tag{2.22}$$

After some calculation, we obtain

$$H(p_{n+1}, q_{n+1}) - H(p_n, q_n)$$

$$= \frac{1}{2}p_n^\mathsf{T}\left(\phi_0^2(V) + V\phi_1^2(V)\right)p_n - \frac{1}{2}p_n^\mathsf{T}p_n + \frac{1}{2}q_n^\mathsf{T}M\left(\phi_0^2(V) + V\phi_1^2(V)\right)q_n - \frac{1}{2}q_n^\mathsf{T}Mq_n$$

$$+ q_n^{\mathsf{T}} V \left(\phi_0(V) \int_0^1 \bar{b}_\tau(V) f(Q_\tau) \, d\tau - \phi_1(V) \int_0^1 b_\tau(V) f(Q_\tau) \, d\tau \right)$$

$$+ h p_n^{\mathsf{T}} \left(\phi_0(V) \int_0^1 b_\tau(V) f(Q_\tau) \, d\tau + V \phi_1(V) \int_0^1 \bar{b}_\tau(V) f(Q_\tau) \, d\tau \right)$$

$$+ \frac{h^2}{2} \left(\int_0^1 b_\tau(V) f(Q_\tau) \, d\tau \right)^{\mathsf{T}} \left(\int_0^1 b_\tau(V) f(Q_\tau) \, d\tau \right) \tag{2.23}$$

$$+ \frac{h^2}{2} \left(\int_0^1 \bar{b}_\tau(V) f(Q_\tau) \, d\tau \right)^{\mathsf{T}} V \left(\int_0^1 \bar{b}_\tau(V) f(Q_\tau) \, d\tau \right)$$

$$+ \int_0^1 [\nabla U(Q_\tau)]^{\mathsf{T}} \mathrm{d} \left(C_\tau(V) q_n + h D_\tau(V) p_n + h^2 \int_0^1 \bar{A}_{\tau\sigma}(V) f(Q_\sigma) \, d\sigma \right).$$

It follows from Proposition 2.1 that

$$H(p_{n+1}, q_{n+1}) - H(p_n, q_n)$$

$$= q_n^{\mathsf{T}} \int_0^1 \left(V \phi_0(V) \bar{b}_\tau(V) - V \phi_1(V) b_\tau(V) - C_\tau'(V) \right) f(Q_\tau) \, d\tau$$

$$+ h p_n^{\mathsf{T}} \int_0^1 \left(\phi_0(V) b_\tau(V) + V \phi_1(V) \bar{b}_\tau(V) - D_\tau'(V) \right) f(Q_\tau) \, d\tau$$

$$+ \frac{h^2}{2} \int_0^1 \int_0^1 f(Q_\tau)^{\mathsf{T}} b_\tau(V) b_\sigma(V) f(Q_\sigma) \, d\tau \, d\sigma$$

$$+ \frac{h^2}{2} \int_0^1 \int_0^1 f(Q_\tau)^{\mathsf{T}} V \bar{b}_\tau(V) \bar{b}_\sigma(V) f(Q_\sigma) \, d\tau \, d\sigma$$

$$+ h^2 \int_0^1 [\nabla U(Y_\tau)]^{\mathsf{T}} \mathrm{d} \left(\int_0^1 \bar{A}_{\tau\sigma}(V) f(Q_\sigma) \, d\sigma \right).$$

Using the first two equations of (2.21) and $f(Q_\tau) = -\nabla U(Q_\tau)$, we obtain

$$H(p_{n+1}, q_{n+1}) - H(p_n, q_n)$$

$$= \frac{h^2}{2} \int_0^1 \int_0^1 f(Q_\tau)^{\mathsf{T}} \left(b_\tau(V) b_\sigma(V) + V \bar{b}_\tau(V) \bar{b}_\sigma(V) - 2 \bar{A}_{\tau\sigma}'(V) \right) f(Q_\sigma) \, d\tau \, d\sigma.$$

Letting $\tau \leftrightarrow \sigma$ and adding the resulting identities gives

$$H(p_{n+1}, q_{n+1}) - H(p_n, q_n) = \frac{h^2}{2} \int_0^1 \int_0^1 f(Q_\tau)^\mathsf{T} \left(b_\tau(V) b_\sigma(V) + V \bar{b}_\tau(V) \bar{b}_\sigma(V) \right.$$
$$\left. - \bar{A}'_{\tau\sigma}(V) - \bar{A}'_{\sigma\tau}(V) \right) f(Q_\sigma) \, \mathrm{d}\tau \mathrm{d}\sigma.$$

It then follows from the third equation of (2.21) that $H(p_{n+1}, q_{n+1}) - H(p_n, q_n) = 0$. The proof is complete. $\qquad\square$

Remark 2.4.1   When $V \to \mathbf{0}$ ($M \to \mathbf{0}$), the CSERKN method (2.15) reduces to CSRKN method. In this case, the energy-preserving conditions of (2.21) reduce to

$$b_\tau = D'_\tau,$$
$$b_\tau b_\sigma = \bar{A}'_{\tau\sigma} + \bar{A}'_{\sigma\tau}, \qquad (2.24)$$

where $D'_\tau I$, $\bar{A}_{\tau\sigma} I_d$, $\bar{b}_\tau I_d$ and $b_\tau I_d$ are the limit values of $D'_\tau(V)$, $\bar{A}_{\tau\sigma}(V)$, $\bar{b}_\tau(V)$ and $b_\tau(V)$ as $V \to \mathbf{0}$. It follows from (2.17) that $D_\tau = \tau$ and $b_\tau = 1$. This result has been shown in [59].

A detailed investigation of the numerical integration of reversible systems has been made in [3], and it has been shown that symmetric integration methods often have excellent long-time behaviour for such systems. Therefore, we turn to the discussion about the symmetry of CSERKN methods.

Definition 2.3 (See [3])   The adjoint method Φ_h^* of a method Φ_h is defined as the inverse map of the original method with reversed time step $-h$, i.e., $\Phi_h^* = \Phi_{-h}^{-1}$. A method with $\Phi_h^* = \Phi_h$ is called symmetric.

The following theorem gives the symmetric conditions of CSERKN methods:

Theorem 2.4 *A CSERKN methods (2.15) is symmetric if and only if the coefficients satisfy following conditions*

$$\phi_1(V) b_\tau(V) - \phi_0(V) \bar{b}_\tau(V) = \bar{b}_{1-\tau}(V),$$
$$\phi_0(V) b_\tau(V) + V \phi_1(V) \bar{b}_\tau(V) = b_{1-\tau}(V),$$
$$C_\tau(V) \phi_0(V) + V D_\tau(V) \phi_1(V) = C_{1-\tau}(V),$$
$$C_\tau(V) \phi_1(V) - D_\tau(V) \phi_0(V) = D_{1-\tau}(V),$$
$$C_\tau(V) \left(\phi_1(V) b_\sigma(V) - \phi_0(V) \bar{b}_\sigma(V) \right)$$
$$- D_\tau(V) \left(\phi_0(V) b_\sigma(V) + V \phi_1(V) \bar{b}_\sigma(V) \right) + \bar{A}_{\tau\sigma}(V) = \bar{A}_{1-\tau, 1-\sigma}(V).$$
$$(2.25)$$

Proof Exchanging $q_{n+1} \leftrightarrow q_n$, $q'_{n+1} \leftrightarrow q'_n$, $t_{n+1} \leftrightarrow t_n$ and replacing h by $-h$ in scheme (2.15) leads to

$$\bar{Q}_\tau = C_\tau(V)q_{n+1} - hD_\tau(V)q'_{n+1} + h^2 \int_0^1 \bar{A}_{\tau\sigma}(V)f\left(\bar{Q}_\sigma\right)d\sigma,$$

$$q_n = \phi_0(V)q_{n+1} - h\phi_1(V)q'_{n+1} + h^2 \int_0^1 \bar{b}_\tau(V)f\left(\bar{Q}_\tau\right)d\tau, \qquad (2.26)$$

$$q'_n = hM\phi_1(V)q_{n+1} + \phi_0(V)q'_{n+1} - h\int_0^1 \bar{b}_\tau(V)f\left(\bar{Q}_\tau\right)d\tau.$$

Using (2.26) and Proposition 2.1, we obtain

$$q_{n+1} = \phi_0(V)q_n + h\phi_1(V)q'_n$$
$$+ h^2 \int_0^1 \left(\phi_1(V)\bar{b}_\tau(V) - \phi_0(V)\bar{b}_\tau(V)\right)f\left(\bar{Q}_\tau\right)d\tau,$$

$$q'_{n+1} = -hM\phi_1(V)q_n + \phi_0(V)q'_n$$
$$+ h\int_0^1 \left(\phi_0(V)\bar{b}_\tau(V) + V\phi_1(V)\bar{b}_\tau(V)\right)f\left(\bar{Q}_\tau\right)d\tau,$$

$$\bar{Q}_\tau = \left(C_\tau(V)\phi_0(V) + VD_\tau(V)\phi_1(V)\right)q_n + \left(C_\tau(V)\phi_1(V) - D_\tau(V)\phi_0(V)\right)hq'_n$$
$$+ h^2 \int_0^1 \left[C_\tau(V)\left(\phi_1(V)\bar{b}_\sigma(V) - \phi_0(V)\bar{b}_\sigma(V)\right)\right.$$
$$\left. - D_\tau(V)\left(\phi_0(V)\bar{b}_\sigma(V) + V\phi_1(V)\bar{b}_\sigma(V)\right) + \bar{A}_{\tau\sigma}\right]f(Q_\sigma)d\sigma.$$

$$(2.27)$$

We replace all indices τ and σ by $1 - \tau$ and $1 - \sigma$, respectively, and denote $\bar{Q}_{1-\tau} = Q_\tau$. It is clear that the scheme defined by (2.27) coincides with the scheme (2.15) if and only if the coefficients satisfy the conditions (2.25). This proves the theorem. □

Remark 2.4.2 When $V \to 0$ ($M \to 0$), the CSERKN method (2.15) reduces to a CSRKN method. In this case, the symmetric conditions reduce to

$$b_\tau - \bar{b}_\tau = \bar{b}_{1-\tau}, \quad b_\tau = b_{1-\tau},$$
$$b_\sigma - \bar{b}_\sigma - \tau b_\sigma + \bar{A}_{\tau\sigma} = \bar{A}_{1-\tau,1-\sigma}. \qquad (2.28)$$

where $\bar{A}_{\tau\sigma}I_d$, $\bar{b}_\tau I_d$ and $b_\tau I_d$ are the limit values of $\bar{A}_{\tau\sigma}(V)$, $\bar{b}_\tau(V)$ and $b_\tau(V)$ as $V \to 0$. This result has been given in [62].

2.5   Linear Stability Analysis

In order to analyse the stability of CSERKN methods, we consider the following
linear scalar test equation

$$q''(t) + \omega^2 q(t) = -\varepsilon q(t), \tag{2.29}$$

where ω represents an estimate of the dominant frequency λ and $\varepsilon = \lambda^2 - \omega^2$ is the
error of the estimate. Applying the CSERKN method (2.15) to (2.29) yields

$$Q_\tau = C_\tau(V)q_n + hD_\tau(V)q_n' - z \int_0^1 \bar{A}_{\tau\sigma}(V)Q_\sigma \,d\sigma,$$

$$q_{n+1} = \phi_0(V)q_n + h\phi_1(V)q_n' - z \int_0^1 \bar{b}_\tau(V)Q_\tau \,d\tau, \tag{2.30}$$

$$hq_{n+1}' = -V\phi_1(V)q_n + h\phi_0(V)q_n' - z \int_0^1 b_\tau(V)Q_\tau \,d\tau,$$

where $V = \omega^2 h^2$ and $z = \varepsilon h^2$. Considering Q_τ is a polynomial of degree s with
respect to τ, we have

$$Q_\tau = \sum_{i=0}^{s} Q_i L_i(\tau), \quad Q_i = Q_{c_i}$$

where $c_0 = 0$, $c_1 = 1$ and $Q_0 = q_n$, $Q_s = q_{n+1}$, and then obtain

$$Q_i = C_i(V)q_n + hD_i(V)q_n' - z \int_0^1 \bar{A}_{i\sigma}(V) \left(\sum_{j=0}^{s} L_j(\sigma)Q_j\right) d\sigma,$$

$$q_{n+1} = \phi_0(V)q_n + h\phi_1(V)q_n' - z \int_0^1 \bar{b}_\tau(V) \left(\sum_{i=0}^{s} L_i(\tau)Q_i\right) d\tau, \tag{2.31}$$

$$hq_{n+1}' = -V\phi_1(V)q_n + h\phi_0(V)q_n' - z \int_0^1 b_\tau(V) \left(\sum_{i=0}^{s} L_i(\tau)Q_i\right) d\tau,$$

where $C_i(V) = C_{c_i}(V)$, $D_i(V) = D_{c_i}(V)$ and $\bar{A}_{i\sigma}(V) = \bar{A}_{c_i\sigma}(V)$. We can express
(2.31) in a vector form

$$Q = C(V)q_n + hD(V)q_n' - z\bar{A}(V)Q,$$

$$q_{n+1} = \phi_0(V)q_n + h\phi_1(V)q_n' - z\bar{B}(V)Q, \tag{2.32}$$

$$hq_{n+1}' = -V\phi_1(V)q_n + h\phi_0(V)q_n' - zB(V)Q,$$

where $Q = (Q_0, \cdots, Q_s)^\mathsf{T}$ and

$$C(V) = (C_0(V), \cdots, C_s(V))^\mathsf{T}, \quad D(V) = (D_0(V), \cdots, D_s(V))^\mathsf{T},$$

$$\bar{A}(V) = \begin{pmatrix} \int_0^1 \bar{A}_{0\sigma}(V)L_0(\sigma)\mathrm{d}\sigma & \cdots & \int_0^1 \bar{A}_{0\sigma}(V)L_s(\sigma)\mathrm{d}\sigma \\ \vdots & \ddots & \vdots \\ \int_0^1 \bar{A}_{s\sigma}(V)L_0(\sigma)\mathrm{d}\sigma & \cdots & \int_0^1 \bar{A}_{s\sigma}(V)L_s(\sigma)\mathrm{d}\sigma \end{pmatrix},$$

$$\bar{B}(V) = \left(\int_0^1 \bar{b}_\sigma(V)L_0(\sigma)\mathrm{d}\sigma, \cdots, \int_0^1 \bar{b}_\sigma(V)L_s(\sigma)\mathrm{d}\sigma \right),$$

$$B(V) = \left(\int_0^1 b_\sigma(V)L_0(\sigma)\mathrm{d}\sigma, \cdots, \int_0^1 b_\sigma(V)L_s(\sigma)\mathrm{d}\sigma \right).$$

The elimination of the vector Q in (2.32) yields the recursion

$$\begin{pmatrix} q_{n+1} \\ hq'_{n+1} \end{pmatrix} = M(V, z) \begin{pmatrix} q_n \\ hq'_n \end{pmatrix}, \tag{2.33}$$

where

$$M = \begin{pmatrix} \phi_0(V) - z\bar{B}(V)N^{-1}C(V) & \phi_1(V) - z\bar{B}(V)N^{-1}D(V) \\ -V\phi_1(V) - zB(V)N^{-1}C(V) & \phi_0(V) - zB(V)N^{-1}D(V) \end{pmatrix}, \tag{2.34}$$

and $N = I + z\bar{A}(V)$. The matrix M is called the stability matrix. The behaviour of the numerical solution will depend on the spectral radius $\rho(M)$. Geometrically, the characterization of stability involves a two-dimensional region in (V, z) space for a CSERKN method.

Definition 2.4 For the CSERKN method (2.15) with the stability matrix $M(V, z)$, the region of the two-dimensional space

$$\Omega := \{(V, z) : V \geqslant 0, \ |\rho(M(V, z))| \leqslant 1\}$$

is called the region of stability. The closed surface defined by $\rho(M(V, z)) = 1$ and $V \geqslant 0$ is the stability boundary of the method.

Definition 2.5 Denoting $\zeta = \sqrt{V + z}$, the two quantities

$$\phi(\zeta) = \zeta - \arccos\left(\frac{\mathrm{tr}(M)}{2\sqrt{\det(M)}}\right), \quad d(\zeta) = 1 - \sqrt{\det(M)}$$

are called the dispersion error and the dissipation error of the underlying CSERKN method, respectively. The method is said to be dispersive of order γ and dissipative of order r, if $\phi(\zeta) = \mathcal{O}(\zeta^{\gamma+1})$ and $d(\zeta) = \mathcal{O}(\zeta^{r+1})$, respectively. If $\phi(\zeta) = 0$ or $d(\zeta) = 0$, then the method is said to be zero dispersive or zero dissipative.

2.6 Construction of CSERKN Methods

In this section, we present second and fourth order symmetric and energy-preserving CSERKN schemes. The derivation process of higher-order methods is completely similar. In the construction of the method, we always choose $D_\tau = \tau$, as described in Remark 2.4.1.

In a CSERKN method, there is a restrictive relation between the internal and final stages for the consistency of the method, because q_{n+1} should coincide with Q_{c_s} while $c_s = 1$. Therefore, $\bar{b}_\sigma(V)$ should be expressed as $\bar{b}_\sigma(V) = \bar{A}_{1\sigma}(V)$.

2.6.1 The Case of Order Two

According to Definition 2.2, a one-degree CSERKN formulation has coefficients with the following form:

$$\bar{A}_{\tau\sigma}(V) = \bar{a}_{11}(V)\tau, \quad \bar{b}_\tau(V) = \bar{b}_1(V), \quad b_\tau(V) = b_1(V). \tag{2.35}$$

On the basis of the energy-preserving conditions (2.21), the coefficients satisfy

$$\bar{a}_{11}(V) = \phi_2((c_1 - c_2)^2 V),$$

$$\bar{b}_1(V) = \frac{(1-c_2)^2\phi_2((1-c_2)^2V) - (1-c_1)^2\phi_2((1-c_1)^2V)}{c_1 - c_2}, \tag{2.36}$$

$$b_1(V) = \frac{(1-c_2)\phi_1((1-c_2)^2V) - (1-c_1)\phi_1((1-c_1)^2V)}{c_1 - c_2}.$$

Under the assumption that the coefficients in (2.36) satisfy the symmetric conditions (2.25), we obtain

$$\bar{a}_{11}(V) = \phi_2\left((2c_1 - 1)^2V\right), \quad c_2 = 1 - c_1,$$

$$\bar{b}_1(V) = \frac{c_1^2\phi_2(c_1^2V) - (1-c_1)^2\phi_2((1-c_1)^2V)}{2c_1 - 1}, \tag{2.37}$$

$$b_1(V) = \frac{c_1\phi_1(c_1^2V) - (1-c_1)\phi_1((1-c_1)^2V)}{2c_1 - 1}.$$

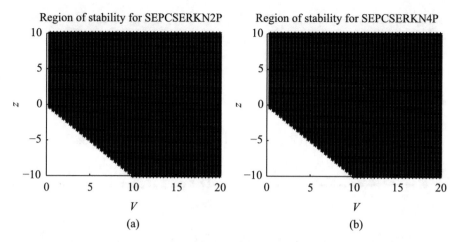

Fig. 2.1 The stability regions of the method SEPCSERKN2P (**a**) and the method SEPCSERKN4P (**b**)

Let $\bar{A}_{1\sigma}(V) = \bar{b}_{\sigma}(V)$ in (2.37), and this gives

$$c_1 = 0, \quad c_2 = 1,$$
$$\bar{a}_{11}(V) = \phi_2(V), \quad \bar{b}_1(V) = \phi_2(V), \quad b_1(V) = \phi_1(V). \tag{2.38}$$

It then can be verified that the coefficients satisfy all the conditions of order two. We denote the CSERKN method determined by (2.38) as SEPCSERKN2P. With regard to the dispersion error and the dissipation error of the method SEPCSERKN2P, we have

$$\phi(\zeta) = \frac{\varepsilon^2 \zeta^3}{12(\varepsilon^2 + w^2)} + \mathcal{O}(\zeta^5), \quad d(\zeta) = 0.$$

This shows that the method is dispersive of order two and zero dissipative, respectively. The stability region of the method SEPCSERKN2P is depicted in Fig. 2.1a.

Remark 2.6.1 Actually, the method SEPCSERKN2P can be expressed as

$$\begin{cases} Q_\tau = \Big(\tau\phi_0(V) + 1 - \tau\Big)q_n + h\tau\phi_1(V)q_n' + h^2\tau \int_0^1 \phi_2(V)f(Q_\sigma)\,d\sigma, \quad \tau \in [0, 1], \\[2mm] q_{n+1} = \phi_0(V)q_n + h\phi_1(V)q_n' + h^2 \int_0^1 \phi_2(V)f(Q_\tau)\,d\tau, \\[2mm] q_{n+1}' = -hM\phi_1(V)q_n + \phi_0(V)q_n' + h\int_0^1 \phi_1(V)f(Q_\tau)\,d\tau. \end{cases} \tag{2.39}$$

Using the first two expressions of (2.39), we write Q_τ as a linear combination of q_n and q_{n+1}.

$$Q_\tau = \tau q_{n+1} + (1 - \tau)q_n. \tag{2.40}$$

The method SEPCSERKN2P then can be expressed as

$$q_{n+1} = \phi_0(V)q_n + h\phi_1(V)q_n' + h^2\phi_2(V) \int_0^1 f\,(\tau q_{n+1} + (1 - \tau)q_n)\,\mathrm{d}\tau,$$

$$q_{n+1}' = -hM\phi_1(V)q_n + \phi_0(V)q_n' + h\phi_1(V) \int_0^1 f\,(\tau q_{n+1} + (1 - \tau)q_n)\,\mathrm{d}\tau. \tag{2.41}$$

This formula (2.41) has been proposed in [63, 64] and is termed the adapted AVF (AAVF) formula in [64]. The authors in [65] studied the application of AAVF formula to Hamiltonian partial differential equations. Therefore, CSERKN methods can be thought of as an extension of the AAVF method (2.41).

2.6.2   The Case of Order Four

A two-degree CSERKN method has the coefficients of the form

$$\bar{A}_{\tau\sigma}(V) = \bar{a}_{11}(V)\tau + \bar{a}_{12}(V)\tau\sigma + \bar{a}_{21}(V)\tau^2 + \bar{a}_{22}(V)\tau^2\sigma,$$
$$\bar{b}_\tau(V) = \bar{b}_1(V) + \bar{b}_2(V)\tau, \; b_\tau(V) = b_1(V) + b_2(V)\tau. \tag{2.42}$$

It then follows from the first two energy-preserving conditions of (2.21) that

$$\bar{b}_1(V) = \Big((-c_2^2 + c_3^2)(1 - c_1)^2\phi_2((1 - c_1)^2 V) + (c_1^2 - c_3^2)(1 - c_2)^2\phi_2((1 - c_2)^2 V)$$

$$+ (-c_1^2 + c_2^2)(1 - c_3)^2\phi_2((1 - c_3)^2 V)\Big)\Big/\Big((c_1 - c_2)(c_1 - c_3)(c_2 - c_3)\Big),$$

$$\bar{b}_2(V) = \Big(2((c_2 - c_3)(1 - c_1)^2\phi_2((1 - c_1)^2 V) + (-c_1 + c_3)(1 - c_2)^2\phi_2((1 - c_2)^2 V)$$

$$+ (c_1 - c_2)(1 - c_3)^2\phi_2((1 - c_3)^2 V))\Big)\Big/\Big((c_1 - c_2)(c_1 - c_3)(c_2 - c_3)\Big),$$

$$b_1(V) = \left((c_2^2 - c_3^2)(1 - c_1)\phi_1((1 - c_1)^2 V) + (-c_1^2 + c_3^2)(1 - c_2)\phi_1((1 - c_2)^2 V) \right.$$

$$\left. + (c_1^2 - c_2^2)(1 - c_3)\phi_1((1 - c_3)^2 V) \right) \Big/ \left((c_1 - c_2)(c_1 - c_3)(c_2 - c_3) \right),$$

$$b_2(V) = 2\left((-c_2 + c_3)(1 - c_1)\phi_1((1 - c_1)^2 V) + (c_1 - c_3)(1 - c_2)\phi_1((1 - c_2)^2 V) \right.$$

$$\left. + (-c_1 + c_2)(1 - c_3)\phi_1((1 - c_3)^2 V) \right) \Big/ \left((c_1 - c_2)(c_1 - c_3)(c_2 - c_3) \right).$$

$$(2.43)$$

Using the last energy-preserving conditions of (2.21), we obtain

$$\bar{a}_{11}(V) = \frac{1}{2}\left(\bar{b}_1^2(V) + \bar{b}_1^2(V)V \right),$$

$$\bar{a}_{21}(V) = \frac{1}{2}\left(-a_{12}(V) + b_1(V)b_2(V) + \bar{b}_1(V)\bar{b}_2(V)V \right), \qquad (2.44)$$

$$\bar{a}_{22}(V) = \frac{1}{4}\left(\bar{b}_2^2(V) + \bar{b}_2^2(V)V \right).$$

Letting the coefficients in (2.6.2) and (2.44) satisfy the symmetric conditions and $\bar{A}_{1\sigma}(V) = \bar{b}_\sigma(V)$, we obtain

$$c_1 = 0, \quad c_2 = \frac{1}{2}, \quad c_3 = 1, \quad \bar{a}_{11}(V) = 4\phi_2\left(\frac{1}{4}V\right) - 3\phi_2(V),$$

$$\bar{a}_{12}(V) = -\phi_1^2\left(\frac{1}{16}V\right)\left(1 + \frac{V}{4}\phi_2\left(\frac{V}{4}\right)\right),$$

$$\bar{a}_{21}(V) = \frac{1}{2}\phi_1^2\left(\frac{1}{16}V\right)\left(1 - \frac{3V}{4}\phi_2\left(\frac{V}{4}\right)\right), \quad \bar{a}_{22}(V) = \frac{V}{4}\phi_1^4\left(\frac{1}{16}V\right),$$

$$\bar{b}_1(V) = 3\phi_2\left(V\right) - \phi_2\left(\frac{1}{4}V\right), \quad \bar{b}_2(V) = 2\phi_2\left(\frac{1}{4}V\right) - 4\phi_2(V),$$

$$b_1(V) = -2\phi_1\left(\frac{1}{4}V\right) + 3\phi_1(V), \quad b_2(V) = 4\phi_1\left(\frac{1}{4}V\right) - 4\phi_1(V).$$

$$(2.45)$$

It can be verified that the coefficients satisfy all the conditions of order four. We denote the CSERKN method (2.15) determined by (2.45) as SEPCSERKN4P. Concerning the dispersion error and the dissipation error of the method SEPCSERKN4P, we have

$$\phi(\zeta) = \frac{\varepsilon^2(4\varepsilon^2 + 3w^2)\zeta^5}{2880(\varepsilon^2 + w^2)^2} + \mathcal{O}(\zeta^7), \quad d(\zeta) = 0,$$

which indicates that the method SEPCSERKN4P is dispersive of order four and zero dissipative, respectively. The stability region of the method SEPCSERKN4P is depicted in Fig. 2.1b.

2.7 Numerical Experiments

In this section, in order to demonstrate the superiority of the continuous-stage ERKN methods in comparison with the existing methods in the literature, we consider three model problems. Since these methods are implicit, iterative solutions are required. We use fixed point iteration with the tolerance 10^{-15}, and the maximum number of iterations is 100. The integrals appearing in the right-hand side of method (2.15) are integrated by using *quad* with the tolerance 10^{-12}. The integrators we select for comparison are

- EPCSRK2P: The energy-preserving CSRK method of order two derived in [48];
- EPCSRK4P: The energy-preserving CSRK method of order four derived in [48];
- SEPCSERKN2P: The symmetric and energy-preserving CSERKN method of order two presented in Sect. 2.6 of this chapter;
- SEPCSERKN4P: The symmetric and energy-preserving CSERKN method of order four presented in Sect. 2.6 of this chapter.

The numerical results are executed on the computer Lenovo M6600 (Inter(R) Pentium(R) CPU 3.00 GHz, 0.99 GB), and the programming language MATLAB is used.

Problem 2.1 We consider the Duffing equation

$$\begin{cases} q'' + \omega^2 q = 2k^2 q^3 - k^2 q, & t \in [0, t_{end}], \\ q(0) = 0, & q'(0) = \omega. \end{cases} \tag{2.46}$$

The Hamiltonian is given by

$$H(p, q) = \frac{1}{2} p^2 + \frac{1}{2} (\omega^2 + k^2) q^2 - \frac{k^2}{2} q^4,$$

where $k = 0.03$. The exact solution of this initial-value problem is $q(t) = sn(\omega t; k/\omega)$, where sn is the so-called Jacobian elliptic function. We choose the frequency $\omega = 50$ in this experiment. Accordingly, this is a highly oscillatory Hamiltonian system.

We first solve this problem on the interval [0, 100] with the stepsizes $h = 1/2^j$ for $j = 4, \cdots, 7$ for each method. We then integrate the problem with a fixed stepsize $h = 1/100$ on the interval [0, 100] to examine the preservation of the Hamiltonian for the four methods. The numerical results are presented in Fig. 2.2.

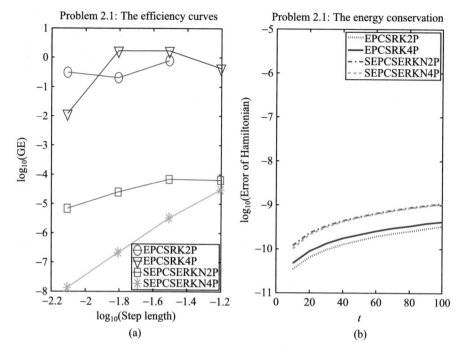

Fig. 2.2 Efficiency curves (**a**) and energy conservation (**b**) for Problem 2.1

Problem 2.2 We consider the two coupled oscillators with different frequen-cies [17]

$$\begin{cases} q_1'' + q_1 = 2\varepsilon q_1 q_2, & q_1(0) = 1, \quad q_1'(0) = 0, \\ q_2'' + 2q_2 = \varepsilon q_1^2 + 4\varepsilon q_2^3, & q_2(0) = 1, \quad q_2'(0) = 0. \end{cases}$$

The Hamiltonian of this system is given by

$$H(p, q) = \frac{1}{2}(p_1^2 + p_2^2) + \frac{1}{2}(q_1^2 + 2q_2^2) - \varepsilon \left(q_1^2 q_2 + q_2^4 \right).$$

In this numerical experiment we choose $\varepsilon = 10^{-3}$. We first solve this problem on the interval [0, 100] with the stepsizes $h = 1/2^j$ for $j = 2, \cdots, 5$ for all the methods. We then integrate the problem with a fixed stepsize $h = 1/10$ on [0, 100] and examine the preservation of the Hamiltonian. The numerical results are presented in Fig. 2.3.

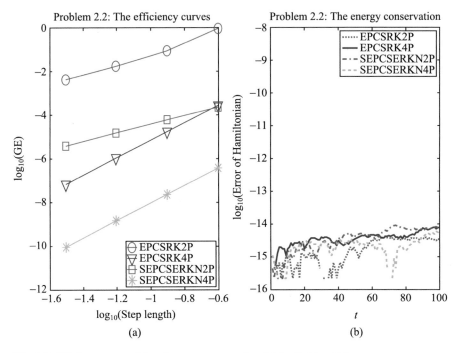

Fig. 2.3 Efficiency curves (**a**) and energy conservation (**b**) for Problem 2.2

Problem 2.3 Consider the semilinear wave equation

$$\begin{cases} \dfrac{\partial^2 u}{\partial t^2} - \dfrac{\partial^2 u}{\partial x^2} = -\dfrac{1}{5} u^3, & 0 < x < 1, \ t > 0, \\[2mm] u(0, t) = u(1, t) = 0, & u(x, 0) = \dfrac{\sin(\pi x)}{2}, \quad u_t(x, 0) = 0. \end{cases}$$

By using second-order symmetric differences, this problem is converted into a system of ODEs in time

$$\begin{cases} \dfrac{d^2 u_i}{dt^2} - \dfrac{u_{i+1} - 2u_i + u_{i-1}}{\Delta x^2} = -\dfrac{1}{5} u_i^3, & 0 < t \leqslant t_{\text{end}}, \\[2mm] u_i(0) = \dfrac{\sin(\pi x_i)}{2}, \quad u_i'(0) = 0, & i = 1, \cdots, N - 1, \end{cases}$$

where $\Delta x = 1/N$ is the spatial mesh stepsize and $x_i = i \Delta x$. Then this semidiscrete oscillatory system has the form

$$\begin{cases} \dfrac{d^2 U}{dt^2} + MU = F(U), & 0 < t \leqslant t_{\text{end}}, \\[2mm] U(0) = \left(\dfrac{\sin(\pi x_1)}{2}, \cdots, \dfrac{\sin(\pi x_{N-1})}{2} \right)^{\mathsf{T}}, & U'(0) = \mathbf{0}, \end{cases}$$

where $U(t) = (u_1(t), \cdots, u_{N-1}(t))^{\mathsf{T}}$ with $u_i(t) \approx u(x_i, t)$ for $i = 1, \cdots, N - 1$, and

$$M = \frac{1}{\Delta x^2} \begin{pmatrix} 2 & -1 & & & \\ -1 & 2 & -1 & & \\ & \ddots & \ddots & \ddots & \\ & & -1 & 2 & -1 \\ & & & -1 & 2 \end{pmatrix}, \qquad (2.47)$$

$$F(U) = \left(-\frac{1}{5} u_1^3, \cdots, -\frac{1}{5} u_{N-1}^3 \right)^{\mathsf{T}}.$$

The Hamiltonian of this system is given by

$$H(p, q) = \frac{1}{2} p^{\mathsf{T}} p + \frac{1}{2} q^{\mathsf{T}} M q + \frac{1}{20} e^{\mathsf{T}} q^4,$$

where $e = (1, \cdots, 1)^{\mathsf{T}}$. In this numerical experiment we choose $N = 100$. We first solve this problem on the interval $[0, 100]$ with the stepsizes $h = 1/2^j$ for $j = 5, \cdots, 8$. We then integrate the problem with a fixed stepsize $h = 1/128$ on $[0, 100]$ and examine the preservation of the Hamiltonian by each code. The numerical results are shown in Fig. 2.4.

It can be observed from Figs. 2.2, 2.3, and 2.4 of the three numerical experiments that the right-hand figures show all the integrators derived in this chapter preserve the Hamiltonian well. The results of the numerical experiments confirm that, for a given stepsize h, the SEPCSERKN integrators are more accurate than EPCSRK methods with the same convergence order.

Remark 2.7.1 In general, the computational cost per step of high order methods is larger than that of low order methods. In order to objectively evaluate these effects, we present in Fig. 2.5 the error versus CPU time for each problem, which indicates that SEPCSERKN4P is the best of these four methods. The related data are the same as those shown in Figs. 2.2a, 2.3a, and 2.4a.

2.8 Conclusions and Discussions

In this chapter, we derived and analysed continuous-stage extended Runge–Kutta–Nyström (CSERKN) methods for (2.1). This class of CSERKN methods is oscillation preserving since the internal stages and the updates exactly integrate the highly oscillatory homogeneous system (2.2) associated with (2.1). Symmetric and energy-preserving conditions for CSERKN methods were derived and analysed for highly oscillatory Hamiltonian systems. In terms of these conditions, two symmetric

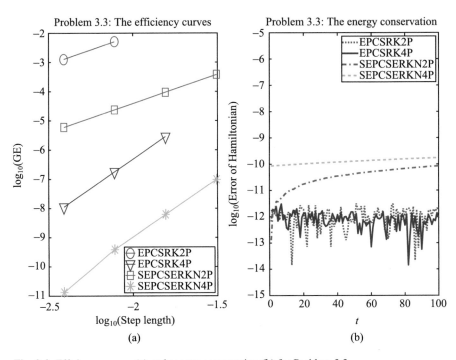

Fig. 2.4 Efficiency curves (**a**) and energy conservation (**b**) for Problem 3.3

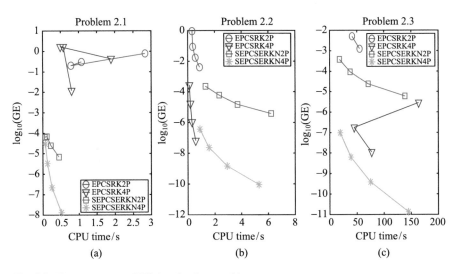

Fig. 2.5 The error versus CPU time for three problems

and energy-preserving CSERKN methods were constructed, of orders two and four respectively. The results of the numerical experiments show that the energy-preserving CSERKN methods preserve the energy well, and are more accurate than EPCSRK methods.

CSERKN methods for semilinear Hamiltonian wave equations could be investigated further. We expect that they may exactly preserve the energy of the underlying Hamiltonian wave equations, including the Klein–Gordon (KG) equation which has received a great deal of attention, both numerical and analytical. We refer the reader to [65] for this topic. A promising approach to the approximation is based on the so-called operator-variation-constants formula (the Duhamel Principle), and we refer the reader to some relevant papers [66–68]. In Chap. 11, symplectic approximations will be derived and analysed in detail for efficiently solving semilinear KG equations. Moreover, continuous-stage leap-frog schemes for semilinear Hamiltonian wave equations will be presented in Chap. 12.

The material in this chapter is based on the work by Li and Wu [69].

References

1. Brugnano, L., Iavernaro, F., Montijano, J.I., et al.: Spectrally accurate space-time solution of Hamiltonian PDEs. Numer. Algor. **81**, 1183–1202 (2019)
2. Butcher, J.C.: Numerical Methods for Ordinary Differential Equations, 2nd edn. Wiley, Chichester (2008)
3. Hairer, E., Nørsett, S.P., Wanner, G.: Solving Ordinary Differential Equations I: Nonstiff Problems, 2nd edn. Springer, Berlin (2002)
4. Gautschi, W.: Numerical integration of ordinary differential equations based on trigonometric polynomials. Numer. Math. **3**, 381–397 (1961)
5. Ixaru, L.G., Vanden, B.G.: Exponential Fitting. Kluwer Academic Publishers, Dordrecht (2004)
6. Kalogiratou, Z., Monovasilis, T.h., Psihoyios, G., et al.: Runge–Kutta type methods with special properties for the numerical integration of ordinary differential equations. Phys. Rep. **536**, 75–146 (2014)
7. Li, J.: Symplectic and symmetric trigonometrically-fitted ARKN methods. Appl. Numer. Math. **135**, 381–395 (2019)
8. Li, J., Gao, Y.: Energy-preserving trigonometrically-fitted continuous stage Runge–Kutta–Nyström methods for oscillatory Hamiltonian systems. Numer. Algor. **81**, 1379–1401 (2019)
9. Li, J., Shi, W., Wu, X.: The existence of explicit symplectic ARKN methods with several stages and algebraic order greater than two. J. Comput. Appl. Math. **353**, 204–209 (2019)
10. Martin-Vaquero, J., Vigo-Aguiar, J.: Exponential fitting BDF algorithms: explicit and implicit 0-stable methods. J. Comput. Appl. Math. **192**, 100–113 (2006)
11. Martin-Vaquero, J., Vigo-Aguiar, J.: On the numerical solution of the heat conduction equations subject to nonlocal conditions. Appl. Numer. Math. **59**, 2507–2514 (2009)
12. Ramos, H., Vigo-Aguiar, J.: On the frequency choice in trigonometrically fitted methods. Appl. Math. Lett. **23**, 1378–1381 (2010)
13. Natesan, S., Jayakumar, J., Vigo-Aguiar, J.: Parameter uniform numerical method for singularly perturbed turning point problems exhibiting boundary layers. J. Comput. Appl. Math. **158**, 121–134 (2003)

14. Simos, T.E., Vigo-Aguiar, J.: A symmetric high order method with minimal phase-lag for the numerical solution of the Schrödinger equation. P. Roy. Soc. A-Math. Phys. **12**, 1035–1042 (2001)
15. Simos, T.E., Vigo-Aguiar, J.: An exponentially-fitted high order method for long-term integration of periodic initial-value problems. Comput. Phys. Commun. **140**, 358–365 (2001)
16. Vigo-Aguiar, J., Simos, T.E.: A family of P-stable eighth algebraic order methods with exponential fitting facilities. J. Math. Chem. **29**, 177–189 (2001)
17. Vigo-Aguiar, J., Simos, T.E., Ferrandiz, J.M.: Controlling the error growth in long-term numerical integration of perturbed oscillations in one or several frequencies. P. Roy. Soc. A-Math. Phys. **460**, 561–567 (2004)
18. Vigo-Aguiar, J., Simos, T.E., Tocino, A.: An adapted symplectic integrator for Hamiltonian problems. Int. J. Mod. Phys. C **12**, 225–234 (2001)
19. Hairer, E., Lubich, C.: Long-time energy conservation of numerical methods for oscillatory differential equations. SIAM J. Numer. Anal. **38**, 414–441 (2000)
20. Hairer, E., Lubich, C., Wanner, G.: Geometric numerical integration illustrated by the Störmer Verlet method. Acta. Numer. **12**, 399–450 (2003)
21. Wu, X., You, X., Wang, B.: Structure-Preserving Algorithms for Oscillatory Differential Equations. Springer, Berlin (2013)
22. Wu, X., You, X., Xia, J.: Order conditions for ARKN methods solving oscillatory systems. Comput. Phys. Commun. **180**, 2250–2257 (2009)
23. Brugnano, L., Montijano, J.I., Rández, L.: On the effectiveness of spectral methods for the numerical solution of multi-frequency highly oscillatory Hamiltonian problems. Numer. Algor. **81**, 345–376 (2019)
24. Li, J.: A family of improved Falkner-type methods for oscillatory systems. Appl. Math. Comput. **293**, 345–357 (2017)
25. Li, J., Deng, S., Wang, X.: Extended explicit pseudo two-step RKN methods for oscillatory systems $y'' + My = f(y)$. Numer. Algor. **78**, 673–700 (2018)
26. Li, J., Wang, B., You, X., et al.: Two-step extended RKN methods for oscillatory systems. Comput. Phys. Commun. **182**, 2486–2507 (2011)
27. Li, J., Wu, X.: Adapted Falkner-type methods solving oscillatory second-order differential equations. Numer. Algor. **62**, 355–381 (2013)
28. Li, J., Wu, X.: Error analysis of explicit TSERKN methods for highly oscillatory systems. Numer. Algor. **65**, 465–483 (2014)
29. Chen, Z.X., You, X., Shi, W., et al.: Symmetric and symplectic ERKN methods for oscillatory Hamiltonian systems. Comput. Phys. Commun. **183**, 86–98 (2012)
30. Wu, X., Wang, B., Xia, J.: Explicit symplectic multidimensional exponential fitting modified Runge–Kutta–Nyström methods. BIT. Numer. Math. **52**, 773–795 (2012)
31. Wu, X., You, X., Shi, W., et al.: ERKN integrators for systems of oscillatory second-order differential equations. Comput. Phys. Commun. **181**, 1873–1887 (2010)
32. Wang, B., Iserles, A., Wu, X.: Arbitrary order trigonometric Fourier collocation methods for second-order ODEs. Found. Comput. Math. **16**, 151–181 (2016)
33. Wang, B., Wu, X., Meng, F.: Trigonometric collocation methods based on Lagrange basis polynomials for multi-frequency oscillatory second-order differential equations. J. Comput. Appl. Math. **313**, 185–201 (2017)
34. Wang, B., Yang, H., Meng, F.: Sixth order symplectic and symmetric explicit ERKN schemes for solving multi-frequency oscillatory nonlinear Hamiltonian equations. Calcolo **54**, 117–140 (2017)
35. Hairer, E., Lubich, C., Wanner, G.: Geometric Numerical Integration: Structure-Preserving Algorithms for Ordinary Differential Equations, 2nd edn. Springer, Berlin (2006)
36. Brugnano, L., Gurioli, G., Iavernaro, F.: Analysis of Energy and QUadratic Invariant Preserving (EQUIP) methods. J. Comput. Appl. Math. **335**, 51–73 (2018)
37. Brugnano, L., Iavernaro, F., Trigiante, D.: Energy and quadratic invariants preserving integrators based upon Gauss collocation formulae. SIAM J. Numer. Anal. **6**, 2897–2916 (2012)

38. Chartier, P., Faou, E., Murua, A.: An algebraic approach to invariant preserving integrators: the case of quadratic and Hamiltonian invariants. Numer. Math. **103**, 575–590 (2006)
39. Zhong, G., Marsden, J.E.: Lie-Poisson Hamilton–Jacobi theory and Lie-Poisson integrators. Phys. Lett. A **133**, 134–139 (1988)
40. Celledoni, E., Grimm, V., Mclachlan, R.I., et al.: Preserving energy resp. dissipation in numerical PDEs using the 'Average Vector Field' method. J. Comput. Phys. **231**, 6770–6789 (2012)
41. Hairer, E.: Variable time step integration with symplectic methods. Appl. Numer. Math. **25**, 219–227 (1997)
42. Simos, J.C.: Assessment of energy-momentum and symplectic schemes for stiff dynamical systems. ASME Winter Annual Meeting, New Orleans, Louisiana (1993)
43. Simos, T.E.: Does variable step size ruin a symplectic integrator? Phys. D Nonl. Phenom. **60**, 311–313 (1992)
44. Owren B, Zennaro M. Derivation of efficient, continuous, explicit Runge–Kutta methods. SIAM J. Sci. Stat. Comput. **13**, 1488–1501 (1992)
45. Owren, B., Zennaro, M.: Order barriers for continuous explicit Runge–Kutta methods. Math. Comput. **56**, 645–661 (1991)
46. Baker, T.S., Dormand, J.R., Gilmore, J.P., et al.: Continuous approximation with embedded Runge–Kutta methods. Appl. Numer. Math. **22**, 51–62 (1996)
47. Verner, J.H., Zennaro, M.: The orders of embedded continuous explicit Runge–Kutta methods. BIT Numer. Math. **35**, 406–416 (1995)
48. Hairer, E.: Energy-preserving variant of collocation methods. J. Numer. Anal. Ind. Appl. Math. **5**, 73–84 (2010)
49. Miyatake, Y., Butcher, J.C.: Characterization of energy-preserving methods and the construction of parallel integrators for Hamiltonian systems. SIAM J. Numer. Anal. **54**, 1993–2013 (2016)
50. Tang, W.: A note on continuous-stage Runge–Kutta methods. Appl. Math. Comput. **339**, 231–241 (2018)
51. Tang, W., Lang, G., Luo, X.: Construction of symplectic (partitioned) Runge–Kutta methods with continuous stage. Appl. Math. Comput. **286**, 279–287 (2016)
52. Tang, W., Sun, Y.: Construction of Runge–Kutta type methods for solving ordinary differential equations. Appl. Math. Comput. **234**, 179–191 (2014)
53. Li, Y., Wu, X.: Functionally fitted energy-preserving methods for solving oscillatory nonlinear Hamiltonian systems. SIAM J. Numer. Anal. **54**, 2036–2059 (2016)
54. Miyatake, Y.: An energy-preserving exponentially-fitted continuous stage Runge–Kutta method for Hamiltonian systems. BIT Numer. Math. **54**, 1–23 (2014)
55. Brugnano, L., Iavernaro, F.: Line integral solution of differential problems. Axioms **7**(2), 36 (2018)
56. Brugnano, L., Iavernaro, F.: Line Integral Methods for Conservative Problems. Chapman and Hall/CRC, Boca Raton (2016)
57. Brugnano, L., Iavernaro, F., Trigiante, D.: Hamiltonian boundary value methods (energy preserving discrete line integral methods). J. Numer. Anal. Ind. Appl. Math. **5**, 17–37 (2010)
58. Tang, W., Zhang, J.: Symplecticity-preserving continuous-stage Runge–Kutta–Nyström methods. Appl. Math. Comput. **323**, 204–219 (2018)
59. Tang, W.: Energy-preserving continuous-stage Runge–Kutta–Nyström methods (2018). arXiv: 1808. 08451
60. Tang, W., Sun, Y., Zhang, J.: High order symplectic integrators based on continuous-stage Runge–Kutta–Nyström methods (2018). arXiv: 1510. 04395
61. Yang, H., Wu, X., You, X., et al.: Extended RKN-type methods for numerical integration of perturbed oscillators. Comput. Phys. Commun. **180**, 1777–1794 (2009)
62. Tang, W.: Continuous-stage Runge–Kutta–Nyström methods (2018). arXiv: 1807. 03393
63. Wang, B., Wu, X.: A new high precision energy-preserving integrator for system of oscillatory second-order differential equations. Phys. Lett. A **376**, 1185–1190 (2012)

64. Wu, X., Wang, B., Shi, W.: Efficient energy-preserving integrators for oscillatory Hamiltonian systems. J. Comput. Phys. **235**, 587–605 (2013)
65. Liu, C., Wu, X.: An energy-preserving and symmetric scheme for nonlinear Hamiltonian wave equations. J. Math. Anal. Appl. **440**, 167–182 (2016)
66. Liu, C., Iserles, A., Wu, X.: Symmetric and arbitrarily high-order Birkhoff–Hermite time integrators and their long-time behavior for solving nonlinear Klein–Gordon equations. J. Comput. Phys. **356**, 1–30 (2018)
67. Liu, C., Wu, X.: Arbitrarily high-order time-stepping schemes based on the operator spectrum theory for high-dimensional nonlinear Klein–Gordon equations. J. Comput. Phys. **340**, 243–275 (2017)
68. Wang, B., Wu, X.: The formulation and analysis of energy-preserving schemes for solving high-dimensional nonlinear Klein–Gordon equations. IMA J. Numer. Anal. **39**, 2016–2044 (2019)
69. Li, J., Wu, X.: Energy-preserving continuous stage extended Runge–Kutta–Nyström method for oscillatory Hamiltonian systems. Appl. Numer. Math. **145**, 469–487 (2019)

Chapter 3
Stability and Convergence Analysis of ERKN Integrators for Second-Order ODEs with Highly Oscillatory Solutions

In this chapter, we commence the nonlinear stability and convergence analysis of ERKN integrators for second-order ODEs with highly oscillatory solutions, depending on a frequency matrix. As one of the most important applications, we also rigorously analyse the global errors of the blend of the ERKN time integrators and the Fourier pseudospectral spatial discretisation (ERKN-FP) when applied to semilinear wave equations. The theoretical results show that the nonlinear stability and the global error bounds are entirely independent of the frequency matrix, and the spatial mesh size. The analysis also provides a new perspective on the class of ERKN time integrators. That is, *the ERKN-FP methods are free from the restriction on the Courant-Friedrichs-Lewy (CFL) condition.*

3.1 Introduction

Nonlinear highly oscillatory problems occur in a variety of fields in science and engineering. The computation of nonlinear highly oscillatory problems contains numerous enduring challenges. In recent years, the investigation of efficient numerical methods for solving such problems has received increasing attention. In this chapter, we consider nonlinear multi-frequency highly oscillatory systems which can be formulated by the following initial value problem of second-order ODEs

$$\begin{cases} \ddot{q}(t) + \kappa^2 A q(t) = g(q(t)), & t \in [t_0, T], \\ q(t_0) = \varphi, \quad \dot{q}(t_0) = \psi, \end{cases} \tag{3.1}$$

where $\kappa^2 > 0$ is a takanami number, $q \in \mathbb{R}^d$, and $A \in \mathbb{R}^{d \times d}$ is a positive semi-definite matrix that implicitly contains the dominant frequencies of the highly oscillatory problem with $\kappa^2 \|A\| \gg \max \left\{ 1, \left\| \dfrac{\partial g}{\partial q} \right\| \right\}$. This type of problem plays

© The Author(s), under exclusive license to Springer Nature Singapore Pte Ltd. 2021
X. Wu, B. Wang, *Geometric Integrators for Differential Equations with Highly Oscillatory Solutions*, https://doi.org/10.1007/978-981-16-0147-7_3

an important role in a wide variety of practical application areas in science and engineering, including nonlinear optics, molecular dynamics, solid state physics and quantum field theory. It is well known that the method of lines is an effective approach for the numerical integration of PDEs such as semilinear wave equations. With suitable spatial discretisation strategies, for example the finite difference method and the pseudospectral or spectral method (see, e.g. [1–6]), semilinear wave equations can be converted into highly oscillatory second-order ODEs (3.1). Therefore, research of the nonlinear multi-frequency highly oscillatory system (3.1) will also be significant for the numerical investigation of semilinear wave equations, including the important Klein–Gordon (KG) equation, in applications.

As is known, if the nonlinear function $g(\cdot)$ satisfies a Lipschitz condition, then the nonlinear highly oscillatory problem (3.1) has a unique solution (see, e.g. [7, 8]) over the interval $[t_0, T]$. Therefore, throughout this chapter we assume that the nonlinear function $g(\cdot)$ is locally Lipschitz continuous in a strip along the exact solution $q(t)$, i.e., there is a positive constant L, s.t.

$$\|g(\alpha(t)) - g(\beta(t))\| \leqslant L\|\alpha(t) - \beta(t)\| \tag{3.2}$$

for all $t \in [t_0, T]$ and

$$\max\{\|\alpha(t) - q(t)\|, \ \|\beta(t) - q(t)\|\} \leqslant R. \tag{3.3}$$

The numerical treatment of the highly oscillatory system (3.1) has received a great deal of attention (see, e.g. [9–15]). Over the last decade, in order to systematically and comprehensively study the nonlinear multi-frequency highly oscillatory second-order ODEs (3.1) from both the analytical and numerical aspects, Wu et al. (see, e.g. [15, 16]) established the following matrix-variation-of-constants formula

$$
\begin{cases}
q(t) = \phi_0\big((t - t_0)^2\kappa^2 A\big)q(t_0) + (t - t_0)\phi_1\big((t - t_0)^2\kappa^2 A\big)\dot{q}(t_0) \\
\qquad + \displaystyle\int_{t_0}^t (t - z)\phi_1\big((t - z)^2\kappa^2 A\big)g(q(z))\mathrm{d}z, \\
\dot{q}(t) = -(t - t_0)\kappa^2 A\phi_1\big((t - t_0)^2\kappa^2 A\big)q(t_0) + \phi_0\big((t - t_0)^2\kappa^2 A\big)\dot{q}(t_0) \\
\qquad + \displaystyle\int_{t_0}^t \phi_0\big((t - z)^2\kappa^2 A\big)g(q(z))\mathrm{d}z,
\end{cases}
\tag{3.4}
$$

where $t \in [t_0, T]$ and the functions $\phi_0(\mathbb{A})$ and $\phi_1(\mathbb{A})$ are defined by the following unconditionally convergent matrix-valued functions:

$$\phi_j(\mathbb{A}) := \sum_{k=0}^{\infty} \frac{(-1)^k \mathbb{A}^k}{(2k + j)!}, \quad j \in \mathbb{N}, \tag{3.5}$$

where $\mathbb{A}$ is a positive semi-definite matrix. Since the matrix A appearing in (3.4) is symmetric and positive semi-definite with $A = \Omega^2$, where Ω is also symmetric and positive semi-definite, (3.4) can also read (see, e.g. [8, 10])

$$
\begin{cases}
q(t) = \cos((t-t_0)\kappa\Omega)q(t_0) + \kappa^{-1}\Omega^{-1}\sin\left((t-t_0)\kappa\Omega\right)\dot{q}(t_0) \\
\qquad + \displaystyle\int_{t_0}^{t} \kappa^{-1}\Omega^{-1}\sin((t-\tau)\kappa\Omega)g\left(q(\tau)\right)\mathrm{d}\tau, \\
\dot{q}(t) = -\kappa\Omega\sin((t-t_0)\kappa\Omega)q(t_0) + \cos((t-t_0)\kappa\Omega)\dot{q}(t_0) \\
\qquad + \displaystyle\int_{t_0}^{t} \cos((t-\tau)\kappa\Omega)g\left(q(\tau)\right)\mathrm{d}\tau.
\end{cases}
\tag{3.6}
$$

It is noted that (3.6) depends on the decomposition of the matrix A, but (3.4) does not. Throughout this chapter we use (3.4).

The matrix-variation-of-constants formula (3.4) has received a lot of attention in the literature over the past decades. In particular, this formula can be used to design and analyse effective and efficient numerical integrators for solving the multi-frequency highly oscillatory system (3.1), such as the Gautschi-type methods of order two (see, e.g. [10, 11, 17, 18]), the exponentially fitted Runge–Kutta (EFRK) method [19], the exponentially fitted Runge–Kutta–Nyström (EFRKN) method [20], the functionally-fitted energy-preserving method [21], the adapted Runge–Kutta–Nyström (ARKN) method (see, e.g. [16, 22]), the extended Runge–Kutta–Nyström (ERKN) method (see, e.g. [23–27]), and arbitrarily high-order time-stepping methods (see, e.g. [28, 29]) and trigonometric Fourier collocation methods (see, e.g. [30, 31]). These methods share the fact that they can exactly integrate the unperturbed multi-frequency highly oscillatory system

$$
q''(t) + \kappa^2 A q(t) = 0. \tag{3.7}
$$

In particular, it is important to note that both the internal stages and updates of an ERKN integrator can solve (3.7) exactly. This property of ERKN method is essential for efficiently solving the nonlinear initial value problem (3.1) with highly oscillatory solutions. Therefore, ERKN integrators are oscillation preserving (see Chap. 1 for details and [32]).

Moreover, we also note that the classical stability analysis for numerical methods deals with the following prototype scalar test equation (see, e.g. [15, 33]):

$$
\ddot{q}(t) + \omega^2 q(t) = -\varepsilon q(t) \quad \text{with} \quad \omega^2 + \varepsilon > 0, \tag{3.8}
$$

where ω represents an estimate of the frequency λ and $\varepsilon = \lambda^2 - \omega^2$ is the error of the estimation. This is a linear system with a single-frequency, and hence this kind of stability analysis is termed as the linear stability analysis. However, it should be pointed out that the original system (3.1) is a nonlinear highly oscillatory system with multiple frequencies. In particular, it may be a large scale system of nonlinear multi-frequency highly oscillatory ODEs yielded by the refinement of spatial discretisations for semilinear wave equations. Therefore, it would be insufficient to apply the linear and single frequency test equation (3.8) to the stability analysis for a numerical method designed for the nonlinear multi-frequency highly oscillatory system (3.1). This implies that it is important to investigate the nonlinear stability for (3.1). Taking into account the special structure brought by the linear term $\kappa^2 A q(t)$ of the system, and approximating the nonlinear integrals appearing in the variation-of-constants formula (3.4) by suitable quadrature formulas, ERKN integrators for solving the nonlinear multi-frequency highly oscillatory system (3.1) have been proposed and developed in the literature. However, in contrast to classical methods, the theoretical analysis associated with ERKN integrators is not sufficient. Therefore, one of our main purposes in this chapter is to analyse the nonlinear stability and convergence for the ERKN integrators based on the matrix-variation-of-constants formula (3.4).

Another important issue in this chapter is to investigate the applications of ERKN time integrators to semilinear wave equations:

$$u_{tt}(x, t) - \varepsilon^2 \Delta u(x, t) + \rho u = f\big(u(x, t)\big), \quad x \in \mathbb{T} = \mathbb{R}/(2\pi\mathbb{Z}), \quad t \in [t_0, T],$$
$$(3.9)$$

where $\varepsilon^2 > 0$ and $\rho > 0$ are parameters, and the function $f(\cdot)$ is smooth and real-valued, satisfying $f(0) = 0$. The wave equation (3.9) is studied with 2π-periodic boundary conditions in one space dimension and its solution is assumed to be real-valued. The initial values at time $t = t_0$ are given by

$$u(x, t_0) = \varphi(x), \quad u_t(x, t_0) = \psi(x). \tag{3.10}$$

In the literature, there exist many numerical strategies for solving the semilinear wave equation, such as the finite difference method [3–5, 34], the pseudospectral or spectral method [2, 6], the radial basis functions methods [35], the dual reciprocity boundary integral equation technique [36] and the He's variational iteration method [37]. In this chapter, using the idea of the so-called operator-variation-of-constants formula described in Chap. 1, we will combine the ERKN time integrators with Fourier pseudospectral spatial discretisation (ERKN-FP) to solve (3.9) with 2π-periodic boundary conditions and initial conditions (3.10), and this leads to a fully discrete scheme. On the basis of energy techniques, which are widely used in the numerical analysis of partial differential equations (see, e.g. [38–44]), we will conclude that the global error bounds of the ERKN-FP schemes are independent of any restriction of the time stepsize and the spatial stepsize. Moreover, it is well known that restriction (CFL) of the time stepsize and the spatial stepsize for the traditional numerical schemes in the literature is required for solving semilinear wave equations. This means that the CFL condition is an essential element associated with numerical PDEs in practice and the traditional schemes for PDEs usually suffer from this crucial condition. Fortunately, however, our analysis of the global errors in this chapter confirms that the ERKN-FP schemes are completely independent of the CFL condition when applied to the semilinear wave equation. This is one of the most essential properties of ERKN time integrators when applied to the semilinear wave equation.

3.2   Nonlinear Stability and Convergence Analysis for ERKN Integrators

This section concerns the study of nonlinear stability and convergence for ERKN integrators over a finite time interval. We begin this study with the nonlinear stability analysis of the matrix-variation-of-constants formula for the nonlinear highly oscillatory system (3.1), possessing multiple frequencies. After completing this, we turn to the nonlinear stability and convergence analysis for ERKN integrators. Throughout this section, $\|\cdot\|$ represents the vector 2-norm or matrix 2-norm (spectral norm).

3.2.1   Nonlinear Stability of the Matrix-Variation-of-Constants Formula

To begin with the stability analysis, we assume that the perturbed problem of (3.1) is

$$\begin{cases} \ddot{p}(t) + \kappa^2 A p(t) = g\big(p(t)\big) + \varepsilon(t), & t \in [t_0, T], \\ p(t_0) = \varphi + \tilde{\varphi}, \quad \dot{p}(t_0) = \psi + \tilde{\psi}, \end{cases} \tag{3.11}$$

where A is symmetric and positive semi-definite, $\tilde{\varphi}$, $\tilde{\psi}$ are perturbations of the initial conditions, and $\varepsilon(t)$ is the perturbation of the nonlinear term. We let $\eta(t) = p(t) - q(t)$. Subtracting (3.1) from (3.11) leads to the following perturbation system:

$$\begin{cases} \ddot{\eta}(t) + \kappa^2 A \eta(t) = g\big(p(t)\big) - g\big(q(t)\big) + \varepsilon(t), & t \in [t_0, T], \\ \eta(t_0) = \tilde{\varphi}, \quad \dot{\eta}(t_0) = \tilde{\psi}. \end{cases} \tag{3.12}$$

We choose the time stepsize $\Delta t = (T - t_0)/N$, where N is a positive integer, and denote the steps as

$$t_n = t_0 + n\Delta t, \quad n = 0, 1, 2, \cdots, N.$$

Applying the matrix-variation-of-constants formula (3.4) to the perturbation system (3.12) yields

$$
\begin{cases}
\eta(t_n + \mu\Delta t) = \phi_0(\mu^2 V)\eta(t_n) + \mu\Delta t\phi_1(\mu^2 V)\dot{\eta}(t_n) \\
\qquad\qquad + \Delta t^2 \displaystyle\int_0^\mu (\mu - z)\phi_1\big((\mu - z)^2 V\big)\Big(g\big(p(t_n + z\Delta t)\big) - g\big(q(t_n + z\Delta t)\big)\Big)\mathrm{d}z \\
\qquad\qquad + \Delta t^2 \displaystyle\int_0^\mu (\mu - z)\phi_1\big((\mu - z)^2 V\big)\varepsilon(t_n + z\Delta t)\mathrm{d}z, \\
\dot{\eta}(t_n + \mu\Delta t) = -\mu\Delta t\kappa^2 A\phi_1(\mu^2 V)\eta(t_n) + \phi_0(\mu^2 V)\dot{\eta}(t_n) \\
\qquad\qquad + \Delta t \displaystyle\int_0^\mu \phi_0\big((\mu - z)^2 V\big)\Big(g\big(p(t_n + z\Delta t)\big) - g\big(q(t_n + z\Delta t)\big)\Big)\mathrm{d}z \\
\qquad\qquad + \Delta t \displaystyle\int_0^\mu \phi_0\big((\mu - z)^2 V\big)\varepsilon(t_n + z\Delta t)\mathrm{d}z,
\end{cases}
$$

(3.13)

where $0 \leqslant \mu \leqslant 1$ and $V = \Delta t^2 \kappa^2 A$. Since the matrix A can be decomposed as $A = \Omega^2$, we denote the matrix $D = \kappa\Omega$, and then the decomposition of matrix $\kappa^2 A$ reads:

$$
\kappa^2 A = D^2,
$$

where D is positive semi-definite matrix. Accordingly, the formula (3.13) can be rewritten as the following compact form:

$$
\begin{bmatrix} D\eta(t_n + \mu\Delta t) \\ \dot{\eta}(t_n + \mu\Delta t) \end{bmatrix} = \Psi(\mu, 0, V) \begin{bmatrix} D\eta(t_n) \\ \dot{\eta}(t_n) \end{bmatrix}
$$

$$
+\Delta t \int_0^\mu \Psi(\mu, z, V) \begin{bmatrix} 0 \\ g\big(p(t_n + z\Delta t)\big) - g\big(q(t_n + z\Delta t)\big) \end{bmatrix} \mathrm{d}z
$$

$$
+\Delta t \int_0^\mu \Psi(\mu, z, V) \begin{bmatrix} 0 \\ \varepsilon(t_n + z\Delta t) \end{bmatrix} \mathrm{d}z,
$$

(3.14)

where

$$
\Psi(\mu, z, V) = \begin{bmatrix} \phi_0((\mu - z)^2 V) & \Delta t(\mu - z)D\phi_1((\mu - z)^2 V) \\ -\Delta t(\mu - z)D\phi_1((\mu - z)^2 V) & \phi_0((\mu - z)^2 V) \end{bmatrix}.
$$

(3.15)

Before going into the details of stability analysis, we summarise some useful properties related to the matrix-valued functions $\phi_j(\mu^2 V)$ for $j \in \mathbb{N}$ and clarify the spectral norm of $\Psi(\mu, z, V)$ for $0 \leqslant \mu, z \leqslant 1$.

Lemma 3.1 (See, e.g. [28, 45]) *The matrix-valued functions defined by (3.5) satisfy*

$$\int_0^1 (1-z)\phi_1\big(\mu^2(1-z)^2V\big)z^j\,dz = \Gamma(j+1)\phi_{j+2}(\mu^2V), \quad j = 0, 1, 2, \cdots,$$

$$\int_0^1 \phi_0\big(\mu^2(1-z)^2V\big)z^j\,dz = \Gamma(j+1)\phi_{j+1}(\mu^2V), \quad j = 0, 1, 2, \cdots,$$

$$(3.16)$$

where $\Gamma(j+1)$ is the Gamma function.

Lemma 3.2 *The matrix-valued functions defined by (3.5) are bounded, i.e.,*

$$\|\phi_j(\mu^2V)\| \leqslant \frac{1}{\Gamma(j+1)}, \quad j \in \mathbb{N}. \tag{3.17}$$

In particular, we have $\|\phi_0(\mu^2V)\| \leqslant 1$ and $\|\phi_1(\mu^2V)\| \leqslant 1$. Moreover, we also have

$$\|\mu\Delta t D\phi_1(\mu^2V)\| \leqslant 1 \quad and \quad \|\mu\Delta t D\phi_j(\mu^2V)\| \leqslant \frac{1}{\Gamma(j)}, \quad j = 2, 3, \cdots, \tag{3.18}$$

where μ is a positive number with $0 \leqslant \mu \leqslant 1$.

Proof The boundedness of $\|\phi_j(\mu^2V)\|$ and $\|\mu\Delta t D\phi_1(\mu^2V)\|$ can be confirmed straightforwardly from the definition of the matrix-valued functions (3.5) and Lemma 3.1. We thus need only to prove the boundedness of $\|\mu\Delta t D\phi_j(\mu^2V)\|$ for $j = 2, 3, \cdots$. Clearly, it follows from the definition of $\phi_j(\cdot)$ in (3.5) that

$$\mu\Delta t D(1-z)\phi_1\big(\mu^2(1-z)^2V\big) = \sin\big(\mu(1-z)\Delta t D\big).$$

Therefore, the conclusion of Lemma 3.1 yields that

$$\mu\Delta t D\phi_j(\mu^2V) = \frac{\mu\Delta t D}{\Gamma(j-1)}\int_0^1 (1-z)\phi_1\big(\mu^2(1-z)^2V\big)z^{j-2}dz$$

$$= \frac{1}{\Gamma(j-1)}\int_0^1 \sin\big(\mu(1-z)\Delta t D\big)z^{j-2}dz. \tag{3.19}$$

Taking the spectral norms on both sides of (3.19) leads to

$$\|\mu\Delta t D\phi_j(\mu^2V)\| \leqslant \frac{1}{\Gamma(j-1)}\int_0^1 \|\sin\big(\mu(1-z)\Delta t D\big)\|z^{j-2}dz \leqslant \frac{1}{\Gamma(j)}. \tag{3.20}$$

The statement of the lemma is confirmed.   □

Lemma 3.3 (See, e.g. [13, 15]) *The bounded matrix-valued functions* $\phi_0(\mu^2 V)$ *and* $\phi_1(\mu^2 V)$ *defined by* (3.5) *satisfy*

$$\phi_0^2(\mu^2 V) + \mu^2 V \phi_1^2(\mu^2 V) = I, \tag{3.21}$$

where V *is any positive semi-definite matrix and* I *is the identity matrix.*

The other conclusions of Lemmas 3.1–3.3 can be proved by direct calculation, see [13, 15, 28, 45], and we here ignore the details of the proof.

Theorem 3.1 *The spectral norms of the matrices* $\Psi(\mu, z, V)$ *satisfy*

$$\|\Psi(\mu, z, V)\| = 1, \quad \forall\, \mu, z \in [0, 1], \tag{3.22}$$

where $V = h^2 \kappa^2 A$ *and* A *is a symmetric and positive semi-definite matrix.*

Proof Obviously, the matrix $\Psi(\mu, z, V)$ is well defined in (3.15) because A is a symmetric and positive semi-definite matrix. Moreover, it is easy to verify that

$$\Psi(\mu, z, V)^\mathsf{T} \Psi(\mu, z, V) = I_{2d \times 2d}.$$

Thus, we have

$$\|\Psi(\mu, z, V)\| = 1, \quad \forall \mu, z \in [0, 1].$$

The conclusion of the lemma is confirmed.   □

According to the assumption of the *finite-energy conditions* (see, e.g. [11, 12, 18])

$$\frac{1}{2}\|\dot{q}(t)\|^2 + \frac{\kappa^2}{2} q(t)^\mathsf{T} A q(t) \leqslant \frac{K^2}{2}, \tag{3.23}$$

where K is a constant, the error bounds of the Gaustchi-type methods of order two were proved to be independent of $\kappa^2 \|A\|$. Here, we observe that Gautschi-type time integrators are special ERKN integrators of order two. Therefore, it seems reasonable to assume that the finite-energy condition (3.23) is also satisfied in a strip along the exact solution. Using this assumption, we will investigate the nonlinear stability and the error bounds of the ERKN integrators. To this end, we also need to quote the following Gronwall's inequality (see, e.g. [29]), which plays an important role for the remainder of our analysis.

Lemma 3.4 *Let* σ *be a positive number and* a_k, b_k $(k = 0, 1, 2, \cdots)$ *be nonnegative and satisfy*

$$a_k \leqslant (1 + \sigma \Delta t) a_{k-1} + \Delta t b_k, \quad k = 1, 2, 3, \cdots,$$

then

$$a_k \leqslant \exp(\sigma k \Delta t)\Big(a_0 + \Delta t \sum_{m=1}^{k} b_m\Big), \quad k = 1, 2, 3, \cdots.$$

In what follows, we first show the nonlinear stability of the matrix-variation-of-constants formula (3.4) whose perturbation formula is given by (3.13).

Theorem 3.2 *Assume that the solution $\eta(t)$ of the perturbation system (3.12) and its derivative $\dot{\eta}(t)$ satisfy the finite-energy condition. If the time stepsize Δt satisfies $\Delta t \leqslant \sqrt{\dfrac{1}{2L}}$, then we have*

$$\|\eta(t_n)\| \leqslant \exp\Big(T(1+4L)\Big)\Big(\|\tilde{\varphi}\| + \sqrt{\|\tilde{\psi}\|^2 + \kappa^2 \tilde{\varphi}^{\mathsf{T}} A \tilde{\varphi}} + 4\Delta t \sum_{k=0}^{n} \max_{0 \leqslant z \leqslant 1} \|\varepsilon(t_k + z\Delta t)\|\Big),$$

$$\|\dot{\eta}(t_n)\| \leqslant \exp\Big(T(1+4L)\Big)\Big(\|\tilde{\varphi}\| + \sqrt{\|\tilde{\psi}\|^2 + \kappa^2 \tilde{\varphi}^{\mathsf{T}} A \tilde{\varphi}} + 4\Delta t \sum_{k=0}^{n} \max_{0 \leqslant z \leqslant 1} \|\varepsilon(t_k + z\Delta t)\|\Big).$$
(3.24)

That is, the matrix-variation-of-constants formula is nonlinearly stable over the time interval $[t_0, T]$.

Proof We take the l_2-norm on both sides of the first formula (3.13) and (3.14), respectively, and obtain

$$\|\eta(t_n + \Delta t)\| \leqslant \|\eta(t_n)\| + \Delta t \|\dot{\eta}(t_n)\| + \Delta t^2 \int_0^1 \|g\big(p(t_n + z\Delta t)\big) - g\big(q(t_n + z\Delta t)\big)\| dz$$

$$+ \Delta t^2 \int_0^1 \|\varepsilon(t_n + z\Delta t)\| dz,$$
(3.25)

and

$$\sqrt{\|\dot{\eta}(t_n + \Delta t)\|^2 + \kappa^2 \eta(t_n + \Delta t)^{\mathsf{T}} A \eta(t_n + \Delta t)} \leqslant \sqrt{\|\dot{\eta}(t_n)\|^2 + \kappa^2 \eta(t_n)^{\mathsf{T}} A \eta(t_n)}$$

$$+ \Delta t \int_0^1 \|g\big(p(t_n + z\Delta t)\big) - g\big(q(t_n + z\Delta t)\big)\| dz + \Delta t \int_0^1 \|\varepsilon(t_n + z\Delta t)\| dz.$$
(3.26)

We then sum up the results in (3.25) and (3.26) and apply the Lipschitz condition. This leads to

$$\|\eta(t_n + \Delta t)\| + \sqrt{\|\dot{\eta}(t_n + \Delta t)\|^2 + \kappa^2 \eta(t_n + \Delta t)^{\mathsf{T}} A \eta(t_n + \Delta t)}$$

$$\leqslant \|\eta(t_n)\| + \sqrt{\|\dot{\eta}(t_n)\|^2 + \kappa^2 \eta(t_n)^{\mathsf{T}} A \eta(t_n)} + \Delta t \|\dot{\eta}(t_n)\| + L \Delta t (1 + \Delta t) \max_{0 \leqslant z \leqslant 1} \|\eta(t_n + z \Delta t)\|$$

$$+ \Delta t (1 + \Delta t) \max_{0 \leqslant z \leqslant 1} \|\varepsilon(t_n + z \Delta t)\|.$$

$$\tag{3.27}$$

It follows from the first equality in (3.13) that

$$\|\eta(t_n + \mu \Delta t)\| \leqslant \|\eta(t_n)\| + \Delta t \|\dot{\eta}(t_n)\| + \Delta t^2 L \max_{0 \leqslant z \leqslant 1} \|\eta(t_n + z \Delta t)\| + \Delta t^2 \max_{0 \leqslant z \leqslant 1} \|\varepsilon(t_n + z \Delta t)\|.$$

Under the assumption that time stepsize Δt satisfies $\Delta t \leqslant \sqrt{\dfrac{1}{2L}}$, we then obtain

$$\max_{0 \leqslant z \leqslant 1} \|\eta(t_n + z \Delta t)\| \leqslant 2 \|\eta(t_n)\| + 2 \Delta t \|\dot{\eta}(t_n)\| + 2 \Delta t^2 \max_{0 \leqslant z \leqslant 1} \|\varepsilon(t_n + z \Delta t)\|.$$

$$\tag{3.28}$$

Inserting (3.28) into (3.27) gives

$$\|\eta(t_n + \Delta t)\| + \sqrt{\|\dot{\eta}(t_n + \Delta t)\|^2 + \kappa^2 \eta(t_n + \Delta t)^{\mathsf{T}} A \eta(t_n + \Delta t)}$$

$$\leqslant \left(1 + \Delta t (1 + 4L)\right) \left(\|\eta(t_n)\| + \sqrt{\|\dot{\eta}(t_n)\|^2 + \kappa^2 \eta(t_n)^{\mathsf{T}} A \eta(t_n)}\right)$$

$$+ 2 \Delta t (1 + \Delta t) \max_{0 \leqslant z \leqslant 1} \|\varepsilon(t_n + z \Delta t)\|.$$

$$\tag{3.29}$$

Applying the Gronwall's inequality (Lemma 3.4) to (3.29) yields

$$\|\eta(t_n + \Delta t)\| + \sqrt{\|\dot{\eta}(t_n + \Delta t)\|^2 + \kappa^2 \eta(t_n + \Delta t)^{\mathsf{T}} A \eta(t_n + \Delta t)}$$

$$\leqslant \exp\left(n \Delta t (1 + 4L)\right) \left(\|\eta(t_0)\| + \sqrt{\|\dot{\eta}(t_0)\|^2 + \kappa^2 \eta(t_0)^{\mathsf{T}} A \eta(t_0)}\right.$$

$$\left. + 4 \Delta t \sum_{k=0}^{n} \max_{0 \leqslant z \leqslant 1} \|\varepsilon(t_k + z \Delta t)\|\right).$$

$$\tag{3.30}$$

We thus obtain the estimations

$$\|\eta(t_n)\| \leqslant \exp\left(T(1+4L)\right)\left(\|\tilde{\varphi}\| + \sqrt{\|\tilde{\psi}\|^2 + \kappa^2\tilde{\varphi}^{\mathsf{T}}A\tilde{\varphi}} + 4\Delta t \sum_{k=0}^{n} \max_{0 \leqslant z \leqslant 1} \|\varepsilon(t_k + z\Delta t)\|\right),$$

$$\|\dot{\eta}(t_n)\| \leqslant \exp\left(T(1+4L)\right)\left(\|\tilde{\varphi}\| + \sqrt{\|\tilde{\psi}\|^2 + \kappa^2\tilde{\varphi}^{\mathsf{T}}A\tilde{\varphi}} + 4\Delta t \sum_{k=0}^{n} \max_{0 \leqslant z \leqslant 1} \|\varepsilon(t_k + z\Delta t)\|\right).$$

Theorem 3.2 is proved.   □

The matrix-variation-of-constants formula (3.4) is fundamental to a true understanding of ERKN integrators for the multi-frequency highly oscillatory system (3.1). Hence, its nonlinear stability is crucial for the nonlinear stability of ERKN integrators for (3.1).

3.2.2 Nonlinear Stability and Convergence of ERKN Integrators

The main theme of this subsection is the nonlinear stability and convergence analysis of ERKN integrators for the nonlinear multi-frequency highly oscillatory system (3.1). Choosing suitable nodes $c_1, c_2, \cdots, c_s$ and approximating the nonlinear integrals appearing in the formula (3.4) by suitable numerical quadrature formulae leads to the following ERKN integrators (see, e.g. [15]).

Definition 3.1 (See [15]) An s-stage multidimensional multi-frequency ERKN integrator with a stepsize Δt for the multidimensional and multi-frequency oscillatory nonlinear system (3.1) is defined as

$$\begin{cases} q_{n+1} = \phi_0(V)q_n + \Delta t\phi_1(V)\dot{q}_n + \Delta t^2 \sum_{i=1}^{s} \bar{B}_i(V)g(Q_{ni}), \\[2mm] \dot{q}_{n+1} = -\Delta t\kappa^2 A\phi_1(V)q_n + \phi_0(V)\dot{q}_n + \Delta t \sum_{i=1}^{s} B_i(V)g(Q_{ni}), \\[2mm] Q_{ni} = \phi_0(c_i^2 V)q_n + c_i\Delta t\phi_1(c_i^2 V)\dot{q}_n + \Delta t^2 \sum_{j=1}^{s} A_{ij}(V)g(Q_{nj}), \quad i = 1, 2, \cdots, s, \end{cases}$$

$$(3.31)$$

where $0 \leqslant c_i \leqslant 1$ for $i = 1, 2, \cdots, s$ are real constants and $\bar{B}_i(V)$, $B_i(V)$ and $A_{ij}(V)$ for $i, j = 1, 2, \cdots, s$ are matrix-valued functions of $V = \Delta t^2\kappa^2 A$.

Using the SSEN-tree set and the corresponding B-series theory (see, e.g. [13]), we now recall the order conditions of ERKN integrators which are summarised in the following theorem. The weights $\bar{B}_i(V)$, $B_i(V)$ and $A_{ij}(V)$ of an ERKN integrator for $i, j = 1, 2, \cdots, s$ can be determined by the following order conditions.

Theorem 3.3 *The ERKN integrator* (3.31) *has order r if and only if the order conditions are satisfied*

$$\sum_{i=1}^{s} \bar{B}_i(V)\Phi_i(\tau) = \frac{\rho(\tau)!}{\tilde{\gamma}(\tau)s(\tau)}\phi_{\rho(\tau)+1}(V) + \mathscr{O}(\Delta t^{r-\rho(\tau)}), \quad \forall \tau \in \mathrm{SSRNT}_m, \quad m \leqslant r-1,$$

$$\sum_{i=1}^{s} B_i(V)\Phi_i(\tau) = \frac{\rho(\tau)!}{\tilde{\gamma}(\tau)s(\tau)}\phi_{\rho(\tau)}(V) + \mathscr{O}(\Delta t^{r-\rho(\tau)+1}), \quad \forall \tau \in \mathrm{SSRNT}_m, \quad m \leqslant r,$$

(3.32)

where the definitions and properties of the order $\rho(\tau)$, the sgn $s(\tau)$, the density $\tilde{\gamma}(\tau)$, and the weight $\Phi_i(\tau)$ are well established and can be found in [13].

The conclusions of Theorem 3.3 indicate that the weights $\bar{B}_i(V)$, $B_i(V)$ and $A_{ij}(V)$ are the linear combination of $\phi_j(V)$. Furthermore, it is evident from the first conditions of (3.32) that the weights $\bar{B}_i(V)$ are independent of $\phi_0(V)$. Therefore, combining this fact with Lemma 3.2, we can establish the uniform boundedness of the weights $\bar{B}_i(V)$, $B_i(V)$ and $A_{ij}(V)$, which will be used in our theoretical analysis.

Lemma 3.5 *The weights $\bar{B}_i(V)$, $B_i(V)$ and $A_{ij}(V)$ are uniformly bounded, i.e.,*

$$\|\bar{B}_i(V)\| \leqslant \bar{B}, \quad \|\Delta t D\bar{B}_i(V)\| \leqslant \hat{B}, \quad \|B_i(V)\| \leqslant B, \quad \|A_{ij}(V)\| \leqslant \beta,$$

(3.33)

for $i, j = 1, 2, \cdots, s$, where $\bar{B}$, $\hat{B}$, B and β are all constants independent of Δt, κ^2 and the matrix V and D.

Applying an ERKN integrator (3.31) to the perturbed system (3.12), we obtain

$$\begin{cases} \eta_{n+1} = \phi_0(V)\eta_n + \Delta t\phi_1(V)\dot{\eta}_n + \Delta t^2 \sum_{i=1}^{s} \bar{B}_i(V)\Big(g(P_{ni}) - g(Q_{ni}) + \varepsilon(t_n + c_i\Delta t)\Big), \\[2mm] \dot{\eta}_{n+1} = -\Delta t\kappa^2 A\phi_1(V)\eta_n + \phi_0(V)\dot{\eta}_n \\[2mm] \qquad\quad + \Delta t \sum_{i=1}^{s} B_i(V)\Big(g(P_{ni}) - g(Q_{ni}) + \varepsilon(t_n + c_i\Delta t)\Big), \\[2mm] \eta_{ni} = \phi_0(c_i^2 V)\eta_n + c_i\Delta t\phi_1(c_i^2 V)\dot{\eta}_n \\[2mm] \qquad\quad + \Delta t^2 \sum_{j=1}^{s} A_{ij}(V)\Big(g(P_{nj}) - g(Q_{nj}) + \varepsilon(t_n + c_j\Delta t)\Big), \\[2mm] i = 1, 2, \cdots, s. \end{cases}$$

(3.34)

The first two equalities of (3.34) can be rewritten as the compact form:

$$
\begin{bmatrix} D\eta_{n+1} \\ \dot{\eta}_{n+1} \end{bmatrix} = \Psi(1, 0, V) \begin{bmatrix} D\eta_n \\ \dot{\eta}_n \end{bmatrix}
$$

$$
+ \Delta t \sum_{i=1}^{s} \begin{bmatrix} \Delta t D \bar{B}_i(V)\Big(g(P_{ni}) - g(Q_{ni}) + \varepsilon(t_n + c_i \Delta t)\Big) \\ B_i(V)\Big(g(P_{ni}) - g(Q_{ni}) + \varepsilon(t_n + c_i \Delta t)\Big) \end{bmatrix}.
$$

$$(3.35)$$

We next present the nonlinear stability analysis for ERKN integrators over the finite time interval $[t_0, T]$.

Theorem 3.4   *It is assumed that the nonlinear function $g(\cdot)$ is locally Lipschitz continuous and the finite-energy condition (3.23) is satisfied. Then, if the time stepsize Δt satisfies the condition $\Delta t \leqslant \sqrt{\dfrac{1}{2sL\beta}}$, we have the following estimates for the perturbation system (3.12)*

$$
\|\eta_n\| \leqslant \exp(C_1 T) \left(\|\tilde{\varphi}\| + \sqrt{\|\tilde{\psi}\|^2 + \kappa^2 \tilde{\varphi}^{\mathsf{T}} A \tilde{\varphi}} + C_2 \Delta t \sum_{k=0}^{n} \sum_{i=1}^{s} \|\varepsilon(t_k + c_i \Delta t)\| \right),
$$

$$
\|\dot{\eta}_n\| \leqslant \exp(C_1 T) \left(\|\tilde{\varphi}\| + \sqrt{\|\tilde{\psi}\|^2 + \kappa^2 \tilde{\varphi}^{\mathsf{T}} A \tilde{\varphi}} + C_2 \Delta t \sum_{k=0}^{n} \sum_{i=1}^{s} \|\varepsilon(t_k + c_i \Delta t)\| \right),
$$

$$(3.36)$$

where C_1 and C_2 are constants independent of Δt, κ^2 and the dominant frequency matrix A.

Proof   Under the hypothesis of the finite-energy condition (3.23), by taking l^2-norm on both sides of the first equality in (3.34) and (3.35), we obtain

$$
\|\eta_{n+1}\| \leqslant \|\eta_n\| + \Delta t \|\dot{\eta}_n\| + \Delta t^2 \bar{B} \sum_{i=1}^{s} \Big(L\|\eta_{ni}\| + \|\varepsilon(t_n + c_i \Delta t)\| \Big),
$$

and

$$
\sqrt{\|\dot{\eta}_{n+1}\|^2 + \kappa^2 \eta_{n+1}^{\mathsf{T}} A \eta_{n+1}} \leqslant \sqrt{\|\dot{\eta}_n\|^2 + \kappa^2 \eta_n^{\mathsf{T}} A \eta_n} + \Delta t \Big(\hat{B}
$$

$$
+ B \Big) \sum_{i=1}^{s} \Big(L\|\eta_{ni}\| + \|\varepsilon(t_n + c_i \Delta t)\| \Big).
$$

Summing up the above inequalities leads to

$$
\|\eta_{n+1}\| + \sqrt{\|\dot{\eta}_{n+1}\|^2 + \kappa^2 \eta_{n+1}^{\mathsf{T}} A \eta_{n+1}} \leqslant \|\eta_n\| + \sqrt{\|\dot{\eta}_n\|^2 + \kappa^2 \eta_n^{\mathsf{T}} A \eta_n} + \Delta t \|\dot{\eta}_n\|
$$

$$
+ \Delta t (\bar{B} + \hat{B} + B) \sum_{i=1}^{s} \Big(L \|\eta_{ni}\| + \|\varepsilon(t_n + c_i \Delta t)\| \Big), \tag{3.37}
$$

where we have used the uniform boundedness of the weights $\bar{B}_i(V)$, $\Delta t D \bar{B}_i(V)$, $B_i(V)$, $A_{ij}(V)$ (see Lemma 3.5). Likewise, it follows from taking norms on both sides of the third equality in (3.34) that

$$
\|\eta_{ni}\| \leqslant \|\eta_n\| + c_i \Delta t \|\dot{\eta}_n\| + \Delta t^2 \beta \sum_{i=1}^{s} \Big(L \|\eta_{ni}\| + \|\varepsilon(t_n + c_i \Delta t)\| \Big). \tag{3.38}
$$

Under the assumption that the time stepsize Δt satisfies $\Delta t \leqslant \sqrt{\dfrac{1}{2s L \beta}}$, it then follows from the inequality (3.38) that

$$
\sum_{i=1}^{s} \|\eta_{ni}\| \leqslant 2s \Big(\|\eta_n\| + \Delta t \|\dot{\eta}_n\| \Big) + \frac{1}{L} \sum_{i=1}^{s} \|\varepsilon(t_n + c_i \Delta t)\|. \tag{3.39}
$$

Inserting (3.39) into (3.37) results in

$$
\|\eta_{n+1}\| + \sqrt{\|\dot{\eta}_{n+1}\|^2 + \kappa^2 \eta_{n+1}^{\mathsf{T}} A \eta_{n+1}}
$$

$$
\leqslant (1 + C_1 \Delta t) \Big(\|\eta_n\| + \sqrt{\|\dot{\eta}_n\|^2 + \kappa^2 \eta_n^{\mathsf{T}} A \eta_n} \Big)
$$

$$
+ C_2 \Delta t \sum_{i=1}^{s} \|\varepsilon(t_n + c_i \Delta t)\|,
$$

where $C_1 = 1 + s L C_2$ and $C_2 = 2(\bar{B} + \hat{B} + B)$ are constants. Thus, using the discrete Gronwall's inequality (Lemma 3.4), we obtain

$$
\|\eta_n\| + \sqrt{\|\dot{\eta}_n\|^2 + \kappa^2 \eta_n^{\mathsf{T}} A \eta_n}
$$

$$
\leqslant \exp(C_1 T) \left(\|\tilde{\varphi}\| + \sqrt{\|\tilde{\psi}\|^2 + \kappa^2 \tilde{\varphi}^{\mathsf{T}} A \tilde{\varphi}} + C_2 \Delta t \sum_{k=0}^{n} \sum_{i=1}^{s} \|\varepsilon(t_k + c_i \Delta t)\| \right).
$$

This completes the proof.   □

We denote

$$\zeta_n = q(t_n) - q_n, \quad \dot{\zeta}_n = \dot{q}(t_n) - \dot{q}_n, \quad \zeta_{ni} = q(t_n + c_i \Delta t) - Q_{ni} \quad \text{for} \quad i = 1, 2, \cdots, s.$$

Subtracting (3.31) from the exact solution (3.4) yields

$$
\begin{cases}
\zeta_{n+1} = \phi_0(V)\zeta_n + \Delta t \phi_1(V)\dot{\zeta}_n + \Delta t^2 \sum_{i=1}^{s} \bar{B}_i(V)\Big(g\big(q(t_n + c_i \Delta t)\big) - g(Q_{ni})\Big) \\
\qquad + \Delta t^2 \int_0^1 (1 - z)\phi_1\big((1 - z)^2 V\big)g\big(q(t_n + z\Delta t)\big)dz - \Delta t^2 \sum_{i=1}^{s} \bar{B}_i(V)g\big(q(t_n + c_i \Delta t)\big), \\[2mm]
\dot{\zeta}_{n+1} = -\Delta t \kappa^2 A\phi_1(V)\zeta_n + \phi_0(V)\dot{\zeta}_n + \Delta t \sum_{i=1}^{s} B_i(V)\Big(g\big(q(t_n + c_i \Delta t)\big) - g(Q_{ni})\Big) \\
\qquad + \Delta t \int_0^1 \phi_0\big((1 - z)^2 V\big)g\big(q(t_n + z\Delta t)\big)dz - \Delta t \sum_{i=1}^{s} B_i(V)g\big(q(t_n + c_i \Delta t)\big), \\[2mm]
\zeta_{ni} = \phi_0(c_i^2 V)\zeta_n + c_i \Delta t \phi_1(c_i^2 V)\dot{\zeta}_n + \Delta t^2 \sum_{j=1}^{s} A_{ij}(V)\Big(g\big(q(t_n + c_j \Delta t)\big) - g(Q_{nj})\Big) \\
\qquad + c_i^2 \Delta t^2 \int_0^1 (1 - z)\phi_1\big((1 - z)^2 c_i^2 V\big)g\big(q(t_n + zc_i \Delta t)\big)dz \\
\qquad - \Delta t^2 \sum_{j=1}^{s} A_{ij}(V)g\big(q(t_n + c_j \Delta t)\big), \\[2mm]
i = 1, 2, \cdots, s.
\end{cases}
$$

$$(3.40)$$

We then expand $g\big(q(t_n + z\Delta t)\big)$ at t_n into a Taylor series with remainder in integral form:

$$g\big(q(t_n + z\Delta t)\big) = \sum_{k=0}^{r-1} \frac{z^k \Delta t^k}{k!} g^{(k)}\big(q(t_n)\big) + \Delta t^r \int_0^z \frac{(z - \tau)^{r-1}}{(r - 1)!} g^{(r)}\big(q(t_n + \tau \Delta t)\big)d\tau.$$

$$(3.41)$$

Inserting the Taylor expression into the right-hand sides of (3.40) leads to

$$
\begin{cases}
\zeta_{n+1} = \phi_0(V)\zeta_n + \Delta t \phi_1(V)\dot{\zeta}_n + \Delta t^2 \sum_{i=1}^{s} \bar{B}_i(V)\Delta g_{ni} + \delta^{n+1}, \\[2mm]
\dot{\zeta}_{n+1} = -\Delta t \kappa^2 A \phi_1(V)\zeta_n + \phi_0(V)\dot{\zeta}_n + \Delta t \sum_{i=1}^{s} B_i(V)\Delta g_{ni} + \dot{\delta}^{n+1}, \\[2mm]
\zeta_{ni} = \phi_0(c_i^2 V)\zeta_n + c_i \Delta t \phi_1(c_i^2 V)\dot{\zeta}_n + \Delta t^2 \sum_{j=1}^{s} A_{ij}(V)\Delta g_{nj} + \Delta_{ni}, \quad i = 1, 2, \cdots, s,
\end{cases}
$$

$$(3.42)$$

where $\Delta g_{ni} = g\big(q(t_n + c_i \Delta t)\big) - g(Q_{ni})$ and the remainders can be explicitly represented as

$$
\begin{aligned}
\delta^{n+1} &= \sum_{k=0}^{r-1} \Delta t^{k+2} \left(\phi_{k+2}(V) - \sum_{i=1}^{s} \bar{B}_i(V)\frac{c_i^k}{k!} \right) g^{(k)}\big(q(t_n)\big) \\
&\quad + \Delta t^{r+2} \int_0^1 (1-z)\phi_1\big((1-z)^2 V\big) \int_0^z \frac{(z-\tau)^{r-1}}{(r-1)!} g^{(r)}\big(q(t_n + \tau \Delta t)\big)\mathrm{d}\tau \mathrm{d}z \\
&\quad - \Delta t^{r+2} \sum_{i=1}^{s} \bar{B}_i(V) \int_0^{c_i} \frac{(c_i - \tau)^{r-1}}{(r-1)!} g^{(r)}\big(q(t_n + \tau \Delta t)\big)\mathrm{d}\tau, \quad\quad (3.43)
\end{aligned}
$$

$$
\begin{aligned}
\dot{\delta}^{n+1} &= \sum_{k=0}^{r-1} \Delta t^{k+1} \left(\phi_{k+1}(V) - \sum_{i=1}^{s} B_i(V)\frac{c_i^k}{k!} \right) g^{(k)}\big(q(t_n)\big) \\
&\quad + \Delta t^{r+1} \int_0^1 \phi_0\big((1-z)^2 V\big) \int_0^z \frac{(z-\tau)^{r-1}}{(r-1)!} g^{(r)}\big(q(t_n + \tau \Delta t)\big)\mathrm{d}\tau \mathrm{d}z \\
&\quad - \Delta t^{r+1} \sum_{i=1}^{s} B_i(V) \int_0^{c_i} \frac{(c_i - \tau)^{r-1}}{(r-1)!} g^{(r)}\big(q(t_n + \tau \Delta t)\big)\mathrm{d}\tau, \quad\quad (3.44)
\end{aligned}
$$

and

$$
\begin{aligned}
\Delta_{ni} =& \sum_{k=0}^{r-1} \Delta t^{k+2} \left(c_i^{k+2} \phi_{k+2}(c_i^2 V) - \sum_{j=1}^{s} A_{ij}(V) \frac{c_j^k}{k!} \right) g^{(k)}(q(t_n)) \\
&+ c_i^{r+2} \Delta t^{r+2} \int_0^1 (1-z) \phi_1\big((1-z)^2 c_i^2 V\big) \int_0^z \frac{(z-\tau)^{r-1}}{(r-1)!} g^{(r)}\big(q(t_n + \tau c_i \Delta t)\big) d\tau dz \\
&- \Delta t^{r+2} \sum_{i=1}^{s} A_{i,j}(V) \int_0^{c_j} \frac{(c_j - \tau)^{r-1}}{(r-1)!} g^{(r)}\big(q(t_n + \tau \Delta t)\big) d\tau.
\end{aligned}
\tag{3.45}
$$

Note that if the weights $\bar{B}_i(V)$, $B_i(V)$ and $A_{ij}(V)$ satisfy the r-th order condition (3.32) in Lemma 3.3 and the exact solution of the multi-frequency highly oscillatory system (3.1) is of a suitable smoothness such that $g^{(r)} \in L^\infty([t_0, T], \mathbb{R}^d)$, then the remainders δ^{n+1}, $\dot{\delta}^{n+1}$ and Δ_{ni} satisfy the following estimates

$$
\|\delta^{n+1}\| \leqslant \tilde{C}_1 \Delta t^{r+2}, \quad \|\dot{\delta}^{n+1}\| \leqslant \tilde{C}_1 \Delta t^{r+1}, \quad \sum_{i=1}^{s} \|\Delta_{ni}\| \leqslant \tilde{C}_1 \Delta t^{r+1}, \quad \|D\delta^{n+1}\| \leqslant \tilde{C}_1 \Delta t^{r+1},
\tag{3.46}
$$

where $\tilde{C}_1$ is constant and obviously independent of Δt, the takanami number κ^2 and the dominant frequency-matrix A. Similarly to the stability analysis, we rewrite the first two equalities of (3.42) as the following matrix-vector form:

$$
\begin{bmatrix} D\zeta_{n+1} \\ \dot{\zeta}_{n+1} \end{bmatrix} = \Psi(1, 0, V) \begin{bmatrix} D\zeta_n \\ \dot{\zeta}_n \end{bmatrix} + \Delta t \sum_{i=1}^{s} \begin{bmatrix} \Delta t D \bar{B}_i(V) \Delta g_{ni} \\ B_i(V) \Delta g_{ni} \end{bmatrix} + \begin{bmatrix} D\delta^{n+1} \\ \dot{\delta}^{n+1} \end{bmatrix}.
\tag{3.47}
$$

Taking norms on both sides of (3.47) and using the estimates in (3.46), we obtain

$$
\sqrt{\|\dot{\zeta}_{n+1}\|^2 + \kappa^2 \zeta_{n+1}^\mathsf{T} A \zeta_{n+1}} \leqslant \sqrt{\|\dot{\zeta}_n\|^2 + \kappa^2 \zeta_n^\mathsf{T} A \zeta_n} + \Delta t L(\hat{B} + B) \sum_{i=1}^{s} \|\zeta_{ni}\| + \tilde{C} \Delta t^{r+1}.
\tag{3.48}
$$

In what follows, we will investigate the convergence of the ERKN integrator (3.31) for solving the system (3.1) of nonlinear multi-frequency highly oscillatory second-order ODEs.

Theorem 3.5 *Assume that the weights $\bar{B}_i(V)$, $B_i(V)$ and $A_{ij}(V)$ satisfy the r-th order conditions (3.32) and the exact solution $q(t)$ of the nonlinear highly oscillatory system (3.1) satisfies suitable smoothness such that $g^{(r)} \in L^\infty([t_0, T], \mathbb{R}^d)$.*

Then, if the time stepsize Δt satisfies $\Delta t \leqslant \sqrt{\dfrac{1}{2sL\beta}}$, we have the estimates

$$\|q(t_n) - q_n\| \leqslant \widetilde{C}\Delta t^r \quad and \quad \|\dot{q}(t_n) - \dot{q}_n\| \leqslant \widetilde{C}\Delta t^r, \tag{3.49}$$

where the constant $\widetilde{C}$ is independent of Δt, κ^2 and the dominant frequency-matrix A.

Proof   In a similar way to the proof of Theorem 3.4, it follows from taking norms on both sides of the first equation in (3.42) and summing up the obtained results with (3.48) that

$$\|\zeta_{n+1}\| + \sqrt{\|\dot{\zeta}_{n+1}\|^2 + \kappa^2\zeta_{n+1}^{\mathsf{T}}A\zeta_{n+1}} \leqslant \|\zeta_n\| + \sqrt{\|\dot{\zeta}_n\|^2 + \kappa^2\zeta_n^{\mathsf{T}}A\zeta_n} + \Delta t\|\dot{\zeta}_n\|$$

$$+ \Delta t L(\bar{B} + \hat{B} + B)\sum_{i=1}^{s}\|\zeta_{ni}\| + 2\widetilde{C}_1\Delta t^{r+1}. \tag{3.50}$$

Taking norms on both sides of the third equation in (3.42) and noting that the time stepsize Δt satisfies $\Delta t \leqslant \sqrt{\dfrac{1}{2sL\beta}}$, we obtain

$$\sum_{i=1}^{s}\|\zeta_{ni}\| \leqslant 2s\left(\|\zeta_n\| + \Delta t\|\dot{\zeta}_n\| + \widetilde{C}_1\Delta t^{r+1}\right). \tag{3.51}$$

Inserting (3.51) into (3.50) yields

$$\|\zeta_{n+1}\| + \sqrt{\|\dot{\zeta}_{n+1}\|^2 + \kappa^2\zeta_{n+1}^{\mathsf{T}}A\zeta_{n+1}}$$

$$\leqslant \left(1 + C_1\Delta t\right)\left(\|\zeta_n\| + \sqrt{\|\dot{\zeta}_n\|^2 + \kappa^2\zeta_n^{\mathsf{T}}A\zeta_n}\right) + \left(sLC_2\widetilde{C}_1\Delta t + 2\widetilde{C}_1\right)\Delta t^{r+1}$$

$$\leqslant \left(1 + C_1\Delta t\right)\left(\|\zeta_n\| + \sqrt{\|\dot{\zeta}_n\|^2 + \kappa^2\zeta_n^{\mathsf{T}}A\zeta_n}\right) + \widetilde{C}_3\Delta t^{r+1},$$

where $\widetilde{C}_3 = sLC_2\widetilde{C}_1 + 2\widetilde{C}_1$ is a constant independent of Δt, κ^2 and $\|A\|$, whereas C_1 and C_2 are given in the proof of Theorem 3.4. Therefore, using the Gronwall's inequality (Lemma 3.4), we obtain

$$\|\zeta_n\| + \sqrt{\|\dot{\zeta}_n\|^2 + \kappa^2\zeta_n^{\mathsf{T}}A\zeta_n} \leqslant \exp\left(C_1n\Delta t\right)\left(\|\zeta_0\| + \sqrt{\|\dot{\zeta}_0\|^2 + \kappa^2\zeta_0^{\mathsf{T}}A\zeta_0} + \widetilde{C}_3n\Delta t^{r+1}\right)$$

$$\leqslant \widetilde{C}_3T\exp\left(C_1T\right)\Delta t^r \leqslant \widetilde{C}\Delta t^r,$$

where $\widetilde{C} = \widetilde{C}_3T\exp\left(C_1T\right)$ is a constant independent of Δt, κ^2 and $\|A\|$. $\square$

Theorem 3.5 shows an important fact that, with the finite-energy condition (3.23), the error bounds of an ERKN integrator for solving the system (3.1) of multi-frequency highly oscillatory second-order ODEs are independent of the takanami number κ^2, and the norm $\|A\|$ of the dominant frequency matrix A. This property is crucial for effectively and efficiently dealing with the system (3.1) of nonlinear multi-frequency highly oscillatory second-order ODEs.

Remark 3.2.1 According to the Theorem 3.4 and Theorem 3.5, the limitation of the time stepsize $\Delta t \leqslant \sqrt{\dfrac{1}{2sL\beta}}$ yields that the ERKN integrator for solving the system (3.1) is unconditionally stable and convergent.

3.3 ERKN Integrators with Fourier Pseudospectral Discretisation for Semilinear Wave Equations

This section presents an effective approach to the numerical solution of semilinear wave equation (3.9) by combining the ERKN time integrators with the Fourier pseudospectral spatial discretisation, which will have better computational efficiency than that of traditional schemes in the literature. The rigorous convergence analysis of the underlying numerical schemes will be based on energy techniques.

To simplify the analysis and practical computation, we truncate the whole space $\mathbb{R} = (-\infty, \infty)$ onto an interval $\Omega = (0, 2\pi)$ with periodic boundary conditions. We will only present and analyse the numerical schemes for the one-dimensional semilinear wave equation:

$$\begin{cases} u_{tt} - \varepsilon^2 \Delta u + \rho u = f(u), & (x, t) \in \Omega \times (t_0, T], \\ u(0, t) = u(2\pi, t), \quad u_t(0, t) = u_t(2\pi, t), & t \in [t_0, T], \\ u(x, t_0) = \varphi_1(x), \quad u_t(x, t_0) = \varphi_2(x), & x \in \bar{\Omega}, \end{cases} \quad (3.52)$$

where $\varepsilon^2 > 0$, $\rho > 0$ are parameters, and 2π is assumed to be the fundamental period. However, the generalisation to higher dimensions is straightforward and the result remains valid without modification.

3.3.1 Time Discretisation: ERKN Time Integrators

We here define $\mathscr{A}$ as the operator:

$$(\mathscr{A}v)(x) = (-\varepsilon^2 \Delta + \rho I)v(x),$$

where $\mathscr{A}$ is a linear, unbounded positive semi-definite operator, whose domain is

$$D(\mathscr{A}) := \left\{ v \in H^1(\Omega) : v(x) = v(x + 2\pi) \right\},$$

and $u(t)$ as the function that maps x to $u(t, x)$, i.e.

$$u(t) = [x \mapsto u(x, t)].$$

Then the semilinear wave equation can be formulated as the following abstract second-order ordinary differential equation:

$$\begin{cases} \ddot{u}(t) + \mathscr{A}u(t) = f\big(u(t)\big), & t_0 < t \leqslant T, \\ u(t_0) = \varphi_1(x), & \dot{u}(t_0) = \varphi_2(x), \end{cases} \tag{3.53}$$

where $\ddot{u}$ denotes the second-order temporal derivatives $\partial_t^2 u$. It follows from the *Duhamel Principle* that the solution of the abstract system (3.53) can be characterised by the following operator-variation-of-constants formula (see [28, 29, 45–49] for details).

Theorem 3.6 *The solution of the abstract ODE (3.53) and its derivative satisfy*

$$\begin{cases} u(t) = \phi_0\big((t - t_0)^2 \mathscr{A}\big)u(t_0) + (t - t_0)\phi_1\big((t - t_0)^2 \mathscr{A}\big)\dot{u}(t_0) \\ \qquad + \displaystyle\int_{t_0}^t (t - \zeta)\phi_1\big((t - \zeta)^2 \mathscr{A}\big) f\big(u(\zeta)\big)d\zeta, \\ \dot{u}(t) = - (t - t_0)\mathscr{A}\phi_1\big((t - t_0)^2 \mathscr{A}\big)u(t_0) + \phi_0\big((t - t_0)^2 \mathscr{A}\big)\dot{u}(t_0) \\ \qquad + \displaystyle\int_{t_0}^t \phi_0\big((t - \zeta)^2 \mathscr{A}\big) f\big(u(\zeta)\big)d\zeta, \end{cases}$$

for $t \geqslant t_0$, where $\phi_0\big((t - t_0)^2 \mathscr{A}\big)$ and $\phi_1\big((t - t_0)^2 \mathscr{A}\big)$ are bounded operator-argument functions of $\mathscr{A}$.

Clearly, the r-th order ERKN integrators (3.31) could be used for the temporal discretisation of the abstract ODE (3.53), i.e.,

$$\begin{cases} u^{n+1} = \phi_0(\mathscr{V})u^n + \Delta t \phi_1(\mathscr{V})\dot{u}^n + \Delta t^2 \displaystyle\sum_{i=1}^s \bar{B}_i(\mathscr{V}) f(u^{ni}), \\ \\ \dot{u}^{n+1} = -\Delta t \mathscr{A}\phi_1(\mathscr{V})u^n + \phi_0(\mathscr{V})\dot{u}^n + \Delta t \displaystyle\sum_{i=1}^s B_i(\mathscr{V}) f(u^{ni}), \\ \\ u^{ni} = \phi_0(c_i^2 \mathscr{V})u^n + c_i \Delta t \phi_1(c_i^2 \mathscr{V})\dot{u}^n + \Delta t^2 \displaystyle\sum_{j=1}^s A_{ij}(\mathscr{V}) f(u^{nj}), \quad i = 1, 2, \cdots, s, \end{cases}$$

$$\tag{3.54}$$

where $\mathscr{V} = \Delta t^2 \mathscr{A}$ and $\phi_0(\mathscr{V})$, $\phi_1(\mathscr{V})$, $B_i(\mathscr{V})$, $\bar{B}_i(\mathscr{V})$ and $A_{ij}(\mathscr{V})$ are bounded operators.

3.3.2   Spatial Discretisation: Fourier Pseudospectral Method

We implement the spatial discretisation based on the Fourier pseudospectral method. To this end, we choose $\Delta x = 2\pi/M$ with the mesh size M, a positive even integer, to discrete the domain $\bar{\Omega}$. The grid points are denoted as $x_j = j\Delta x$ for $j = 0, 1, \cdots, M$. We define

$$X_M = \mathrm{span}\left\{\mathrm{e}^{\mathrm{i}kx}, \quad k = -M/2, \cdots, M/2 - 1\right\}$$

and

$$Y_M = \left\{v = (v_0, v_1, \cdots, v_M)^{\mathsf{T}} \in \mathbb{R}^{M+1} : v_0 = v_M\right\}.$$

For a periodic function $v(x)$ defined on $\bar{\Omega}$ and a vector $v \in Y_M$, let $P_M : L^2(\bar{\Omega}) \to X_M$ be the standard L^2-projection operator, and $I_M : C(\bar{\Omega}) \to X_M$ or $Y_M \to X_M$ be the interpolation operator, i.e.

$$\left(P_M v\right)(x) = \sum_{k=-M/2}^{M/2-1} \hat{v}_k \mathrm{e}^{\mathrm{i}kx}, \qquad \left(I_M v\right)(x) = \sum_{k=-M/2}^{M/2-1} \tilde{v}_k \mathrm{e}^{\mathrm{i}kx}, \quad 0 \leqslant x \leqslant 2\pi,$$

where $\hat{v}_k$ and $\tilde{v}_k$ are the Fourier and discrete Fourier transform coefficients of the periodic function $v(x)$ and vector v, respectively, defined as

$$\hat{v}_k = \frac{1}{2\pi} \int_0^{2\pi} v(x)\mathrm{e}^{-\mathrm{i}kx}\mathrm{d}x \qquad \text{and} \qquad \tilde{v}_k = \frac{1}{M} \sum_{j=0}^{M-1} v_j \mathrm{e}^{-\mathrm{i}kx_j}.$$

To obtain the fully discrete scheme, the Fourier spectral method is used to discretise the ERKN integrators (3.54). This is described as follows. Find $u_M^{n+1}(x)$, $\dot{u}_M^{n+1}(x)$, $u_M^{ni}(x) \in X_M$, i.e.,

$$u_M^{n+1}(x) = \sum_{k=-M/2}^{M/2-1} \widehat{u}_k^{n+1} \mathrm{e}^{\mathrm{i}kx}, \quad \dot{u}_M^{n+1}(x) = \sum_{k=-M/2}^{M/2-1} \widehat{\dot{u}}_k^{n+1} \mathrm{e}^{\mathrm{i}kx}, \quad u_M^{ni}(x) = \sum_{k=-M/2}^{M/2-1} \widehat{u}_k^{ni} \mathrm{e}^{\mathrm{i}kx},$$

$$\tag{3.55}$$

such that

$$
\begin{cases}
\widehat{u}_k^{n+1} = \phi_0(\lambda_k^2)\widehat{u}_k^n + \Delta t \phi_1(\lambda_k^2)\widehat{\dot{u}}_k^n + \Delta t^2 \sum_{i=1}^{s} \bar{B}_i(\lambda_k^2) f\left(\widehat{u_M^{ni}(x)}\right)_k, \\[2mm]
\widehat{\dot{u}}_k^{n+1} = -\Delta t \beta_k^2 \phi_1(\lambda_k^2)\widehat{u}_k^n + \phi_0(\lambda_k^2)\widehat{\dot{u}}_k^n + \Delta t \sum_{i=1}^{s} B_i(\lambda_k^2) f\left(\widehat{u_M^{ni}(x)}\right)_k, \\[2mm]
\widehat{u}_k^{ni} = \phi_0(c_i^2\lambda_k^2)\widehat{u}_k^n + c_i \Delta t \phi_1(c_i^2\lambda_k^2)\widehat{\dot{u}}_k^n + \Delta t^2 \sum_{j=1}^{s} A_{ij}(\lambda_k^2) f\left(\widehat{u_M^{nj}(x)}\right)_k, \\[2mm]
i = 1, 2, \cdots, s, \quad k = -M/2, \cdots, M/2 - 1,
\end{cases}
\tag{3.56}
$$

where $\lambda_k^2 = \Delta t^2 \beta_k^2$ with $\beta_k^2 = \rho + \varepsilon^2 k^2$ and $\phi_0(\lambda_k^2) = \cos(\lambda_k)$, $\phi_1(\lambda_k^2) = \dfrac{\sin(\lambda_k)}{\lambda_k}$.
The blend of the ERKN time integrator and the Fourier spectral discretisation (ERKN-FS) can be represented by the Butcher tableau:

$$
\begin{array}{c|c}
\begin{array}{c|c}
c & A(\lambda_k^2) \\ \hline
 & \bar{B}^{\mathsf{T}}(\lambda_k^2) \\ \hline
 & B^{\mathsf{T}}(\lambda_k^2)
\end{array}
=
\begin{array}{c|cccc}
c_1 & A_{11}(\lambda_k^2) & A_{12}(\lambda_k^2) & \cdots & A_{1s}(\lambda_k^2) \\
c_2 & A_{21}(\lambda_k^2) & A_{22}(\lambda_k^2) & \cdots & A_{2s}(\lambda_k^2) \\
\vdots & \vdots & \vdots & & \vdots \\
c_s & A_{s1}(\lambda_k^2) & A_{s2}(\lambda_k^2) & \cdots & A_{ss}(\lambda_k^2) \\ \hline
 & \bar{B}_1(\lambda_k^2) & \bar{B}_2(\lambda_k^2) & \cdots & \bar{B}_s(\lambda_k^2) \\ \hline
 & B_1(\lambda_k^2) & B_2(\lambda_k^2) & \cdots & B_s(\lambda_k^2)
\end{array}
\end{array}.
$$

However, the computation of the Fourier coefficient defined in integral form is unsuitable in practice. In order to achieve an efficient implementation, we usually use the interpolation to replace the integral. Thus, the ERKN time integrator with the Fourier pseudospectral spatial discretisation (ERKN-FP) for the semilinear wave equation (3.52) can be formulated as follows.

Let

$$
u_j^n \approx u(x_j, t_n), \quad \dot{u}_j^n \approx \partial_t u(x_j, t_n), \quad u_j^{ni} \approx u(x_j, t_n + c_i \Delta t), \quad j = 0, 1, \cdots, M,
$$

and choose $u_j^0 = \varphi_1(x_j)$, $\dot{u}_j^0 = \varphi_2(x_j)$, we then have

$$
u_j^{n+1} = \sum_{k=-M/2}^{M/2-1} \tilde{u}_k^{n+1} e^{ikx_j}, \quad \dot{u}_j^{n+1} = \sum_{k=-M/2}^{M/2-1} \tilde{\dot{u}}_k^{n+1} e^{ikx_j}, \quad u_j^{ni} = \sum_{k=-M/2}^{M/2-1} \tilde{u}_k^{ni} e^{ikx_j},
\tag{3.57}
$$

for $j = 0, 1, \cdots, M$, where

$$
\begin{cases}
\tilde{u}_k^{n+1} = \phi_0(\lambda_k^2)\tilde{u}_k^n + \Delta t\phi_1(\lambda_k^2)\tilde{u}_k'^n + \Delta t^2 \sum_{i=1}^{s} \bar{B}_i(\lambda_k^2)\widetilde{f(u^{ni})}_k, \\[2ex]
\tilde{u}_k'^{n+1} = -\Delta t\beta_k^2\phi_1(\lambda_k^2)\tilde{u}_k^n + \phi_0(\lambda_k^2)\tilde{u}_k'^n + \Delta t \sum_{i=1}^{s} B_i(\lambda_k^2)\widetilde{f(u^{ni})}_k, \\[2ex]
\tilde{u}_k^{ni} = \phi_0(c_i^2\lambda_k^2)\tilde{u}_k^n + c_i\Delta t\phi_1(c_i^2\lambda_k^2)\tilde{u}_k'^n + \Delta t^2 \sum_{j=1}^{s} A_{ij}(\lambda_k^2)\widetilde{f(u^{nj})}_k, \\[2ex]
i = 1, 2, \cdots, s, \quad k = -M/2, \cdots, M/2 - 1.
\end{cases}
\tag{3.58}
$$

It is obvious that the ERKN-FP method (3.57)–(3.58) can be efficiently implemented due to the fast Fourier transform (FFT). Its memory cost is $\mathcal{O}(M)$ and the computational cost per time step is $\mathcal{O}(M\log(M))$.

3.3.3 Error Bounds of the ERKN-FP Method (3.57)–(3.58)

Before dealing with the error estimation of the ERKN-FP method (3.57)–(3.58), we clarify some notations and assumptions:

• Denote the Soblev space

$$
H_p^m(\Omega) = \left\{ u(x) \in H^m(\Omega) \mid \partial_x^l u(0) = \partial_x^l u(2\pi),\ l = 0, 1, \cdots, m \right\}
$$

and the L^2-norm and the H^1-norm as:

$$
\|v\|_{L^2}^2 = \sum_{k \in \mathbb{Z}} |\hat{v}_k|^2 \quad \text{and} \quad \|v\|_{H^1}^2 = \sum_{k \in \mathbb{Z}} (1+k^2)|\hat{v}_k|^2 \quad \text{with} \quad v(x) = \sum_{k \in \mathbb{Z}} \hat{v}_k e^{ikx}.
$$

• The solutions $(u(x, t), \partial_t u(x, t))$ of the semilinear wave equation are studied in the space $H_p^1(\Omega) \times L^2(\Omega)$ with the energy norm:

$$
|||(u, \partial_t u)||| = \sqrt{\|u\|_{H^1}^2 + \|\partial_t u\|_{L^2}^2}.
$$

• Assume that the nonlinear function $f(\cdot)$ and the exact solution of the semilinear wave equation (3.52) satisfy

$$
f(\cdot) \in C^r(\mathbb{R}), \quad u \in C^r\left([0, T], H_p^{m_0+1}(\Omega)\right) \quad (m_0 \geqslant 1, r \geqslant 0).
$$

Under this assumption, we denote

$$K_1 = \max \left\{ \|u(\cdot, t)\|_{L^\infty([0,T];H^1)}, \ \|\partial_t u(\cdot, t)\|_{L^\infty([0,T];L^2)} \right\} \lesssim 1$$

and

$$K_2 = \max_{0 \leqslant l \leqslant r} \max_{\|w\|_{L^2} \leqslant K_1} \left\| f^{(l)}(w) \right\|_{L^2} \lesssim 1.$$

With u_k^n, $\dot{u}_k^n$ and u_k^{ni}, which are obtained from (3.57)–(3.58), define the error functions as

$$e^n(x) := u(x, t_n) - \left(I_M u^n \right)(x),$$

$$\dot{e}^n(x) := u(x, t_n) - \left(I_M \dot{u}^n \right)(x),$$

$$e^{ni}(x) := u(x, t_n + c_i \Delta t) - \left(I_M u^{ni} \right)(x).$$

To proceed to the proof of the error bound for the ERKN-FP method, we define the projected error as

$$e_M^n(x) := P_M u(x, t_n) - u_M^n(x),$$

$$\dot{e}_M^n(x) := P_M \partial_t u(x, t_n) - \dot{u}_M^n(x),$$

$$e_M^{ni}(x) := P_M u(x, t_n + c_i \Delta t) - u_M^{ni}(x),$$

where $u_M^n(x)$, $\dot{u}_M^n(x)$ and $u_M^{ni}(x)$ are yielded from the ERKN-FS method (3.55)–(3.56). It then follows from the triangle inequality and estimates on the projection error [6, 50] that

$$\|e^n\|_{H^1} + \|\dot{e}^n\|_{L^2} \leqslant \|e_M^n\|_{H^1} + \|\dot{e}_M^n\|_{L^2} + \|u_M^n(\cdot) - \left(I_M u^n \right)(\cdot)\|_{H^1} + \|\dot{u}_M^n(\cdot) - \left(I_M \dot{u}^n \right)(\cdot)\|_{L^2}$$

$$+ \|u(\cdot, t_n) - P_M u(\cdot, t_n)\|_{H^1} + \|\partial_t u(\cdot, t_n) - P_M \partial_t u(\cdot, t_n)\|_{L^2}$$

$$\lesssim \|e_M^n\|_{H^1} + \|\dot{e}_M^n\|_{L^2} + \Delta x^{m_0},$$

$$(3.59)$$

and

$$\|e^{ni}\|_{L^2} \leqslant \|e_M^{ni}\|_{L^2} + \|u_M^{ni}(\cdot) - \left(I_M u^{ni} \right)(\cdot)\|_{L^2} + \|u(\cdot, t_{ni}) - P_M u(\cdot, t_{ni})\|_{L^2}$$

$$\lesssim \|e_M^n\|_{L^2} + \Delta x^{m_0+1}.$$

Hence, the error estimates for the ERKN-FP methods can be converted to the estimates for the ERKN-FS methods. Moreover, the theoretical analysis for PDEs is quite different from that for ODEs. In particular, the assumption for the nonlinear function $f(\cdot)$ satisfying the Lipschitz condition will not be the same. Fortunately, the boundedness of the numerical solutions will be helpful to the theoretical analysis. In what follows, we will first analyse the boundedness of the numerical solutions for the ERKN-FS methods (3.55)–(3.56). We then will deduce the convergence of the ERKN-FP methods (3.57)–(3.58).

With regard to the boundedness of the numerical methods, we start with the explicit ERKN-FS methods (3.55)–(3.56), which can be expressed as

$$
\begin{cases}
\widehat{u}_k^{n+1} = \phi_0(\lambda_k^2)\widehat{u}_k^n + \Delta t\phi_1(\lambda_k^2)\widehat{\dot{u}}_k^n + \Delta t^2 \sum_{i=1}^{s} \bar{B}_i(\lambda_k^2) f\big(\widehat{u_M^{ni}(x)}\big)_k, \\[2mm]
\widehat{\dot{u}}_k^{n+1} = -\Delta t\beta_k^2\phi_1(\lambda_k^2)\widehat{u}_k^n + \phi_0(\lambda_k^2)\widehat{\dot{u}}_k^n + \Delta t \sum_{i=1}^{s} B_i(\lambda_k^2) f\big(\widehat{u_M^{ni}(x)}\big)_k, \\[2mm]
\widehat{u}_k^{ni} = \phi_0(c_i^2\lambda_k^2)\widehat{u}_k^n + c_i\Delta t\phi_1(c_i^2\lambda_k^2)\widehat{\dot{u}}_k^n + \Delta t^2 \sum_{j=1}^{i-1} A_{ij}(\lambda_k^2) f\big(\widehat{u_M^{nj}(x)}\big)_k, \\[2mm]
i = 1, 2, \cdots, s, \quad k = -M/2, \cdots, M/2 - 1,
\end{cases}
$$

$$(3.60)$$

with the Butcher tableau:

$$
\begin{array}{c|ccccc}
c_1 & 0 & 0 & \cdots & 0 & 0 \\
c_2 & A_{21}(\lambda_k^2) & 0 & \cdots & 0 & 0 \\
\vdots & \vdots & \vdots & & \vdots & \vdots \\
c_s & A_{s1}(\lambda_k^2) & A_{s2}(\lambda_k^2) & \cdots & A_{s,s-1}(\lambda_k^2) & 0 \\
\hline
 & \bar{B}_1(\lambda_k^2) & \bar{B}_2(\lambda_k^2) & \cdots & \bar{B}_{s-1}(\lambda_k^2) & \bar{B}_s(\lambda_k^2) \\
\hline
 & B_1(\lambda_k^2) & B_2(\lambda_k^2) & \cdots & B_{s-1}(\lambda_k^2) & B_s(\lambda_k^2)
\end{array}
$$

$$
\frac{c \;\Big|\; A(\lambda_k^2)}{\dfrac{\bar{B}^{\mathsf{T}}(\lambda_k^2)}{B^{\mathsf{T}}(\lambda_k^2)}} \;=\;
$$

Theorem 3.7 (Boundedness for a Single Time Step: Explicit ERKN-FS Method) *Let the weights* $\bar{B}_i(\lambda_k^2)$, $B_i(\lambda_k^2)$ *and* $A_{ij}(\lambda_k^2)$ *of the explicit ERKN-FS method* (3.55)–(3.60) *satisfy the r-th order conditions* (3.32). *There exists a sufficiently small* $0 < \tau_0 \leqslant 1$ *such that the time stepsize* $\Delta t \leqslant \tau_0$. *If the numerical solution* $(u_M^n, \dot{u}_M^n) \in H_p^1(\Omega) \times L^2(\Omega)$ *of the explicit ERKN-FS method satisfies* $|||(u_M^n, \dot{u}_M^n)||| \leqslant K$, *then we have*

$$
\|u_M^{ni}\| \lesssim 1, \quad i = 1, 2, \cdots, s,
$$

and $(u_M^{n+1}, \dot{u}_M^{n+1}) \in H_p^1(\Omega) \times L^2(\Omega)$ *with*

$$
\|u_M^{n+1}\|_{H^1} \leqslant C_K \quad \text{and} \quad \|\dot{u}_M^{n+1}\|_{L^2} \leqslant C_K,
$$

where C_K *is independent of the time stepsize* Δt *and spatial mesh size* M.

Proof   1. *"Estimations for update stage procedures: $u_M^{n+1}(x)$ and $\dot{u}_M^{n+1}(x)$".*

For the first equality in (3.60), an application of the triangle inequality results in

$$|\widehat{u}_k^{n+1}| \leqslant |\widehat{u}_k^n| + \Delta t |\widehat{\dot{u}}_k^n| + \Delta t^2 \bar{B} \sum_{i=1}^s |f\big(\widehat{u_M^{ni}(x)}\big)_k|$$

and

$$\sqrt{1+k^2}|\widehat{u}_k^{n+1}| \leqslant \sqrt{1+k^2}|\widehat{u}_k^n| + \frac{\sqrt{1+k^2}}{\sqrt{\rho + \varepsilon^2 k^2}}|\widehat{\dot{u}}_k^n| + \Delta t \hat{B} \sum_{i=1}^s |\widehat{f(u^{nj})}_k|.$$

Then, applying Minkowski's inequality and Parseval's identity to the above inequalities, we have

$$\|u_M^{n+1}\|_{L^2} \leqslant \|u_M^n\|_{L^2} + \Delta t \|\dot{u}_M^n\|_{L^2} + \Delta t^2 \bar{B} \sum_{i=1}^s \big\|P_M f\big(u_M^{ni}\big)\big\|_{L^2} \qquad (3.61)$$

and

$$\|u_M^{n+1}\|_{H^1} \leqslant \|u_M^n\|_{H^1} + \varsigma \|\dot{u}_M^n\|_{L^2} + \Delta t \hat{B} \sum_{i=1}^s \big\|P_M f\big(u_M^{ni}\big)\big\|_{L^2}, \qquad (3.62)$$

where $\varsigma = 1/\min\big\{\sqrt{\rho}, \varepsilon\big\} > 0$ is a constant parameter. Likewise, it follows from the second equality in (3.60) that

$$|\widehat{\dot{u}}_k^{n+1}| \leqslant \sqrt{\rho + \varepsilon^2 k^2}|\widehat{u}_k^n| + |\widehat{\dot{u}}_k^n| + \Delta t B \sum_{i=1}^s |f\big(\widehat{u_M^{ni}(x)}\big)_k|$$

$$\leqslant \varrho\sqrt{1+k^2}|\widehat{u}_k^n| + |\widehat{\dot{u}}_k^n| + \Delta t B \sum_{i=1}^s |f\big(\widehat{u_M^{ni}(x)}\big)_k|,$$

where $\varrho = \max\{\sqrt{\rho}, \varepsilon\} > 0$ is also a constant parameter. Similarly, we have the following estimate

$$\|\dot{u}_M^{n+1}\|_{L^2} \leqslant \varrho\|u_M^n\|_{H^1} + \|\dot{u}_M^n\|_{L^2} + \Delta t B \sum_{i=1}^s \big\|P_M f\big(u_M^{ni}\big)\big\|_{L^2}. \qquad (3.63)$$

Consequently, summing up (3.62) and (3.63) leads to

$$\|u_M^{n+1}\|_{H^1} + \|\dot{u}_M^{n+1}\|_{L^2} \leqslant \hat{C}_1 \left(\|u_M^n\|_{H^1} + \|\dot{u}_M^n\|_{L^2} \right) + \Delta t \left(\hat{B} + B \right) \sum_{i=1}^{s} \left\| f\left(u_M^{ni}\right) \right\|_{L^2},$$

$$(3.64)$$

where $\hat{C}_1 = \max \{ \varrho, \varsigma \} + 1$ is a constant parameter.

2. *"Estimations for internal stage procedures: $u_M^{ni}(x)$"*.
Using the third equality in (3.60), we obtain

$$\left| \widehat{u}_k^{ni} \right| \leqslant \left| \widehat{u}_k^n \right| + c_i \Delta t \left| \widehat{\dot{u}}_k^n \right| + \Delta t^2 \beta \sum_{j=1}^{i-1} \left| \widehat{f(u^{nj})}_k \right|.$$

Applying Minkowski's inequality and Parseval's identity to the above inequality yields

$$\|u_M^{n1}\|_{L^2} \leqslant \|u_M^n\|_{L^2} + c_i \Delta t \|\dot{u}_M^n\|_{L^2},$$

$$\|u_M^{ni}\|_{L^2} \leqslant \|u_M^n\|_{L^2} + c_i \Delta t \|\dot{u}_M^n\|_{L^2} + \Delta t^2 \beta \sum_{j=1}^{i-1} \left\| P_M f(u_M^{nj}) \right\|_{L^2}, \quad i = 2, \cdots, s.$$

$$(3.65)$$

3. *"Boundedness of the numerical solutions"*.
According to the inequalities of (3.65), if the solution $(u_M^n, \dot{u}_M^n) \in H_p^1(\Omega) \times L^2(\Omega)$ of the explicit ERKN-FS method (3.55)–(3.60) satisfies

$$\||(u_M^n, \dot{u}_M^n)\|| \leqslant K,$$

then, the following approximations can be obtained by recursion:

$$\|u_M^{ni}\|_{L^2} \leqslant (1 + \Delta t)^i K \lesssim 1, \qquad i = 1, 2, \cdots, s, \qquad (3.66)$$

where we have used the fact that $\|u_M^n\|_{L^2} \leqslant \|u_M^n\|_{H^1} \leqslant K$ and the sufficiently small time stepsize Δt such that $\Delta t \beta K_2 \leqslant 1$. Inserting the result (3.66) into (3.64) yields

$$\|u_M^{n+1}\|_{H^1} + \|\dot{u}_M^{n+1}\|_{L^2} \leqslant \hat{C}_1\left(\|u_M^n\|_{H^1} + \|\dot{u}_M^n\|_{L^2}\right)$$

$$+ \Delta t(\hat{B} + B) \sum_{i=1}^s \left\|\int_0^1 f'(\tau u_M^{ni})\mathrm{d}\tau \cdot u_M^{ni}\right\|_{L^2}$$

$$\leqslant \hat{C}_1\left(\|u_M^n\|_{H^1} + \|\dot{u}_M^n\|_{L^2}\right) + \Delta t(\hat{B} + B)K_2 \sum_{i=1}^s \|u_M^{ni}\|_{L^2}$$

$$\leqslant \hat{C}_2\left(1 + \Delta t + \Delta t \sum_{i=1}^s (1 + \Delta t)^i\right) \leqslant C_K,$$

where $\hat{C}_2 = K_1 \max\left\{\hat{C}_1, (\hat{B} + B)K_2\right\}$ is a constant and C_K is obviously independent of time stepsize Δt and spatial mesh size M. □

For the implicit ERKN-FS method (3.55)–(3.56):

$$
\begin{cases}
\widehat{u}_k^{n+1} = \phi_0(\lambda_k^2)\widehat{u}_k^n + \Delta t\phi_1(\lambda_k^2)\widehat{\dot{u}}_k^n + \Delta t^2 \sum_{i=1}^s \bar{B}_i(\lambda_k^2) f\widehat{\left(u_M^{ni}(x)\right)}_k, \\[2mm]
\widehat{\dot{u}}_k^{n+1} = -\Delta t\beta_k^2\phi_1(\lambda_k^2)\widehat{u}_k^n + \phi_0(\lambda_k^2)\widehat{\dot{u}}_k^n + \Delta t \sum_{i=1}^s B_i(\lambda_k^2) f\widehat{\left(u_M^{ni}(x)\right)}_k, \\[2mm]
\widehat{u}_k^{ni} = \phi_0(c_i^2\lambda_k^2)\widehat{u}_k^n + c_i\Delta t\phi_1(c_i^2\lambda_k^2)\widehat{\dot{u}}_k^n + \Delta t^2 \sum_{j=1}^s A_{ij}(\lambda_k^2) f\widehat{\left(u_M^{nj}(x)\right)}_k, \\[2mm]
i = 1, 2, \cdots, s, \quad k = -M/2, \cdots, M/2 - 1,
\end{cases}
$$

(3.67)

with the Butcher tableau:

$$
\begin{array}{c|c}
\mathbf{c} & A(\lambda_k^2) \\
\hline
& \bar{B}^{\mathsf{T}}(\lambda_k^2) \\
\hline
& B^{\mathsf{T}}(\lambda_k^2)
\end{array}
=
\begin{array}{c|cccc}
c_1 & A_{11}(\lambda_k^2) & A_{12}(\lambda_k^2) & \cdots & A_{1s}(\lambda_k^2) \\
c_2 & A_{21}(\lambda_k^2) & A_{22}(\lambda_k^2) & \cdots & A_{2s}(\lambda_k^2) \\
\vdots & \vdots & \vdots & & \vdots \\
c_s & A_{s1}(\lambda_k^2) & A_{s2}(\lambda_k^2) & \cdots & A_{ss}(\lambda_k^2) \\
\hline
& \bar{B}_1(\lambda_k^2) & \bar{B}_2(\lambda_k^2) & \cdots & \bar{B}_s(\lambda_k^2) \\
\hline
& B_1(\lambda_k^2) & B_2(\lambda_k^2) & \cdots & B_s(\lambda_k^2)
\end{array},
$$

iteration is needed for practical application. In this chapter, we use the waveform relaxation iteration, which can be split into the following two phases.

I. Iteration procedure:

$$
\begin{cases}
\widehat{u}_{k,[0]}^{ni} = \phi_0(c_i^2\lambda_k^2)\widehat{u}_k^n + c_i\Delta t\phi_1(c_i^2\lambda_k^2)\widehat{\dot{u}}_k^n, \\[2mm]
\widehat{u}_{k,[l+1]}^{ni} = \phi_0(c_i^2\lambda_k^2)\widehat{u}_k^n + c_i\Delta t\phi_1(c_i^2\lambda_k^2)\widehat{\dot{u}}_k^n + \Delta t^2\sum_{j=1}^{s}A_{ij}(\lambda_k^2)\widehat{f\left(u_{M,[l]}^{nj}(x)\right)}_k, \\[2mm]
i = 1,2,\cdots,s, \quad k = -M/2,\cdots,M/2-1, \quad l = 1,2,\cdots.
\end{cases}
\tag{3.68}
$$

For any error tolerance $\varepsilon > 0$, if the condition

$$
\|u_{M,[l+1]}^{ni} - u_{M,[l]}^{ni}\|_{L^2} \leqslant \varepsilon
$$

is satisfied, we define

$$
u_M^{ni}(x) := u_{M,[l+1]}^{ni}(x) = \sum_{k=-M/2}^{M/2-1}\widehat{u}_{k,[l+1]}^{ni}e^{ikx}.
$$

II. Output procedure:

$$
\begin{cases}
\widehat{u}_k^{n+1} = \phi_0(\lambda_k^2)\widehat{u}_k^n + \Delta t\phi_1(\lambda_k^2)\widehat{\dot{u}}_k^n + \Delta t^2\sum_{i=1}^{s}\bar{B}_i(\lambda_k^2)\widehat{f\left(u_M^{ni}(x)\right)}_k, \\[2mm]
\widehat{\dot{u}}_k^{n+1} = -\Delta t\beta_k^2\phi_1(\lambda_k^2)\widehat{u}_k^n + \phi_0(\lambda_k^2)\widehat{\dot{u}}_k^n + \Delta t\sum_{i=1}^{s}B_i(\lambda_k^2)\widehat{f\left(u_M^{ni}(x)\right)}_k,
\end{cases}
\tag{3.69}
$$

and define

$$
u_M^{n+1}(x) = \sum_{k=-M/2}^{M/2-1}\widehat{u}_k^{n+1}e^{ikx}, \quad \dot{u}_M^{n+1}(x) = \sum_{k=-M/2}^{M/2-1}\widehat{\dot{u}}_k^{n+1}e^{ikx}.
$$

In practice, the application of procedure (3.68)–(3.69) of the implicit ERKN-FS method could be understood as an explicit method. Therefore, if the solution $(u_M^n, \dot{u}_M^n) \in H_p^1(\Omega) \times L^2(\Omega)$ of the implicit ERKN-FS method satisfies

$$
|||(u_M^n, \dot{u}_M^n)||| \leqslant K,
$$

we then obtain

$$
\|u_{M,[l]}^{ni}\|_{L^2} \leqslant (1 + \Delta t)^{l+1}K \lesssim 1, \quad i = 1,2,\cdots,s. \tag{3.70}
$$

In a similar way to the proof of Theorem 3.7, we can deduce the boundedness for the implicit ERKN-FS methods (3.55)–(3.67).

Theorem 3.8 (Boundedness for a Single Time Step: Implicit ERKN-FS Methods) *Let the weights $\bar{B}_i(\lambda_k^2)$, $B_i(\lambda_k^2)$ and $A_{ij}(\lambda_k^2)$ of the implicit ERKN-FS method (3.55)–(3.67) satisfy the r-th order conditions (3.32). There exists a sufficiently small $0 < \tau_0 \leqslant 1$ such that $\Delta t \leqslant \tau_0$. If the numerical solution $(u_M^n, \dot{u}_M^n) \in H_p^1(\Omega) \times L^2(\Omega)$ of the implicit ERKN-FS method satisfies $|||(u_M^n, \dot{u}_M^n)||| \leqslant K$, then we have*

$$\|u_M^{ni}\| \lesssim 1, \quad i = 1, 2, \cdots, s,$$

and $(u_M^{n+1}, \dot{u}_M^{n+1}) \in H_p^1(\Omega) \times L^2(\Omega)$ with

$$\|u_M^{n+1}\|_{H^1} \leqslant \widetilde{C}_K \quad and \quad \|\dot{u}_M^{n+1}\|_{L^2} \leqslant \widetilde{C}_K,$$

where $\widetilde{C}_K$ is independent of the time stepsize Δt and spatial mesh size M.

According to the conclusion in Theorem 3.7 and Theorem 3.8 and using mathematical induction, suitable smoothness assumptions for the initial values $\varphi_1(\cdot)$ and $\varphi_2(\cdot)$ yield the boundedness of numerical solutions over a long-time interval $[t_0, T]$.

Theorem 3.9 *Assume that the weights $\bar{B}_i(\lambda_k^2)$, $B_i(\lambda_k^2)$ and $A_{ij}(\lambda_k^2)$ of the ERKN-FS method (3.55)–(3.56) satisfy the r-th order conditions (3.32). There exists a sufficiently small $0 < \tau_0 \leqslant 1$ such that $\Delta t \leqslant \tau_0$. If the initial conditions $\big(\varphi_1(x), \varphi_2(x)\big) \in H_p^1(\Omega) \times L^2(\Omega)$ satisfy $|||(\varphi_1, \varphi_2)||| \leqslant K_0$, then we have $(u_M^n, \dot{u}_M^n) \in H_p^1(\Omega) \times L^2(\Omega)$ with*

$$|||(u_M^{n+1}, \dot{u}_M^{n+1})||| \leqslant C_{K_0} \quad and \quad \|u_M^{ni}\|_{L^2} \leqslant C_{K_0}, \quad i = 1, 2, \cdots, s,$$

where C_{K_0} is independent of the time stepsize Δt and spatial mesh size M.

Proof By mathematical induction, the proof of the theorem is quite similar to Theorem 3.7 and Theorem 3.8, we omit the details here for brevity. $\qquad\square$

Using the boundedness of numerical solutions, we will analyse the error bounds for the ERKN-FS methods. To this end, we introduce the modified H^1-norm and modified energy norm:

$$|[u_M]|_{H^1} = \left(\sum_{k=-M/2}^{M/2-1} (\rho + \varepsilon^2 k^2)|\hat{u}_k|^2 \right)^{1/2} \quad and \quad |||[(u_M, \dot{u}_M)]||| = \sqrt{|[u_M]|_{H^1}^2 + \|\dot{u}_M\|_{L^2}^2}.$$

Obviously, the modified H^1-norm is equivalent to the normal H^1-norm, namely

$$\min\{\sqrt{\rho}, \varepsilon\} \|u_M\|_{H^1} \leqslant |[u_M]|_{H^1} \leqslant \max\{\sqrt{\rho}, \varepsilon\} \|u_M\|_{H^1}. \tag{3.71}$$

We also assume that the weights $\bar{B}_i(\lambda_k^2)$, $B_i(\lambda_k^2)$ and $A_{ij}(\lambda_k^2)$ satisfy the r-th order conditions and the nonlinear function $f(\cdot)$ satisfies $\partial_t^r f \in L^\infty([t_0, T], L^2(\Omega))$. Then the error system of ERKN-FS methods is to find $e_M^n(x)$, $\dot{e}_M^n(x)$ and $e_M^{ni}(x)$ in the space X_M, i.e.,

$$e_M^{n+1}(x) = \sum_{k=-M/2}^{M/2-1} \widehat{e}_k^{n+1} e^{ikx}, \quad \dot{e}_M^{n+1}(x) = \sum_{k=-M/2}^{M/2-1} \widehat{\dot{e}}_k^{n+1} e^{ikx}, \quad e_M^{ni}(x) = \sum_{k=-M/2}^{M/2-1} \widehat{e}_k^{ni} e^{ikx}$$

$$\tag{3.72}$$

such that

$$\begin{cases} \widehat{e}_k^{n+1} = \phi_0(\lambda_k^2)\widehat{e}_k^n + \Delta t \phi_1(\lambda_k^2)\widehat{\dot{e}}_k^n + \Delta t^2 \sum_{i=1}^s \bar{B}_i(\lambda_k^2)\Delta \widehat{f}_k^{ni} + \widehat{\delta}_k^{n+1}, \\[3mm] \widehat{\dot{e}}_k^{n+1} = -\Delta t \beta_k^2 \phi_1(\lambda_k^2)\widehat{e}_k^n + \phi_0(\lambda_k^2)\widehat{\dot{e}}_k^n + \Delta t \sum_{i=1}^s B_i(\lambda_k^2)\Delta \widehat{f}_k^{ni} + \widehat{\dot{\delta}}_k^{n+1}, \\[3mm] \widehat{e}_k^{ni} = \phi_0(c_i^2\lambda_k^2)\widehat{e}_k^n + c_i \Delta t \phi_1(c_i^2\lambda_k^2)\widehat{\dot{e}}_k^n + \Delta t^2 \sum_{j=1}^s A_{ij}(\lambda_k^2)\Delta \widehat{f}_k^{nj} + \widehat{\Delta}_k^{ni}, \\[3mm] i = 1, 2, \cdots, s, \quad k = -M/2, \cdots, M/2 - 1. \end{cases}$$

$$\tag{3.73}$$

where $\Delta \widehat{f}_k^{ni} = \widehat{f(u)_k}(t_n + c_i\Delta t) - \widehat{f(u_M^{ni})_k}$ and the remainders $\widehat{\delta}_k^{n+1}$, $\widehat{\dot{\delta}}_k^{n+1}$ and $\widehat{\Delta}_k^{ni}$ can be represented as

$$\widehat{\delta}_k^{n+1} = \sum_{l=0}^{r-1} \Delta t^{k+2} \left(\phi_{k+2}(\lambda_k^2) - \sum_{i=1}^s \bar{B}_i(\lambda_k^2) \frac{c_i^k}{k!} \right) \frac{d^l}{dt^l} \widehat{f(u)_k}(t_n)$$

$$+ \Delta t^{r+2} \int_0^1 (1-z)\phi_1\big((1-z)^2\lambda_k^2\big) \int_0^z \frac{(z-\tau)^{r-1}}{(r-1)!} \frac{d^r}{dt^r} \widehat{f(u)_k}(t_n + \tau \Delta t) d\tau dz$$

$$- \Delta t^{r+2} \sum_{i=1}^s \bar{B}_i(\lambda_k^2) \int_0^{c_i} \frac{(c_i-\tau)^{r-1}}{(r-1)!} \frac{d^r}{dt^r} \widehat{f(u)_k}(t_n + \tau \Delta t) d\tau,$$

$$\tag{3.74}$$

$$\widehat{\dot{\delta}}_k^{n+1} = \sum_{k=0}^{r-1} \Delta t^{k+1} \left(\phi_{k+1}(\lambda_k^2) - \sum_{i=1}^s B_i(\lambda_k^2) \frac{c_i^k}{k!} \right) \frac{d^l}{dt^l} \widehat{f(u)_k}(t_n)$$

$$+ \Delta t^{r+1} \int_0^1 \phi_0\big((1-z)^2\lambda_k^2\big) \int_0^z \frac{(z-\tau)^{r-1}}{(r-1)!} \frac{d^r}{dt^r} \widehat{f(u)_k}(t_n + \tau \Delta t) d\tau dz$$

$$- \Delta t^{r+1} \sum_{i=1}^s B_i(\lambda_k^2) \int_0^{c_i} \frac{(c_i-\tau)^{r-1}}{(r-1)!} \frac{d^r}{dt^r} \widehat{f(u)_k}(t_n + \tau \Delta t) d\tau, \tag{3.75}$$

and

$$
\begin{aligned}
\widehat{\Delta}_k^{ni} = {} & \sum_{k=0}^{r-1} \Delta t^{k+2} \left(c_i^{k+2} \phi_{k+2}(c_i^2 \lambda_k^2) - \sum_{j=1}^{s} A_{ij}(\lambda_k^2) \frac{c_j^k}{k!} \right) \frac{\mathrm{d}^l}{\mathrm{d}t^l} \widehat{f(u)}_k(t_n) \\
& + c_i^{r+2} \Delta t^{r+2} \int_0^1 (1-z)\phi_1 \big((1-z)^2 c_i^2 \lambda_k^2\big) \int_0^z \frac{(z-\tau)^{r-1}}{(r-1)!} \frac{\mathrm{d}^r}{\mathrm{d}t^r} \widehat{f(u)}_k(t_n + \tau c_i \Delta t) \mathrm{d}\tau \mathrm{d}z \\
& - \Delta t^{r+2} \sum_{i=1}^{s} A_{i,j}(\lambda_k^2) \int_0^{c_j} \frac{(c_j - \tau)^{r-1}}{(r-1)!} \frac{\mathrm{d}^r}{\mathrm{d}t^r} \widehat{f(u)}_k(t_n + \tau \Delta t) \mathrm{d}\tau.
\end{aligned}
$$

$$(3.76)$$

Using energy techniques, we can obtain the convergence result for the ERKN-FP methods.

Theorem 3.10 ($H^1 \times L^2$ **Error Bounds of the ERKN-FP Method**) *Let u^n, $\dot{u}^n$ and u^{ni} be the approximations obtained from the ERKN-FP method* (3.57)–(3.58), *and the weights $\bar{B}_i(\lambda_k^2)$, $B_i(\lambda_k^2)$ and $A_{ij}(\lambda_k^2)$ satisfy the r-th order conditions* (3.32). *Then, under the assumption of Theorem 3.9, there exist two sufficiently small constants $0 < \tau_0 \leqslant 1$ and $0 < h_0 \leqslant 1$, such that*

$$
\|u(\cdot, t_n) - (I_M u^n)(\cdot)\|_{H^1} + \|u_t(\cdot, t_n) - (I_M \dot{u}^n)(\cdot)\|_{L^2} \lesssim \Delta t^r + \Delta x^{m_0},
$$

when $0 < \Delta t \leqslant \tau_0$ and $0 < \Delta x \leqslant h_0$.

Proof According to Lemma 3.5, it is easy to obtain the following estimates for the remainders $\widehat{\delta}_k^{n+1}$, $\widehat{\dot{\delta}}_k^{n+1}$ and $\widehat{\Delta}_k^{ni}$:

$$
\|[\delta_M^{n+1}]\|_{H^1} \leqslant K_3 \Delta t^{r+1}, \quad \|\dot{\delta}_M^{n+1}\|_{L^2} \leqslant K_3 \Delta t^{r+1}, \quad \sum_{i=1}^{s} \|\Delta_M^{ni}\|_{L^2} \leqslant K_3 \Delta t^{r+2},
$$

where the constant K_3 is dependent on K_2, B, $\hat{B}$, $\bar{B}$ and β, but independent of the time stepsize Δt and the spatial mesh size M. Rewriting the first two equations in (3.73) as

$$
\begin{bmatrix} \beta_k \widehat{e}_k^{n+1} \\ \widehat{\dot{e}}_k^{n+1} \end{bmatrix} = \Omega(1, 0, \lambda_k^2) \begin{bmatrix} \beta_k \widehat{e}_k^{n} \\ \widehat{\dot{e}}_k^{n} \end{bmatrix} + \Delta t \sum_{i=1}^{s} \begin{bmatrix} \beta_k \bar{B}_i(\lambda_k^2) \Delta \widehat{f}_k^{ni} \\ B_i(\lambda_k^2) \Delta \widehat{f}_k^{ni} \end{bmatrix} + \begin{bmatrix} \beta_k \widehat{\delta}_k^{n+1} \\ \widehat{\dot{\delta}}_k^{n+1} \end{bmatrix},
$$

by taking the l_2 inner product on both sides and using the Cauchy inequality, we have

$$\beta_k^2|\widehat{e}_k^{n+1}|^2 + |\widehat{\dot{e}}_k^{n+1}|^2 \leqslant (1 + \Delta t(s+1))\left(\beta_k^2|\widehat{e}_k^n|^2 + |\widehat{\dot{e}}_k^n|^2\right) + 3\Delta t(\hat{B}^2 + B^2)\sum_{i=1}^s |\Delta \widehat{f}_k^{ni}|^2$$

$$+ (2 + \frac{1}{\Delta t})\left(\beta_k^2|\widehat{\delta}_k^{n+1}|^2 + |\widehat{\dot{\delta}}_k^{n+1}|^2\right).$$

Summing up the above inequality for k from $-M/2$ to $M/2-1$ and using Parseval's identity yields

$$\||(e_M^{n+1}, \dot{e}_M^{n+1})\||^2 \leqslant (1 + \Delta t(1+s))\||(e_M^n, \dot{e}_M^n)\||^2$$

$$+ 3\Delta t(\hat{B}^2 + B^2)\sum_{i=1}^s \left\| f\left(u(\cdot, t_n + c_i\Delta t)\right) - f\left(u_M^{ni}\right)\right\|_{L^2}^2 + 4K_3\Delta t^{2r+1}.$$

$$(3.77)$$

It then follows from the conclusion of Theorem 3.9 and the assumptions for $f(\cdot)$ that

$$\left\| f\left(u(\cdot, t_n + c_i\Delta t)\right) - f\left(u_M^{ni}\right)\right\|_{L^2}$$

$$= \left\|\int_0^1 f'\left(\tau u_M^{ni} + (1-\tau)u(\cdot, t_n + c_i\Delta t)\right)d\tau \cdot \left(u_M^{ni}(\cdot) - u(\cdot, t_n + c_i\Delta t)\right)\right\|_{L^2}$$

$$\leqslant K_2\left\| u_M^{ni}(\cdot) - u(\cdot, t_n + c_i\Delta t)\right\|_{L^2} \leqslant K_2\left(\left\| e_M^{ni}\right\|_{L^2} + \Delta x^{m_0+1}\right).$$

Hence, inserting the above inequality into (3.77), we have

$$\||(e_M^{n+1}, \dot{e}_M^{n+1})\||^2 \leqslant (1 + (1+s)\Delta t)\||(e_M^n, \dot{e}_M^n)\||^2 + K_4\Delta t\sum_{i=1}^s \left\| e_M^{ni}\right\|_{L^2}^2$$

$$+ K_5\Delta t\left(\Delta t^{2r} + \Delta x^{2m_0+2}\right),$$

$$(3.78)$$

where K_4 and K_5 are constants and independent of Δt and Δx. Clearly, to show the required error bounds, we need to estimate the term $\sum_{i=1}^s \|e_M^{ni}\|_{L^2}$. It follows from taking the L^2 norm on both sides of the third equation in (3.73) that

$$\left\| e_M^{ni}\right\|_{L^2} \leqslant \left\| e_M^n\right\|_{L^2} + \left\| \dot{e}_M^n\right\|_{L^2} + \Delta t^2\beta\sum_{i=1}^s \left\| f\left(u(\cdot, t_n + c_i\Delta t)\right) - f\left(u_M^{ni}\right)\right\|_{L^2} + \sum_{i=1}^s \left\| \Delta_M^{ni}\right\|_{L^2}$$

$$\leqslant \||(e_M^n, \dot{e}_M^n)\||_{L^2} + \Delta t^2\beta K_2\sum_{i=1}^s \left(\left\| e_M^{ni}\right\|_{L^2} + \Delta x^{m_0+1}\right) + \sum_{i=1}^s \left\| \Delta_M^{ni}\right\|_{L^2}.$$

This, together with the assumption that the time stepsize satisfies $\Delta t \leqslant \sqrt{\dfrac{1}{2s\beta K_2}}$, implies

$$\sum_{i=1}^{s} \left\| e_M^{ni} \right\|_{L^2} \leqslant 2s \||[(e_M^n, \dot{e}_M^n)]\|_{L^2} + 2s K_3 \Delta t^{r+2} + s\Delta x^{m_0+1}.$$

Inserting the result into (3.78) leads to

$$\||[(e_M^{n+1}, \dot{e}_M^{n+1})]\||^2 \leqslant \left(1 + K_6 \Delta t\right)\||[(e_M^n, \dot{e}_M^n)]\||^2 + K_7 \Delta t \left(\Delta t^{2r} + \Delta x^{2m_0+2}\right).$$

$$(3.79)$$

Applying Gronwall's inequality to (3.79) results in

$$\||[(e_M^{n+1}, \dot{e}_M^{n+1})]\|| \lesssim \Delta t^r + \Delta x^{m_0+1}.$$

Since the modified H^1-norm is equivalent to the normal H^1-norm and the relation (3.59), we obtain

$$\|e^n\|_{H^1} + \|\dot{e}^n\|_{L^2} \lesssim \Delta t^r + \Delta x^{m_0}.$$

Theorem 3.10 is proved.   □

Remark 3.3.1   It follows from the convergence analysis stated above that we gain an insight into the significance of the ERKN-FP methods. That is, the ERKN-FP methods are independent of the restriction between the time stepsize Δt and the spatial stepsize Δx. In other words, the ERKN-FP methods are free from the CFL condition. This is another important property of ERKN integrators when applied to the semilinear wave equation, which, unfortunately, is not shared by traditional schemes for PDEs in the literature.

3.4 Numerical Experiments

In this section, we present results of numerical experiments to verify our theoretical analysis for the ERKN time integrators. In order to demonstrate the superiority of ERKN time integrators, we select the following time integrators for comparison:

- ISV: the improved explicit symplectic Störmer–Verlet formula of order two given in [13];
- ERKN3s4: the three-stage symmetric and symplectic explicit ERKN method of order four (see [15]);
- IERKN2s4: the two-stage implicit symplectic ERKN method of order four;
- IERKN3s6: the three-stage implicit symplectic ERKN method of order six;

- **SV**: the classical explicit symplectic Strömer–Verlet formula of order two (see [8]);
- **RKN3s4**: the three-stage explicit symplectic RKN method of order four (see [8]);
- **IRKN2s4**: the two-stage implicit symplectic RKN method of order four (see [51]);
- **IRKN3s6**: the three-stage implicit symplectic RKN method of order six (see [51]).

Using the established mapping between the ERKN group and the RKN group (see [52]), it is known that ERKN methods with an arbitrarily high order can be obtained from the corresponding RKN methods. Hence, the IERKN2s4 method and the IERKN3s6 method are yielded by the well-known IRKN2s4 method and IRKN3s6 method, respectively. For implicit time integrators, we use fixed-point iteration and choose the tolerance as 10^{-15} and the maximum iteration number as 100. Here, it is noted that when the error of a method under consideration is very large for some Δt, we do not plot the corresponding points in the efficiency curves in the numerical experiments. The efficiency curves are given as the log-log plots of the errors.

Problem 3.1 We consider the Duffing equation

$$\begin{cases} \ddot{q} + \omega^2 q = k^2 (2q^3 - q), \\ q(0) = 0, \quad \dot{q}(0) = \omega, \end{cases}$$

where $0 \leqslant k < \omega$. This is a Hamiltonian system with the conservation of the following Hamiltonian

$$H\big(q(t), \dot{q}(t)\big) = \frac{1}{2}\dot{q}(t)^2 + \frac{1}{2}\omega^2 q(t)^2 + \frac{k^2}{2}\big(q(t)^2 - q(t)^4\big).$$

The analytic solution of the Duffing equation is well known, and given by

$$q(t) = sn(\omega t, k/\omega),$$

where sn means the Jacobian elliptic function. Obviously, the analytic solution $q(t)$ satisfies $|q(t)| \leqslant 1$, i.e., $q^2 \geqslant q^4$. Therefore, for each $\omega > 0$ (no matter how big ω is) there exists a constant K such that

$$\frac{1}{2}\dot{q}(t)^2 + \frac{1}{2}\omega^2 q(t)^2 \leqslant H\big(q(0), \dot{q}(0)\big) \leqslant \frac{K^2}{2}.$$

Then, the finite-energy condition (3.23) is verified. We choose $k = 0.03$ and different frequencies $\omega = 5, 10$ and 20 which are similar to those in [53]. We integrate the Problem 3.1 on the interval $[0, 1000]$ to verify our error estimates for the ISV method, the ERKN3s4 method, the IERKN2s4 method and the IERKN3s6

Table 3.1 Temporal accuracy of the "ISV" method for solving Problem 3.1 with different ω and Δt up to $T = 1000$

ω	Δt				
	$\Delta t_0 = 0.08$	$\Delta t_0/2$	$\Delta t_0/2^2$	$\Delta t_0/2^3$	$\Delta t_0/2^4$
$\omega = 5$	3.6227E − 7	8.4317E − 8	2.0714E − 8	5.1545E − 9	1.2843E − 9
Rate	*	2.1032	2.0252	2.0067	2.0048
$\omega = 10$	4.0951E − 7	7.1460E − 8	1.6455E − 8	4.0277E − 9	9.9857E − 10
Rate	*	2.5187	2.1186	2.0305	2.0120
$\omega = 20$	1.2973E − 5	9.1229E − 8	1.5503E − 8	3.5447E − 9	8.5937E − 10
Rate	*	−	2.5569	2.1288	2.0443

Table 3.2 Temporal accuracy of the "ERKN3s4" method for solving Problem 3.1 with different ω and Δt up to $T = 1000$

ω	Δt			
	$\Delta t_0 = 0.08$	$\Delta t_0/2$	$\Delta t_0/2^2$	$\Delta t_0/2^3$
$\omega = 5$	4.4048E − 8	2.7520E − 9	1.7404E − 10	1.3915E − 11
Rate	*	4.0005	3.9830	3.6447
$\omega = 10$	1.1427E − 7	6.5012E − 9	4.0112E − 10	3.0784E − 11
Rate	*	4.1356	4.0186	3.7038
$\omega = 20$	6.2331E − 6	2.0477E − 8	1.0892E − 9	7.1579E − 11
Rate	*	8.2498	4.2327	3.9276

Table 3.3 Temporal precision of the "IERKN2s4" method for solving Problem 3.1 with different ω and Δt up to $T = 1000$

ω	Δt				
	$\Delta t_0 = 0.1$	$\Delta t_0/2$	$\Delta t_0/2^2$	$\Delta t_0/2^3$	$\Delta t_0/2^4$
$\omega = 5$	1.9339E − 6	1.2139E − 7	7.6054E − 9	4.7731E − 10	3.0324E − 11
Rate	*	3.9938	3.9964	3.9940	3.9764
$\omega = 10$	1.5263E − 5	9.6938E − 7	6.0899E − 8	3.8101E − 9	2.4023E − 10
Rate	*	3.9768	3.9926	3.9985	3.9873
$\omega = 20$	1.1468E − 4	7.6411E − 6	4.8518E − 7	3.0467E − 8	1.9073E − 9
Rate	*	3.9077	3.9772	3.9932	3.9976

method with the different frequencies. The results in Tables 3.1 and 3.2 indicate that the convergence order of the ISV method and the ERKN3s4 method are of order two and order four, respectively. Tables 3.3 and 3.4 demonstrate that the IERKN2s4 method and the IERKN3s6 method are of order four and order six, respectively. The computational results are coincide with our theoretical analysis results.

The logarithm of the global errors GE$= \|q_N - q(1000)\|_2$ against different stepsizes for Problem 3.1 are plotted in Fig. 3.1. The logarithm of the global errors against different frequencies ω are displayed in Fig. 3.2. It can be observed from Fig. 3.2 that the ERKN integrators are independent of the frequency ω, whereas other traditional integrators depend on the frequency. In conclusion, Figs. 3.1

Table 3.4 Temporal precision of the "IERKN3s6" method for solving Problem 3.1 with different ω and Δt up to $T = 1000$

ω	Δt				
	$\Delta t_0 = 0.4$	$\Delta t_0/2$	$\Delta t_0/2^2$	$\Delta t_0/2^3$	$\Delta t_0/2^4$
$\omega = 5$	1.3355E $-$ 5	2.2286E $-$ 7	3.4864E $-$ 9	5.6017E $-$ 11	3.4615E $-$ 12
Rate	*	5.9051	5.9983	5.9597	4.0164
$\omega = 10$	3.3252E $-$ 4	6.5535E $-$ 6	1.0957E $-$ 7	1.7381E $-$ 9	2.8857E $-$ 10
Rate	*	5.6650	5.9024	5.9782	5.9124
$\omega = 20$	4.0302E $-$ 3	1.6588E $-$ 4	3.2996E $-$ 6	5.4632E $-$ 8	8.6855E $-$ 9
Rate	*	4.6026	5.6517	5.9164	5.9750

and 3.2 demonstrate that the ERKN time integrators are much more superior to the traditional numerical methods in the literature.

Problem 3.2 We consider the nonlinear KG equation (see, e.g. [5, 29])

$$u_{tt}(x, t) - a^2 \Delta u(x, t) + au(x, t) - bu^3(x, t) = 0,$$

in the region $(x, t) \in [-20, 20] \times [0, 10]$ with the initial conditions

$$u(x, 0) = \sqrt{\frac{2a}{b}} \operatorname{sech}(\lambda x), \qquad u_t(x, 0) = c\lambda \sqrt{\frac{2a}{b}} \operatorname{sech}(\lambda x) \tanh(\lambda x),$$

where $\lambda = \sqrt{a/(a^2 - c^2)}$ and $a, b, a^2 - c^2 > 0$. The exact solution of Problem 3.2 is given by

$$u(x, t) = \sqrt{\frac{2a}{b}} \operatorname{sech}(\lambda(x - ct)).$$

The real parameter $\sqrt{2a/b}$ represents the amplitude of a soliton which travels with velocity c. We use the parameters $a = 0.3$, $b = 1$ and $c = 0.25$ which are similar to those in [5, 29]. We integrate Problem 3.2 by using the IERKN3s6 time integrator with Fourier pseudospectral spatial discretisation (IERKN3s6-FP). The error graphs are shown in Fig. 3.3, with fixed time stepsize $\Delta t = 0.01$ and several values of spatial mesh size M. Numerical results demonstrate the spectral accuracy of the spatial discretisation.

In Tables 3.5 and 3.6, we fixed the spatial mesh size $M = 800$ and integrate the Problem 3.2 with different time stepsizes Δt to compute the temporal convergence order. The results demonstrate that the temporal accuracy is completely consistent with our theoretical analysis. In Fig. 3.4, we plot the logarithms of the global error $GE = \|U(\Delta t; T) - u(\cdot, T)\|_2$ against different time stepsizes, where $U(\Delta t; T)$ denotes the numerical solution at time T with the time stepsize Δt. The results illustrate that the ERKN time integrators have much better precision than the RKN time integrators.

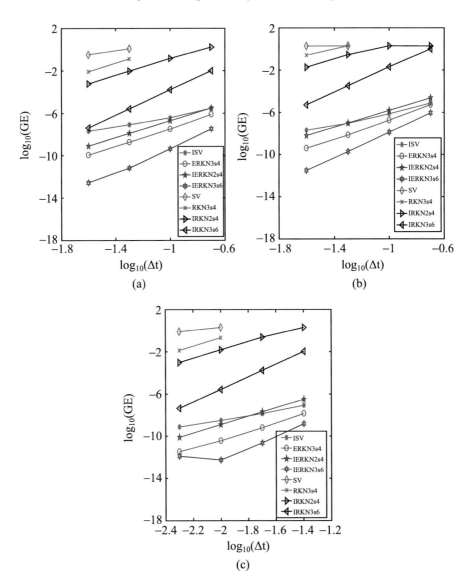

Fig. 3.1 Efficiency curves for Problem 3.1: The logarithm of the global errors GE $= \|q_N - q(1000)\|_2$ against different time stepsizes with frequencies $\omega = 5$ (**a**), 10 (**b**) and 20 (**c**)

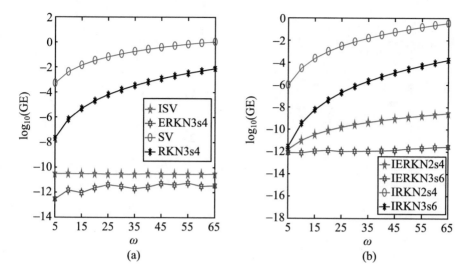

Fig. 3.2 Results of Problem 3.1: The logarithm of the global errors (GE) against different frequencies ω. (a) $\Delta t = 0.001$. (b) $\Delta t = 0.005$

Problem 3.3 Consider the nonlinear KG equation in the nonrelativistic limit regime (see [54, 55]):

$$\begin{cases} \varepsilon^2 u_{tt}(x,t) - \Delta u(x,t) + \dfrac{1}{\varepsilon^2} u(x,t) + f\big(u(x,t)\big) = 0, \\[2mm] u(x,0) = \psi_1(x), \quad u_t(x,0) = \dfrac{1}{\varepsilon^2}\psi_2(x), \end{cases} \tag{3.80}$$

in the region $(x,t) \in [-30, 30] \times [0, T]$ with the initial functions

$$\psi_1(x) = 2e^{-x^2}, \quad \psi_2(x) = 3e^{-x^2}$$

and the cubic nonlinearity, i.e. $f(u) = u^3$. Here $0 < \varepsilon \ll 1$ is a dimensionless parameter which is inversely proportional to the speed of light, ψ_1 and ψ_2 are two given pieces of real-valued initial data which are independent of ε. We simulate the experiment by using the IERKN3s6-FP method with the time stepsize $\Delta t = 10^{-4}$ and spatial mesh size $M = 1200$. The simulation results are displayed in Figs. 3.5 and 3.6. Obviously, the problem is highly oscillatory in time with respect to different values of parameter ε.

To test the temporal accuracy of the time integrators "ISV", "ERKN3s4", "IERKN2s4" and "IERKN3s6", we fixed the spatial mesh size as $M = 1200$. As is known, the exact solution of the Problem 3.3 cannot be represented explicitly. Therefore, we use a posterior error estimate, i.e. RE $= \|U(\Delta t; T) - U(\Delta t/2; T)\|_2$, to compute the convergence order. The computational results are

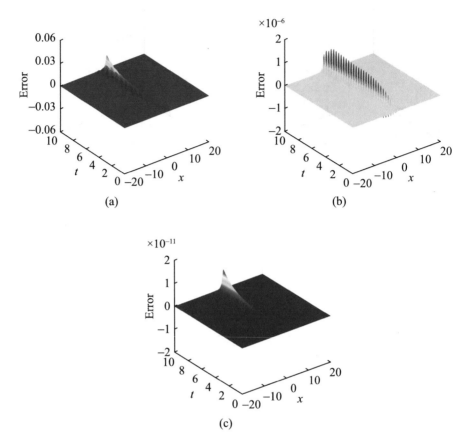

Fig. 3.3 The errors for Problem 3.2 obtained by using the IERKN3s6-FP method for $\Delta t = 0.01$ with (**a**) $M = 200$, (**b**) $M = 400$, and (**c**) $M = 800$

Table 3.5 Temporal precision of "ISV" and "ERKN3s4" methods for solving Problem 3.2 with different Δt up to $T = 10$ ($\Delta t_0 = 0.1$)

	ISV		ERKN3s4	
	Global error	Rate	Global error	Rate
Δt_0	1.2884E − 1	*	6.9442E − 4	*
$\Delta t_0/2$	3.3095E − 2	1.9609	4.3965E − 5	3.9814
$\Delta t_0/2^2$	8.3311E − 3	1.9900	2.7567E − 6	3.9953
$\Delta t_0/2^3$	2.0864E − 3	1.9975	1.7244E − 7	3.9988

Table 3.6 Temporal precision of "IERKN2s4" and "IERKN3s6" methods for solving Problem 3.2 with different Δt up to $T = 10$ ($\Delta t_0 = 0.4$)

	IERKN2s4		IERKN3s6	
	Global error	Rate	Global error	Rate
Δt_0	1.4215E − 3	*	9.0669E − 6	*
$\Delta t_0/2$	9.0773E − 5	3.9690	1.4335E − 7	5.9830
$\Delta t_0/2^2$	5.7039E − 6	3.9922	2.2590E − 9	5.9877
$\Delta t_0/2^3$	3.5696E − 7	3.9981	1.7062E − 11	7.0488

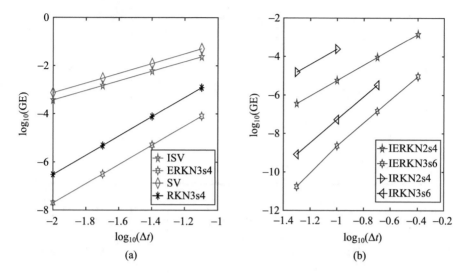

Fig. 3.4 Efficiency curves for Problem 3.2: The logarithm of the errors GE $= \|U(\Delta t; T) - u(\cdot, T)\|_2$ against different time stepsizes. (**a**) Explicit methods. (**b**) Implicit methods

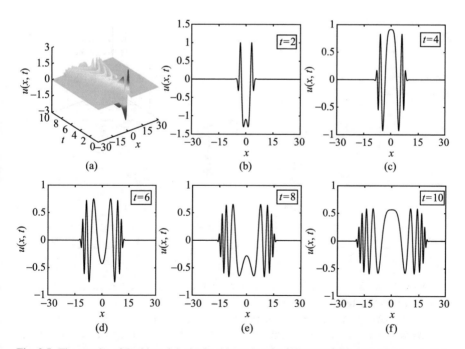

Fig. 3.5 The graphs of Problem 3.3 obtained by using the IERKN3s6-FP method for $\varepsilon = 0.5$, $\Delta t = 10^{-4}$ and $\Delta x = 1/20$. (**a**) $\varepsilon = 0.5$, (**b**) $t = 2$, (**c**) $t = 4$, (**d**) $t = 6$, (**e**) $t = 8$, (**f**) $t = 10$

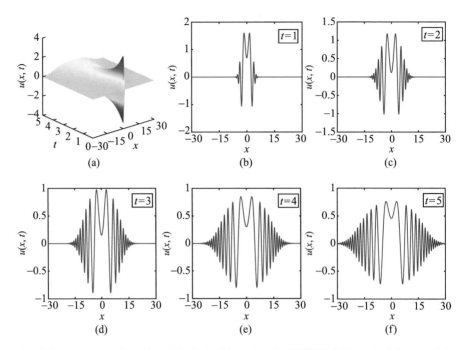

Fig. 3.6 The graphs of Problem 3.3 obtained by using the IERKN3s6-FP method for $\varepsilon = 0.1$, $\Delta t = 10^{-4}$ and $\Delta x = 1/20$. (a) $\varepsilon = 0.1$, (b) $t = 1$, (c) $t = 2$, (d) $t = 3$, (e) $t = 4$, (f) $t = 5$

Table 3.7 Temporal accuracy of the "ISV" method for solving Problem 3.3 with different ε and Δt at time $T = 2$

ε	Δt				
	$\Delta t_0 = 0.1$	$\Delta t_0/2$	$\Delta t_0/2^2$	$\Delta t_0/2^3$	$\Delta t_0/2^4$
$\varepsilon = 1$	4.1772E − 2	9.8027E − 3	2.4133E − 3	6.0104E − 4	1.5012E − 4
Rate	*	2.0913	2.0222	2.0055	2.0014
	$\Delta t_0 = 0.04$	$\Delta t_0/2$	$\Delta t_0/2^2$	$\Delta t_0/2^3$	$\Delta t_0/2^4$
$\varepsilon = 0.5$	3.8736E − 2	9.5530E − 3	2.3790E − 3	5.9416E − 4	1.4850E − 4
Rate	*	2.0196	2.0056	2.0014	2.0004
$\varepsilon = 0.1$	1.9373E − 2	5.1113E − 3	1.2885E − 3	3.2271E − 4	8.0712E − 5
Rate	*	1.9223	1.9880	1.9974	1.9994

listed in Tables 3.7, 3.8, 3.9 and 3.10 demonstrating that the temporal accuracy is completely consistent with our theoretical analysis.

In comparison with the corresponding time integrators "SV", "RKN3s4", "IRKN2s4" and "IRKN3s6", we fix the spatial mesh size as $M = 1200$ and integrate with different time stepsizes at time $T = 2$. The logarithms of the relative errors $\log_{10}(RE)$ are plotted in Fig. 3.7. It can be observed from Fig. 3.7 that the ERKN time integrators are more accurate than these traditional methods.

Table 3.8 Temporal accuracy of the "ERKN3s4" method for solving Problem 3.3 with different ε and Δt at time $T = 2$

ε	Δt				
	$\Delta t_0 = 0.1$	$\Delta t_0/2$	$\Delta t_0/2^2$	$\Delta t_0/2^3$	$\Delta t_0/2^4$
$\varepsilon = 1$	$4.0049E - 2$	$2.0521E - 3$	$1.2269E - 4$	$7.5856E - 6$	$4.7283E - 7$
Rate	*	4.2866	4.0641	4.0156	4.0039
	$\Delta t_0 = 0.04$	$\Delta t_0/2$	$\Delta t_0/2^2$	$\Delta t_0/2^3$	$\Delta t_0/2^4$
$\varepsilon = 0.5$	$4.3176E - 2$	$2.3654E - 3$	$1.4380E - 4$	$8.9274E - 6$	$5.5704E - 7$
Rate	*	4.1900	4.0400	4.0096	4.0024
$\varepsilon = 0.1$	$3.4882E - 2$	$2.2846E - 3$	$1.4431E - 4$	$9.0431E - 6$	$5.6556E - 7$
Rate	*	3.9325	3.9847	3.9962	3.9991

Table 3.9 Temporal precision of the "IERKN2s4" method for solving Problem 3.3 with different ε and Δt at time $T = 2$

ε	Δt				
	$\Delta t_0 = 0.1$	$\Delta t_0/2$	$\Delta t_0/2^2$	$\Delta t_0/2^3$	$\Delta t_0/2^4$
$\varepsilon = 1$	$2.3594E - 4$	$1.4193E - 5$	$8.8175E - 7$	$5.5037E - 8$	$3.4388E - 9$
Rate	*	4.0551	4.0087	4.0019	4.0004
	$\Delta t_0 = 0.05$	$\Delta t_0/2$	$\Delta t_0/2^2$	$\Delta t_0/2^3$	$\Delta t_0/2^4$
$\varepsilon = 0.5$	$6.5929E - 4$	$3.4212E - 5$	$2.0537E - 6$	$1.2721E - 7$	$7.9337E - 9$
Rate	*	4.2683	4.0582	4.0129	4.0031
	$\Delta t_0 = 0.005$	$\Delta t_0/2$	$\Delta t_0/2^2$	$\Delta t_0/2^3$	$\Delta t_0/2^4$
$\varepsilon = 0.1$	$2.1318E - 3$	$2.0379E - 4$	$1.3783E - 5$	$8.7673E - 7$	$5.5029E - 8$
Rate	*	3.3869	3.8861	3.9746	3.9939

Table 3.10 Temporal precision of the "IERKN3s6" method for solving Problem 3.3 with different ε and Δt at time $T = 2$

ε	Δt				
	$\Delta t_0 = 0.1$	$\Delta t_0/2$	$\Delta t_0/2^2$	$\Delta t_0/2^3$	$\Delta t_0/2^4$
$\varepsilon = 1$	$3.0107E - 6$	$3.5378E - 8$	$5.2866E - 10$	$8.1239E - 12$	$2.6356E - 13$
Rate	*	6.4111	6.0644	6.0240	–
	$\Delta t_0 = 0.05$	$\Delta t_0/2$	$\Delta t_0/2^2$	$\Delta t_0/2^3$	$\Delta t_0/2^4$
$\varepsilon = 0.5$	$2.3162E - 5$	$2.9908E - 7$	$4.5292E - 9$	$7.0507E - 11$	$1.2447E - 12$
Rate	*	6.2751	6.0451	6.0053	5.8239
	$\Delta t_0 = 0.005$	$\Delta t_0/2$	$\Delta t_0/2^2$	$\Delta t_0/2^3$	$\Delta t_0/2^4$
$\varepsilon = 0.1$	$4.3967E - 4$	$2.0432E - 6$	$3.0716E - 8$	$4.7584E - 10$	$6.6287E - 12$
Rate	*	7.7494	6.0557	6.0124	6.1656

In Fig. 3.8, we use the numerical solution obtained by the sixth-order IERKN3s6-FP method with the very small time stepsize $\Delta t = 10^{-4}$ and the spatial mesh size $M = 1200$, as the reference solution of the exact solution. The logarithms of the global errors against different parameters are plotted in Fig. 3.8. The results again show that the ERKN time integrators for solving the highly oscillatory problems are much superior to the RKN time integrators.

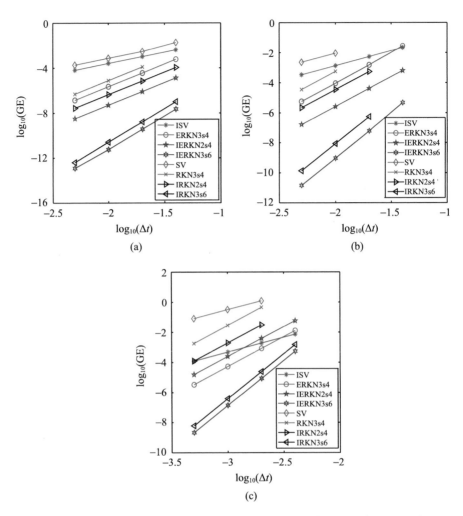

Fig. 3.7 Efficiency curves for Problem 3.3: The logarithm of the relative errors RE $=$ $\|U(\Delta t; T) - U(\Delta t/2; T)\|_2$ against different time stepsizes with parameters $\varepsilon = 1$ (**a**), 0.5 (**b**) and 0.1 (**c**)

3.5   Conclusions

In this chapter, we have made a comprehensive investigation on the nonlinear stability and convergence of ERKN integrators for solving the system of nonlinear multi-frequency highly oscillatory second-order ODEs (3.1) with a takanami number. On the basis of the finite-energy condition, it turns out that the nonlinear stability and the global error bounds are independent of the dominant frequency-matrix and the takanami number. Employing the energy technique, we also analysed

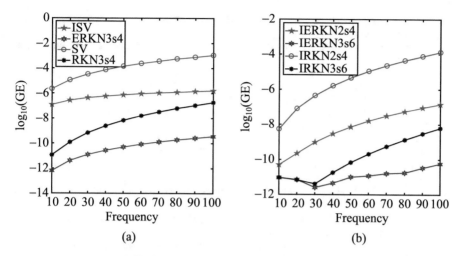

Fig. 3.8 Results of Problem 3.3: The logarithm of the global errors (GE) against different parameters $1/\varepsilon^2$. (**a**) $\Delta t = 0.001$. (**b**) $\Delta t = 0.005$

the convergence of the ERKN time integrators with the Fourier pseudospectral spatial discretisation when applied to semilinear wave equations. Another important issue is that the ERKN-FP method eliminates necessity for the CFL restriction, when applied to semilinear wave equations, whereas traditional schemes for solving PDEs suffer from this crucial restriction which greatly affects the efficiency of these schemes. This outstanding property of ERKN integrators ensures that an ERKN-type time integrator can use a larger time stepsize in comparison with the traditional methods for numerical solution of semilinear wave equations. This point is significant in the long-time numerical simulation of nonlinear phenomena in a wide variety of practical application areas in Science and Engineering.

The material in this chapter is based on the work by Liu and Wu [56].

References

1. Bátkai, A., Farkas, B., Csomós, P., et al.: Operator semigroups for numerical analysis. 15th Internet Seminar 2011/12 (2011)
2. Hesthaven, J.S., Gottlieb, S., Gottlieb, D.: Spectral methods for time-dependent problems. Cambridge Monographs on Applied and Computational Mathematics. Cambridge University Press, Cambridge (2007)
3. Lakestani, M., Dehghan, M.: Collocation and finite difference-collocation methods for the solution of nonlinear Klein-Gordon equation. Comput. Phys. Commun. **181**, 1392–1401 (2010)
4. Li, S., Vu-Quoc, L.: Finite difference calculus invariant structure of a class of algorithms for the nonlinear Klein-Gordon equation. SIAM J. Numer. Anal. **32**, 1839–1875 (1995)
5. Liu, C., Shi, W., Wu, X.: An efficient high-order explicit scheme for solving Hamiltonian nonlinear wave equations. Appl. Math. Comput. **246**, 696–710 (2014)

6. Shen, J., Tang, T., Wang, L.L.: Spectral Methods: Algorithms, Analysis, Applications. Springer, Berlin (2011)
7. Butcher, J.C.: Numerical Methods for Ordinary Differential Equations, 2nd edn. Wiley, New York (2008)
8. Hairer, E., Lubich, C., Wanner, G.: Geometric Numerical Integration: Structure-preserving Algorithms for Ordinary Differental Equations, 2nd edn. Springer, Berlin (2006)
9. Cohen, D., Jahnke, T., Lorenz, K.C., et al.: Numerical integrators for highly oscillatory Hamiltonian systems: a review. In: Mielke A. Analysis, Modeling and Simulation of Multiscale Problems. Springer, Berlin (2006)
10. García-Archilla, B., Sanz-Serna, J.M., Skeel, R.D.: Long-time-step methods for oscillatory differential equations. SIAM J. Sci. Comput. **20**, 930–963 (1998)
11. Grimm, V.: On error bounds for the Gautschi-type exponential integrator applied to oscillatory second-order differential equations. Numer. Math. **100**, 71–89 (2005)
12. Hochbruck, M., Lubich, C.: A Gautschi-type method for oscillatory second-order differential equations. Numer. Math. **83**, 403–426 (1999)
13. Wu, X., Liu, K., Shi, W.: Structure-Preserving Algorithms for Oscillatory Differential Equations II. Springer, Heidelberg (2015)
14. Wu, X., Wang, B.: Recent Developments in Structure-Preserving Algorithms for Oscillatory Differential Equations. Springer Nature Singapore Pte Ltd., Singapore (2018)
15. Wu, X., You, X., Wang, B.: Structure-Preserving Algorithms for Oscillatory Differential Equations. Springer, Heidelberg (2013)
16. Wu, X., You, X., Xia, J.: Order conditions for ARKN methods solving oscillatory systems. Comput. Phys. Commun. 180: 2250–2257 (2009)
17. Gautschi, W.: Numerical integration of ordinary differential equations based on trigonometric polynomials. Numer. Math. **3**, 381–397 (1961)
18. Hairer, E., Lubich, C.: Long-time energy conservation of numerical methods for oscillatory differential equations. SIAM J. Numer. Anal. **38**, 414–441 (2000)
19. Vanden Berghe, G., de Meyer, H., van Daele, M., et al.: Exponentially fitted Runge–Kutta methods. J. Comput. Appl. Math. **125**, 107–115 (2000)
20. Franco, J.M.: Exponentially fitted explicit Runge–Kutta–Nyström methods. J. Comput. Appl. Math. **167**, 1–19 (2004)
21. Li, Y.W., Wu, X.: Functionally-fitted energy-preserving methods for solving oscillatory nonlinear Hamiltonian systems. SIAM J. Numer. Anal. **54**, 2036–2059 (2016)
22. Franco, J.M.: New methods for oscillatory systems based on ARKN methods. Appl. Numer. Math. **56**, 1040–1053 (2006)
23. Mei, L., Liu, C., Wu, X.: An essential extension of the finite-energy condition for extended Runge–Kutta–Nyström integrators when applied to nonlinear wave equations. Commun. Comput. Phys. **22**, 742–764 (2017)
24. Shi, W., Wu, X., Xia, J.: Explicit Multi-symplectic extended leap-frog methods for Hamiltonian wave equations. J. Comput. Phys. **231**, 7671–7694 (2012)
25. Wang, B., Wu, X.: Long-time momentum and actions behaviour of energy-preserving methods for semilinear wave equations via spatial spectral semi-discretizations. Adv. Comput. Math. **45**, 2921–2952 (2019)
26. Wu, X., Wang, B., Shi, W.: Efficient energy-preserving integrators for oscillatory Hamiltonian systems. J. Comput. Phys. **235**, 587–605 (2013)
27. Wu, X., You, X., Shi, W., et al.: ERKN integrators for systems of oscillatory second-order differential equations. Comput. Phys. Commun. **181**, 1873–1887 (2010)
28. Liu, C., Iserles, A., Wu, X.: Symmetric and arbitrarily high-order Birkhoff-Hermite time integrators and their long-time behaviour for solving nonlinear Klein-Gordon equations. J. Comput. Phys. **356**, 1–30 (2018)
29. Liu, C., Wu, X.: Arbitrarily high-order time-stepping schemes based on the operator spectrum theory for high-dimensional nonlinear Klein-Gordon equations. J. Comput. Phys. **340**, 243–275 (2017)

30. Wang, B., Iserles, A., Wu, X.: Arbitrary-order trigonometric Fourier collocation methods for multi-frequency oscillatory systems. Found. Comput. Math. **16**, 151–181 (2016)
31. Wang, B., Wu, X., Meng, F.: Trigonometric collocation methods based on Lagrange basis polynomials for multi-frequency oscillatory second-order differential equations. J. Comput. Appl. Math. **313**, 185–201 (2017)
32. Wu, X., Wang, B., Mei, L.: Oscillation-preserving algorithms for efficiently solving highly oscillatory second-order ODEs. Numer. Algor. **86**, 693–727 (2021)
33. van der Houwen, P.J., Sommeijer, B.P.: Explicit Runge–Kutta(–Nyström) methods with reduced phase errors for computing oscillating solutions. SIAM J. Numer. Anal. **24**, 595–617 (1987)
34. Dehghan, M., Mohebbi, A., Asgari, Z.: Fourth-order compact solution of the nonlinear Klein–Gordon equation. Numer. Algor. **52**, 523–540 (2009)
35. Dehghan, M., Shokri, A.: Numerical solution of the nonlinear Klein-Gordon equation using radial basis functions. J. Comput. Appl. Math. **230**, 400–410 (2009)
36. Dehghan, M., Ghesmati, A.: Application of the dual reciprocity boundary integral equation technique to solve the nonlinear Klein–Gordon equation. Comput. Phys. Commun. **181**, 1410–1418 (2010)
37. Shakeri, F., Dehghan, M.: Numerical solution of the Klein–Gordon equation via He's variational iteration method. Nonlinear Dynam. **51**, 89–97 (2008)
38. Cano, B.: Conservation of invariants by symmetric multistep cosine methods for second-order partial differential equations. BIT Numer. Math. **53**, 29–56 (2013)
39. Cohen, D., Hairer, E., Lubich, C.: Conservation of energy, momentum and actions in numerical discretizations of non-linear wave equations. Numer. Math. **110**, 113–143 (2008)
40. Dong, X.: Stability and convergence of trigonometric integrator pseudo spectral discretization for N-coupled nonlinear Klein-Gordon equations. Appl. Math. Comput. **232**, 752–765 (2014)
41. Gauckler, L.: Error analysis of trigonometric integrators for semilinear wave equations. SIAM J. Numer. Anal. **53**, 1082–1106 (2015)
42. Hochbruck, M., Pažur, T.: Error analysis of implicit Euler methods for quasilinear hyperbolic evolution equations. Numer. Math. **135**, 547–569 (2017)
43. Wang, B., Wu, X.: The formulation and analysis of energy-preserving schemes for solving high-dimensional nonlinear Klein-Gordon equations. IMA J. Numer. Anal. **39**(4), 2016–2044 (2019)
44. Wang, B., Wu, X.: Globel error bounds of one-stage extended RKN integrators for semilinear wave equations. Numer. Algor. **81**, 1203–1218 (2019)
45. Liu, C., Wu, X.: The boundness of the operator-valued functions for multidimensional nonlinear wave equations with applications. Appl. Math. Lett. **74**, 60–67 (2017)
46. Liu, C., Wu, X.: An energy-preserving and symmetric scheme for nonlinear Hamiltonian wave equations. J. Math. Anal. Appl. **440**, 167–182 (2016)
47. Wu, X., Liu, C.: An integral formula adapted to different boundary conditions for arbitrarily high-dimensional nonlinear Klein-Gordon equations with its applications. J. Math. Phys. **57**(2), 3239–3249 (2016)
48. Wu, X., Liu, C., Mei, L.: A new framework for solving partial differential equations using semi-analytical explicit RK(N)-type integrators. J. Comput. Appl. Math. **301**, 74–90 (2016)
49. Wu, X., Mei, L., Liu, C.: An analytical expression of solutions to nonlinear wave equations in higher dimensions with Robin boundary conditions. J. Math. Anal. Appl. **426**, 1164–1173 (2015)
50. Gottlieb, D., Orszag, S.: Numerical Analysis of Spectral Methods: Theory and Applications. Society for Industrial and Applied Mathematics, Philadelphia (1993)
51. Tang, W.S., Ya, Y.J., Zhang, J.J.: High order symplectic integrators based on continuous-stage Runge–Kutta–Nyström methods. Appl. Math. Comput. **361**, 670–679 (2019)
52. Mei, L., Wu, X.: The construction of arbitrary order ERKN methods based on group theory for solving oscillatory Hamiltonian systems with applications. J. Comput. Phys. **323**, 171–190 (2016)

53. Wang, B., Liu, K., Wu, X.: A Filon-type asymptotic approach to solving highly oscillatory second-order initial value problems. J. Comput. Phys. **243**, 210–223 (2013)
54. Bao, W.Z., Dong, X.C.: Analysis and comparison of numerical methods for the Klein-Gordon equation in the nonrelativistic limit regime. Numer. Math. **120**, 189–229 (2012)
55. Wang, Y., Zhao, X.F.: Symmetric high order Gautschi-type exponential wave integrators pseudospectral method for the nonlinear Klein-Gordon equation in the nonrelativistic limit regime. Int. J. Numer. Anal. Mod. **15**(3), 405–427 (2018)
56. Liu, C., Wu, X.: Nonlinear stability and convergence of ERKN integrators for solving nonlinear multi-frequency highly oscillatory second-order ODEs with applications to semi-linear wave equations. Appl. Numer. Math. **153**, 352–380 (2020)

Chapter 4
Functionally-Fitted Energy-Preserving Integrators for Poisson Systems

This chapter presents a class of energy-preserving integrators for Poisson systems based on the functionally-fitted strategy, and these energy-preserving integrators can have arbitrarily high order. This approach permits us to obtain the energy-preserving methods proposed in [1] by Cohen and Hairer and [2] by Brugnano et al. for Poisson systems.

4.1 Introduction

It is well known that Poisson systems arise in many applications. Moreover, it is noted that Poisson systems often have periodic or oscillatory solutions. This chapter is devoted to efficient numerical integrators for solving Poisson systems (non-canonical Hamiltonian systems)

$$y'(t) = B(y(t))\nabla H(y(t)), \quad y(0) = y_0 \in \mathbb{R}^d, \quad t \in [0, T], \tag{4.1}$$

where the prime denotes $\dfrac{\mathrm{d}}{\mathrm{d}t}$, $B(y)$ is a skew-symmetric matrix, $H(y)$ is a scalar function, and both are sufficiently smooth. It is assumed that the system (4.1) has a unique solution $y = y(t)$ defined for all $t \in [0, T]$. An important feature of (4.1) is that the energy $H(y)$ is preserved along the exact solution $y(t)$, since we have

$$\frac{\mathrm{d}}{\mathrm{d}t}H(y(t)) = \nabla H(y(t))^{\mathsf{T}} y'(t) = \nabla H(y(t))^{\mathsf{T}} B(y(t))\nabla H(y(t)) = 0.$$

Numerical integrators that preserve $H(y)$ are termed energy-preserving (EP) integrators. The main aim of this chapter is to formulate and analyse some EP integrators for efficiently solving Poisson systems. Other geometric properties of the Poisson systems such as the preservation of Casimir functions and the Poisson map of the flow will not be considered in this chapter.

© The Author(s), under exclusive license to Springer Nature Singapore Pte Ltd. 2021
X. Wu, B. Wang, *Geometric Integrators for Differential Equations with Highly Oscillatory Solutions*, https://doi.org/10.1007/978-981-16-0147-7_4

If the matrix $B(y)$ is independent of y, d is an even number and

$$B = J = \begin{pmatrix} 0_{\frac{d}{2}} & I_{\frac{d}{2}} \\ -I_{\frac{d}{2}} & 0_{\frac{d}{2}} \end{pmatrix},$$

then the system (4.1) is a canonical Hamiltonian system. Much effort has been made for solving this system, and we refer the reader to [3–13] and references therein. For canonical Hamiltonian systems, EP integrators are important and efficient methods and a variety of EP methods have been derived and studied in the past few decades, such as the average vector field (AVF) method (see, e.g. [14–16]), discrete gradient methods (see, e.g. [17, 18]), Hamiltonian Boundary Value Methods (HBVMs) (see, e.g. [19, 20]), EP collocation methods (see, e.g. [21]) and exponential/trigonometric EP methods (see, e.g. [22–26]).

Among typical EP methods for solving $\dot{y} = J\nabla H(y)$ is the well-known AVF method given by Quispel and McLaren [16] as follows:

$$y_1 = y_0 + h \int_0^1 J\nabla H(y_0 + \sigma(y_1 - y_0))\mathrm{d}\sigma. \tag{4.2}$$

Quispel and McLaren in [16] pointed out that this method is a B-series method. Hairer extended this second-order method to higher-order schemes by introducing the so-called continuous-stage Runge–Kutta methods [21]. On noticing the fact that the dependence of the matrix $B(y)$ should be discretised in a different manner, an additional strategy will be required for Poisson systems. This means that it is necessary to design and analyse the EP methods specifically for Poisson systems. As is known, McLachlan et al. [18] discussed DG methods for various kinds of ODEs including Poisson systems, and Cohen et al. [1] succeeded in constructing arbitrary high-order EP schemes for Poisson systems and the following second-order EP scheme for (4.1) was derived

$$y_1 = y_0 + hB\big((1/2)(y_1 + y_0)\big) \int_0^1 \nabla H(y_0 + \sigma(y_1 - y_0))\mathrm{d}\sigma. \tag{4.3}$$

In the light of HBVMs, Brugnano et al. gave an alternative derivation of such methods and presented a new proof of their orders in [27]. EP exponentially-fitted integrators for Poisson systems were researched by Miyatake [28]. Using discrete gradients, Dahlby et al. [29] constructed useful methods that simultaneously preserve several invariants in systems of type (4.1). With regard to other multiple invariant-preserving integrators we refer the reader to [2, 20, 30–32].

We note that the functionally-fitted (FF) technique is a very useful approach to the construction of effective numerical methods for solving differential equations. In general, an FF method can be derived by requiring it to integrate members of a given finite-dimensional function space X exactly. The corresponding methods are termed trigonometrically-fitted (TF) or exponentially-fitted (EF) methods if

X is generated by trigonometrical or exponential functions. Using the FF/TF/EF technique, many efficient methods have been constructed for canonical Hamiltonian systems including the symplectic methods (see, e.g. [33–40]) and EP methods (see, e.g. [23, 41]). This technique has also been used successfully for Poisson systems in [28] and second- and fourth-order schemes were derived. In this chapter, using the functionally-fitted technology, we will design and analyse efficient EP integrators for Poisson systems. These integrators of arbitrarily high order can be derived in a routine and convenient manner, and different EP schemes can be obtained by considering different function spaces. We will show that choosing a special function space allows us to obtain the EP schemes proposed by Cohen and Hairer [1] and Brugnano et al. [27].

4.2 Functionally-Fitted EP Integrators

In order to derive the EP integrators for Poisson systems (4.1), we first define a vector function space $Y=\text{span}\{\Phi_0(t), \cdots, \Phi_{r-1}(t)\}$ on $[0, T]$ by (see [41])

$$Y = \left\{ w : w(t) = \sum_{i=0}^{r-1} \Phi_i(t) W_i, \ t \in [0, T], \ W_i \in \mathbb{R}^d \right\},$$

where the real functions $\{\Phi_i(t)\}_{i=0}^{r-1}$ are linearly independent and C^1 $([0, T] \to \mathbb{R})$. In this chapter, we choose a stepsize $h > 0$ and consider the following two function spaces

$$Y_h = \text{span}\{\Phi_0(\tau h), \cdots, \Phi_{r-1}(\tau h)\}, \quad X_h = \text{span}\left\{ 1, \int_0^{\tau h} \Phi_0(s)\mathrm{d}s, \cdots, \int_0^{\tau h} \Phi_{r-1}(s)\mathrm{d}s \right\},$$
$$(4.4)$$

where τ is a variable with $\tau \in [0, 1]$, the stepsize h is a positive parameter with $0 < h \leqslant h_0 \leqslant T$, and h_0 depends on the underlying problem.

We now introduce a projection (see [41]), which will be used in this chapter and we summarise its definition as follows.

Definition 4.1 (See [41]) Let $\mathscr{P}_h$ be a linear operator that maps d-vector valued real functions defined on $[0, h]$ into the finite dimensional space Y_h. The definition of $\mathscr{P}_h w(\tau h)$ is given by

$$\langle v(\tau h), \mathscr{P}_h w(\tau h) \rangle = \langle v(\tau h), w(\tau h) \rangle, \quad \text{for any } v \in Y_h, \qquad (4.5)$$

where $w(\tau h)$ be a continuous $\mathbb{R}^d$-valued function for $\tau \in [0, 1]$ and $\mathscr{P}_h w(\tau h)$ is a projection of w onto Y_h. The scalar product $\langle \cdot, \cdot \rangle$ is defined as an inner product in

$C([0, 1] \to \mathbb{R}^d)$ so that for

$$u = u(\tau h) = (u_1(\tau h), \cdots, u_d(\tau h))^{\mathsf{T}}, \quad v = v(\tau h) = (v_1(\tau h), \cdots, v_d(\tau h))^{\mathsf{T}},$$

$\langle u, v \rangle$ is a d-dimensional vector with components $\displaystyle\int_0^1 u_i(\tau h)v_i(\tau h) \, d\tau$ for $i = 1, \cdots, d$.

The following property of $\mathscr{P}_h$ is also needed, which has been proved in [41].

Lemma 4.1 (See [41]) *The projection* $\mathscr{P}_h w(\tau h)$ *can be explicitly expressed as*

$$\mathscr{P}_h w(\tau h) = \langle P_{\tau,\sigma}, w(\sigma h) \rangle_\sigma,$$

where

$$P_{\tau,\sigma} = \sum_{i=0}^{r-1} \psi_i(\tau h)\psi_i(\sigma h),$$

and $\{\psi_0, \cdots, \psi_{r-1}\}$ *is a standard orthonormal basis of* Y_h *under the inner product* $\langle \cdot, \cdot \rangle$.

With these preliminaries, we first present the definition of the integrators and then show that they exactly preserve the energy of Poisson system (4.1).

Definition 4.2 Let $u = u(\tau h)$ be the unique solution of the following initial value problem

$$\frac{1}{h}\frac{du(\tau h)}{d\tau} = B(u(\tau h))\mathscr{P}_h\big(\nabla H(u(\tau h))\big), \quad u(0) = y_0, \quad \tau \in [0, 1]. \tag{4.6}$$

If $u \in X_h$, then the numerical solution after one step is defined by $y_1 = u(h)$. In this chapter, the integrator is termed a functionally-fitted EP (FFEP) integrator.

Remark 4.1 It is important to note that the exact solution of the Poisson system (4.1) may not belong to the function space X_h. In this definition, the function $u \in X_h$ is considered as a numerical approximation of the exact solution. This approach is similar to that given by Cohen and Hairer in [1], where they consider a polynomial function as the approximation of the exact solution. In particular, we remark that, for the Euler equation considered as a numerical experiment in Sect. 4.7, the solution of (4.6) belongs to X_h.

Theorem 4.1 *The FFEP integrator* (4.6) *exactly preserves the energy, i.e.,*

$$H(y_1) = H(y_0).$$

Proof It follows from $u \in X_h$ that $u' \in Y_h$. Using the definition of $\mathscr{P}_h$ yields

$$\int_0^1 u'(\tau h)_i \left(\mathscr{P}_h \big(\nabla H(u(\tau h)) \big) \right)_i d\tau = \int_0^1 u'(\tau h)_i \big(\nabla H(u(\tau h)) \big)_i d\tau, \quad i = 1, 2, \cdots, d,$$

where $(\cdot)_i$ denotes the i-th entry of a vector. We then obtain

$$\int_0^1 u'(\tau h)^\mathsf{T} \mathscr{P}_h \big(\nabla H(u(\tau h)) \big) d\tau = \int_0^1 u'(\tau h)^\mathsf{T} \nabla H(u(\tau h)) d\tau.$$

Hence, we have

$$H(y_1) - H(y_0) = \int_0^1 \frac{d}{d\tau} H(u(\tau h)) d\tau$$

$$= h \int_0^1 u'(\tau h)^\mathsf{T} \nabla H(u(\tau h)) d\tau$$

$$= h \int_0^1 u'(\tau h)^\mathsf{T} \mathscr{P}_h \big(\nabla H(u(\tau h)) \big) d\tau.$$

Inserting (4.6) into this formula gives

$$H(y_1) - H(y_0) = h \int_0^1 \mathscr{P}_h \big(\nabla H(u(\tau h)) \big)^\mathsf{T} B(u(\tau h))^\mathsf{T} \mathscr{P}_h \big(\nabla H(u(\tau h)) \big) d\tau.$$

This proves the result by considering that $B(u)$ is a skew-symmetric matrix. □

Remark 4.2 If $B(y)$ is a constant skew-symmetric matrix, (4.1) is a canonical Hamiltonian system. In this case, the FFEP integrator (4.6) is identical to the functionally-fitted EP method derived in Li and Wu [41]. Apart from this, if Y_h is generated by the shifted Legendre polynomials on [0, 1], then the FFEP integrator (4.6) reduces to the EP collocation method given by Cohen and Hairer [21] and Brugnano et al. [27].

4.3 Implementation Issues

We next pay attention to practical implementation issues of the FFEP integrator. We choose the generalised Lagrange interpolation functions $\hat{l}_i(\tau) \in Y_h$ with respect to

r distinct points $\hat{d}_i \in [0, 1]$ for $i = 1, \cdots, r$ as follows:

$$
(\hat{l}_1(\tau), \cdots, \hat{l}_r(\tau)) = (\Phi_0(\tau h), \Phi_1(\tau h), \cdots, \Phi_{r-1}(\tau h))
$$

$$
\cdot
\begin{pmatrix}
\Phi_0(\hat{d}_1 h) & \Phi_1(\hat{d}_1 h) & \cdots & \Phi_{r-1}(\hat{d}_1 h) \\
\Phi_0(\hat{d}_2 h) & \Phi_1(\hat{d}_2 h) & \cdots & \Phi_{r-1}(\hat{d}_2 h) \\
\vdots & \vdots & & \vdots \\
\Phi_0(\hat{d}_r h) & \Phi_1(\hat{d}_r h) & \cdots & \Phi_{r-1}(\hat{d}_r h)
\end{pmatrix}^{-1}
. \tag{4.7}
$$

Clearly, $\{\hat{l}_i(\tau)\}_{i=1}^r$ provides another basis of Y_h, satisfying $\hat{l}_i(\hat{d}_j) = \delta_{ij}$. As $u' \in Y_h$, u' can be expressed in

$$
u'(\tau h) = \sum_{i=1}^r \hat{l}_i(\tau) u'(\hat{d}_i h).
$$

It follows from Lemma 4.1 that the FFEP integrator (4.6) is identical to

$$
u'(\tau h) = B(u(\tau h)) \int_0^1 P_{\tau,\sigma} \nabla H(u(\sigma h)) d\sigma,
$$

which leads to

$$
u'(\hat{d}_i h) = B(u(\hat{d}_i h)) \int_0^1 P_{\hat{d}_i,\sigma} \nabla H(u(\sigma h)) d\sigma.
$$

We then obtain

$$
u'(\tau h) = \sum_{i=1}^r \hat{l}_i(\tau) u'(\hat{d}_i h) = \sum_{i=1}^r \hat{l}_i(\tau) \left(B(u(\hat{d}_i h)) \int_0^1 P_{\hat{d}_i,\sigma} \nabla H(u(\sigma h)) d\sigma \right).
$$

$$
\tag{4.8}
$$

Integrating (4.8) gives

$$
u(\tau h) = y_0 + \int_0^{\tau h} u'(x) dx = y_0 + h \int_0^\tau u'(\alpha h) d\alpha
$$

$$
= y_0 + h \int_0^\tau \sum_{i=1}^r \hat{l}_i(\alpha) d\alpha B(u(\hat{d}_i h)) \int_0^1 P_{\hat{d}_i,\sigma} \nabla H(u(\sigma h)) d\sigma.
$$

Let $y_\sigma = u(\sigma h)$, and we are now in a position to present the FFEP integrator (4.6) for Poisson system (4.1).

Definition 4.3 An FFEP integrator (4.6) for Poisson system (4.1) is defined by

$$
\begin{cases}
y_\tau = y_0 + h \sum_{i=1}^{r} \int_0^1 \left(P_{\hat{d}_i,\sigma} \int_0^\tau \hat{l}_i(\alpha)d\alpha \right) B(y_{\hat{d}_i}) \nabla H(y_\sigma) d\sigma, \quad 0 < \tau < 1, \\[4mm]
y_1 = y_0 + h \sum_{i=1}^{r} \int_0^1 \left(P_{\hat{d}_i,\sigma} \int_0^1 \hat{l}_i(\alpha)d\alpha \right) B(y_{\hat{d}_i}) \nabla H(y_\sigma) d\sigma.
\end{cases}
$$

$$(4.9)$$

Remark 4.3 It can be observed from (4.9) that this integrator has a pattern similar to the formula (2.4) given by Cohen and Hairer in [1]. We need the first formula of (4.9) only for $\tau = \hat{d}_1, \cdots, \hat{d}_r$ and this leads to a nonlinear system of equations for the unknowns $y_{\hat{d}_1}, \cdots, y_{\hat{d}_r}$ which can be solved by a fixed-point iteration method.

Remark 4.4 It is noted that the integrals $\int_0^\tau \hat{l}_i(\alpha)d\alpha$ and $\int_0^1 \hat{l}_i(\alpha)d\alpha$ can be calculated exactly if Y_h is generated by many kinds of functions such as polynomials, exponential and trigonometrical functions. The integral $\int_0^1 P_{\hat{d}_i,\sigma} \nabla H(y_\sigma)d\sigma$ appearing in (4.9) can also be calculated exactly for many cases. If these integrals cannot be explicitly calculated, they can be approximated by quadrature to any desired degree of accuracy.

4.4 The Existence, Uniqueness and Smoothness

We note that the FFEP integrator (4.6) fails to be well defined unless its existence and uniqueness is shown. This section is devoted to this issue.

In what follows, we assume that the solution $y = y(t)$ of (4.1) is contained in the following ball for $t \in [0, h_0]$

$$
\bar{B}(y_0, R) = \left\{ y \in \mathbb{R}^d : ||y - y_0|| \leqslant R \right\},
$$

where R is a positive constant and $|| \cdot ||$ is a fixed norm in $\mathbb{R}^d$ which is the same as that stated in Assumption 4.1 below. Besides, it has been shown in [41] that $P_{\tau,\sigma}$ is a smooth function of h. In this setting, we assume that $A_n = \max_{\tau,\sigma,h\in[0,1]} \left| \dfrac{\partial^n P_{\tau,\sigma}}{\partial h^n} \right|$ for $n = 0, 1$. Furthermore, the nth-order derivative of ∇H at y is a multilinear map from $\underbrace{\mathbb{R}^d \times \cdots \times \mathbb{R}^d}_{n-fold}$ to $\mathbb{R}^d$ defined by

$$
\nabla H^{(n)}(y)(z_1, \cdots, z_n) = \sum_{1 \leqslant \alpha_1, \cdots, \alpha_n \leqslant d} \frac{\partial^n \nabla H}{\partial y_{\alpha_1} \cdots \partial y_{\alpha_n}}(y) z_1^{\alpha_1} \cdots z_n^{\alpha_n},
$$

where $y = (y_1, \cdots, y_d)^\mathsf{T}$ and $z_i = (z_i^1, \cdots, z_i^d)^\mathsf{T}$ for $i = 1, \cdots, n$. The same notation is used for $B(y)$. Before presenting the result, we also need the following assumption.

Assumption 4.1 Denote $D_0 = \max_{y \in \bar{B}(y_0, R)} \|\nabla H(y)\|$ and $C_0 = \max_{y \in \bar{B}(y_0, R)} \|B(y)\|$. It is assumed that ∇H and $\nabla H^{(1)}$ are Lipschitz-continuous, i.e., there exist $D_1, D_2 > 0$ such that

$$\|\nabla H(y_1) - \nabla H(y_2)\| \leqslant D_1 \|y_1 - y_2\|, \quad \|\nabla H^{(1)}(y_1) - \nabla H^{(1)}(y_2)\| \leqslant D_2 \|y_1 - y_2\|$$

for all $y_1, y_2 \in \bar{B}(y_0, R)$. The same assumption is required for $B(y)$ and $B^{(1)}(y)$, and the corresponding Lipschitz constants are denoted by C_1 and C_2, respectively.

Theorem 4.2 *Under the assumptions stated above, the FFEP integrator* (4.6) *has a unique solution* $u(\tau h)$ *provided the stepsize h satisfies*

$$0 \leqslant h \leqslant \delta < \min\left\{ \frac{1}{A_0 C_0 D_1 + A_0 C_1 D_0}, \frac{R}{A_0 C_0 D_0}, h_0, 1 \right\}. \qquad (4.10)$$

Moreover, $u(\tau h)$ *is smoothly dependent on h for any fixed* $\tau \in (0, 1]$.

Proof Existence and uniqueness. It follows from Lemma 4.1 that the FFEP integrator (4.6) can be written as

$$u'(\tau h) = B(u(\tau h)) \int_0^1 P_{\tau,\sigma} \nabla H(u(\sigma h)) d\sigma.$$

By integration we obtain

$$u(\tau h) = y_0 + h \int_0^\tau B(u(\alpha h)) \int_0^1 P_{\alpha,\sigma} \nabla H(u(\sigma h)) d\sigma\, d\alpha. \qquad (4.11)$$

The formula (4.11) generates a function series $\{u_n(\tau h)\}_{n=0}^\infty$ by the following recursive definition

$$u_{n+1}(\tau h) = y_0 + h \int_0^1 \left(\int_0^\tau B(u_n(\alpha h)) P_{\alpha,\sigma} d\alpha \right) \nabla H(u_n(\sigma h)) d\sigma, \quad n = 0, 1, \cdots,$$
$$\qquad (4.12)$$

which will be shown to be uniformly convergent by proving the uniform convergence of the infinite series $\sum_{n=0}^\infty (u_{n+1}(\tau h) - u_n(\tau h))$. Then the integrator (4.6) has a solution $\lim_{n \to \infty} u_n(\tau h)$.

We next prove the uniform convergence of $\sum_{n=0}^\infty (u_{n+1}(\tau h) - u_n(\tau h))$. First, it is clear that $\|u_0(\tau h) - y_0\| = 0 \leqslant R$. We assume that $\|u_n(\tau h) - y_0\| \leqslant R$ for $n = 0, \cdots, m$. It then follows from (4.10) and (4.12) that

$$\|u_{m+1}(\tau h) - y_0\| \leqslant h A_0 C_0 D_0 \leqslant R,$$

which implies that $u_n(\tau h)$ are uniformly bounded by $||u_n(\tau h) - y_0|| \leqslant R$ for $n = 0, 1, \cdots$. Then using (4.12) and the Lipschitz conditions, we obtain

$$||u_{n+1}(\tau h) - u_n(\tau h)||_c$$

$$\leqslant h \int_0^1 \int_0^\tau \left|\left|\left[B(u_n(\alpha h)) P_{\alpha,\sigma} \nabla H(u_n(\sigma h)) - B(u_{n-1}(\alpha h)) P_{\alpha,\sigma} \nabla H(u_{n-1}(\sigma h)) \right]\right|\right|_c d\alpha d\sigma$$

$$\leqslant h \int_0^1 \int_0^\tau \left|\left|\left[B(u_n(\alpha h)) P_{\alpha,\sigma} \nabla H(u_n(\sigma h)) - B(u_n(\alpha h)) P_{\alpha,\sigma} \nabla H(u_{n-1}(\sigma h)) \right.\right.\right.$$

$$\left.\left.\left. + B(u_n(\alpha h)) P_{\alpha,\sigma} \nabla H(u_{n-1}(\sigma h)) - B(u_{n-1}(\alpha h)) P_{\alpha,\sigma} \nabla H(u_{n-1}(\sigma h)) \right]\right|\right|_c d\alpha d\sigma$$

$$\leqslant h(A_0 C_0 D_1 + A_0 C_1 D_0) \int_0^1 ||u_n(\sigma h) - u_{n-1}(\sigma h)|| d\sigma \leqslant \beta ||u_n(\tau h) - u_{n-1}(\tau h)||_c,$$

where $\beta = \delta(A_0 C_0 D_1 + A_0 C_1 D_0)$ and $||w||_c = \max_{\tau \in [0,1]} ||w(\tau h)||$ for a continuous $\mathbb{R}^d$-valued function w on $[0, 1]$. This implies that

$$||u_{n+1} - u_n||_c \leqslant \beta ||u_n - u_{n-1}||_c$$

and then

$$||u_{n+1} - u_n||_c \leqslant \beta^n ||u_1 - y_0||_c \leqslant \beta^n R, \quad n = 0, 1, \cdots .$$

Using the Weierstrass M-test and the fact that $\beta < 1$, we confirm that $\sum_{n=0}^{\infty} (u_{n+1}(\tau h) - u_n(\tau h))$ is uniformly convergent.

With regard to the uniqueness, we suppose that the integrator has another solution $v(\tau h)$. We then have

$$||u(\tau h) - v(\tau h)|| \leqslant \beta ||u(\tau h) - v(\tau h)|| \leqslant \beta ||u - v||_c,$$

and

$$||u - v||_c \leqslant \beta ||u - v||_c.$$

Hence, we obtain $||u - v||_c = 0$ and $u(\tau h) \equiv v(\tau h)$. Therefore, the solution of the FFEP integrator (4.6) exists and is unique.

Smoothness We next prove the result that $u(\tau h)$ is smoothly dependent on h for any fixed $\tau \in (0, 1]$. This is true if the series $\left\{ \dfrac{\partial^k u_n}{\partial h^k}(\tau h) \right\}_{n=0}^{\infty}$ is uniformly convergent for $k \geqslant 1$. We note that the analysis of this part needs the bounds on $\nabla H^{(1)}(y)$ and $B^{(1)}(y)$, which are also denoted by D_1 and C_1, respectively.

Differentiating (4.12) with respect to h yields

$$
\frac{\partial u_{n+1}}{\partial h}(\tau h) = \int_0^1 \left(\int_0^\tau B(u_n(\alpha h)) P_{\alpha,\sigma} \, d\alpha \right) \nabla H(u_n(\sigma h)) d\sigma
$$

$$
+ h \int_0^1 \left(\int_0^\tau B^{(1)}(u_n(\alpha h)) \frac{\partial u_n(\alpha h)}{\partial h} P_{\alpha,\sigma} \, d\alpha \right) \nabla H(u_n(\sigma h)) d\sigma
$$

$$
+ h \int_0^1 \left(\int_0^\tau B(u_n(\alpha h)) \frac{\partial P_{\alpha,\sigma}}{\partial h} \, d\alpha \right) \nabla H(u_n(\sigma h)) d\sigma
$$

$$
+ h \int_0^1 \left(\int_0^\tau B(u_n(\alpha h)) P_{\alpha,\sigma} \, d\alpha \right) \nabla H^{(1)}(u_n(\sigma h)) \frac{\partial u_n(\sigma h)}{\partial h} d\sigma. \quad (4.13)
$$

We then have

$$
\left\| \frac{\partial u_{n+1}}{\partial h} \right\|_c \leqslant \alpha + \beta \left\| \frac{\partial u_n}{\partial h} \right\|_c \qquad \text{with } \alpha = A_0 C_0 D_0 + \delta A_1 C_0 D_0,
$$

which yields that $\left\{ \dfrac{\partial u_n}{\partial h}(\tau h) \right\}_{n=0}^{\infty}$ is uniformly bounded as follows:

$$
\left\| \frac{\partial u_n}{\partial h} \right\|_c \leqslant \alpha(1 + \beta + \cdots + \beta^{n-1}) \leqslant \frac{\alpha}{1-\beta} =: C^*, \quad n = 0, 1, \cdots.
$$

It follows from (4.13) that

$$
\frac{\partial u_{n+1}}{\partial h} - \frac{\partial u_n}{\partial h}
$$

$$
= \int_0^1 \int_0^\tau \Big[B(u_n(\alpha h)) P_{\alpha,\sigma} \nabla H(u_n(\sigma h)) - B(u_{n-1}(\alpha h)) P_{\alpha,\sigma} \nabla H(u_{n-1}(\sigma h)) \Big] d\alpha d\sigma
$$

$$
+ h \int_0^1 \int_0^\tau \Big[B^{(1)}(u_n(\alpha h)) \frac{\partial u_n(\alpha h)}{\partial h} P_{\alpha,\sigma} \nabla H(u_n(\sigma h))
$$

$$
- B^{(1)}(u_{n-1}(\alpha h)) \frac{\partial u_{n-1}(\alpha h)}{\partial h} P_{\alpha,\sigma} \nabla H(u_{n-1}(\sigma h)) \Big] d\alpha d\sigma
$$

$$
+ h \int_0^1 \int_0^\tau \Big[B(u_n(\alpha h)) \frac{\partial P_{\alpha,\sigma}}{\partial h} \nabla H(u_n(\sigma h))
$$

$$
- B(u_{n-1}(\alpha h)) \frac{\partial P_{\alpha,\sigma}}{\partial h} \nabla H(u_{n-1}(\sigma h)) \Big] d\alpha d\sigma
$$

$$
+ h \int_0^1 \int_0^\tau \Big[B(u_n(\alpha h)) P_{\alpha,\sigma} \nabla H^{(1)}(u_n(\sigma h)) \frac{\partial u_n(\sigma h)}{\partial h}
$$

$$
- B(u_{n-1}(\alpha h)) P_{\alpha,\sigma} \nabla H^{(1)}(u_{n-1}(\sigma h)) \frac{\partial u_{n-1}(\sigma h)}{\partial h} \Big] d\alpha d\sigma.
$$

Adding and removing some expressions with careful simplifications gives

$$\left\| \frac{\partial u_{n+1}}{\partial h} - \frac{\partial u_n}{\partial h} \right\|_c \leqslant \gamma \beta^{n-1} + \beta \left\| \frac{\partial u_n}{\partial h} - \frac{\partial u_{n-1}}{\partial h} \right\|_c,$$

where

$$\gamma = (C_0 A_0 D_1 + C_1 A_0 D_0 + \delta C_0 A_1 D_1 + \delta C_1 A_1 D_0 + 2\delta C_1 A_0 D_1 C^*$$
$$+ \delta A_0 D_0 C^* C_2 + \delta C_0 A_0 C^* D_2) R.$$

Hence, by induction, it is true that

$$\left\| \frac{\partial u_{n+1}}{\partial h} - \frac{\partial u_n}{\partial h} \right\|_c \leqslant n\gamma \beta^{n-1} + \beta^n C^*, \quad n = 1, 2, \cdots,$$

which confirms the uniform convergence of $\sum_{n=0}^{\infty} \left(\frac{\partial u_{n+1}}{\partial h}(\tau h) - \frac{\partial u_n}{\partial h}(\tau h) \right)$.

Thus, $\left\{ \frac{\partial u_n}{\partial h}(\tau h) \right\}_{n=0}^{\infty}$ is uniformly convergent.

Likewise, the uniform convergence of other function series $\left\{ \frac{\partial^k u_n}{\partial h^k}(\tau h) \right\}_{n=0}^{\infty}$ for $k \geqslant 2$ can be shown as well. Therefore, $u(\tau h)$ is smoothly dependent on h. $\square$

4.5 Algebraic Order

We consider the algebraic order of the FFEP integrator in this section. For this purpose, we begin with the regularity of the integrators. Following [41], if an h-dependent function $w(\tau h)$ can be expanded as

$$w(\tau h) = \sum_{n=0}^{r-1} w^{[n]}(\tau h) h^n + \mathcal{O}(h^r),$$

then $w(\tau h)$ is termed regular, where $w^{[n]}(\tau h) = \frac{1}{n!} \frac{\partial^n w(\tau h)}{\partial h^n}\Big|_{h=0}$ is a vector-valued function with polynomial entries of degrees $\leqslant n$.

Lemma 4.2 *The FFEP integrator* (4.6) *gives a regular h-dependent function $u(\tau h)$.*

Proof It has been proved in Theorem 4.2 that $u(\tau h)$ is smoothly dependent on h. We then can expand $u(\tau h)$ with respect to h at zero as follows:

$$u(\tau h) = \sum_{m=0}^{r-1} u^{[m]}(\tau h) h^m + \mathcal{O}(h^r).$$

Let $\Delta = u(\tau h) - y_0$ and it is clear that $\Delta = \mathcal{O}(h)$. Expanding $\nabla H(u(\tau h))$ at $h = 0$ and inserting the above equalities into (4.11) leads to

$$\sum_{m=0}^{r-1} u^{[m]}(\tau h) h^m$$

$$= y_0 + h \int_0^1 \int_0^\tau P_{\alpha,\sigma} B(u(\alpha h)) d\alpha \sum_{n=0}^{r-1} \frac{1}{n!} \nabla H^{(n)}(y_0) (\underbrace{\Delta, \cdots, \Delta}_{n-fold}) d\sigma + \mathcal{O}(h^r). \quad (4.14)$$

In what follows, we prove the following result by induction

$$u^{[m]}(\tau h) \in P_m^d = \underbrace{P_m([0, 1]) \times \cdots \times P_m([0, 1])}_{d-fold} \quad \text{for} \quad m = 0, 1, \cdots, r - 1,$$

where $P_m([0, 1])$ consists of polynomials with degrees $\leqslant m$ on $[0, 1]$.

First, it is clear that $u^{[0]}(\tau h) = y_0 \in P_0^d$. Assume that $u^{[n]}(\tau h) \in P_n^d$ for $n = 0, 1, \cdots, m$. Compare the coefficients of h^{m+1} on both sides of (4.14) and then we have

$$u^{[m+1]}(\tau h) = \sum_{k+n=m} \int_0^1 \int_0^\tau \left[P_{\alpha,\sigma} B(u(\alpha h)) \right]^{[k]} d\alpha h_n(\sigma h) d\sigma, \quad h_n(\sigma h) \in P_n^d.$$

Because $P_{\alpha,\sigma}$ is regular (see [41]) and $u^{[n]}(\tau h) \in P_n^d$, it can be verified that $\left[P_{\alpha,\sigma} B(u(\alpha h)) \right]^{[k]} \in P_k^{d \times d}$. Hence, with the condition $k + n = m$, we have

$$\sum_{k+n=m} \int_0^1 \int_0^\tau \left[P_{\alpha,\sigma} B(u(\alpha h)) \right]^{[k]} d\alpha h_n(\sigma h) d\sigma \in P_{m+1}^d.$$

Thus, it is true that

$$u^{[m+1]}(\tau h) \in P_{m+1}^d.$$

$\square$

Let us now quote a result which is needed in the analysis of algebraic order.

Lemma 4.3 (See [41]) *Given a regular function w and an h-independent sufficiently smooth function g, the composition (if it exists) is regular. Moreover, one has*

$$\mathscr{P}_h g(w(\tau h)) - g(w(\tau h)) = \mathcal{O}(h^r).$$

Before presenting the algebraic order of the integrators, we recall the following elementary theory of ordinary differential equations. Denoting by $y(\cdot, \tilde{t}, \tilde{y})$ the solution of $y'(t) = B(y(t))\nabla H(y(t))$ satisfying the initial condition $y(\tilde{t}, \tilde{t}, \tilde{y}) = \tilde{y}$ [1] for any given $\tilde{t} \in [0, h]$ and setting

$$\Phi(s, \tilde{t}, \tilde{y}) = \frac{\partial y(s, \tilde{t}, \tilde{y})}{\partial \tilde{y}},$$

one has the standard result

$$\frac{\partial y(s, \tilde{t}, \tilde{y})}{\partial \tilde{t}} = -\Phi(s, \tilde{t}, \tilde{y})B(\tilde{y})\nabla H(\tilde{y}).$$

Theorem 4.3 *The FFEP integrator* (4.6) *is of order 2r, which implies*

$$u(h) - y(t_0 + h) = \mathcal{O}(h^{2r+1}).$$

Moreover, we have

$$u(\tau h) - y(t_0 + \tau h) = \mathcal{O}(h^{r+1}), \quad 0 < \tau < 1.$$

Proof On the basis of the previous preliminaries, we obtain

$u(h) - y(t_0 + h)$

$= y(t_0 + h, t_0 + h, u(h)) - y(t_0 + h, t_0, y_0)$

$= \displaystyle\int_0^1 \frac{d}{d\alpha} y(t_0 + h, t_0 + \alpha h, u(\alpha h))d\alpha$

$= \displaystyle\int_0^1 \left(h\frac{\partial y}{\partial \tilde{t}}(t_0 + h, t_0 + \alpha h, u(\alpha h)) + \frac{\partial y}{\partial \tilde{y}}(t_0 + h, t_0 + \alpha h, u(\alpha h))hu'(\alpha h) \right)d\alpha$

$= \displaystyle\int_0^1 \left(-h\frac{\partial y}{\partial \tilde{y}}(t_0 + h, t_0 + \alpha h, u(\alpha h))B(u(\alpha h))\nabla H(u(\alpha h)) \right.$

$\left. + \frac{\partial y}{\partial \tilde{y}}(t_0 + h, t_0 + \alpha h, u(\alpha h))hB(u(\alpha h))\mathscr{P}_h\nabla H(u(\alpha h)) \right)d\alpha$

$= -h\displaystyle\int_0^1 \Phi^1(\alpha)B(u(\alpha h))\left(\nabla H(u(\alpha h)) - \mathscr{P}_h\nabla H(u(\alpha h))\right)d\alpha,$

---

[0] 1 Clearly, since the problem is autonomous, then $y(t, \tilde{t}, \tilde{y}) = y(t - \tilde{t}, 0, \tilde{y})$.

where

$$\Phi^1(\alpha) = \frac{\partial y}{\partial \tilde{y}}(t_0 + h, t_0 + \alpha h, u(\alpha h)).$$

It follows from Lemmas 4.2 and 4.3 that

$$\mathscr{P}_h \nabla H(u(\tau h)) - \nabla H(u(\tau h)) = \mathcal{O}(h^r).$$

Partition the matrix-valued function $\Phi^1(\alpha)$ as $\Phi^1(\alpha) = (\Phi_1^1(\alpha), \cdots, \Phi_d^1(\alpha))^\mathsf{T}$ and then it follows from Lemma 4.2 that

$$\Phi_i^1(\alpha) = \mathscr{P}_h \Phi_i^1(\alpha) + \mathcal{O}(h^r), \quad i = 1, 2, \cdots, d.$$

As $\mathscr{P}_h \Phi_i^1(\alpha) \in Y_h$, we obtain

$$\int_0^1 (\mathscr{P}_h \Phi_i^1(\alpha))^\mathsf{T} \nabla H(u(\alpha h)) d\alpha = \int_0^1 (\mathscr{P}_h \Phi_i^1(\alpha))^\mathsf{T} \mathscr{P}_h \nabla H(u(\alpha h)) d\alpha, \quad i = 1, 2, \cdots, d.$$

Hence, we have

$$u(h) - y(t_0 + h)$$

$$= -h \int_0^1 \left(\begin{pmatrix} (\mathscr{P}_h \Phi_1^1(\alpha))^\mathsf{T} \\ \vdots \\ (\mathscr{P}_h \Phi_d^1(\alpha))^\mathsf{T} \end{pmatrix} + \mathcal{O}(h^r) \right) (\nabla H(u(\alpha h)) - \mathscr{P}_h \nabla H(u(\alpha h))) \, d\alpha$$

$$= -h \int_0^1 \begin{pmatrix} (\mathscr{P}_h \Phi_1^1(\alpha))^\mathsf{T} (\nabla H(u(\alpha h)) - \mathscr{P}_h \nabla H(u(\alpha h))) \\ \vdots \\ (\mathscr{P}_h \Phi_d^1(\alpha))^\mathsf{T} (\nabla H(u(\alpha h)) - \mathscr{P}_h \nabla H(u(\alpha h))) \end{pmatrix} d\alpha$$

$$\quad - h \int_0^1 \mathcal{O}(h^r) \times \mathcal{O}(h^r) d\alpha$$

$$= 0 + \mathcal{O}(h^{2r+1}) = \mathcal{O}(h^{2r+1}).$$

Similarly, we deduce that

$$u(\tau h) - y(t_0 + \tau h) = y(t_0 + \tau h, t_0 + \tau h, u(\tau h)) - y(t_0 + \tau h, t_0, y_0)$$

$$= -h \int_0^\tau \Phi^\tau(\alpha) B(u(\alpha h)) (\nabla H(u(\alpha h)) - \mathscr{P}_h \nabla H(u(\alpha h))) d\alpha$$

$$= -h \int_0^\tau \Phi^\tau(\alpha) B(u(\alpha h)) \mathcal{O}(h^r) d\alpha = \mathcal{O}(h^{r+1}).$$

$$\square$$

4.6 Practical FFEP Integrators

In what follows, we consider two illustrative examples of FFEP integrators.

Example 1 We choose

$$\Phi_k(\tau h) = (\tau h)^k, \quad k = 0, 1, \cdots, r - 1,$$

for the function spaces X_h and Y_h, and then we have

$$\hat{l}_i(\tau) = \prod_{j=1, j \neq i}^{r} \frac{\tau - \hat{d}_j}{\hat{d}_i - \hat{d}_j},$$

for $i = 1, 2, \cdots, r$. The Gram–Schmidt process leads to the standard orthonormal basis of Y_h as follows:

$$\hat{p}_j(\tau h) = (-1)^j \sqrt{2j + 1} \sum_{k=0}^{j} \binom{j}{k} \binom{j + k}{k} (-\tau)^k, \quad t \in [0, 1],$$

for $j = 0, 1, \cdots, r - 1$, which are the shifted Legendre polynomials on $[0, 1]$. Consequently, $P_{\tau, \sigma}$ can be determined by

$$P_{\tau, \sigma} = \sum_{i=0}^{r-1} \hat{p}_i(\tau h) \hat{p}_i(\sigma h).$$

Here it is important to note that all the above functions are independent of h. In this situation, the FFEP integrator (4.6) is identical to the EP method given by Cohen and Hairer [21] and Brugnano et al. [27].

If we choose $r = 1$ and $\hat{d}_1 = 1/2$, we obtain

$$\hat{l}_1(\tau) = 1, \quad P_{\tau, \sigma} = 1.$$

Accordingly, the integrator (4.9) yields

$$\begin{cases} y_\tau = y_0 + h\tau B(y_{\frac{1}{2}}) \int_0^1 \nabla H(y_\sigma) d\sigma, \\ \\ y_1 = y_0 + h B(y_{\frac{1}{2}}) \int_0^1 \nabla H(y_\sigma) d\sigma, \end{cases} \tag{4.15}$$

which gives

$$y_\tau = y_0 + h\tau B(y_{\frac{1}{2}}) \int_0^1 \nabla H(y_\sigma) d\sigma = y_0 + \tau(y_1 - y_0).$$

Let $\tau = 1/2$ for the first equality of (4.15), and then we have

$$y_{\frac{1}{2}} = y_0 + \frac{1}{2}hB(y_{\frac{1}{2}})\int_0^1 \nabla H(y_\sigma)d\sigma = y_0 + \frac{1}{2}(y_1 - y_0) = \frac{1}{2}(y_0 + y_1).$$

This leads to

$$y_1 = y_0 + hB\left(\frac{1}{2}(y_0 + y_1)\right)\int_0^1 \nabla H(y_0 + \sigma(y_1 - y_0))d\sigma.$$

This second-order integrator has been given by Cohen and Hairer in [1].

Example 2 Let us consider another choice for Y_h by

$$Y_h = \text{span}\{\cos(\omega\tau h)\},$$

and this gives

$$\hat{l}_1(\tau) = \frac{\cos(\tau v)}{\cos(\hat{d}_1 v)}, \quad P_{\tau,\sigma} = \frac{4v\cos(\sigma v)\cos(\tau v)}{2v + \sin(2v)},$$

where $v = \omega h$. With this choice, the integrator (4.9) becomes

$$\begin{cases} y_\tau = y_0 + h\int_0^\tau \hat{l}_1(\alpha)d\alpha\, B(y_{\hat{d}_1})\int_0^1 P_{\hat{d}_1,\sigma}\nabla H(y_\sigma)d\sigma, \\ y_1 = y_0 + h\int_0^1 \hat{l}_1(\alpha)d\alpha\, B(y_{\hat{d}_1})\int_0^1 P_{\hat{d}_1,\sigma}\nabla H(y_\sigma)d\sigma. \end{cases} \tag{4.16}$$

Let $\tau = \hat{d}_1 = \dfrac{1}{2}$ in (4.16). We then obtain

$$y_{1/2} = y_0 + h\frac{\tan(v/2)}{v}B(y_{1/2})\int_0^1 P_{1/2,\sigma}\nabla H(y_\sigma)d\sigma,$$

$$y_1 = y_0 + h\frac{2\sin(v/2)}{v}B(y_{1/2})\int_0^1 P_{1/2,\sigma}\nabla H(y_\sigma)d\sigma.$$

It follows from these two equalities that

$$y_{1/2} = y_0 + \frac{\tan(v/2)}{v}\frac{v(y_1 - y_0)}{2\sin(v/2)} = y_0 + \frac{1}{2\cos(v/2)}(y_1 - y_0),$$

$$y_\tau = y_0 + \frac{\sin(v\tau)}{v\cos(v/2)}\frac{v(y_1 - y_0)}{2\sin(v/2)} = y_0 + \frac{\sin(v\tau)}{\sin(v)}(y_1 - y_0).$$

This then results in

$$y_1 = y_0 + h \frac{2 \sin(v/2)}{v} B\left(y_0 + \frac{y_1 - y_0}{2 \cos(v/2)}\right) \int_0^1 P_{1/2,\sigma} \nabla H\left(y_0 + \frac{\sin(v\sigma)}{\sin(v)}(y_1 - y_0)\right) d\sigma.$$

(4.17)

Clearly, this integrator reduces to (4.3) when $v = 0$. We denote the second-order scheme by FFEP1.

Example 3 We now consider

$$Y_h = \text{span}\left\{\cos(\omega \tau h), \sin(\omega \tau h)\right\}.$$

This choice of Y_h leads to

$$\hat{l}_1(\tau) = \frac{\sin((\tau - \hat{d}_2)v)}{\sin((\hat{d}_1 - \hat{d}_2)v)}, \quad \hat{l}_2(\tau) = \frac{\sin((\tau - \hat{d}_1)v)}{\sin((\hat{d}_2 - \hat{d}_1)v)}$$

and

$$P_{\tau,\sigma} = \frac{2v(2v \cos((\sigma - \tau)v) + \sin((-2 + \sigma + \tau)v) - \sin((\sigma + \tau)v))}{-1 + 2v^2 + \cos(2v)}.$$

We here choose $\tau = \hat{d}_1$ and $\hat{d}_2$ for the integrator (4.9). We then obtain

$$\begin{cases} y_{\hat{d}_1} = y_0 + h \int_0^1 \left(\bar{A}_{11}(\sigma)B(y_{\hat{d}_1}) + \bar{A}_{12}(\sigma)B(y_{\hat{d}_2})\right)\nabla H(y_\sigma)d\sigma, \\[2mm] y_{\hat{d}_2} = y_0 + h \int_0^1 \left(\bar{A}_{21}(\sigma)B(y_{\hat{d}_1}) + \bar{A}_{22}(\sigma)B(y_{\hat{d}_2})\right)\nabla H(y_\sigma)d\sigma, \\[2mm] y_1 = y_0 + h \int_0^1 \left(\bar{b}_1(\sigma)B(y_{\hat{d}_1}) + \bar{b}_2(\sigma)B(y_{\hat{d}_2})\right)\nabla H(y_\sigma)d\sigma, \end{cases}$$

(4.18)

where

$$\bar{A}_{ij}(\sigma) = P_{\hat{d}_j,\sigma} \int_0^{\hat{d}_i} \hat{l}_j(\alpha)d\alpha, \quad \bar{b}_j(\sigma) = P_{\hat{d}_j,\sigma} \int_0^1 \hat{l}_j(\alpha)d\alpha \quad i, j = 1, 2.$$

We denote this fourth-order integrator (4.18) by FFEP2. It is worth noting that when $v = 0$ and $\hat{d}_{1,2} = 1/2 \mp \sqrt{3}/6$, this scheme becomes

$$
\begin{cases}
y_{\hat{d}_1} = y_0 + h \displaystyle\int_0^1 \left(\frac{1}{2} l_1(\sigma) B(y_{\hat{d}_1}) + \left(\frac{1}{2} - \frac{\sqrt{3}}{3} \right) l_2(\sigma) B(y_{\hat{d}_2}) \right) \nabla H(y_\sigma) \mathrm{d}\sigma, \\[3mm]
y_{\hat{d}_2} = y_0 + h \displaystyle\int_0^1 \left(\left(\frac{1}{2} + \frac{\sqrt{3}}{3} \right) l_1(\sigma) B(y_{\hat{d}_1}) + \frac{1}{2} l_2(\sigma) B(y_{\hat{d}_2}) \right) \nabla H(y_\sigma) \mathrm{d}\sigma, \\[3mm]
y_1 = y_0 + h \displaystyle\int_0^1 \left(l_1(\sigma) B(y_{\hat{d}_1}) + l_2(\sigma) B(y_{\hat{d}_2}) \right) \nabla H(y_\sigma) \mathrm{d}\sigma,
\end{cases}
$$

where

$$
l_1(\sigma) = \frac{\sigma - \hat{d}_2}{\hat{d}_1 - \hat{d}_2}
$$

and

$$
l_2(\sigma) = \frac{\sigma - \hat{d}_1}{\hat{d}_2 - \hat{d}_1}.
$$

This fourth-order integrator has been proposed by Cohen and Hairer in [1].

Remark 4.5 We remark that different choices of Y_h and X_h will derive different practical integrators. We do not pursue this point further for brevity.

4.7 Numerical Experiments

To illustrate the efficiency and robustness of the integrators derived in this chapter, we apply our integrators FFEP1 and FFEP2 to the Euler equation. For comparison, we consider the second-order and fourth-order EP collocation methods given in [1] and denote them by EPCM1 and EPCM2, respectively. We also choose the following second-order trigonometrically-fitted EP method (see [28])

$$
y_1 = y_0 + h \frac{2 \sinh(v/2)}{v \cosh(v/2)} B((1/2)(y_0 + y_1)) \int_0^1 \nabla H(y_0 + \sigma(y_1 - y_0)) \mathrm{d}\sigma,
$$

$$
\tag{4.19}
$$

which is denoted by TFEP1. Since these five methods are all implicit, we use fixed-point iteration. We set 10^{-16} as the error tolerance and 10 as the maximum number of each iteration.

We will use as a test problem the following Euler equations (see [28, 33]) given by

$$\dot{y} = \left((\alpha - \beta)y_2 y_3, (1 - \alpha)y_3 y_1, (\beta - 1)y_1 y_2\right)^\mathsf{T}, \quad t \in [0, T],$$

which describes the motion of a rigid body under no forces. This system can be written as a Poisson system

$$\dot{y} = \begin{pmatrix} 0 & \alpha y_3 & -\beta y_2 \\ -\alpha y_3 & 0 & y_1 \\ \beta y_2 & -y_1 & 0 \end{pmatrix} \nabla H(y)$$

with

$$H(y) = \frac{y_1^2 + y_2^2 + y_3^2}{2}.$$

Following [28, 33], the initial value is chosen as $y(0) = (0, 1, 1)$, and the parameters are given by

$$\alpha = 1 + \frac{1}{\sqrt{1.51}}, \quad \beta = 1 - \frac{0.51}{\sqrt{1.51}}.$$

The exact solution is given by

$$y(t) = (\sqrt{1.51}sn(t, 0.51), cn(t, 0.51), dn(t, 0.51))^\mathsf{T},$$

where sn, cn, dn are the Jacobi elliptic functions. This solution is periodic with the period

$$T_p = 7.450563209330954,$$

and thence we consider choosing $\omega = 2\pi/T_p$ for the methods FFEP1 and TFEP1. We integrate this problem with the stepsizes $h = 0.5$ and $h = 0.2$ on the interval $[0, 10000]$. The energy conservation for different methods is shown in Fig. 4.1. We then solve the problem on the interval $[0, T]$ with different stepsizes $h = 1/2^i$ for $i = 4, 5, 6, 7$. The global errors are presented in Fig. 4.2 for $T = 10, 100$.

We also consider another case. As mentioned in [28], when $\beta \approx 1$, it is expected that $\dot{y}_3 \approx 0$ and thus $y_3(t) \approx 1$. Therefore, the variables y_1 and y_2 seem to behave like the harmonic oscillator with the period $T_p = 2\pi/(\alpha - 1)$. We choose $\alpha = 2$ and $\beta = 1.01$. We integrate this problem with $h = 0.5$ and $h = 0.2$ on the interval $[0, 10000]$. The energy conservation for different methods is shown in Fig. 4.3.

Then the problem is solved on the interval $[0, T]$ with $h = 1/2^i$ for $i = 4, 5, 6, 7$, and see Fig. 4.4 for the global errors of $T = 10, 100$.

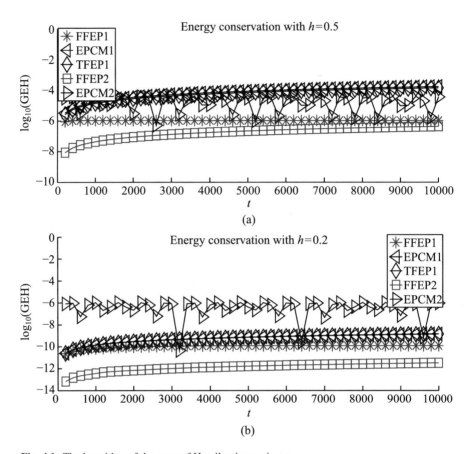

Fig. 4.1 The logarithm of the error of Hamiltonian against t

It is very clear from the numerical results that our FFEP methods when applied to the underlying Euler equations show remarkable numerical behaviour compared with the existing EP methods in the literature.

4.8 Conclusions

The Poisson system is an important model in applications. It is well known that the energy of Poisson system is preserved along its exact solution. This chapter paid attention to the analysis of preserving the energy exactly in the numerical treatment, so that we can obtain $H(y_1) = H(y_0)$ after one step of the method starting from y_0 with a time stepsize h. In this chapter, we presented functionally-fitted energy-preserving integrators for Poisson systems by using a functionally-fitted strategy. It has been shown that these integrators preserve exactly the energy of

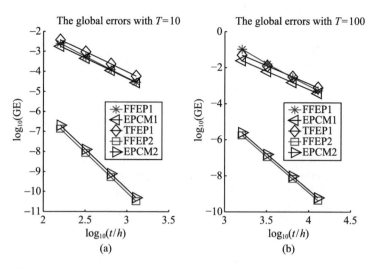

Fig. 4.2 The logarithm of the global error against the logarithm of t/h

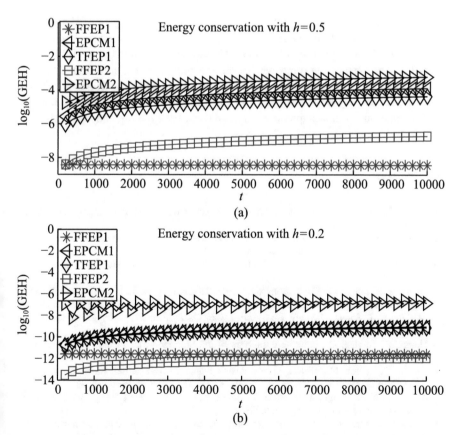

Fig. 4.3 The logarithm of the error of Hamiltonian against t

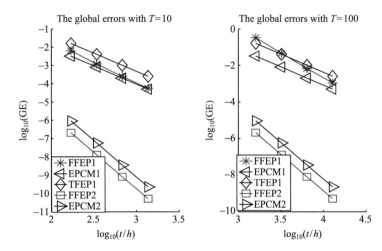

Fig. 4.4 The logarithm of the global error against the logarithm of t/h

Poisson systems and can be of arbitrarily high order by choosing a sufficiently large integer r for the function spaces Y_h and X_h. These integrators contain the energy-preserving schemes given by Cohen and Hairer [1] and Brugnano et al. [27]. The remarkable efficiency and robustness of the integrators were demonstrated through the numerical experiments for the Euler equations. In a similar way, it is possible to develop functionally-fitted energy-diminishing integrators for gradient systems.

The materials in this chapter are based on the work by Wang and Wu [42].

References

1. Cohen, D., Hairer, E.: Linear energy-preserving integrators for Poisson systems. BIT Numer. Math. **51**, 91–101 (2011)
2. Brugnano, L., Iavernaro, F.: Line integral methods which preserve all invariants of conservative problems. J. Comput. Appl. Math. **236**, 3905–3919 (2012)
3. Feng, K., Qin, M.: Symplectic Geometric Algorithms for Hamiltonian Systems. Springer, Berlin (2010)
4. Hairer, E., Lubich, C.: Oscillations over Long Times in Numerical Hamiltonian Systems. Highly Oscillatory Problems. In: Engquist, B., Fokas, A., Hairer, E., et al. (eds.) London Mathematical Society Lecture Note Series 366. Cambridge University Press, Cambridge (2009)
5. Hairer, E., Lubich, C.: Long-term analysis of the Störmer-Verlet method for Hamiltonian systems with a solution-dependent high frequency. Numer. Math. **134**, 119–138 (2016)
6. Hairer, E., Lubich, C., Wanner, G.: Geometric Numerical Integration: Structure-Preserving Algorithms for Ordinary Differential Equations, 2nd edn. Springer, Berlin (2006)
7. Mei, L., Wu, X.: Symplectic exponential Runge–Kutta methods for solving nonlinear Hamiltonian systems. J. Comput. Phys. **338**, 567–584 (2017)
8. Wang, B., Iserles, A., Wu, X.: Arbitrary-order trigonometric Fourier collocation methods for multi-frequency oscillatory systems. Found. Comput. Math. **16**, 151–181 (2016)

9. Wang, B., Wu, X.: Improved Filon type asymptotic methods for highly oscillatory differential equations with multiple time scales. J. Comput. Phys. **276**, 62–73 (2014)

10. Wang, B., Wu, X., Meng, F.: Trigonometric collocation methods based on Lagrange basis polynomials for multi-frequency oscillatory second-order differential equations. J. Comput. Appl. Math. **313**, 185–201 (2017)

11. Wang, B., Wu, X., Meng, F., et al.: Exponential Fourier collocation methods for solving first-order differential equations. J. Comput. Math. **35**, 711–736 (2017)

12. Wu, X., Liu, K., Shi, W.: Structure-preserving Algorithms for Oscillatory Differential Equations II. Springer, Heidelberg (2015)

13. Wu, X., You, X., Wang, B.: Structure-preserving algorithms for oscillatory differential equations. Springer, Berlin (2013)

14. Celledoni, E., Mclachlan, R.I., Owren, B., et al.: Energy-preserving integrators and the structure of B-series. Found. Comput. Math. **10**, 673–693 (2010)

15. Celledoni, E., Owren, B., Sun, Y.: The minimal stage, energy preserving Runge–Kutta method for polynomial Hamiltonian systems is the averaged vector field method. Math. Comput. **83**, 1689–1700 (2014)

16. Quispel, G.R.W., McLaren, D.I.: A new class of energy-preserving numerical integration methods. J. Phys. A Math. Ther. **41**, 045206 (2008)

17. McLachlan, R.I., Quispel, G.R.W.: Discrete gradient methods have an energy conservation law. Discrete Contin. Dyn. Syst. **34**, 1099–1104 (2014)

18. McLachlan, R.I., Quispel, G.R.W., Robidoux, N.: Geometric integration using discrete gradient. Philos. Trans. R. Soc. Lond. A **357**, 1021–1045 (1999)

19. Brugnano, L., Iavernaro, F., Trigiante, D.: Hamiltonian boundary value methods (energy preserving discrete line integral methods). J. Numer. Anal. Ind. Appl. Math. **5**, 13–17 (2010)

20. Brugnano, L., Iavernaro, F., Trigiante, D.: Energy- and quadratic invariants-preserving integrators based upon Gauss-collocation formulae. SIAM J. Numer. Anal. **50**, 2897–2916 (2012)

21. Hairer, E.: Energy-preserving variant of collocation methods. J. Numer. Anal. Ind. Appl. Math. **5**, 73–84 (2010)

22. Li, Y.W., Wu, X.: Exponential integrators preserving first integrals or Lyapunov functions for conservative or dissipative systems. SIAM J. Sci. Comput. **38**, 1876–1895 (2016)

23. Miyatake, Y.: An energy-preserving exponentially-fitted continuous stage Runge–Kutta method for Hamiltonian systems. BIT Numer. Math. **54**, 777–799 (2014)

24. Wang, B., Wu, X.: A new high precision energy-preserving integrator for system of oscillatory second-order differential equations. Phys. Lett. A **376**, 1185–1190 (2014)

25. Wang, B., Wu, X.: Exponential collocation methods for conservative or dissipative systems. J. Comput. Appl. Math. **360**, 99–116 (2019)

26. Wu, X., Wang, B., Shi, W.: Efficient energy preserving integrators for oscillatory Hamiltonian systems. J. Comput. Phys. **235**, 587–605 (2013)

27. Brugnano, L., Calvo, M., Montijano, J.I., et al.: Energy-preserving methods for Poisson systems. J. Comput. Appl. Math. **236**, 3890–3904 (2012)

28. Miyatake, Y.: A derivation of energy-preserving exponentially-fitted integrators for Poisson systems. Comput. Phys. Commun. **187**, 156–161 (2015)

29. Dahlby, M., Owren, B., Yaguchi, T.: Preserving multiple first integrals by discrete gradients. J. Phys. A Math. Theor. **44**, 1651–1659 (2012)

30. Brugnano, L., Sun, Y.: Multiple invariants conserving Runge–Kutta type methods for Hamiltonian problems. Numer. Algor. **65**, 611–632 (2014)

31. Chartier, P., Faou, E., Murua, A.: An algebraic approach to invariant preserving integrators: the case of quadratic and Hamiltonian invariants. Numer. Math. **103**, 575–590 (2006)

32. Iserles, A., Zanna, A.: Preserving algebraic invariants with Runge–Kutta methods. J. Comput. Appl. Math. **125**, 69–81 (2000)

33. Calvo, M., Franco, J.M., Montijano, J.I., et al.: Sixth-order symmetric and symplectic exponentially fitted Runge–Kutta methods of the Gauss type. J. Comput. Appl. Math. **223**, 387–398 (2009)

34. Calvo, M., Franco, J.M., Montijano, J.I., et al.: On high order symmetric and symplectic trigonometrically fitted Runge–Kutta methods with an even number of stages. BIT Numer. Math. **50**, 3–21 (2010)
35. Calvo, M., Franco, J.M., Montijano, J.I., et al.: Symmetric and symplectic exponentially fitted Runge–Kutta methods of high order. Comput. Phys. Commun. **181**, 2044–2056 (2010)
36. Franco, J.M.: Exponentially fitted symplectic integrators of RKN type for solving oscillatory problems. Comput. Phys. Commun. **177**, 479–492 (2007)
37. Vanden Berghe, G.: Exponentially-fitted Runge–Kutta methods of collocation type: fixed or variable knots? J. Comput. Appl. Math. **159**, 217–239 (2003)
38. van de Vyver, H.: A fourth order symplectic exponentially fitted integrator. Comput. Phys. Commun. **176**, 255–262 (2006)
39. Wang, B., Yang, H., Meng, F.: Sixth order symplectic and symmetric explicit ERKN schemes for solving multi-frequency oscillatory nonlinear Hamiltonian equations. Calcolo **54**, 117–140 (2017)
40. Wu, X., Wang, B., Xia, J.: Explicit symplectic multidimensional exponential fitting modified Runge–Kutta–Nyström methods. BIT Numer. Math. **52**, 773–795 (2012)
41. Li, Y.W., Wu, X.: Functionally fitted energy-preserving methods for solving oscillatory nonlinear Hamiltonian systems. SIAM J. Numer. Anal. **54**, 2036–2059 (2016)
42. Wang, B., Wu, X.: Functionally-fitted energy-preserving integrators for Poisson systems. J. Comput. Phys. **364**, 137–152 (2018)

Chapter 5
Exponential Collocation Methods for Conservative or Dissipative Systems

The main purpose of this chapter is to present exponential collocation methods (ECMs) for solving conservative or dissipative systems. ECMs can be of arbitrarily high order and preserve exactly or approximately first integrals or Lyapunov functions. In particular, the application of ECMs to stiff gradient systems is discussed in detail, and it turns out that ECMs are unconditionally energy-diminishing and strongly damped even for very stiff gradient systems. As a consequence of this discussion, arbitrary-order trigonometric/RKN collocation methods are also presented and analysed for second-order highly oscillatory/general systems. The chapter is accompanied by numerical results that demonstrate the potential value of this research.

5.1 Introduction

In this chapter, we consider systems of ordinary differential equations (ODEs) of the form

$$y'(t) = Q\nabla H(y(t)), \quad y(0) = y_0 \in \mathbb{R}^d, \quad t \in [0, T], \tag{5.1}$$

where Q is an invertible and $d \times d$ real matrix, and $H : \mathbb{R}^d \to \mathbb{R}$ is defined by

$$H(y) = \frac{1}{2}y^{\mathsf{T}}My + V(y). \tag{5.2}$$

Here M is a $d \times d$ symmetric real matrix, and $V : \mathbb{R}^d \to \mathbb{R}$ is a differentiable function.

© The Author(s), under exclusive license to Springer Nature Singapore Pte Ltd. 2021
X. Wu, B. Wang, *Geometric Integrators for Differential Equations with Highly Oscillatory Solutions*, https://doi.org/10.1007/978-981-16-0147-7_5

It is important to note that the system (5.1) exhibits remarkable geometrical/ physical structures, which should be preserved by a numerical method in the spirit of geometric numerical integration. In fact, if the matrix Q is skew symmetric, then (5.1) is a conservative system with the first integral H: i.e.,

$$H(y(t)) \equiv H(y_0) \qquad \text{for any } t \geqslant 0.$$

If the matrix Q is negative semi-definite, then (5.1) is a dissipative system with the Lyapunov function H: i.e.,

$$H(y(t_2)) \leqslant H(y(t_1)) \quad \text{if } t_2 \geqslant t_1.$$

Throughout this chapter, we call H energy for both cases in a broad sense. The objective of this chapter is to design and analyse a class of arbitrary-order exponential energy-preserving collocation methods which can preserve first integrals or Lyapunov functions of the underlying conservative/dissipative system (5.1).

It is convenient to express

$$A = QM, \quad g(y(t)) = Q\nabla V(y(t)).$$

We then rewrite the system (5.1) as

$$y'(t) = Ay(t) + g(y(t)), \quad y(0) = y_0 \in \mathbb{R}^d. \tag{5.3}$$

As is known, the exact solution of (5.1) or (5.3) can be represented by the variation-of-constants formula (the Duhamel Principle)

$$y(t) = e^{tA}y_0 + t\int_0^1 e^{(1-\tau)tA}g(y(\tau t))d\tau. \tag{5.4}$$

The system (5.1) or (5.3) plays a prominent role in a wide range of applications in physics and engineering, including mechanics, astronomy, molecular dynamics, and in problems of wave propagation in classical and quantum physics (see, e.g. [1–4]). Some highly oscillatory problems and semidiscrete PDEs such as semilinear Schrödinger equations fit this pattern. Among typical examples are the multi-frequency highly oscillatory Hamiltonian systems with the Hamiltonian

$$H(p, q) = \frac{1}{2}p^{\mathsf{T}}\bar{M}^{-1}p + \frac{1}{2}q^{\mathsf{T}}\bar{K}q + U(q), \tag{5.5}$$

where $\bar{K}$ is a symmetric positive semi-definite stiffness matrix, $\bar{M}$ is a symmetric positive definite mass matrix, and $U(q)$ is a smooth potential with moderately bounded derivatives.

As an interesting class of numerical methods for (5.3), exponential integrators have been widely investigated and developed in recent decades, and we refer the reader to [5–17] for example. Exponential integrators make good use of the variation-of-constants formula (5.4), and their performance has been evaluated on a range of test problems. A systematic survey of exponential integrators is presented in [2]. However, apart from symplectic exponential integrators (see, e.g. [18]), most existing publications dealing with exponential integrators focus on the construction and analysis of the schemes and pay little attention to energy-preserving exponential integrators which can preserve the first integrals/Lyapunov functions. Energy-preserving exponential integrators, especially higher-order schemes have not been well researched yet in the literature.

On the other hand, various effective energy-preserving methods have been proposed and researched for (5.3) in the special case of $A = 0$, such as the average vector field (AVF) method [19–21], discrete gradient (DG) methods [22–24], Hamiltonian Boundary Value Methods (HBVMs) [25–28], the Runge–Kutta-type energy-preserving collocation (RKEPC) methods [29, 30], time finite elements (TFE) methods [31–35], and energy-preserving exponentially-fitted (EPEF) methods [36, 37]. Some numerical methods preserving Lyapunov functions have also been studied for (5.3) with $A = 0$ (see, e.g. [38–40]). It is noted that all these methods are constructed and studied for the special case $A = 0$ and thus they do not take advantage of the structure brought by the linear term Ay in the system (5.3). These methods could be applied to (5.3) with $A \neq 0$ if the right-hand side of (5.3) is considered as a whole (function), i.e., $y' = f(y) \equiv Ay + g(y)$.

Recently, in order to take advantage of the structure of the underlying system and preserve its energy simultaneously, a novel energy-preserving method has been studied in [41, 42] for second-order ODEs and a new energy-preserving exponential scheme for the conservative or dissipative system has been researched in [43]. However, those two kinds of methods are both based on the AVF methods and thence they are only of order two, in general. This may not be sufficient to deal with some practical problems for high-precision numerical simulations in sciences and engineering.

On noting the above observation, we are concerned in this chapter with deriving and analysing structure-preserving exponential collocation methods. To this end we make good use of the variation-of-constants formula and the structure introduced by the underlying system. These exponential integrators can in such a way exactly or approximately preserve the first integral or the Lyapunov function of (5.1). Very recently, there have been some publications on the numerical solution of Hamiltonian PDEs, and the analysis is related to the approach of this chapter (see, e.g. [44–48]).

5.2   Formulation of Methods

Following [34], we begin by defining the finite-dimensional function spaces Y_h as follows:

$$Y_h = \text{span}\,\{\tilde{\varphi}_0(\tau), \cdots, \tilde{\varphi}_{r-1}(\tau)\}$$

$$= \left\{\tilde{w} : \tilde{w}(\tau) = \sum_{i=0}^{r-1} \tilde{\varphi}_i(\tau) W_i, \ \tau \in [0, 1], \ W_i \in \mathbb{R}^d\right\}, \tag{5.6}$$

where $\{\tilde{\varphi}_i\}_{i=0}^{r-1}$ are supposed to be linearly independent on $I = [0, T]$ and sufficiently smooth. We use $\tilde{\varphi}_i(\tau)$ to denote $\varphi_i(\tau h)$ for all the functions φ_i throughout this chapter and $h > 0$ is the stepsize. With this definition, we consider another finite-dimensional function space X_h such that $\tilde{w}' \in Y_h$ for any $\tilde{w} \in X_h$.

We introduce the idea of the formulation of methods. Find $\tilde{u}(\tau)$ with $\tilde{u}(0) = y_0$, satisfying

$$\tilde{u}'(\tau) = A\tilde{u}(\tau) + \mathscr{P}_h g(\tilde{u}(\tau)), \tag{5.7}$$

where the projection operation $\mathscr{P}_h$ is given by (see [34])

$$\langle \tilde{v}(\tau), \mathscr{P}_h \tilde{w}(\tau)\rangle = \langle \tilde{v}(\tau), \tilde{w}(\tau)\rangle \quad \text{for any } \tilde{v}(\tau) \in Y_h \tag{5.8}$$

and the inner product $\langle \cdot, \cdot \rangle$ is defined by (see [34])

$$\langle w_1, w_2 \rangle = \langle w_1(\tau), w_2(\tau)\rangle_\tau = \int_0^1 w_1(\tau) \cdot w_2(\tau) \mathrm{d}\tau.$$

With regard to the projection operation $\mathscr{P}_h$, we have the following property (see [34]) which is needed in this chapter.

Lemma 5.1   *The projection $\mathscr{P}_h \tilde{w}$ can be explicitly expressed as*

$$\mathscr{P}_h \tilde{w}(\tau) = \langle P_{\tau,\sigma}, \tilde{w}(\sigma)\rangle_\sigma,$$

where

$$P_{\tau,\sigma} = (\tilde{\varphi}_0(\tau), \cdots, \tilde{\varphi}_{r-1}(\tau))\Theta^{-1}(\tilde{\varphi}_0(\sigma), \cdots, \tilde{\varphi}_{r-1}(\sigma))^\mathsf{T},$$

$$\Theta = (\langle \tilde{\varphi}_i(\tau), \tilde{\varphi}_j(\tau)\rangle)_{0 \leqslant i,j \leqslant r-1}. \tag{5.9}$$

When h tends to 0, the limit of $P_{\tau,\sigma}$ exists. If $P_{\tau,\sigma}$ is computed by a standard orthonormal basis $\left\{\tilde{\psi}_0, \cdots, \tilde{\psi}_{r-1}\right\}$ of Y_h under the inner product $\langle \cdot, \cdot \rangle$, then Θ

is an identity matrix and $P_{\tau,\sigma}$ has a simpler expression:

$$P_{\tau,\sigma} = \sum_{i=0}^{r-1} \tilde{\psi}_i(\tau)\tilde{\psi}_i(\sigma).$$ (5.10)

As $\tilde{u}(\tau) = u(\tau h)$, (5.7) can be expressed in

$$u'(\tau h) = Au(\tau h) + \langle P_{\tau,\sigma}, g(u(\sigma h))\rangle_\sigma.$$

Applying the variation-of-constants formula (5.4) to (5.7), we obtain

$$\tilde{u}(\tau) = u(\tau h) = e^{\tau h A} y_0 + \tau h \int_0^1 e^{(1-\xi)\tau h A} \langle P_{\xi\tau,\sigma}, g(u(\sigma h))\rangle_\sigma \, d\xi$$

$$= e^{\tau h A} y_0 + \tau h \int_0^1 e^{(1-\xi)\tau h A} \langle P_{\xi\tau,\sigma}, g(\tilde{u}(\sigma))\rangle_\sigma \, d\xi$$ (5.11)

Inserting (5.10) into (5.11) yields

$$\tilde{u}(\tau) = e^{\tau h A} y_0 + \tau h \int_0^1 e^{(1-\xi)\tau h A} \int_0^1 \sum_{i=0}^{r-1} \tilde{\psi}_i(\xi\tau)\tilde{\psi}_i(\sigma) g(\tilde{u}(\sigma)) d\sigma \, d\xi$$

$$= e^{\tau h A} y_0 + \tau h \int_0^1 \sum_{i=0}^{r-1} \int_0^1 e^{(1-\xi)\tau h A} \tilde{\psi}_i(\xi\tau) d\xi \, \tilde{\psi}_i(\sigma) g(\tilde{u}(\sigma)) d\sigma.$$

We are now in a position to define exponential collocation methods.

Definition 5.1   An exponential collocation method for solving the system (5.1) or (5.3) is defined as follows:

$$\tilde{u}(\tau) = e^{\tau h A} y_0 + \tau h \int_0^1 \bar{A}_{\tau,\sigma}(A) g(\tilde{u}(\sigma)) d\sigma, \qquad y_1 = \tilde{u}(1),$$ (5.12)

where h is a stepsize,

$$\bar{A}_{\tau,\sigma}(A) = \int_0^1 e^{(1-\xi)\tau h A} P_{\xi\tau,\sigma} \, d\xi = \sum_{i=0}^{r-1} \int_0^1 e^{(1-\xi)\tau h A} \tilde{\psi}_i(\xi\tau) d\xi \, \tilde{\psi}_i(\sigma),$$ (5.13)

and $\left\{\tilde{\psi}_0, \cdots, \tilde{\psi}_{r-1}\right\}$ is a standard orthonormal basis of Y_h. We denote the method as ECr.

Remark 5.1   Once the stepsize h is chosen, the method (5.12) approximates the solution of (5.1) in the time interval I_0. Obviously, the obtained result can be considered as the initial condition for a new initial value problem and it can be approximated in the next time interval I_1. In general, the method can be extended to the approximation of the solution in the interval $[0, T]$.

Remark 5.2   It can be observed that the ECr method (5.12) exactly integrates the homogeneous linear system $y' = Ay$. The scheme (5.12) can be classified into the category of exponential integrators (which can be thought of as continuous-stage exponential integrators). This is an interesting and important class of numerical methods for first-order ODEs (see, e.g. [2, 13, 14, 49, 50]). In [43], the authors researched a new energy-preserving exponential scheme for the conservative or dissipative system. Here we note that its order is only two since this scheme combines the ideas of DG and AVF methods. We have proposed a kind of arbitrary-order exponential Fourier collocation methods in [16]. However, those methods cannot preserve energy exactly. Fortunately, we will show that the ECr method (5.12) can be of arbitrarily high order and can preserve energy exactly or approximately, and which is different from the existing exponential integrators in the literature. This feature is significant and makes the methods more efficient and robust.

Remark 5.3   In the case of $M = 0$ and $Q = \begin{pmatrix} O_{d_1 \times d_1} & -I_{d_1 \times d_1} \\ I_{d_1 \times d_1} & O_{d_1 \times d_1} \end{pmatrix}$, (5.1) is a Hamiltonian system. In this special case, if we choose X_h and Y_h as

$$Y_h = \mathrm{span}\left\{\tilde{\varphi}_0(\tau), \cdots, \tilde{\varphi}_{r-1}(\tau)\right\},$$

$$X_h = \mathrm{span}\left\{1, \int_0^\tau \tilde{\varphi}_0(s)\mathrm{d}s, \cdots, \int_0^\tau \tilde{\varphi}_{r-1}(s)\mathrm{d}s\right\},$$

then the ECr method (5.12) becomes the following energy-preserving Runge–Kutta type collocation methods

$$\tilde{u}(\tau) = y_0 + \tau h \int_0^1 \int_0^1 P_{\xi\tau,\sigma}\mathrm{d}\xi g(\tilde{u}(\sigma))\mathrm{d}\sigma, \qquad y_1 = \tilde{u}(1),$$

which yields the functionally-fitted TFE method derived in [34]. Moreover, under the above choices of M and Q, if Y_h is particularly generated by the shifted Legendre polynomials on $[0, 1]$, then the ECr method (5.12) reduces to the RKEPC method of order $2r$ given in [30] or HBVM(∞, r) presented in [26]. Consequently, the ECr method (5.12) can be regarded as a generalisation of these existing methods in the literature.

5.3 Methods for Second-Order ODEs with Highly Oscillatory Solutions

We first consider the following systems of second-order ODEs with highly oscillatory solutions

$$q''(t) - Nq'(t) + \Upsilon q(t) = -\nabla U(q(t)), \qquad q(0) = q_0, \quad q'(0) = q'_0, \qquad t \in [0, T],$$
(5.14)

where N is a symmetric negative semi-definite matrix, Υ is a symmetric positive semi-definite matrix, and $U : \mathbb{R}^d \to \mathbb{R}$ is a differentiable function. By introducing $p = q'$, (5.14) can be transformed into

$$\begin{pmatrix} q \\ p \end{pmatrix}' = \begin{pmatrix} 0 & I \\ -I & N \end{pmatrix} \nabla H(p, q)$$
(5.15)

with

$$H(p, q) = \frac{1}{2} p^{\mathsf{T}} p + \frac{1}{2} q^{\mathsf{T}} \Upsilon q + U(q).$$
(5.16)

This is exactly the same as the problem (5.1). Since N is symmetric negative semi-definite, (5.15) is a dissipative system with the Lyapunov function (5.16). In the particular case $N = 0$, (5.15) becomes a conservative Hamiltonian system with the first integral (5.16). This is an important highly oscillatory system which has been investigated by many researchers (see, e.g. [4, 51–58]).

Applying the ECr method (5.12) to (5.15) yields the trigonometric collocation method for second-order highly oscillatory systems. In particular, for Hamiltonian systems

$$q''(t) + \Upsilon q(t) = -\nabla U(q(t)),$$
(5.17)

the case where $N = 0$ in (5.14), the ECr method (5.12) leads to the following form.

Definition 5.2 The trigonometric collocation (denoted by TCr) method for (5.17) is defined as:

$$\begin{cases} \tilde{q}(\tau) = \phi_0(K) q_0 + \tau h \phi_1(K) p_0 - \tau^2 h^2 \int_0^1 \mathscr{A}_{\tau,\sigma}(K) f(\tilde{q}(\sigma)) d\sigma, & q_1 = \tilde{q}(1), \\[2mm] \tilde{p}(\tau) = -\tau h \Upsilon \phi_1(K) q_0 + \phi_0(K) p_0 - \tau h \int_0^1 \mathscr{B}_{\tau,\sigma}(K) f(\tilde{q}(\sigma)) d\sigma, & p_1 = \tilde{p}(1), \end{cases}$$
(5.18)

where $K = \tau^2 h^2 \Upsilon$, $f(q) = \nabla U(q)$,

$$\phi_i(K) := \sum_{l=0}^{\infty} \frac{(-1)^l K^l}{(2l+i)!},$$

for $i = 0, 1, \cdots$, and

$$\mathscr{A}_{\tau,\sigma}(K) = \sum_{i=0}^{r-1} \int_0^1 (1-\xi)\phi_1\big((1-\xi)^2 K\big)\tilde{\psi}_i(\xi\tau)d\xi\,\tilde{\psi}_i(\sigma),$$

$$\mathscr{B}_{\tau,\sigma}(K) = \sum_{i=0}^{r-1} \int_0^1 \phi_0\big((1-\xi)^2 K\big)\tilde{\psi}_i(\xi\tau)d\xi\,\tilde{\psi}_i(\sigma). \tag{5.19}$$

Remark 5.4   In [59], the authors developed and researched a type of trigonometric Fourier collocation methods for second-order ODEs $q''(t) + Mq(t) = f(q(t))$. However, as shown in [59], those methods cannot preserve the energy exactly. From the analysis to be presented in this chapter, it turns out that the trigonometric collocation scheme (5.18) derived here can attain arbitrary algebraic order and can preserve the energy of (5.16) exactly or approximately.

Remark 5.5   It is remarked that the multi-frequency highly oscillatory Hamiltonian system (5.5) is a kind of second-order system $q''(t) + \bar{M}^{-1}\bar{K}q(t) = -\bar{M}^{-1}\nabla U(q(t))$ and applying the ECr method (5.12) to it leads to the TCr method (5.18) with $K = \tau^2 h^2 \bar{M}^{-1}\bar{K}$ and $f(q) = \bar{M}^{-1}\nabla U(q)$.

In the special case where $N = 0$ and $\Upsilon = 0$, the system (5.14) reduces to the conventional second-order ODEs

$$q''(t) = -\nabla U(q(t)), \qquad q(0) = q_0, \ \ q'(0) = q_0', \qquad t \in [0, T]. \tag{5.20}$$

Then the TCr method has the following form.

Definition 5.3   A TCr method for solving (5.20) is defined as

$$\begin{cases} \tilde{q}(\tau) = q_0 + \tau h p_0 - \tau^2 h^2 \displaystyle\int_0^1 \bar{\mathscr{A}}_{\tau,\sigma} \nabla U(\tilde{q}(\sigma))d\sigma, & q_1 = \tilde{q}(1), \\[2mm] \tilde{p}(\tau) = p_0 - \tau h \displaystyle\int_0^1 \bar{\mathscr{B}}_{\tau,\sigma} \nabla U(\tilde{q}(\sigma))d\sigma, & p_1 = \tilde{p}(1), \end{cases} \tag{5.21}$$

where

$$\bar{\mathscr{A}}_{\tau,\sigma} = \sum_{i=0}^{r-1} \int_0^1 (1-\xi)\tilde{\psi}_i(\xi\tau)d\xi\,\tilde{\psi}_i(\sigma), \quad \bar{\mathscr{B}}_{\tau,\sigma} = \sum_{i=0}^{r-1} \int_0^1 \tilde{\psi}_i(\xi\tau)d\xi\,\tilde{\psi}_i(\sigma). \tag{5.22}$$

This scheme looks like a continuous-stage RKN method, and is denoted by RKNCr in this chapter.

5.4   Energy-Preserving Analysis

In this section, we analyse the energy-preserving property of the ECr methods.

Theorem 5.1   *If Q is skew symmetric and $\tilde{u}(\tau) \in X_h$, the first integral H determined by (5.2) of the conservative system (5.1) can be preserved exactly by the ECr method (5.12): i.e., $H(y_1) = H(y_0)$. If $\tilde{u}(\tau) \notin X_h$, the ECr method (5.12) approximately preserves the energy H with the following accuracy $H(y_1) = H(y_0) + \mathcal{O}(h^{2r+1})$.*

Proof   We begin with the first part of this proof under the assumption that Q is skew symmetric and $\tilde{u}(\tau) \in X_h$. It follows from $\tilde{u}(\tau) \in X_h$ that $\tilde{u}'(\tau) \in Y_h$ and $Q^{-1}\tilde{u}'(\tau) \in Y_h$. Then, in the light of (5.8), we obtain

$$\int_0^1 \tilde{u}'(\tau)^{\mathsf{T}}(Q^{-1})^{\mathsf{T}}\tilde{u}'(\tau)\mathrm{d}\tau = \int_0^1 \tilde{u}'(\tau)^{\mathsf{T}}(Q^{-1})^{\mathsf{T}}\big(A\tilde{u}(\tau) + \mathscr{P}_h g(\tilde{u}(\tau))\big)\mathrm{d}\tau$$

$$= \int_0^1 \tilde{u}'(\tau)^{\mathsf{T}}(Q^{-1})^{\mathsf{T}}\big(A\tilde{u}(\tau) + g(\tilde{u}(\tau))\big)\mathrm{d}\tau.$$

Here Q is skew symmetric, so is Q^{-1}. We then have

$$0 = \int_0^1 \tilde{u}'(\tau)^{\mathsf{T}}(Q^{-1})^{\mathsf{T}}\tilde{u}'(\tau)\mathrm{d}\tau = -\int_0^1 \tilde{u}'(\tau)^{\mathsf{T}}Q^{-1}\big(A\tilde{u}(\tau) + g(\tilde{u}(\tau))\big)\mathrm{d}\tau.$$

On the other hand, it is clear that

$$H(y_1) - H(y_0) = \int_0^1 \frac{\mathrm{d}}{\mathrm{d}\tau}H(\tilde{u}(\tau))\mathrm{d}\tau = h\int_0^1 \tilde{u}'(\tau)^{\mathsf{T}}\nabla H(\tilde{u}(\tau))\mathrm{d}\tau.$$

It follows from (5.1) and (5.3) that

$$\nabla H(\tilde{u}(\tau)) = Q^{-1}\big(A\tilde{u}(\tau) + g(\tilde{u}(\tau))\big).$$

Therefore, we obtain

$$H(y_1) - H(y_0) = h\int_0^1 \tilde{u}'(\tau)^{\mathsf{T}}Q^{-1}\big(A\tilde{u}(\tau) + g(\tilde{u}(\tau))\big)\mathrm{d}\tau = h \cdot 0 = 0.$$

We next prove the second part of this theorem under the assumption that $\tilde{u}(\tau) \notin X_h$. With the above analysis for the first part of the proof, we have

$$
H(y_1) - H(y_0)
$$

$$
= h \int_0^1 \tilde{u}'(\tau)^{\mathsf{T}} Q^{-1} \big(A\tilde{u}(\tau) + g(\tilde{u}(\tau)) \big) d\tau
$$

$$
= h \int_0^1 \tilde{u}'(\tau)^{\mathsf{T}} Q^{-1} \big(A\tilde{u}(\tau) + \mathscr{P}_h g(\tilde{u}(\tau)) + g(\tilde{u}(\tau)) - \mathscr{P}_h g(\tilde{u}(\tau)) \big) d\tau
$$

$$
= -h \int_0^1 \tilde{u}'(\tau)^{\mathsf{T}} (Q^{-1})^{\mathsf{T}} \tilde{u}'(\tau) d\tau + h \int_0^1 \tilde{u}'(\tau)^{\mathsf{T}} Q^{-1} \big(g(\tilde{u}(\tau)) - \mathscr{P}_h g(\tilde{u}(\tau)) \big) d\tau
$$

$$
= h \int_0^1 \tilde{u}'(\tau)^{\mathsf{T}} Q^{-1} \big(g(\tilde{u}(\tau)) - \mathscr{P}_h g(\tilde{u}(\tau)) \big) d\tau.
$$

Exploiting Lemma 3.4 presented in [34] and Lemma 5.2 proved in Sect. 5.6, we obtain $\tilde{u}'(\tau) = \mathscr{P}_h \tilde{u}'(\tau) + \mathcal{O}(h^r)$. Therefore, one arrives at

$$
H(y_1) - H(y_0)
$$

$$
= h \int_0^1 \big(\mathscr{P}_h \tilde{u}'(\tau) + \mathcal{O}(h^r) \big)^{\mathsf{T}} Q^{-1} \big(g(\tilde{u}(\tau)) - \mathscr{P}_h g(\tilde{u}(\tau)) \big) d\tau
$$

$$
= h \int_0^1 \big(\mathscr{P}_h \tilde{u}'(\tau) \big)^{\mathsf{T}} Q^{-1} \big(g(\tilde{u}(\tau)) - \mathscr{P}_h g(\tilde{u}(\tau)) \big) d\tau + \mathcal{O}(h^{2r+1})
$$

$$
= h \int_0^1 \big(\mathscr{P}_h \tilde{u}'(\tau) \big)^{\mathsf{T}} Q^{-1} \big(g(\tilde{u}(\tau)) - g(\tilde{u}(\tau)) \big) d\tau + \mathcal{O}(h^{2r+1}) = \mathcal{O}(h^{2r+1}),
$$

where the result (5.28) in Sect. 5.6 is used.

The proof is complete. $\square$

Remark 5.6   It is noted that for the special case $g(y) = 0$ or $A = 0$, it is easy to choose Y_h and X_h such that $\tilde{u}(\tau) \in X_h$. For the case $A \neq 0$ and $g(y) \equiv C$, if we consider $Y_h = \mathrm{span}\{1, e^{\tau h A}\}$ and $X_h = \mathrm{span}\{1, \tau h, e^{\tau h A}\}$, it follows from (5.12) that $\tilde{u}(\tau) = e^{\tau h A} y_0 + A^{-1}(e^{\tau h A} - I)C$. This also leads to $\tilde{u}(\tau) \in X_h$. However, for the general situation, it is usually not easy to check whether the fact $\tilde{u}(\tau) \in X_h$ is true or not for the considered Y_h and X_h. Therefore, we present the results for two different cases $\tilde{u}(\tau) \in X_h$ and $\tilde{u}(\tau) \notin X_h$ in Theorem 5.1.

Remark 5.7   For the result of $\tilde{u}(\tau) \notin X_h$, we only present the local error of the energy conservation, which is a direct consequence of Theorem 5.4. For the long-time energy conservation, we have proved the result for exponential integrators in [60]. It is possible to perform the long-time analysis for the methods presented in this chapter by using modulated Fourier expansions.

Theorem 5.2 *If Q is negative semi-definite and $\tilde{u}(\tau) \in X_h$, then H, the Lyapunov function of the dissipative system (5.1), given by (5.2), can be preserved by the ECr method (5.12); i.e., $H(y_1) \leqslant H(y_0)$. If $\tilde{u}(\tau) \notin X_h$, it is true that $H(y_1) \leqslant H(y_0) + \mathcal{O}(h^{2r+1})$.*

Proof Applying the fact that $\int_0^1 \tilde{u}'(\tau)^{\mathsf{T}} Q^{-1} \tilde{u}'(\tau) d\tau \leqslant 0$, this theorem can be proved in a similar way to the proof of Theorem 5.1. □

5.5 Existence, Uniqueness and Smoothness of the Solution

In this section, we focus on the study of the existence and uniqueness of $\tilde{u}(\tau)$ associated with the ECr method (5.12).

According to Lemma 3.1 given in [50], it is easily verified that the coefficients $e^{\tau hA}$ and $\bar{A}_{\tau,\sigma}(A)$ of the methods for $0 \leqslant \tau \leqslant 1$ and $0 \leqslant \sigma \leqslant 1$ are uniformly bounded. We begin by assuming that

$$M_k = \max_{\tau,\sigma,h\in[0,1]} \left\| \frac{\partial^k \bar{A}_{\tau,\sigma}}{\partial h^k} \right\|, \quad C_k = \max_{\tau,h\in[0,1]} \left\| \frac{\partial^k e^{\tau hA}}{\partial h^k} y_0 \right\|, \quad k = 0, 1, \cdots.$$

Furthermore, denoting n-th-order derivative of g at y by $g^{(n)}(y)$, we then have the following result about the existence and uniqueness of the methods.

Theorem 5.3 *Let $B(\bar{y}_0, R) = \left\{ y \in \mathbb{R}^d : \|y - \bar{y}_0\| \leqslant R \right\}$ and*

$$D_n = \max_{y \in B(\bar{y}_0, R)} \|g^{(n)}(y)\|, \; n = 0, 1, \cdots,$$

where R is a positive constant, $\bar{y}_0 = e^{\tau hA} y_0$, $\|\cdot\| = \|\cdot\|_\infty$ is the maximum norm for vectors in $\mathbb{R}^d$ or the corresponding induced norm for the multilinear maps $g^{(n)}(y)$. If h satisfies

$$0 \leqslant h \leqslant \kappa < \min \left\{ \frac{1}{M_0 D_1}, \frac{R}{M_0 D_0}, 1 \right\}, \tag{5.23}$$

then the ECr method (5.12) has a unique solution $\tilde{u}(\tau)$ which is smoothly dependent on h.

Proof Set $\tilde{u}_0(\tau) = \bar{y}_0$ and define

$$\tilde{u}_{n+1}(\tau) = e^{\tau hA} y_0 + \tau h \int_0^1 \bar{A}_{\tau,\sigma}(A) g(\tilde{u}_n(\sigma)) d\sigma, \quad n = 0, 1, \cdots, \tag{5.24}$$

which leads to a function sequence $\{\tilde{u}_n(\tau)\}_{n=0}^\infty$. We note that $\lim_{n\to\infty} \tilde{u}_n(\tau)$ is a solution of the TCr method (5.12) if $\{\tilde{u}_n(\tau)\}_{n=0}^\infty$ is uniformly convergent, which will be shown by proving the uniform convergence of the infinite series $\sum_{n=0}^{\infty}(\tilde{u}_{n+1}(\tau) - \tilde{u}_n(\tau))$.

By induction and according to (5.23) and (5.24), we obtain $||\tilde{u}_n(\tau) - \bar{y}_0|| \leqslant R$ for $n = 0, 1, \cdots$. It then follows from (5.24) that

$$||\tilde{u}_{n+1}(\tau) - \tilde{u}_n(\tau)||$$

$$\leqslant \tau h \int_0^1 M_0 D_1 ||\tilde{u}_n(\sigma) - \tilde{u}_{n-1}(\sigma)|| \mathrm{d}\sigma$$

$$\leqslant h \int_0^1 M_0 D_1 ||\tilde{u}_n(\sigma) - \tilde{u}_{n-1}(\sigma)|| \mathrm{d}\sigma \leqslant \beta ||\tilde{u}_n - \tilde{u}_{n-1}||_c, \quad \beta = \kappa M_0 D_1,$$

where $|| \cdot ||_c$ is the maximum norm for continuous functions defined as $||w||_c = \max_{\tau \in [0,1]} ||w(\tau)||$ for a continuous $\mathbb{R}^d$-valued function w on $[0, 1]$. Hence, we obtain

$$||\tilde{u}_{n+1} - \tilde{u}_n||_c \leqslant \beta ||\tilde{u}_n - \tilde{u}_{n-1}||_c$$

and

$$||\tilde{u}_{n+1} - \tilde{u}_n||_c \leqslant \beta^n ||\tilde{u}_1 - y_0||_c \leqslant \beta^n R, \quad n = 0, 1, \cdots.$$

It then immediately follows from Weierstrass M-test and the fact of $\beta < 1$ that $\sum_{n=0}^{\infty}(\tilde{u}_{n+1}(\tau) - \tilde{u}_n(\tau))$ is uniformly convergent.

If the ECr method (5.12) has another solution $\tilde{v}(\tau)$, we obtain the following inequalities

$$||\tilde{u}(\tau) - \tilde{v}(\tau)|| \leqslant h \int_0^1 ||\bar{A}_{\tau,\sigma}(A)\big(g(\tilde{u}(\sigma)) - g(\tilde{v}(\sigma))\big)|| \mathrm{d}\sigma \leqslant \beta ||\tilde{u} - \tilde{v}||_c,$$

and $||\tilde{u} - \tilde{v}||_c \leqslant \beta ||\tilde{u} - \tilde{v}||_c$. This yields $||\tilde{u} - \tilde{v}||_c = 0$ and $\tilde{u}(\tau) \equiv \tilde{v}(\tau)$. The existence and uniqueness have been proved.

With respect to the result that $\tilde{u}(\tau)$ is smoothly dependent of h, since each $\tilde{u}_n(\tau)$ is a smooth function of h, we need only to prove that the sequence $\left\{ \dfrac{\partial^k \tilde{u}_n}{\partial h^k}(\tau) \right\}_{n=0}^{\infty}$ is uniformly convergent for $k \geqslant 1$. Differentiating (5.24) with respect to h gives

$$\frac{\partial \tilde{u}_{n+1}}{\partial h}(\tau) = \tau A e^{\tau h A} y_0 + \tau \int_0^1 \left(\bar{A}_{\tau,\sigma}(A) + h \frac{\partial \bar{A}_{\tau,\sigma}}{\partial h} \right) g(\tilde{u}_n(\sigma)) \mathrm{d}\sigma$$

$$+ \tau h \int_0^1 \bar{A}_{\tau,\sigma}(A) g^{(1)}(\tilde{u}_n(\sigma)) \frac{\partial \tilde{u}_n}{\partial h}(\sigma) \mathrm{d}\sigma, \tag{5.25}$$

which yields

$$\left\|\frac{\partial \tilde{u}_{n+1}}{\partial h}\right\|_c \leqslant \alpha + \beta \left\|\frac{\partial \tilde{u}_n}{\partial h}\right\|_c, \qquad \alpha = C_1 + (M_0 + \kappa M_1)D_0.$$

By induction, it is easy to show that $\left\{\dfrac{\partial \tilde{u}_n}{\partial h}(\tau)\right\}_{n=0}^{\infty}$ is uniformly bounded:

$$\left\|\frac{\partial \tilde{u}_n}{\partial h}\right\|_c \leqslant \alpha(1 + \beta + \cdots + \beta^{n-1}) \leqslant \frac{\alpha}{1-\beta} = C^*, \qquad n = 0, 1, \cdots. \tag{5.26}$$

It follows from (5.25)–(5.26) that

$$\left\|\frac{\partial \tilde{u}_{n+1}}{\partial h} - \frac{\partial \tilde{u}_n}{\partial h}\right\|_c$$

$$\leqslant \tau \int_0^1 (M_0 + hM_1) \left\| g(\tilde{u}_n(\sigma)) - g(\tilde{u}_{n-1}(\sigma)) \right\| d\sigma$$

$$+ \tau h \int_0^1 M_0 \left(\left\| (g^{(1)}(\tilde{u}_n(\sigma)) - g^{(1)}(\tilde{u}_{n-1}(\sigma))) \frac{\partial \tilde{u}_n}{\partial h}(\sigma) \right\| \right.$$

$$+ \left. \left\| g^{(1)}(\tilde{u}_{n-1}(\sigma)) \left(\frac{\partial \tilde{u}_n}{\partial h}(\sigma) - \frac{\partial \tilde{u}_{n-1}}{\partial h}(\sigma) \right) \right\| \right) d\sigma \leqslant \gamma \beta^{n-1} + \beta \left\| \frac{\partial \tilde{u}_n}{\partial h} - \frac{\partial \tilde{u}_{n-1}}{\partial h} \right\|_c,$$

where $\gamma = (M_0 D_1 + \kappa M_1 D_1 + \kappa M_0 L_2 C^*)R$, and L_2 is a constant satisfying

$$\|g^{(1)}(y) - g^{(1)}(z)\| \leqslant L_2\|y - z\|, \qquad \text{for } y, z \in B(\bar{y}_0, R).$$

Therefore, the following result is obtained by induction

$$\left\|\frac{\partial \tilde{u}_{n+1}}{\partial h} - \frac{\partial \tilde{u}_n}{\partial h}\right\|_c \leqslant n\gamma\beta^{n-1} + \beta^n C^*, \qquad n = 1, 2, \cdots.$$

This shows the uniform convergence of $\sum_{n=0}^{\infty} \left(\dfrac{\partial \tilde{u}_{n+1}}{\partial h}(\tau) - \dfrac{\partial \tilde{u}_n}{\partial h}(\tau) \right)$ and then $\left\{\dfrac{\partial \tilde{u}_n}{\partial h}(\tau)\right\}_{n=0}^{\infty}$ is uniformly convergent.

Likewise, it can be shown that other function series $\left\{\dfrac{\partial^k \tilde{u}_n}{\partial h^k}(\tau)\right\}_{n=0}^{\infty}$ for $k \geqslant 2$ are uniformly convergent as well. Therefore, $\tilde{u}(\tau)$ is smoothly dependent on h. $\qquad\square$

5.6   Algebraic Order

In this section, we analyse the algebraic order of the ECr method (5.12). To express the dependence of the solutions of $y'(t) = Ay(t) + g(y(t))$ on the initial values, we denote by $y(\cdot, \tilde{t}, \tilde{y})$ the solution satisfying the initial condition $y(\tilde{t}, \tilde{t}, \tilde{y}) = \tilde{y}$ for any given $\tilde{t} \in [0, h]$ and set $\Phi(s, \tilde{t}, \tilde{y}) = \dfrac{\partial y(s, \tilde{t}, \tilde{y})}{\partial \tilde{y}}$. Recalling the elementary theory of ODEs, we have the following standard result

$$\frac{\partial y(s, \tilde{t}, \tilde{y})}{\partial \tilde{t}} = -\Phi(s, \tilde{t}, \tilde{y})\big(A\tilde{y} + g(\tilde{y})\big).$$

Throughout this section, for convenience, an h-dependent function $w(\tau)$ is called regular if it can be expanded as $w(\tau) = \sum_{n=0}^{r-1} w^{[n]}(\tau)h^n + \mathcal{O}(h^r)$, where $w^{[n]}(\tau) = \dfrac{1}{n!} \dfrac{\partial^n w(\tau)}{\partial h^n}\big|_{h=0}$ is a vector-valued function with polynomial entries of degrees $\leqslant n$.

It can be deduced from Proposition 3.3 in [34] that $P_{\tau,\sigma}$ is regular. Moreover, we can prove the following result.

Lemma 5.2   *The ECr method (5.12) generates a regular h-dependent function $\tilde{u}(\tau)$.*

Proof   By the result given in [34], we know that $P_{\tau,\sigma}$ can be smoothly extended to $h = 0$ by setting $P_{\tau,\sigma}|_{h=0} = \lim\limits_{h \to 0} P_{\tau,\sigma}(h)$. Furthermore, it follows from Theorem 5.3 that $\tilde{u}(\tau)$ is smoothly dependent on h. Therefore, $\tilde{u}(\tau)$ and $\bar{A}_{\tau,\sigma}(A)$ can be expanded with respect to h at zero as follows:

$$\tilde{u}(\tau) = \sum_{m=0}^{r-1} \tilde{u}^{[m]}(\tau)h^m + \mathcal{O}(h^r), \quad \bar{A}_{\tau,\sigma}(A) = \sum_{m=0}^{r-1} \bar{A}_{\tau,\sigma}^{[m]}(A)h^m + \mathcal{O}(h^r).$$

Then let $\delta = \tilde{u}(\sigma) - y_0$ and we have

$$\delta = \tilde{u}^{[0]}(\sigma) - y_0 + \mathcal{O}(h) = y_0 - y_0 + \mathcal{O}(h) = \mathcal{O}(h).$$

We expand $f(\tilde{u}(\sigma))$ at y_0 and insert the above equalities into the first equation of the ECr method (5.12). This manipulation yields

$$\sum_{m=0}^{r-1} \tilde{u}^{[m]}(\tau)h^m = \sum_{m=0}^{r-1} \frac{\tau^m A^m y_0}{m!}h^m$$

$$+ \tau h \int_0^1 \sum_{k=0}^{r-1} \bar{A}_{\tau,\sigma}^{[k]}(A)h^k \sum_{n=0}^{r-1} \frac{1}{n!} g^{(n)}(y_0)\underbrace{(\delta, \cdots, \delta)}_{n-fold}\mathrm{d}\sigma + \mathcal{O}(h^r). \quad (5.27)$$

In order to show that $\tilde{u}(\tau)$ is regular, we need only to prove that

$$\tilde{u}^{[m]}(\tau) \in P_m^d = \underbrace{P_m([0,1]) \times \cdots \times P_m([0,1])}_{d-fold} \quad \text{for} \quad m = 0, 1, \cdots, r-1,$$

where $P_m([0,1])$ consists of polynomials of degree $\leq m$ on $[0,1]$. This can be confirmed by induction as follows.

Firstly, it is clear that $\tilde{u}^{[0]}(\tau) = y_0 \in P_0^d$. We assume that $\tilde{u}^{[n]}(\tau) \in P_n^d$ for $n = 0, 1, \cdots, m$. Comparing the coefficients of h^{m+1} on both sides of (5.27) and using (5.13) lead to

$$\tilde{u}^{[m+1]}(\tau)$$

$$= \frac{\tau^{m+1} A^{m+1}}{(m+1)!} y_0 + \sum_{k+n=m} \tau \int_0^1 \bar{A}_{\tau,\sigma}^{[k]}(A) h_n(\sigma) d\sigma$$

$$= \frac{\tau^{m+1} A^{m+1}}{(m+1)!} y_0 + \sum_{k+n=m} \tau \int_0^1 \int_0^1 \left[e^{(1-\xi)\tau hA} P_{\xi\tau,\sigma} \right]^{[k]} h_n(\sigma) d\sigma d\xi,$$

$$h_n(\sigma) \in P_n^d.$$

Since $P_{\xi\tau,\sigma}$ is regular, it is easy to check that $e^{(1-\xi)\tau hA} P_{\xi\tau,\sigma}$ is also regular. Thus, under the condition $k+n=m$, we have

$$\int_0^1 \left[e^{(1-\xi)\tau hA} P_{\xi\tau,\sigma} \right]^{[k]} h_n(\sigma) d\sigma := \check{p}_m^k(\xi\tau) \in P_m^d([0,1]).$$

Then, the above result can be simplified as

$$\tilde{u}^{[m+1]}(\tau) = \frac{\tau^{m+1} A^{m+1}}{(m+1)!} y_0 + \sum_{k+n=m} \tau \int_0^1 \check{p}_m^k(\xi\tau) d\xi$$

$$= \frac{\tau^{m+1} A^{m+1}}{(m+1)!} y_0 + \sum_{k+n=m} \int_0^\tau \check{p}_m^k(\alpha) d\alpha \in P_{m+1}^d.$$

$\square$

According to Lemma 3.4 presented in [34] and the above lemma, we obtain

$$\mathscr{P}_h g(\tilde{u}(\tau)) - g(\tilde{u}(\tau)) = \mathcal{O}(h^r), \tag{5.28}$$

which will be used in the analysis of algebraic order. We are now ready to present the result about the algebraic order of the ECr method (5.12).

Theorem 5.4   *About the stage order and order of the ECr method* (5.12), *we have*

$$\tilde{u}(\tau) - y(t_0 + \tau h) = \mathcal{O}(h^{r+1}), \quad 0 < \tau < 1,$$

$$\tilde{u}(1) - y(t_0 + h) = \mathcal{O}(h^{2r+1}).$$

Proof   According to the previous preliminaries, we obtain

$$\tilde{u}(\tau) - y(t_0 + \tau h)$$

$$= y(t_0 + \tau h, t_0 + \tau h, \tilde{u}(\tau)) - y(t_0 + \tau h, t_0, y_0)$$

$$= \int_0^\tau \frac{\mathrm{d}}{\mathrm{d}\alpha} y(t_0 + \tau h, t_0 + \alpha h, \tilde{u}(\alpha))\mathrm{d}\alpha$$

$$= \int_0^\tau (h\frac{\partial y}{\partial t}(t_0 + \tau h, t_0 + \alpha h, \tilde{u}(\alpha)) + \frac{\partial y}{\partial \tilde{y}}(t_0 + \tau h, t_0 + \alpha h, \tilde{u}(\alpha))h\tilde{u}'(\alpha))\mathrm{d}\alpha$$

$$= \int_0^\tau \Big(-h\frac{\partial y}{\partial \tilde{y}}(t_0 + \tau h, t_0 + \alpha h, \tilde{u}(\alpha))\big(A\tilde{u}(\alpha) + g(\tilde{u}(\alpha))\big)$$

$$+ \frac{\partial y}{\partial \tilde{y}}(t_0 + \tau h, t_0 + \alpha h, \tilde{u}(\alpha))\big(hA\tilde{u}(\alpha) + h\langle P_{\tau,\sigma}, g(\tilde{u}(\alpha))\rangle_\alpha\big)\Big)\mathrm{d}\alpha$$

$$= -h\int_0^\tau \Phi^\tau(\alpha)\big(g(\tilde{u}(\alpha)) - \mathscr{P}_h(g \circ \tilde{u})(\alpha)\big)\mathrm{d}\alpha = \mathcal{O}(h^{r+1}), \tag{5.29}$$

where $\Phi^\tau(\alpha) = \dfrac{\partial y}{\partial \tilde{y}}(t_0 + \tau h, t_0 + \alpha h, \tilde{u}(\alpha))$. Letting $\tau = 1$ in (5.29) yields

$$\tilde{u}(1) - y(t_0 + h) = -h\int_0^1 \Phi^1(\alpha)\big(g(\tilde{u}(\alpha)) - \mathscr{P}_h(g \circ \tilde{u})(\alpha)\big)\mathrm{d}\alpha. \tag{5.30}$$

We partition the matrix-valued function $\Phi^1(\alpha)$ as $\Phi^1(\alpha) = (\Phi_1^1(\alpha), \cdots, \Phi_d^1(\alpha))^\mathsf{T}$. It follows from Lemma 5.2 that

$$\Phi_i^1(\alpha) = \mathscr{P}_h\Phi_i^1(\alpha) + \mathcal{O}(h^r), \quad i = 1, \cdots, d. \tag{5.31}$$

On the other hand, we have

$$\int_0^1 (\mathscr{P}_h\Phi_i^1(\alpha))^\mathsf{T} g(\tilde{u}(\alpha))\mathrm{d}\alpha = \int_0^1 (\mathscr{P}_h\Phi_i^1(\alpha))^\mathsf{T} \mathscr{P}_h(g \circ \tilde{u})(\alpha)\mathrm{d}\alpha, \quad i = 1, \cdots, d. \tag{5.32}$$

Therefore, it follows from (5.30), (5.31) and (5.32) that

$$\tilde{u}(1) - y(t_0 + h)$$

$$= -h \int_0^1 \left(\begin{pmatrix} (\mathscr{P}_h \Phi_1^1(\alpha))^{\mathsf{T}} \\ \vdots \\ (\mathscr{P}_h \Phi_d^1(\alpha))^{\mathsf{T}} \end{pmatrix} + \mathscr{O}(h^r) \right) \left(g(\tilde{u}(\alpha)) - \mathscr{P}_h(g \circ \tilde{u})(\alpha) \right) d\alpha$$

$$= -h \int_0^1 \begin{pmatrix} (\mathscr{P}_h \Phi_1^1(\alpha))^{\mathsf{T}} \left(g(\tilde{u}(\alpha)) - \mathscr{P}_h(g \circ \tilde{u})(\alpha) \right) \\ \vdots \\ (\mathscr{P}_h \Phi_d^1(\alpha))^{\mathsf{T}} \left(g(\tilde{u}(\alpha)) - \mathscr{P}_h(g \circ \tilde{u})(\alpha) \right) \end{pmatrix} d\alpha - h \int_0^1 \mathscr{O}(h^r) \times \mathscr{O}(h^r) d\alpha$$

$$= 0 + \mathscr{O}(h^{2r+1}) = \mathscr{O}(h^{2r+1}).$$

$\square$

5.7 Application in Stiff Gradient Systems

When the matrix Q in (5.1) is the identity matrix, the system (5.1) is a stiff gradient system as follows:

$$y' = -\nabla U(y), \quad y(0) = y_0 \in \mathbb{R}^d, \quad t \in [0, T], \tag{5.33}$$

where the potential U has the form

$$U(y) = \frac{1}{2} y^{\mathsf{T}} My + V(y). \tag{5.34}$$

Such problems arise from the spatial discretisation of Allen–Cahn and Cahn–Hilliard PDEs (see, e.g. [61]). Along every exact solution, it is true that

$$\frac{\mathrm{d}}{\mathrm{d}t} U(y(t)) = \nabla U(y(t))^{\mathsf{T}} y'(t) = -y'(t)^{\mathsf{T}} y'(t) \leqslant 0,$$

which implies that $U(y(t))$ is monotonically decreasing.

For solving this stiff gradient system, it follows from Theorem 5.2 that the practical ECr method (5.40) is unconditionally energy-diminishing. For a quadratic potential (i.e., $V(y) = 0$ in (5.34)), the numerical solution of the method is given by

$$y_1 = R(-hA)y_0 = \mathrm{e}^{-hA} y_0.$$

The importance of the damping property $|R(\infty)| < 1$ for the approximation properties of Runge–Kutta methods has been studied and well understood in

[62, 63] for solving semilinear parabolic equations. The role of the condition $|R(\infty)| < 1$ in the approximation of stiff differential equations has been researched in Chapter VI of [64]. It has been shown in [39] that for each Runge–Kutta method the energy decreases once the stepsize satisfies some conditions. Discrete-gradient methods, AVF methods and AVF collocation methods derived in [39] are unconditionally energy-diminishing methods but they show no damping for very stiff gradient systems. However, it is clear that the methods are unconditionally energy-diminishing methods and they have

$$|R(\infty)| = |e^{-\infty}| = 0.$$

This implies that the methods are strongly damped even for very stiff gradient systems and this is a significant feature.

5.8   Practical Examples of Exponential Collocation Methods

In this section, we present practical examples of exponential collocation methods. Choosing $\tilde{\varphi}_k(\tau) = (\tau h)^k$ for $k = 0, 1, \cdots, r - 1$ and using the Gram–Schmidt process, we obtain the standard orthonormal basis of Y_h as follows:

$$\hat{p}_j(\tau) = (-1)^j \sqrt{2j+1} \sum_{k=0}^{j} \binom{j}{k}\binom{j+k}{k}(-\tau)^k,$$

$$j = 0, 1, \cdots, r-1, \quad \tau \in [0, 1],$$

which are the shifted Legendre polynomials on $[0, 1]$. Therefore, $P_{\tau,\sigma}$ can be determined by (5.10) as follows $P_{\tau,\sigma} = \sum_{i=0}^{r-1} \hat{p}_i(\tau)\hat{p}_i(\sigma)$.

5.8.1   An Example of ECr Methods

For the ECr method (5.12), we need to calculate $\bar{A}_{\tau,\sigma}(A)$ appearing in the methods. It follows from (5.13) that

$$\bar{A}_{\tau,\sigma}(A) = \int_0^1 e^{(1-\xi)\tau h A} P_{\xi\tau,\sigma} d\xi = \sum_{i=0}^{r-1} \int_0^1 e^{(1-\xi)\tau h A} \hat{p}_i(\xi\tau) d\xi \, \hat{p}_i(\sigma)$$

$$= \sum_{i=0}^{r-1} \int_0^1 e^{(1-\xi)\tau h A}(-1)^i \sqrt{2i+1} \sum_{k=0}^{i} \binom{i}{k}\binom{i+k}{k}(-\xi\tau)^k d\xi \, \hat{p}_i(\sigma)$$

$$= \sum_{i=0}^{r-1} \sqrt{2i+1} \sum_{k=0}^{i}(-1)^{i+k}\frac{(i+k)!}{k!(i-k)!}\bar{\varphi}_{k+1}(\tau h A)\hat{p}_i(\sigma). \tag{5.35}$$

Here the $\bar{\varphi}$-functions (see, e.g. [2, 14, 49, 50]) are defined by:

$$\bar{\varphi}_0(z) = e^z, \quad \bar{\varphi}_k(z) = \int_0^1 e^{(1-\sigma)z} \frac{\sigma^{k-1}}{(k-1)!} d\sigma, \quad k = 1, 2, \cdots.$$

It is noted that a number of approaches have been developed which work with the application of the φ-functions on a vector (see [2, 65, 66], for example).

5.8.2 An Example of TCr Methods

For the TCr method (5.18) solving $q''(t) + \Omega q(t) = -\nabla U(q(t))$, we need to compute $\mathscr{A}_{\tau,\sigma}$ and $\mathscr{B}_{1,\sigma}$. It follows from (5.19) that

$$\mathscr{A}_{\tau,\sigma}(K)$$

$$= \sum_{j=0}^{r-1} \int_0^1 (1-\xi)\phi_1\big((1-\xi)^2 K\big)\hat{p}_j(\xi\tau)d\xi\,\hat{p}_j(\sigma)$$

$$= \sum_{j=0}^{r-1} \sqrt{2j+1} \sum_{l=0}^{\infty} (-1)^j \sum_{k=0}^{j} \binom{j}{k}\binom{j+k}{k} \int_0^1 (-\xi)^k (1-\xi)^{2l+1}d\xi \frac{(-1)^l K^l}{(2l+1)!}\tau^k \hat{p}_j(\sigma)$$

$$= \sum_{j=0}^{r-1} \sqrt{2j+1} \sum_{l=0}^{\infty} \sum_{k=0}^{j} (-1)^{j+k}\binom{j}{k}\binom{j+k}{k} \frac{k!(2l+1)!}{(2l+k+2)!}\frac{(-1)^l K^l}{(2l+1)!}\tau^k \hat{p}_j(\sigma)$$

$$= \sum_{j=0}^{r-1} \sqrt{2j+1}\hat{p}_j(\sigma) \sum_{l=0}^{\infty} \sum_{k=0}^{j} \frac{(-1)^{j+k+l}(j+k)!}{k!(j-k)!(2l+k+2)!}\tau^k K^l.$$

Recall that the generalised hypergeometric function $_mF_n$ is defined by

$$_mF_n\begin{bmatrix} \alpha_1, \alpha_2, \cdots, \alpha_m; \\ \beta_1, \beta_2, \cdots, \beta_n; \end{bmatrix} x = \sum_{l=0}^{\infty} \frac{\prod_{i=1}^{m}(\alpha_i)_l}{\prod_{i=1}^{n}(\beta_i)_l}\frac{x^l}{l!}, \tag{5.36}$$

where α_i and β_i are arbitrary complex numbers, except that β_i can be neither zero nor a negative integer, and $(z)_l$ is the Pochhammer symbol which is defined as

$$(z)_0 = 1, \quad (z)_l = z(z+1)\cdots(z+l-1), \quad l \in \mathbb{N}.$$

Then, $\mathscr{A}_{\tau,\sigma}$ can be expressed by

$$\mathscr{A}_{\tau,\sigma}(K) = \sum_{j=0}^{r-1} \sqrt{2j+1}\,\hat{p}_j(\sigma) \sum_{l=0}^{\infty} \frac{(-1)^{j+l}}{(2l+2)!} {}_2F_1\left[\begin{array}{c} -j,\ j+1;\\ 2l+3; \end{array}\tau\right] K^l. \tag{5.37}$$

Likewise, we can obtain

$$\mathscr{B}_{1,\sigma}(K) = \sum_{j=0}^{r-1} \sqrt{2j+1}\,\hat{p}_j(\sigma) S_j(K), \tag{5.38}$$

where $S_j(K)$ are

$$S_{2j}(K) = (-1)^j \frac{(2j)!}{(4j+1)!} K^j {}_0F_1\left[\begin{array}{c} -;\\ \frac{1}{2}; \end{array} -\frac{K}{16}\right] {}_0F_1\left[\begin{array}{c} -;\\ 2j+\frac{3}{2}; \end{array} -\frac{K}{16}\right],$$

$$S_{2j+1}(K) = (-1)^j \frac{(2j+2)!}{(4j+4)!} K^{j+1} {}_0F_1\left[\begin{array}{c} -;\\ \frac{3}{2}; \end{array} -\frac{K}{16}\right] {}_0F_1\left[\begin{array}{c} -;\\ 2j+\frac{5}{2}; \end{array} -\frac{K}{16}\right], \quad j=0,1,\cdots.$$

$$\tag{5.39}$$

5.8.3   An Example of RKNCr Methods

By letting $K = 0$ in the above analysis, we obtain an example of RKNCr methods for solving the general second-order ODEs (5.20) as

$$\begin{cases} q_{d_i} = q_0 + d_i h p_0 - d_i^2 h^2 \displaystyle\int_0^1 \bar{\mathscr{A}}_{d_i,\sigma} \nabla U\left(\sum_{m=1}^r q_{d_m} l_m(\sigma)\right) \mathrm{d}\sigma, \quad i=1,\cdots,r, \\[3mm] q_1 = q_0 + h p_0 - h^2 \displaystyle\int_0^1 \bar{\mathscr{A}}_{1,\sigma} \nabla U\left(\sum_{m=1}^r q_{d_m} l_m(\sigma)\right) \mathrm{d}\sigma, \\[3mm] p_1 = p_0 - h \displaystyle\int_0^1 \bar{\mathscr{B}}_{1,\sigma} \nabla U\left(\sum_{m=1}^r q_{d_m} l_m(\sigma)\right) \mathrm{d}\sigma, \end{cases}$$

where $\bar{\mathscr{A}}_{\tau,\sigma} = \sum_{i=0}^{r-1} \int_0^1 (1-\xi)\hat{p}_i(\xi\tau)\mathrm{d}\xi\,\hat{p}_i(\sigma)$ and $\bar{\mathscr{B}}_{1,\sigma} = \sum_{i=0}^{r-1} \int_0^1 \hat{p}_i(\xi)\mathrm{d}\xi\,\hat{p}_i(\sigma)$.

Remark 5.8 It is noted that one can make different choices of Y_h and X_h and the whole analysis presented in this chapter still holds. Different choices will produce different practical methods, and in this chapter, we do not pursue this point for brevity.

5.9 Numerical Experiments

Applying the r-point Gauss–Legendre quadrature to the integral of (5.12) yields

$$
\begin{cases}
y_{c_i} = e^{c_i h A} y_0 + c_i h \sum_{j=1}^{r} b_j \bar{A}_{c_i, c_j}(A) g(y_{c_j}), & i = 1, \cdots, r, \\[3mm]
y_1 = e^{hA} y_0 + h \sum_{j=1}^{r} b_j \bar{A}_{1, c_j}(A) g(y_{c_j}),
\end{cases}
\tag{5.40}
$$

where c_j and b_j for $j = 1, \cdots, r$ are the nodes and weights of the quadrature, respectively. It is shown that the quadrature formula used here is not exact in general for arbitrary g. According to Theorem 5.4 and the order of Gauss–Legendre quadrature, it is obtained that this scheme approximately preserves the energy H with the accuracy $H(y_1) = H(y_0) + \mathcal{O}(h^{2r+1})$.

In this section, we use fixed-point iteration in practical computation. Concerning the convergence of the fixed-point iteration for the above scheme (5.40), we have the following result.

Theorem 5.5 *Assume that g satisfies a Lipschitz condition in the variable y, i.e., there exists a constant L with the property that $\|g(y_1) - g(y_2)\| \leqslant L \|y_1 - y_2\|$. If the stepsize h satisfies*

$$
0 < h < \frac{1}{LC \max\limits_{i=1,\cdots,r} c_i \max\limits_{j=1,\cdots,r} |b_j|},
\tag{5.41}
$$

then the fixed-point iteration for the scheme (5.40) is convergent, where the constant C depends on r but is independent of A.

Proof We rewrite the first formula of (5.40) as

$$
Y = e^{chA} y_0 + h\bar{K}(A) g(Y),
\tag{5.42}
$$

where $c = (c_1, c_2, \cdots, c_r)^\mathsf{T}$, $Y = (y_1, y_2, \cdots, y_r)^\mathsf{T}$, $\bar{K}(A) = (\bar{K}_{ij}(A))_{r \times r}$ and $\bar{K}_{ij}(A)$ are defined by

$$\bar{K}_{ij}(A) := c_i b_j \bar{A}_{c_i, c_j}(A).$$

It then follows from (5.35) that

$$\left\| \bar{K}_{ij}(A) \right\| \leqslant c_i |b_j| \sum_{l=0}^{r-1} \sqrt{2l+1} \sum_{k=0}^{l} \frac{(l+k)!}{k!(l-k)!} \left\| \bar{\varphi}_{k+1}(c_i h A) \right\| \left| \hat{p}_l(c_j) \right| \leqslant C c_i |b_j|,$$

where the constant C depends on r but is independent of A. It then follows that $\left\| \bar{K}(A) \right\| \leqslant C \max\limits_{i=1,\cdots,r} c_i \max\limits_{j=1,\cdots,r} |b_j|$. Letting

$$\varphi(x) = e^{chA} y_0 + h \bar{K}(A) g(x),$$

we obtain that

$$\left\| \varphi(x) - \varphi(y) \right\| = \left\| h \bar{K}(A) g(x) - h \bar{K}(A) g(y) \right\| \leqslant hL \left\| \bar{K}(A) \right\| \left\| x - y \right\|$$
$$\leqslant hLC \max\limits_{i=1,\cdots,r} c_i \max\limits_{j=1,\cdots,r} |b_j| \left\| x - y \right\|.$$

The proof is complete by the Contraction Mapping Theorem.   □

Remark 5.9   It can be concluded from this theorem that the convergence of the method (5.40) is independent of $\|A\|$. However, it can be checked easily that the convergence of some other methods such as RKEPC methods given in [30] depends on $\|A\|$. This fact confirms the efficiency of the method (5.40) and is demonstrated numerically by the experiments presented in this section. This is also a reason why the RKEPC2 formula does not precisely conserve the energy of Problem 5.1.

In order to show the efficiency and robustness of the methods, we take $r = 2$ and denote the corresponding method by EC2P. Then we choose the same Y_h and X_h for the functionally fitted energy-preserving method developed in [34], and by this choice, the method becomes the $2r$th order RKEPC method given in [30]. For this method, we choose $r = 2$ and approximate the integral by the Lobatto quadrature of order eight, which is precisely the "extended Labatto IIIA method of order four" in [67]. We denote this corresponding method as RKEPC2. Another integrator we select for comparison is the explicit three-stage exponential integrator of order four derived in [14] which is denoted by EEI3s4. It is noted that the first two methods

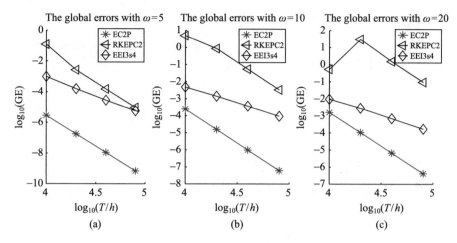

Fig. 5.1 The logarithm of the global error against the logarithm of T/h

are implicit and we set 10^{-16} as the error tolerance and 5 as the maximum number[1]
demonstrate the efficiency of ECr methods when applied to first-order systems, for
brevity.

Problem 5.1 Consider the Duffing equation defined by

$$\begin{pmatrix} q \\ p \end{pmatrix}' = \begin{pmatrix} 0 & 1 \\ -\omega^2 - k^2 & 0 \end{pmatrix} \begin{pmatrix} q \\ p \end{pmatrix} + \begin{pmatrix} 0 \\ 2k^2 q^3 \end{pmatrix}, \quad \begin{pmatrix} q(0) \\ p(0) \end{pmatrix} = \begin{pmatrix} 0 \\ \omega \end{pmatrix}.$$

It is a Hamiltonian system with the Hamiltonian:

$$H(p, q) = \frac{1}{2}p^2 + \frac{1}{2}(\omega^2 + k^2)q^2 - \frac{k^2}{2}q^4.$$

The exact solution of this system is $q(t) = sn(\omega t; k/\omega)$ with the Jacobi elliptic
function sn. Choose $k = 0.07$, $\omega = 5, 10, 20$ and solve the problem on the interval
$[0, 1000]$ with different stepsizes $h = 0.1/2^i$ for $i = 0, \cdots, 3$. The global errors are
presented in Fig. 5.1. Then, we integrate this problem with the stepsize $h = 1/100$
on the interval $[0, 10000]$. See Fig. 5.2 for the energy conservation for different
methods. Finally, we solve this problem on the interval $[0, 10]$ with $\omega = 20$, $h =
0.01$ and different error tolerances in the fixed-point iteration. See Table 5.1 for the
total numbers of iterations for the implicit methods EC2P and RKEPC2.

---

[1] It is noted that in order to show that the methods can perform well even for few iterations, a
low maximum number 5 of fixed-point iterations is used in this section. It is possible to increase to
other bigger maximum number of fixed-point iterations, but we do not go further here for brevity.

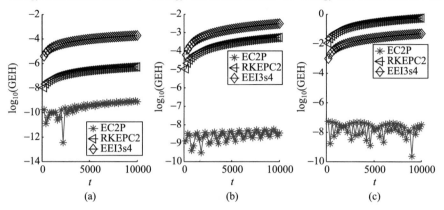

Fig. 5.2 The logarithm of the error of Hamiltonian against t

Table 5.1 Results for Problem 5.1: the total numbers of iterations for different error tolerances (tol)

Methods	tol $= 1.0 \times 10^{-6}$	tol $= 1.0 \times 10^{-8}$	tol $= 1.0 \times 10^{-10}$	tol $= 1.0 \times 10^{-12}$
EC2P	859	992	1000	1651
RKEPC2	6886	8907	10, 647	11, 899

Problem 5.2   Consider the following averaged system in wind-induced oscillation (see [40])

$$\begin{pmatrix} x_1 \\ x_2 \end{pmatrix}' = \begin{pmatrix} -\zeta & -\lambda \\ \lambda & -\zeta \end{pmatrix}\begin{pmatrix} x_1 \\ x_2 \end{pmatrix} + \begin{pmatrix} x_1 x_2 \\ \frac{1}{2}(x_1^2 - x_2^2) \end{pmatrix},$$

where $\zeta \geqslant 0$ is a damping factor and λ is a detuning parameter. By setting

$$\zeta = r\cos\theta, \qquad \lambda = r\sin\theta, \qquad r \geqslant 0, \qquad 0 \leqslant \theta \leqslant \pi/2,$$

this system can be transformed into the scheme (5.1) with

$$Q = \begin{pmatrix} -\cos\theta & -\sin\theta \\ \sin\theta & -\cos\theta \end{pmatrix}, \quad M = \begin{pmatrix} r & 0 \\ 0 & r \end{pmatrix},$$

$$V = -\frac{1}{2}\sin\theta\left(x_1 x_2^2 - \frac{1}{3}x_1^3\right) + \frac{1}{2}\cos\theta\left(-x_1^2 x_2 + \frac{1}{3}x_2^3\right).$$

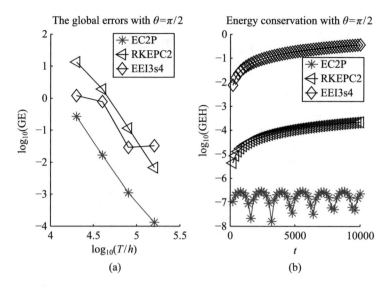

Fig. 5.3 (a) The logarithm of the global error against the logarithm of T/h. (b) The logarithm of the error of Hamiltonian against t

Its first integral (conservative case, when $\theta = \pi/2$) or Lyapunov function (dissipative case, when $\theta < \pi/2$) is

$$H = \frac{1}{2}r(x_1^2 + x_2^2) - \frac{1}{2}\sin\theta\left(x_1x_2^2 - \frac{1}{3}x_1^3\right) + \frac{1}{2}\cos\theta\left(-x_1^2x_2 + \frac{1}{3}x_2^3\right).$$

The initial values are given by $x_1(0) = 0$, $x_2(0) = 1$. Firstly we consider the conservative case and choose $\theta = \pi/2$, $r = 20$. The problem is integrated on $[0, 1000]$ with the stepsize $h = 0.1/2^i$ for $i = 1, \cdots, 4$ and the global errors are given in Fig. 5.3a. Then we solve this system with the stepsize $h = 1/200$ on the interval $[0, 10000]$ and Fig. 5.3b shows the results of the energy preservation. Secondly we choose $\theta = \pi/2 - 10^{-4}$ and this gives a dissipative system. The system is solved on $[0, 1000]$ with $h = 0.1/2^i$ for $i = 1, \cdots, 4$ and the errors are presented in Fig. 5.4a. See Fig. 5.4b for the results of the Lyapunov function with $h = 1/20$. Here we consider the results given by EC2P with a smaller stepsize $h = 1/1000$ as the 'exact' values of the Lyapunov function. Table 5.2 gives the total numbers of iterations when applying the methods to this problem on $[0, 10]$ with $\theta = \pi/2$, $r = 20$, $h = 0.01$ and different error tolerances.

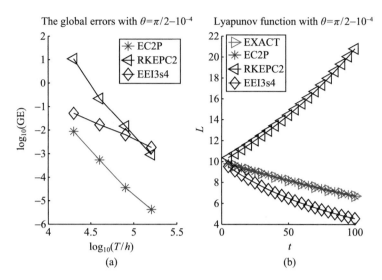

Fig. 5.4   (a) The logarithm of the global error against the logarithm of T/h. (b) The results for the Lyapunov function against t

Table 5.2   Results for Problem 5.2: the total number of iterations for different error tolerances (tol)

Methods	tol $= 1.0 \times 10^{-6}$	tol $= 1.0 \times 10^{-8}$	tol $= 1.0 \times 10^{-10}$	tol $= 1.0 \times 10^{-12}$
EC2P	2000	3000	3434	4000
RKEPC2	6000	8000	9999	11, 000

Problem 5.3   Consider the nonlinear Schrödinger equation (see [68])

$$i\psi_t + \psi_{xx} + 2|\psi|^2\psi = 0, \quad \psi(x, 0) = 0.5 + 0.025 \cos(\mu x),$$

with the periodic boundary condition $\psi(0, t) = \psi(L, t)$. Following [68], we choose $L = 4\sqrt{2}\pi$ and $\mu = 2\pi/L$. The initial condition chosen here is in the vicinity of the homoclinic orbit. Using $\psi = p + iq$, this equation can be rewritten as a pair of real-valued equations

$$p_t + q_{xx} + 2(p^2 + q^2)q = 0,$$
$$q_t - p_{xx} - 2(p^2 + q^2)p = 0.$$

Discretising the spatial derivative ∂_{xx} by the pseudospectral method given in [68], this problem is converted into the following system:

$$\begin{pmatrix} p \\ q \end{pmatrix}' = \begin{pmatrix} 0 & -D_2 \\ D_2 & 0 \end{pmatrix} \begin{pmatrix} p \\ q \end{pmatrix} + \begin{pmatrix} -2(p^2 + q^2) \cdot q \\ 2(p^2 + q^2) \cdot p \end{pmatrix}, \tag{5.43}$$

where $p = (p_0, p_1, \cdots, p_{N-1})^\mathsf{T}$, $q = (q_0, q_1, \cdots, q_{N-1})^\mathsf{T}$ and $D_2 = (D_2)_{0 \leqslant j,k \leqslant N-1}$ is the pseudospectral differentiation matrix defined by:

$$(D_2)_{jk} = \begin{cases} \dfrac{1}{2}\mu^2(-1)^{j+k+1}\dfrac{1}{\sin^2(\mu(x_j - x_k)/2)}, & j \neq k, \\ -\mu^2 \dfrac{2(N/2)^2 + 1}{6}, & j = k, \end{cases}$$

with $x_j = j\dfrac{L}{N}$ for $j = 0, 1, \cdots, N - 1$. The Hamiltonian of (5.43) is

$$H(p, q) = \frac{1}{2}p^\mathsf{T} D_2 p + \frac{1}{2}q^\mathsf{T} D_2 q + \frac{1}{2}\sum_{i=0}^{N-1}(p_i^2 + q_i^2)^2.$$

We choose $N = 128$ and first solve the problem on the interval $[0, 10]$ with $h = 0.1/2^i$ for $i = 3, \cdots, 6$. See Fig. 5.5a for the global errors. Then, this problem is integrated with $h = 1/200$ on $[0, 1000]$ and the energy conservation is presented in Fig. 5.5b. The total numbers of iterations when solving this problem on $[0, 10]$ with $N = 32$, $h = 0.1$ and different error tolerances are shown in Table 5.3.

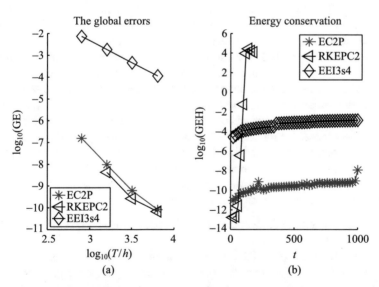

Fig. 5.5 (a) The logarithm of the global error against the logarithm of T/h. (b) The logarithm of the Hamiltonian error against t

Table 5.3 Results for Problem 5.3: the total number of iterations for different error tolerances (tol)

Methods	tol $= 1.0 \times 10^{-6}$	tol $= 1.0 \times 10^{-8}$	tol $= 1.0 \times 10^{-10}$	tol $= 1.0 \times 10^{-12}$
EC2P	488	632	796	963
RKEPC2	2558	4229	6991	8551

It can be concluded from these numerical experiments that the EC2P method definitely shows higher accuracy, better invariant-preserving property, and good long-term behaviour in the numerical simulations, compared to the other effective methods in the literature.

5.10 Concluding Remarks and Discussions

For several decades, exponential integrators have constituted an important class of methods for the numerical simulation of first-order ODEs, including the semi-discrete nonlinear Schrödinger equation etc. Finite element methods for ODEs can be traced back to the early 1960s and they have been investigated by many researchers. In this chapter, combining the ideas of these two types of effective methods, we derived and analysed a type of exponential collocation method for the conservative or dissipative system (5.1). We have also rigorously analysed its properties including existence and uniqueness, and algebraic order. It has been proved that the exponential collocation methods can achieve an arbitrary order of accuracy as well as preserve first integrals or Lyapunov functions exactly or approximately. The application of the methods to stiff gradient systems was discussed. The efficiency and superiority of exponential collocation methods were demonstrated by numerical results. By the analysis of this chapter, arbitrary-order energy-preserving methods were presented for second-order highly oscillatory/general systems.

Last, but not least, it is noted that the application of the methodology presented in this chapter to other ODEs such as general gradient systems (see [69]) and Poisson systems (see [70]) has been presented recently. We also note that there are some further issues of these methods to be considered.

- The error bounds and convergence properties of exponential collocation methods can be investigated.
- Another issue for exploration is the application of the methodology to PDEs such as nonlinear Schrödinger equations and wave equations (see, e.g. [71]).
- The long-time energy conservation of exponential collocation methods as well as its analysis by modulated Fourier expansion is another point which can be researched.

The material in this chapter is based on the work by Wang and Wu [72].

References

1. Hairer E, Lubich C, Wanner G. Geometric Numerical Integration: Structure-Preserving Algorithms for Ordinary Differential Equations. 2nd ed. Berlin, Heidelberg: Springer-Verlag, 2006.
2. Hochbruck M, Ostermann A. Exponential integrators. Acta Numer., 2010, 19: 209–286.
3. Wu X, Wang B. Recent Developments in Structure-Preserving Algorithms for Oscillatory Differential Equations. Singapore: Springer Nature Singapore Pte Ltd., 2018.
4. Wu X, You X, Wang B. Structure-preserving Algorithms for Oscillatory Differential Equations. Berlin, Heidelberg: Springer-Verlag, 2013.
5. Berland H, Owren B, Skaflestad B. B-series and order conditions for exponential integrators. SIAM J. Numer. Anal., 2005, 43: 1715–1727.
6. Butcher J C. Trees, B-series and exponential integrators. IMA J. Numer. Anal., 2009, 30: 131–140.
7. Caliari M, Ostermann A. Implementation of exponential Rosenbrock-type integrators. Appl. Numer. Math., 2009, 59: 568–581.
8. Calvo M, Palencia C. A class of explicit multistep exponential integrators for semilinear problems. Numer. Math., 2006, 102: 367–381.
9. Cano B, Gonzalez-Pachon A. Projected explicit Lawson methods for the integration of Schrödinger equation. Numer. Methods Partial Differ. Eq., 2015, 31: 78–104.
10. Celledoni E, Cohen D, Owren B. Symmetric exponential integrators with an application to the cubic Schrödinger equation. Found. Comput. Math., 2008, 8: 303–317.
11. Einkemmer L, Tokman M, Loffeld J. On the performance of exponential integrators for problems in magnetohydrodynamics. J. Comput. Phys., 2017, 330: 550–565.
12. Grimm V, Hochbruck M. Error analysis of exponential integrators for oscillatory second order differential equations. J. Phys. A: Math. Gen., 2006, 39: 5495–5507.
13. Hochbruck M, Ostermann A. Exponential Runge–Kutta methods for parabolic problems. Appl. Numer. Math., 2005, 53: 323–339.
14. Hochbruck M, Ostermann A, Schweitzer J. Exponential rosenbrock-type methods. SIAM J. Numer. Anal., 2009, 47: 786–803.
15. Ostermann A, Thalhammer M, Wright W M. A class of explicit exponential general linear methods. BIT Numer. Math., 2006, 46: 409–431.
16. Wang B, Wu X, Meng F, et al. Exponential Fourier collocation methods for solving first-order differential equations. J. Comput. Math., 2017, 35: 711–736.
17. Wu X, Wang B, Xia J. Explicit symplectic multidimensional exponential fitting modified Runge–Kutta–Nyström methods. BIT Numer. Math., 2012, 52: 773–795.
18. Mei L, Wu X. Symplectic exponential Runge–Kutta methods for solving nonlinear Hamiltonian systems. J. Comput. Phys., 2017, 338: 567–584.
19. Celledoni E, Mclachlan R I, Owren B, et al. Energy-preserving integrators and the structure of B-series. Found. Comput. Math., 2010, 10: 673–693.
20. Celledoni E, Owren B, Sun Y. The minimal stage, energy preserving Runge–Kutta method for polynomial Hamiltonian systems is the averaged vector field method. Math. Comput., 2014, 83: 1689–1700.
21. Quispel G R W, McLaren D I. A new class of energy-preserving numerical integration methods. J. Phys. A: Math. Theor., 2008, 41: 045206.
22. McLachlan R I, Quispel G R W. Discrete gradient methods have an energy conservation law. Disc. Contin. Dyn. Syst., 2014, 34: 1099–1104.
23. McLachlan, R.I., Quispel, G.R.W., Robidoux, N.: Geometric integration using discrete gradient. Philos. Trans. R. Soc. Lond. A **357**, 1021–1045 (1999)
24. Wang, B., Wu, X.: The formulation and analysis of energy-preserving schemes for solving high-dimensional nonlinear Klein-Gordon equations. SIMA J. Numer. Anal. **39**, 2016–2044 (2019)

25. Brugnano, L., Iavernaro, F.: Line Integral Methods for Conservative Problems. Chapman and Hall/CRC Press, Boca Raton (2016)
26. Brugnano, L., Iavernaro, F., Trigiante, D.: Hamiltonian boundary value methods (energy preserving discrete line integral methods). J. Numer. Anal. Ind. Appl. Math. **5**, 13–17 (2010)
27. Brugnano, L., Iavernaro, F., Trigiante, D.: A simple framework for the derivation and analysis of effective one-step methods for ODEs. Appl. Math. Comput. **218**, 8475–8485 (2012)
28. Brugnano, L., Iavernaro, F., Trigiante, D.: Energy- and quadratic invariants-preserving integrators based upon Gauss-collocation formulae. SIAM J. Numer. Anal. **50**, 2897–2916 (2012)
29. Cohen, D., Hairer, E.: Linear energy-preserving integrators for Poisson systems. BIT Numer. Math. **51**, 91–101 (2011)
30. Hairer, E.: Energy-preserving variant of collocation methods. J. Numer. Anal. Ind. Appl. Math. **5**, 73–84 (2010)
31. Betsch, P., Steinmann, P.: Inherently energy conserving time finite elements for classical mechanics. J. Comput. Phys. **160**, 88–116 (2000)
32. Betsch, P., Steinmann, P.: Conservation properties of a time FE method, I. Time-stepping schemes for N-body problems. Int. J. Numer. Meth. Eng. **49**, 599–638 (2000)
33. Hansbo, P.: A note on energy conservation for Hamiltonian systems using continuous time finite elements. Commun. Numer. Methods Eng. **7**, 863–869 (2001)
34. Li, Y.W., Wu, X.: Functionally fitted energy-preserving methods for solving oscillatory nonlinear Hamiltonian systems. SIAM J. Numer. Anal. **54**, 2036–2059 (2016)
35. Tang, W., Sun, Y.: Time finite element methods: a unified framework for the numerical discretizations of ODEs. Appl. Math. Comput. **219**, 2158–2179 (2012)
36. Miyatake, Y.: An energy-preserving exponentially-fitted continuous stage Runge–Kutta method for Hamiltonian systems. BIT Numer. Math. **54**, 777–799 (2014)
37. Miyatake, Y.: A derivation of energy-preserving exponentially-fitted integrators for Poisson systems. Comput. Phys. Commun. **187**, 156–161 (2015)
38. Calvo, M., Laburta, M.P., Montijano, J.I., et al.: Projection methods preserving Lyapunov functions. BIT Numer. Math. **50**, 223–241 (2010)
39. Hairer, E., Lubich, C.: Energy-diminishing integration of gradient systems. IMA J. Numer. Anal. **34**, 452–461 (2014)
40. Mclachlan, R.I., Quispel, G.R.W., Robidoux, N.: A unified approach to Hamiltonian systems, Poisson systems, gradient systems, and systems with Lyapunov functions or first integrals. Phys. Rev. Lett. **81**, 2399–2411 (1998)
41. Wang, B., Wu, X.: A new high precision energy-preserving integrator for system of oscillatory second-order differential equations. Phys. Lett. A **376**, 1185–1190 (2012)
42. Wu, X., Wang, B., Shi, W.: Efficient energy preserving integrators for oscillatory Hamiltonian systems. J. Comput. Phys. **235**, 587–605 (2013)
43. Li, Y.W., Wu, X.: Exponential integrators preserving first integrals or Lyapunov functions for conservative or dissipative systems. SIAM J. Sci. Comput. **38**, 1876–1895 (2016)
44. Brugnano, L., Gurioli, G., Sun, Y.: Energy-conserving Hamiltonian boundary value methods for the numerical solution of the Korteweg-de Vries equation. J. Comput. Appl. Math. **351**, 117–135 (2019)
45. Brugnano, L., Gurioli, G., Zhang, C.: Spectrally accurate energy-preserving methods for the numerical solution of the "Good" Boussinesq equation. Numer. Methods Partial Differ. Eq. **35**, 1343–1362 (2019)
46. Brugnano, L., Iavernaro, F., Montijano, J.I., et al.: Spectrally accurate space-time solution of Hamiltonian PDEs. Numer. Algor. **81**, 1183–1202 (2019)
47. Brugnano, L., Montijano, J.I., Rández, L.: On the effectiveness of spectral methods for the numerical solution of multi-frequency highly-oscillatory Hamiltonian problems. Numer. Algor. **81**, 345–376 (2019)
48. Brugnano, L., Zhang, C., Li, D.: A class of energy-conserving Hamiltonian boundary value methods for nonlinear Schrödinger equation with wave operator. Commun. Nonlinear Sci. Numer. Simul. **60**, 33–49 (2018)

49. Hochbruck, M., Lubich, C., Selhofer, H.: Exponential integrators for large systems of differential equations. SIAM J. Sci. Comput. **19**, 1552–1574 (1998)
50. Hochbruck, M., Ostermann, A.: Explicit exponential Runge–Kutta methods for semilinear parabolic problems. SIAM J. Numer. Anal. **43**, 1069–1090 (2005)
51. Cohen, D., Hairer, E., Lubich, C.: Numerical energy conservation for multi-frequency oscillatory differential equations. BIT Numer. Math. **45**, 287–305 (2005)
52. Franco, J.M.: New methods for oscillatory systems based on ARKN methods. Appl. Numer. Math. **56** 1040–1053 (2006)
53. García-Archilla, B., Sanz-Serna, J.M., Skeel, R.D.: Long-time-step methods for oscillatory differential equations. SIAM J. Sci. Comput. **20**, 930–963 (1999)
54. Hochbruck, M., Lubich, C.: A Gautschi-type method for oscillatory second-order differential equations. Numer. Math. **83**, 403–426 (1999)
55. Iserles, A.: Think globally, act locally: solving highly-oscillatory ordinary differential equations. Appl. Numer. Math. **43**, 145–160 (2002)
56. Wang, B., Meng, F., Fang, Y.: Efficient implementation of RKN-type Fourier collocation methods for second-order differential equations. Appl. Numer. Math. **119**, 164–178 (2017)
57. Wang, B., Wu, X., Meng, F.: Trigonometric collocation methods based on Lagrange basis polynomials for multi-frequency oscillatory second-order differential equations. J. Comput. Appl. Math. **313**, 185–201 (2017)
58. Wu, X., You, X., Shi, W., et al.: ERKN integrators for systems of oscillatory second-order differential equations. Comput. Phys. Commun. **181**, 1873–1887 (2010)
59. Wang, B., Iserles, A., Wu, X.: Arbitrary-order trigonometric Fourier collocation methods for multi-frequency oscillatory systems. Found. Comput. Math. **16**, 151–181 (2016)
60. Wang, B., Li, J., Fang, Y.: Long-term analysis of exponential integrators for highly oscillatory conservative systems (2018). arXiv: 1809. 07268
61. Barrett, J., Blowey, J.: Finite element approximation of an Allen-Cahn/Cahn-Hilliard system. IMA J. Numer. Anal. **22**, 11–71 (2002)
62. Lubich, C., Ostermann, A.: Runge–Kutta methods for parabolic equations and convolution quadrature. Math. Comput. **60**, 105–131 (1993)
63. Lubich, C., Ostermann, A.: Runge–Kutta time discretization of reaction-diffusion and Navier Stokes equations: nonsmooth-data error estimates and applications to long-time behaviour. Appl. Numer. Math. **22**, 279–292 (1996)
64. Hairer, E., Wanner, G.: Solving ordinary differential equations II. Stiff and Differential Algebraic Problems, Springer Series in Computational Mathematics 14, 2nd edn. Springer, Berlin (1996)
65. Al-Mohy, A.H., Higham, N.J.: Computing the action of the matrix exponential, with an application to exponential integrators. SIAM J. Sci. Comput. **33**, 488–511 (2011)
66. Hochbruck, M., Lubich, C.: On Krylov subspace approximations to the matrix exponential operator. SIAM J. Numer. Anal. **34**, 1911–1925 (1997)
67. Iavernaro, F., Trigiante, D.: High-order symmetric schemes for the energy conservation of polynomial Hamiltonian problems. J. Numer. Anal. Ind. Appl. Math. **4**, 87–101 (2009)
68. Chen, J.B., Qin, M.Z.: Multi-symplectic Fourier pseudospectral method for the nonlinear Schrödinger equation. Electron. Trans. Numer. Anal. **12**, 193–204 (2001)
69. Wang, B., Li, T., Wu, Y.: Arbitrary-order functionally fitted energy-diminishing methods for gradient systems. Appl. Math. Lett. **83**, 130–139 (2018)
70. Wang, B., Wu, X.: Functionally-fitted energy-preserving integrators for Poisson systems. J. Comput. Phys. **364**, 137–152 (2018)
71. Wang, B., Wu, X.: Exponential collocation methods based on continuous finite element approximations for efficiently solving the cubic Schrödinger equation. Numer. Methods Partial Differ. Eq. **36**, 1735–1757 (2020)
72. Wang, B., Wu, X.: Exponential collocation methods for conservative or dissipative systems. J. Comput. Appl. Math. **360**, 99–116 (2019)

Chapter 6
Volume-Preserving Exponential Integrators

Since various dynamical systems preserve volume in phase space, such as all Hamiltonian systems, this qualitative geometrical property of the analytical solution should be preserved within the framework of Geometric Integration. This chapter considers the volume-preserving exponential integrators for different vector fields. We first analyse a necessary and sufficient condition of volume preservation for exponential integrators. We then discuss volume-preserving exponential integrators for four kinds of vector fields. It turns out that symplectic exponential integrators can be volume preserving for a much larger class of vector fields than Hamiltonian systems. On the basis of this profound analysis, the applications of volume-preserving exponential integrators are demonstrated. For solving highly oscillatory second-order systems, efficient volume-preserving exponential integrators are derived, and for separable partitioned systems, volume-preserving ERKN integrators are presented. Moreover, volume-preserving RKN methods are also investigated.

6.1 Introduction

Geometric numerical integrators (also known as structure-preserving algorithms) have been an active area of great interest in recent decades. A remarkable advantage of such integrators for solving ordinary differential equations (ODEs) is that they can exactly preserve some qualitative geometrical property of the analytical solution, such as the symplecticity, preservation of energy, momentum, angular momentum, phase-space volume, and symmetries. These geometric properties are of crucial importance in physical applications. Various geometric integrators have been designed and researched recently and we refer the reader to [1–11]. For a good theoretical foundation in connection with geometric numerical integration for ODEs, we refer the reader to [12, 13]. Highly oscillatory differential equations are

© The Author(s), under exclusive license to Springer Nature Singapore Pte Ltd. 2021
X. Wu, B. Wang, *Geometric Integrators for Differential Equations with Highly Oscillatory Solutions*, https://doi.org/10.1007/978-981-16-0147-7_6

currently a subject of great interest, and surveys of structure-preserving algorithms for them are presented in [14–17].

As is known, volume preservation is an important property of numerous dynamical systems. It is clear from the classical theorem due to Liouville that all Hamiltonian systems are volume preserving. In the sense of geometric integrators, the structure of volume preservation should also be respected when the underlying dynamical system is discretised. Some numerical methods have been proposed and shown to be (or not to be) volume preserving (see, e.g. [18–25] and references therein). It has been shown that all symplectic methods are volume preserving for Hamiltonian systems. However, it is important to recognise that this result does not hold for the non-Hamiltonian systems (see [19, 20, 22]). The authors in [26] have pointed out that the derivation of efficient volume-preserving (VP) methods is still an open problem in geometric numerical integration. Recently, various VP methods have been constructed and analysed, such as splitting methods (see [23, 25]), Runge–Kutta (RK) methods (see [18]) and the methods based on generating functions (see [24, 27]).

On the other hand, exponential integrators have been developed and researched as an efficient approach to the numerical integration of ODEs/PDEs. The reader is referred to [28–31] for some examples of exponential integrators. In comparison with RK methods, exponential integrators exactly solve the linear system associated with the underlying ODEs. Accordingly, exponential integrators can be expected to perform better than RK methods when solving highly oscillatory systems and the results of many numerical experiments have demonstrated this point (see [29, 31]). Hence, the study of this chapter focuses on volume-preserving exponential integrators. More precisely, we are concerned with systems of ODEs of the form

$$y'(t) = Ky(t) + g(y(t)) := f(y(t)), \quad y(0) = y_0 \in \mathbb{R}^n, \tag{6.1}$$

where K is an $n \times n$ matrix which is assumed to satisfy $\left|e^{hK}\right| \neq -1$ for $0 < h < 1$, and $g : \mathbb{R}^n \to \mathbb{R}^n$ is a differentiable nonlinear function. In this chapter, $|\cdot|$ denotes the determinant. The function f is assumed to be divergence free so that this system is volume preserving. It is well known that the exact solution of (6.1) can be represented by the variation-of-constants formula (or Volterra integral equation)

$$y(t) = e^{tK}y_0 + t \int_0^1 e^{(1-\tau)tK} g(y(\tau t)) d\tau. \tag{6.2}$$

The focus of our attention in this chapter is to derive the volume-preserving condition for exponential integrators and analyse the qualitative feature of volume preservation for larger classes of vector fields than Hamiltonian systems. Furthermore, on the basis of the analysis, volume-preserving adapted exponential integrators are formulated for highly oscillatory systems of second-order ODEs and volume-preserving extended Runge–Kutta–Nyström (ERKN) integrators are derived for separable partitioned systems. We also discuss the volume preservation of Runge–Kutta–Nyström (RKN) methods by considering them as a special class of ERKN integrators.

6.2 Exponential Integrators

We approximate the integral appearing in (6.2) by a quadrature formula with suitable nodes $c_1, c_2, \cdots, c_s$. This leads to the following definition of exponential integrators for (6.1).

Definition 6.1 (See [29]) An s-stage exponential integrator applied with stepsize h for numerically solving (6.1) is defined by

$$\begin{cases} k_i = e^{c_i h K} y_n + h \sum_{j=1}^{s} \bar{a}_{ij}(hK) g(k_j), \quad i = 1, 2, \cdots, s, \\ y_{n+1} = e^{hK} y_n + h \sum_{i=1}^{s} \bar{b}_i(hK) g(k_i), \end{cases} \quad (6.3)$$

where c_i for $i = 1, \cdots, s$ are real constants, and $\bar{b}_i(hK)$ and $\bar{a}_{ij}(hK)$ for $i, j = 1, \cdots, s$ are matrix-valued functions of hK.

The exponential integrator can be represented briefly in Butcher's notation by the following block tableau of coefficients:

$$\frac{c \mid \bar{A}}{\mid \bar{b}^{\mathsf{T}}} = \begin{array}{c|ccc} c_1 & \bar{a}_{11} & \cdots & \bar{a}_{1s} \\ \vdots & \vdots & \ddots & \vdots \\ c_s & \bar{a}_{s1} & \cdots & \bar{a}_{ss} \\ \hline & \bar{b}_1 & \cdots & \bar{b}_s \end{array},$$

where (hK) is suppressed for brevity. This kind of exponential integrator has been successfully used for solving different kinds of ODEs/PDEs (see [28–31]). Clearly, when $K = 0$, an s-stage exponential integrator reduces to a classical s-stage RK method represented by the Butcher tableau

$$\frac{c \mid A}{\mid b^{\mathsf{T}}} = \begin{array}{c|ccc} c_1 & a_{11} & \cdots & a_{1s} \\ \vdots & \vdots & \ddots & \vdots \\ c_s & a_{s1} & \cdots & a_{ss} \\ \hline & b_1 & \cdots & b_s \end{array}.$$

In what follows, we consider an important and special kind of exponential integrators which was first proposed in [32].

Definition 6.2 (See [32]) A kind of special s-stage exponential integrator applied with stepsize h is defined by

$$\bar{a}_{ij}(hK) = a_{ij}e^{(c_i-c_j)hK}, \quad \bar{b}_i(hK) = b_i e^{(1-c_i)hK}, \quad i, j = 1, \cdots, s, \qquad (6.4)$$

where

$$c = (c_1, \cdots, c_s)^\mathsf{T}, \quad b = (b_1, \cdots, b_s)^\mathsf{T}, \quad A = (a_{ij})_{s \times s} \qquad (6.5)$$

are the coefficients of an s-stage Runge–Kutta (RK) method.

Two useful properties of this kind of exponential integrator are shown in [32] and are summarised below.

Theorem 6.1 (See [32]) *If a Runge–Kutta method with the coefficients (6.5) is of order m, then the exponential integrator given by (6.4) is also of order m.*

Theorem 6.2 (See [32]) *The exponential integrator defined by (6.4) is symplectic if the corresponding Runge–Kutta method (6.5) is symplectic.*

In this chapter, we supplement an additional requirement for b and use the following two abbreviations.

Definition 6.3 An s-stage exponential integrator (6.4) is called as a symplectic exponential integrator (SEI) if the RK method (6.5) is symplectic. Moreover, we call the integrator (6.4) a special symplectic exponential integrator (SSEI) if $b_j \neq 0$ for all $j = 1, \cdots, s$ and $BA + A^\mathsf{T}B - bb^\mathsf{T} = 0$ with $B = \mathrm{diag}(b)$.

Remark 6.2.1 It is important to note that a kind of special symplectic RK (SSRK) methods has been considered in [18] and the SSEI integrators reduce to the SSRK methods when $K = \mathbf{0}$.

6.3 VP Condition of Exponential Integrators

This section is devoted to VP condition of exponential integrators. To this end, we denote the s-stage exponential integrator (6.3) applied with stepsize h by a map $\varphi_h : \mathbb{R}^n \to \mathbb{R}^n$, which is

$$\begin{cases} \varphi_h(y) = e^{hK}y + h\displaystyle\sum_{i=1}^{s} \bar{b}_i(hK)g(k_i(y)), \\ k_i(y) = e^{c_i hK}y + h\displaystyle\sum_{j=1}^{s} \bar{a}_{ij}(hK)g(k_j(y)), \quad i = 1, 2, \cdots, s. \end{cases} \qquad (6.6)$$

We first derive the Jacobian matrix of φ_h and then present the result of its determinant.

Lemma 6.1 *The Jacobian matrix of the exponential integrator* (6.6) *can be expressed as*

$$\varphi_h'(y) = e^{hK} + h\bar{b}^\mathsf{T} F(I_s \otimes I - h\bar{A}F)^{-1} e^{chK},$$

where $F = \mathrm{diag}(g'(k_1), \cdots, g'(k_s))$, I_s *and* I *are the* $s \times s$ *and* $n \times n$ *identity matrices, respectively, and* $e^{chK} = (e^{c_1 hK}, \cdots, e^{c_s hK})^\mathsf{T}$. *Its determinant reads*

$$|\varphi_h'(y)| = \frac{|e^{hK}| \, |I_s \otimes I - h(\bar{A} - e^{(c-1)hK}\bar{b}^\mathsf{T})F|}{|I_s \otimes I - h\bar{A}F|}, \tag{6.7}$$

where $e^{(c-1)hK} = (e^{(c_1-1)hK}, \cdots, e^{(c_s-1)hK})^\mathsf{T}$. *Here we make use of the Kronecker product* $\otimes$ *throughout this chapter.*

Proof The proof is similar to that of Lemma 2.1 in [18] but with some modifications. According to the first formula of (6.6), we obtain

$$\varphi_h'(y) = e^{hK} + h\sum_{i=1}^{s} \bar{b}_i g'(k_i(y))k_i'(y) = e^{hK} + h\bar{b}^\mathsf{T} F(k_1', \cdots, k_s')^\mathsf{T}. \tag{6.8}$$

Likewise, it follows from $k_i(y)$ in (6.6) that

$$\begin{pmatrix} I - h\bar{a}_{11}g'(k_1) & -h\bar{a}_{12}g'(k_2) & \cdots & -h\bar{a}_{1s}g'(k_s) \\ -h\bar{a}_{21}g'(k_1) & I - h\bar{a}_{22}g'(k_2) & \cdots & -h\bar{a}_{2s}g'(k_s) \\ \vdots & \vdots & & \vdots \\ -h\bar{a}_{s1}g'(k_1) & -h\bar{a}_{s2}g'(k_2) & \cdots & I - h\bar{a}_{ss}g'(k_s) \end{pmatrix} \begin{pmatrix} k_1' \\ k_2' \\ \vdots \\ k_s' \end{pmatrix} = e^{chK},$$

which can be rewritten as

$$(I_s \otimes I - h\bar{A}F)(k_1', \cdots, k_s')^\mathsf{T} = e^{chK}. \tag{6.9}$$

Substituting (6.9) into (6.8) yields the first statement of this lemma.

For the second statement, we will use the following block determinant identity (see [13, 18]):

$$|U| \, |X - WU^{-1}V| = \begin{vmatrix} U & V \\ W & X \end{vmatrix} = |X| \, |U - VX^{-1}W|,$$

which is yielded by Gaussian elimination. Let

$$X = e^{hK}, \quad W = -h\bar{b}^\mathsf{T} F, \quad U = I_s \otimes I - h\bar{A}F, \quad V = e^{chK}.$$

It is clear that

$$
\left| I_s \otimes I - h \bar{A} F \right| \left| \varphi_h'(y) \right| = \left| e^{hK} \right| \left| I_s \otimes I - h \bar{A} F + h e^{chK} e^{-hK} \bar{b}^{\mathsf{T}} F \right|
$$

$$
= \left| e^{hK} \right| \left| I_s \otimes I - h(\bar{A} - e^{(c-1)hK} \bar{b}^{\mathsf{T}}) F \right|,
$$

which leads to the result (6.7).   □

By Lemma 6.1, a necessary and sufficient condition for the SSEI methods to be volume preserving is shown in the following lemma.

Lemma 6.2 *An s-stage SSEI method defined in Definition 6.3 is volume preserving if and only if the following VP condition is satisfied*

$$
\left| I_s \otimes I - h(A \otimes I. * E(hK)) F \right| = \left| e^{hK} \right| \left| I_s \otimes I + h(A^{\mathsf{T}} \otimes I. * E(hK)) F \right|,
$$

$$
(6.10)
$$

where $E(hK)$ is a block matrix defined by

$$
E(hK) = (E_{i,j}(hK))_{s \times s} = \begin{pmatrix} I & e^{(c_1-c_2)hK} & \cdots & e^{(c_1-c_s)hK} \\ e^{(c_2-c_1)hK} & I & \cdots & e^{(c_2-c_s)hK} \\ \vdots & \vdots & & \vdots \\ e^{(c_s-c_1)hK} & e^{(c_s-c_2)hK} & \cdots & I \end{pmatrix}, \quad (6.11)
$$

and $. *$ denotes the element-wise multiplication of two matrices.

Proof From the choice (6.4) of the coefficients, we calculate

$$
\bar{A} - e^{(c-1)hK} \bar{b}^{\mathsf{T}}
$$

$$
= (A \otimes I). * E(hK) - (e^{(c_1-1)hK}, \cdots, e^{(c_s-1)hK})^{\mathsf{T}} (b_1 e^{(1-c_1)hK}, \cdots, b_s e^{(1-c_s)hK})
$$

$$
= (A \otimes I). * E(hK) - (\mathbf{1} b^{\mathsf{T}} \otimes I). * E(hK)
$$

$$
= (A - \mathbf{1} b^{\mathsf{T}}) \otimes I. * E(hK).
$$

We then obtain

$$
\left| \varphi_h'(y) \right| = \frac{\left| e^{hK} \right| \left| I_s \otimes I - h(A - \mathbf{1} b^{\mathsf{T}}) \otimes I. * E(hK) F \right|}{\left| I_s \otimes I - h \bar{A} F \right|}. \quad (6.12)
$$

Furthermore, it can be verified that for $B = \text{diag}(b_1, \cdots, b_s)$, the following result holds

$$
\left| I_s \otimes I - h(A - \mathbf{1} b^{\mathsf{T}}) \otimes I. * E(hK) F \right|
$$

$$
= \left| I_s \otimes I - h(B \otimes I)(A - \mathbf{1} b^{\mathsf{T}}) \otimes I. * E(hK) F(B^{-1} \otimes I) \right|
$$

$$= \left| I_s \otimes I - h(B \otimes I)(A - \mathbf{1}b^\mathsf{T}) \otimes I. * E(hK)(B^{-1} \otimes I)F \right|$$

$$= \left| I_s \otimes I - hB(A - \mathbf{1}b^\mathsf{T})B^{-1} \otimes I. * E(hK)F \right|. \tag{6.13}$$

because the RK method is symplectic, we have $BA + A^\mathsf{T}B - bb^\mathsf{T} = 0$ (see [13]), which leads to $B(A - \mathbf{1}b^\mathsf{T})B^{-1} = -A^\mathsf{T}$. The result (6.13) then can be simplified as

$$|I_s \otimes I - h(A - \mathbf{1}b^\mathsf{T}) \otimes I. * E(hK)F| = |I_s \otimes I + h(A^\mathsf{T} \otimes I. * E(hK))F|.$$

The proof is complete by considering (6.12).   □

Remark 6.3.1 Clearly, when $K = \mathbf{0}$, the VP condition (6.10) reduces to the condition of RK methods presented in [18]. This implies that the condition (6.10) can be regarded as a generalisation of that of RK methods.

6.4   VP Results for Different Vector Fields

This section concerns the volume-preserving properties of exponential integrators for the following four kinds of vector fields.

Definition 6.4 (See [18]) Define the following four classes of vector fields on Euclidean space using vector fields $f(y)$

$$\mathscr{H} = \{f(y)| \text{ there exists } a \; matrix \; P \text{ such that for all } y, \; Pf'(y)P^{-1} = -f'(y)^\mathsf{T}\},$$

$$\mathscr{S} = \{f(y)| \text{ there exists } a \; matrix \; P \text{ such that for all } y, \; Pf'(y)P^{-1} = -f'(y)\},$$

$$\mathscr{F}^{(\infty)} = \{f(y_1, y_2) = (u(y_1), v(y_1, y_2))^\mathsf{T} \text{ where } u \in \mathscr{H} \cup \mathscr{F}^{(\infty)}| \text{ there exists}$$

$$a \; matrix \; P \text{ such that for all } y_1, y_2, \; P\partial_{y_2}v(y_1, y_2)P^{-1} = -\partial_{y_2}v(y_1, y_2)^\mathsf{T}\},$$

$$\mathscr{F}^{(2)} = \{f(y_1, y_2) = (u(y_1), v(y_1, y_2))^\mathsf{T} \text{ where } u \in \mathscr{H} \cup \mathscr{S} \cup \mathscr{F}^{(2)}| \text{ there exists } a$$

$$matrix \; P \text{ such that for all } y_1, y_2, \text{ either } P\partial_{y_2}v(y_1, y_2)P^{-1} = -\partial_{y_2}v(y_1, y_2)^\mathsf{T},$$

$$\text{or } P\partial_{y_2}v(y_1, y_2)P^{-1} = -\partial_{y_2}v(y_1, y_2)\}.$$

Remark 6.4.1 As shown in [18], all these fields are equal to divergence free vector fields. The relationships of these vector fields are also given in [18] as

$$\mathscr{H} \subset \mathscr{F}^{(\infty)} \subset \mathscr{F}^{(2)} \text{ and } \mathscr{S} \subset \mathscr{F}^{(\infty)} \subset \mathscr{F}^{(2)}.$$

It follows from Lemma 3.2 of [18] that the set $\mathscr{H}$ contains all Hamiltonian systems. We denote the set of Hamiltonian systems by H and see Fig. 6.1 for the venn diagram illustrating the relationships. It can be seen from this figure that the sets $\mathscr{H}$, $\mathscr{F}^{(\infty)}$ and $\mathscr{F}^{(2)}$ are larger classes of vector fields than Hamiltonian systems. It

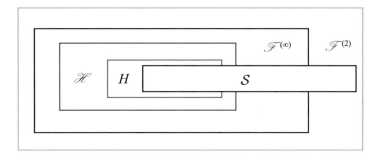

Fig. 6.1 Venn diagram illusting the relationships

is noted that the volume-preserving properties of RK methods for these vector fields
have been researched in [18]. Following this work and in what follows, we consider
extending those results for exponential integrators.

6.4.1 Vector Fields in $\mathscr{H}$

The following theorem shows the volume-preserving feature of SSEI methods for
vector fields in $\mathscr{H}$.

Theorem 6.3 *All SSEI methods for solving* (6.1) *are volume preserving for vector
fields f and g in* $\mathscr{H}$ *with the same P.*

Proof For vector fields f and g in $\mathscr{H}$ with the same P, we obtain that
$Pf'(y)P^{-1} = -f'(y)^\mathsf{T}$ and $Pg'(y)P^{-1} = -g'(y)^\mathsf{T}$. According to these conditions
and the expression $f(y) = Ky + g(y)$, one has that $PKP^{-1} = -K^\mathsf{T}$. Thus it is
clear that $Pe^{hK}P^{-1} = e^{-hK^\mathsf{T}}$. In the light of this result, we have

$$\left|Pe^{hK}P^{-1}\right| = \left|e^{hK}\right| = \left|e^{-hK^\mathsf{T}}\right| = \left|e^{-hK}\right| = \left|(e^{hK})^{-1}\right| = \frac{1}{\left|e^{hK}\right|},$$

which yields $\left|e^{hK}\right| = 1$ (it is assumed that $\left|e^{hK}\right| \neq -1$ in the introduction of this
chapter). We then compute the left-hand side of (6.10) as follows:

$$|I_s \otimes I - h(A \otimes I. * E(hK))F|$$
$$= \left|(I_s \otimes P)(I_s \otimes P^{-1}) - h(I_s \otimes P)(A \otimes I. * E(hK))(I_s \otimes P^{-1})(I_s \otimes P)F(I_s \otimes P^{-1})\right|$$
$$= \left|I_s \otimes I + h(I_s \otimes P)(A \otimes I. * E(hK))(I_s \otimes P^{-1})F^\mathsf{T}\right|$$
$$= \left|I_s \otimes I + h(A \otimes I. * E(-hK^\mathsf{T}))F^\mathsf{T}\right|$$
$$= \left|I_s \otimes I + hF(A^\mathsf{T} \otimes I. * E(-hK^\mathsf{T})^\mathsf{T})\right| \text{ (transpose)}$$
$$= \left|I_s \otimes I + h(A^\mathsf{T} \otimes I. * E(-hK^\mathsf{T})^\mathsf{T})F\right| \text{ (Sylvester's law).}$$

It follows from the definition (6.11) that

$$E(-hK^\mathsf{T})^\mathsf{T} = (E_{i,j}(-hK^\mathsf{T}))_{s\times s}^\mathsf{T} = (E_{j,i}(-hK))_{s\times s} = (E_{i,j}(hK))_{s\times s} = E(hK).$$
$$(6.14)$$

Thus, we have

$$\left|I_s \otimes I - h(A \otimes I.*E(hK))F\right| = \left|I_s \otimes I + h(A^\mathsf{T} \otimes I.*E(hK))F\right|.$$

This shows the statement of this theorem by considering Lemma 6.2.   □

6.4.2   Vector Fields in $\mathscr{S}$

Theorem 6.4 *All one-stage SSEI methods and all two-stage SSEI methods with*

$$e^{(c_2-c_1)hK}g'(k_2)e^{(c_1-c_2)hK}g'(k_1) = e^{(c_1-c_2)hK}g'(k_2)e^{(c_2-c_1)hK}g'(k_1) \qquad (6.15)$$

(and any composition of such methods) are volume preserving for vector fields f and g in $\mathscr{S}$ with the same P.

Proof In a similar way to the proof of the previous theorem, we obtain that $PKP^{-1} = -K$. Thus it is true that $Pe^{hK}P^{-1} = e^{-hK}$ and $\left|e^{hK}\right| = 1$.

For the one-stage SSEI methods, according to Lemma 6.2, they are volume preserving if and only if

$$\left|I - ha_{11}g'(k_1)\right| = \left|I + ha_{11}g'(k_1)\right|,$$

which can be verified by considering

$$\left|I - ha_{11}g'(k_1)\right| = \left|PP^{-1} - ha_{11}Pg'(k_1)P^{-1}\right| = \left|I + ha_{11}g'(k_1)\right|.$$

For a two-stage SSEI method, according to Lemma 6.2 again, this two-stage SSEI method is volume preserving if and only if

$$\begin{vmatrix} I - ha_{11}g'(k_1) & -ha_{12}e^{(c_1-c_2)hK}g'(k_2) \\ -ha_{21}e^{(c_2-c_1)hK}g'(k_1) & I - ha_{22}g'(k_2) \end{vmatrix}$$
$$= \begin{vmatrix} I + ha_{11}g'(k_1) & ha_{21}e^{(c_1-c_2)hK}g'(k_2) \\ ha_{12}e^{(c_2-c_1)hK}g'(k_1) & I + ha_{22}g'(k_2) \end{vmatrix},$$

which gives the necessary & sufficient condition

$$
\begin{aligned}
&|I - ha_{11}g'(k_1) - ha_{22}g'(k_2) + h^2a_{11}a_{22}g'(k_1)g'(k_2) \\
&\quad - h^2a_{12}a_{21}e^{(c_1-c_2)hK}g'(k_2)e^{(c_2-c_1)hK}g'(k_1)| \\
&= |I + ha_{11}g'(k_1) + ha_{22}g'(k_2) + h^2a_{11}a_{22}g'(k_1)g'(k_2) \\
&\quad - h^2a_{12}a_{21}e^{(c_1-c_2)hK}g'(k_2)e^{(c_2-c_1)hK}g'(k_1)|.
\end{aligned} \tag{6.16}
$$

Next, we show that (6.16) is satisfied by all two-stage SSEI methods obeying (6.15). Now, the left-hand side of (6.16)

$$
\begin{aligned}
&= |\, PP^{-1} - ha_{11}Pg'(k_1)P^{-1} - ha_{22}Pg'(k_2)P^{-1} + h^2a_{11}a_{22}Pg'(k_1)g'(k_2)P^{-1} \\
&\quad - h^2a_{12}a_{21}Pe^{(c_1-c_2)hK}g'(k_2)e^{(c_2-c_1)hK}g'(k_1)P^{-1}\,| \\
&= |\, I + ha_{11}g'(k_1) + ha_{22}g'(k_2) + h^2a_{11}a_{22}Pg'(k_1)P^{-1}Pg'(k_2)P^{-1} \\
&\quad - h^2a_{12}a_{21}Pe^{(c_1-c_2)hK}P^{-1}Pg'(k_2)P^{-1}Pe^{(c_2-c_1)hK}P^{-1}Pg'(k_1)P^{-1}\,| \\
&= |\, I + ha_{11}g'(k_1) + ha_{22}g'(k_2) + h^2a_{11}a_{22}g'(k_1)g'(k_2) \\
&\quad - h^2a_{12}a_{21}e^{(c_2-c_1)hK}g'(k_2)e^{(c_1-c_2)hK}g'(k_1)\,|\,.
\end{aligned}
$$

Under the assumption (6.15), the last line becomes

$$
\begin{aligned}
&|\, I + ha_{11}g'(k_1) + ha_{22}g'(k_2) + h^2a_{11}a_{22}g'(k_1)g'(k_2) \\
&\quad - h^2a_{12}a_{21}e^{(c_1-c_2)hK}g'(k_2)e^{(c_2-c_1)hK}g'(k_1)\,|\,.
\end{aligned}
$$

Thus (6.16) is obtained immediately, and then all two-stage SSEI methods with (6.15) are volume preserving.   □

Remark 6.4.2 It is noted that for the vector fields in $\mathscr{S}$ and two-stage SSEI methods, the condition (6.15) can be true for many special cases such as for some special matrix K or some special function g. The same situation will happen in the analysis of Sect. 6.4.4.

6.4.3 Vector Fields in $\mathscr{F}^{(\infty)}$

For vector fields in $\mathscr{F}^{(\infty)}$, if the function $f(y) := Ky + g(y)$ has the pattern $(u(y_1), v(y_1, y_2))^{\mathsf{T}}$, this means that K and g can be expressed in blocks as

$$
K = \begin{pmatrix} K_{11} & 0 \\ 0 & K_{22} \end{pmatrix}, \quad g(y) = \begin{pmatrix} g_1(y_1) \\ g_2(y_1, y_2) \end{pmatrix}. \tag{6.17}
$$

Then the following relation is true

$$u(y_1) = K_{11}y_1 + g_1(y_1), \quad v(y_1, y_2) = K_{22}y_2 + g_2(y_1, y_2). \tag{6.18}$$

Theorem 6.5 *Consider an s-stage SSEI method for solving* $y_1' = u(y_1)$ *that is volume preserving for the vector field* $u(y_1) : \mathbb{R}^m \to \mathbb{R}^m$. *Let* $v(y_1, y_2) : \mathbb{R}^{m+n} \to \mathbb{R}^{m+n}$ *and assume that there exists an invertible matrix* P *such that for all* y_1, y_2,

$$P\partial_{y_2} v(y_1, y_2) P^{-1} = -\partial_{y_2} v(y_1, y_2)^\mathsf{T}, \quad P\partial_{y_2} g_2(y_1, y_2) P^{-1} = -\partial_{y_2} g_2(y_1, y_2)^\mathsf{T}.$$

Then the SSEI method is volume preserving for vector fields $f(y_1, y_2) = (u(y_1), v(y_1, y_2))^\mathsf{T}$ *in* $\mathscr{F}^{(\infty)}$.

Proof It follows from the property of v that $PK_{22}P^{-1} = -K_{22}^\mathsf{T}$ and $\left|e^{hK_{22}}\right| = 1$. Thus $\left|e^{hK}\right| = \left|e^{hK_{11}}\right|\left|e^{hK_{22}}\right| = \left|e^{hK_{11}}\right|$. The Jacobian matrix of $g(y)$ is block triangular as follows

$$g'(y_1, y_2) = \begin{pmatrix} \partial_{y_1} g_1(y_1) & 0 \\ * & \partial_{y_2} g_2(y_1, y_2) \end{pmatrix}.$$

In what follows, we prove the condition (6.10). Using the block transformation, we can bring the left-hand side of (6.10) to the block form

$$|I_s \otimes I - h(A \otimes I. * E(hK))F| = \begin{pmatrix} \Phi_1 & 0 \\ * & \Phi_2 \end{pmatrix},$$

where

$$\Phi_1 = \begin{pmatrix} I - h\bar{a}_{11}(hK_{11})\partial_{y_1} g_1(k_1) & \cdots & -h\bar{a}_{1s}(hK_{11})\partial_{y_1} g_1(k_s) \\ \vdots & \ddots & \vdots \\ -h\bar{a}_{s1}(hK_{11})\partial_{y_1} g_1(k_1) & \cdots & I - h\bar{a}_{ss}(hK_{11})\partial_{y_1} g_1(k_s) \end{pmatrix},$$

$$\Phi_2 = \begin{pmatrix} I - h\bar{a}_{11}(hK_{22})\partial_{y_2} g_2(k_1) & \cdots & -h\bar{a}_{1s}(hK_{22})\partial_{y_2} g_2(k_s) \\ \vdots & \ddots & \vdots \\ -h\bar{a}_{s1}(hK_{22})\partial_{y_2} g_2(k_1) & \cdots & I - h\bar{a}_{ss}(hK_{22})\partial_{y_2} g_2(k_s) \end{pmatrix}.$$

Let $F_1 = \mathrm{diag}(\partial_{y_1} g_1(k_1), \cdots, \partial_{y_1} g_1(k_s))$ and $F_2 = \mathrm{diag}(\partial_{y_2} g_2(k_1), \cdots, \partial_{y_2} g_2(k_s))$. The above result can be simplified as

$$|I_s \otimes I - h(A \otimes I. * E(hK))F|$$
$$= |I_s \otimes I - h(A \otimes I. * E(hK_{11}))F_1| \, |I_s \otimes I - h(A \otimes I. * E(hK_{22}))F_2|.$$

Since the SSEI method is volume preserving for the vector field $u(y_1)$, the following condition is true

$$|I_s \otimes I - h(A \otimes I. * E(hK_{11}))F_1| = \left|e^{hK_{11}}\right| \left|I_s \otimes I + h(A^\mathsf{T} \otimes I. * E(hK_{11}))F_1\right|.$$

On the other hand, we compute

$$|I_s \otimes I - h(A \otimes I. * E(hK_{22}))F_2|$$

$$= \left|(I_s \otimes P)(I_s \otimes P^{-1}) - h(I_s \otimes P)(A \otimes I. * E(hK_{22}))(I_s \otimes P^{-1})(I_s \otimes P)F(I_s \otimes P^{-1})\right|$$

$$= \left|I_s \otimes I + h(I_s \otimes P)(A \otimes I. * E(hK_{22}))(I_s \otimes P^{-1})F_2^\mathsf{T}\right|$$

$$= \left|I_s \otimes I + h(A \otimes I. * E(-hK_{22}^\mathsf{T}))F_2^\mathsf{T}\right|$$

$$= \left|I_s \otimes I + hF_2(A^\mathsf{T} \otimes I. * E(-hK_{22}^\mathsf{T})^\mathsf{T})\right| \quad \text{(transpose)}$$

$$= \left|I_s \otimes I + h(A^\mathsf{T} \otimes I. * E(-hK_{22}^\mathsf{T})^\mathsf{T})F_2\right| \quad \text{(Sylvester's law)}$$

$$= \left|I_s \otimes I + h(A^\mathsf{T} \otimes I. * E(hK_{22}))F_2\right| \quad \text{(property (6.14))}.$$

Hence, the VP condition (6.10) holds and the SSEI method is volume preserving for vector fields in $\mathscr{F}^{(\infty)}$. □

6.4.4 Vector Fields in $\mathscr{F}^{(2)}$

It is assumed that the function $f(y)$ of (6.1) falls into $\mathscr{F}^{(2)}$. Under this situation, (6.17) and (6.18) are still true. This leads to the following result about the VP property of SSEI methods.

Theorem 6.6 *Consider a one-stage or two-stage SSEI with (6.15) (or a composition of such method) that is volume preserving for the vector field $u(y_1) : \mathbb{R}^m \to \mathbb{R}^m$. Letting $v(y_1, y_2) : \mathbb{R}^{m+n} \to \mathbb{R}^{m+n}$, we assume that there exists an invertible matrix P such that for all y_1, y_2,*

$$P\partial_{y_2}v(y_1, y_2)P^{-1} = -\partial_{y_2}v(y_1, y_2), \quad P\partial_{y_2}g_2(y_1, y_2)P^{-1} = -\partial_{y_2}g_2(y_1, y_2).$$

Then the SSEI method is volume preserving for the vector fields $f(y_1, y_2) = (u(y_1), v(y_1, y_2))^\mathsf{T}$ in $\mathscr{F}^{(2)}$.

Proof It follows from the conditions of this theorem that $PK_{22}P^{-1} = -K_{22}$ and $\left|e^{hK_{22}}\right| = 1$.

For the one-stage SSEI, the condition for volume preservation is

$$\left|I - ha_{11}g'(k_1)\right| = \left|I + ha_{11}g'(k_1)\right|,$$

which can be rewritten as

$$\left|I - ha_{11}\partial_{y_1}g_1\right|\left|I - ha_{11}\partial_{y_2}g_2\right| = \left|I + ha_{11}\partial_{y_1}g_1\right|\left|I + ha_{11}\partial_{y_2}g_2\right|. \qquad (6.19)$$

Since the method is volume preserving for the vector field $u(y_1)$, we have

$$\left|I - ha_{11}\partial_{y_1}g_1\right| = \left|I + ha_{11}\partial_{y_1}g_1\right|.$$

On the other hand,

$$\left|I - ha_{11}\partial_{y_2}g_2\right| = \left|PP^{-1} - ha_{11}P\partial_{y_2}g_2P^{-1}\right| = \left|I + ha_{11}\partial_{y_2}g_2\right|.$$

Thus (6.19) is proved.

For the two-stage SSEI, it is volume preserving if and only if (6.16) is true. Using the special result of g', we obtain

the left hand side of (6.16)

$$= |I - ha_{11}\partial_{y_1}g_1(k_1) - ha_{22}\partial_{y_1}g_1(k_2) + h^2a_{11}a_{22}\partial_{y_1}g_1(k_1)\partial_{y_1}g_1(k_2)$$
$$-h^2a_{12}a_{21}\partial_{y_1}g_1(k_2)\partial_{y_1}g_1(k_1)|$$
$$|I - ha_{11}\partial_{y_2}g_2(k_1) - ha_{22}\partial_{y_2}g_2(k_2) + h^2a_{11}a_{22}\partial_{y_2}g_2(k_1)\partial_{y_2}g_2(k_2)$$
$$-h^2a_{12}a_{21}\partial_{y_2}g_2(k_2)\partial_{y_2}g_2(k_1)|$$
$$= |I + ha_{11}\partial_{y_1}g_1(k_1) + ha_{22}\partial_{y_1}g_1(k_2) + h^2a_{11}a_{22}\partial_{y_1}g_1(k_1)\partial_{y_1}g_1(k_2)$$
$$-h^2a_{12}a_{21}\partial_{y_1}g_1(k_2)\partial_{y_1}g_1(k_1)|$$
$$|I - ha_{11}\partial_{y_2}g_2(k_1) - ha_{22}\partial_{y_2}g_2(k_2) + h^2a_{11}a_{22}\partial_{y_2}g_2(k_1)\partial_{y_2}g_2(k_2)$$
$$-h^2a_{12}a_{21}\partial_{y_2}g_2(k_2)\partial_{y_2}g_2(k_1)|.$$

It then can be verified that

$$|I - ha_{11}\partial_{y_2}g_2(k_1) - ha_{22}\partial_{y_2}g_2(k_2) + h^2a_{11}a_{22}\partial_{y_2}g_2(k_1)\partial_{y_2}g_2(k_2)$$
$$-h^2a_{12}a_{21}\partial_{y_2}g_2(k_2)\partial_{y_2}g_2(k_1)|$$
$$= |PP^{-1} - ha_{11}P\partial_{y_2}g_2(k_1)P^{-1} - ha_{22}P\partial_{y_2}g_2(k_2)P^{-1} + h^2a_{11}a_{22}P\partial_{y_2}g_2(k_1)P^{-1}$$
$$P\partial_{y_2}g_2(k_2)P^{-1} - h^2a_{12}a_{21}P\partial_{y_2}g_2(k_2)P^{-1}P\partial_{y_2}g_2(k_1)P^{-1}|$$
$$= |I + ha_{11}\partial_{y_2}g_2(k_1) + ha_{22}\partial_{y_2}g_2(k_2) + h^2a_{11}a_{22}\partial_{y_2}g_2(k_1)\partial_{y_2}g_2(k_2)$$
$$-h^2a_{12}a_{21}\partial_{y_2}g_2(k_2)\partial_{y_2}g_2(k_1)|.$$

Hence,

the left hand side of (6.16)

$$= |I + ha_{11}\partial_{y_1}g_1(k_1) + ha_{22}\partial_{y_1}g_1(k_2) + h^2a_{11}a_{22}\partial_{y_1}g_1(k_1)\partial_{y_1}g_1(k_2)$$

$$-h^2a_{12}a_{21}\partial_{y_1}g_1(k_2)\partial_{y_1}g_1(k_1)|$$

$$|I + ha_{11}\partial_{y_2}g_2(k_1) + ha_{22}\partial_{y_2}g_2(k_2) + h^2a_{11}a_{22}\partial_{y_2}g_2(k_1)\partial_{y_2}g_2(k_2)$$

$$-h^2a_{12}a_{21}\partial_{y_2}g_2(k_2)\partial_{y_2}g_2(k_1)|$$

$$= \text{the right hand side of (6.16).}$$

Consequently, all two-stage SSEI methods with (6.15) are volume preserving.   □

Remark 6.4.3 We here remark that when $K = \mathbf{0}$, all the results given in this section reduce to those proposed in [18], which demonstrates the wider applications of the analysis. Moreover, using these results of exponential integrators, we will formulate and study different volume-preserving methods for different problems in the next section.

6.5   Applications to Various Problems

In this section, our sole goal is to demonstrate the applications of the SSEI methods to various problems. Using the analysis given in Sect. 6.4, we will show the volume preservation of different integrators.

6.5.1   Highly Oscillatory Second-Order Systems

Consider the following first-order systems

$$y'(t) = J^{-1}My(t) + J^{-1}\nabla V(y(t)), \tag{6.20}$$

where the matrix J is constant and invertible, M is a symmetric matrix and V is a differentiable function.

Corollary 6.1 *All SSEI methods are volume preserving for solving the system (6.20).*

Proof This system is the exact pattern (6.1) with

$$K = J^{-1}M, \quad g(y(t)) = J^{-1}\nabla V(y(t)). \tag{6.21}$$

It can be verified that

$$Jg'(y)J^{-1} = JJ^{-1}\nabla^2 V(y)J^{-1} = \nabla^2 V(y)J^{-1} = -g'(y)^\mathsf{T}$$

and

$$J(K + g'(y))J^{-1} = -(K + g'(y))^\mathsf{T}.$$

This shows that the set $\mathscr{H}$ contains all vector fields of (6.20) with the same $P = J$. Consequently, according to Theorem 6.3, the result is proved.   □

Remark 6.5.1 When $J = \begin{pmatrix} 0 & I \\ -I & 0 \end{pmatrix}$, the system (6.20) is a Hamiltonian system $y'(t) = J^{-1}\nabla H(y(t))$ with the Hamiltonian $H(y) = \dfrac{1}{2}y^\mathsf{T}My + V(y)$. Corollary 6.1 shows that all SSEI methods are volume preserving for this Hamiltonian system. This is another explanation of the fact that symplectic exponential integrators are volume preserving for Hamiltonian systems.

Consider another special and important case of (6.20) by choosing

$$y = \begin{pmatrix} q \\ p \end{pmatrix}, \quad J^{-1} = \begin{pmatrix} 0 & I \\ -I & N \end{pmatrix}, \quad M = \begin{pmatrix} \Omega & 0 \\ 0 & I \end{pmatrix}, \quad V(y) = V_1(q),$$

which gives the following second-order ODE

$$q'' - Nq' + \Omega q = -\nabla V_1(q). \tag{6.22}$$

This system stands for highly oscillatory problems and many problems fall into this kind of equation such as the dissipative molecular dynamics, the (damped) Duffing, charged-particle dynamics in a constant magnetic field and semidiscrete nonlinear wave equations. Applying the SSEI methods to (6.22) and considering Theorem 6.3, we obtain the following corollary.

Corollary 6.2 *The following s-stage adapted exponential integrator applied with stepsize h*

$$\begin{cases} k_i = \exp^{11}(c_i hK)q_n + \exp^{12}(c_i hK)q'_n - h\sum_{j=1}^{s} a_{ij}\exp^{12}((c_i - c_j)hK)\nabla V_1(k_j), \\ \qquad\qquad\qquad\qquad\qquad\qquad\qquad\qquad\qquad i = 1, 2, \cdots, s, \\ q_{n+1} = \exp^{11}(hK)q_n + \exp^{12}(hK)q'_n - h\sum_{i=1}^{s} b_i \exp^{12}((1 - c_i)hK)\nabla V_1(k_i), \\ q'_{n+1} = \exp^{21}(hK)q_n + \exp^{22}(hK)q'_n - h\sum_{i=1}^{s} b_i \exp^{22}((1 - c_i)hK)\nabla V_1(k_i) \end{cases}$$

$$\tag{6.23}$$

are volume preserving for the second-order highly oscillatory equation (6.22), where $\exp(hK)$ *is partitioned into*

$$\begin{pmatrix} \exp^{11}(hK) \ \exp^{12}(hK) \\ \exp^{21}(hK) \ \exp^{22}(hK) \end{pmatrix}$$

and the same denotations are used for other matrix-valued functions. Here $(c_1, \cdots, c_s)^{\mathsf{T}}$, $(b_1, \cdots, b_s)^{\mathsf{T}}$ *and* $(a_{ij})_{s \times s}$ *are given in Definition 6.2. If* N *commutes with* Ω, *the results of* $\exp^{ij}$ *for* $i, j = 1, 2$ *can be expressed explicitly:*

$$\exp^{11}(hK) = e^{\frac{h}{2}N}\left(\cosh\left(\frac{h}{2}\sqrt{N^2 - 4\Omega}\right) - N\sinh\left(\frac{h}{2}\sqrt{N^2 - 4\Omega}\right)(\sqrt{N^2 - 4\Omega})^{-1}\right),$$

$$\exp^{12}(hK) = 2e^{\frac{h}{2}N}\sinh\left(\frac{h}{2}\sqrt{N^2 - 4\Omega}\right)(\sqrt{N^2 - 4\Omega})^{-1},$$

$$\exp^{21}(hK) = -\Omega\exp^{12}(hK),$$

$$\exp^{22}(hK) = e^{\frac{h}{2}N}\left(\cosh\left(\frac{h}{2}\sqrt{N^2 - 4\Omega}\right) + N\sinh\left(\frac{h}{2}\sqrt{N^2 - 4\Omega}\right)(\sqrt{N^2 - 4\Omega})^{-1}\right).$$

$$(6.24)$$

These results are still true if we replace h *by* kh *with any* $k \in \mathbb{R}$.

If we further assume that $\Omega = \mathbf{0}$, equation (6.22) becomes

$$q'' = Nq' - \nabla V_1(q). \tag{6.25}$$

One typical example of this type of system is charged-particle dynamics in a constant magnetic field (see [33])

$$x'' = x' \times B + F(x). \tag{6.26}$$

Here $x(t) \in \mathbb{R}^3$ describes the position of a particle moving in an electro-magnetic field, $F(x) = -\nabla_x U(x)$ is an electric field with the scalar potential $U(x)$, and $B = \nabla_x \times A(x)$ is a constant magnetic field with the vector potential $A(x) = -\frac{1}{2}x \times B$. Under the condition that $\Omega = \mathbf{0}$, the formula (6.24) can be rewritten more succinctly as:

$$\exp^{11}(hK) = I, \ \ \exp^{12}(hK) = h\varphi_1(hN), \ \ \exp^{21}(hK) = 0, \ \ \exp^{22}(hK) = \varphi_0(hN),$$

where the φ-functions are defined by (see [29, 30])

$$\varphi_0(z) = e^z, \ \ \varphi_k(z) = \int_0^1 e^{(1-\sigma)z}\frac{\sigma^{k-1}}{(k-1)!}d\sigma, \ \ k = 1, 2, \cdots. \tag{6.27}$$

We then obtain the following volume-preserving methods for the special and important second-order system (6.25).

Corollary 6.3 *The following s-stage integrator applied with stepsize h*

$$
\begin{cases}
k_i = q_n + c_i h \varphi_1(c_i h N) q_n' - h^2 \sum_{j=1}^{s} a_{ij}(c_i - c_j)\varphi_1((c_i - c_j)hN)\nabla V_1(k_j), \\
\hspace{9cm} i = 1, 2, \cdots, s, \\
q_{n+1} = q_n + h\varphi_1(hN)q_n' - h^2 \sum_{i=1}^{s} b_i(1 - c_i)\varphi_1((1 - c_i)hN)\nabla V_1(k_i), \\
q_{n+1}' = \varphi_0(hN)q_n' - h \sum_{i=1}^{s} b_i \varphi_0((1 - c_i)hN)\nabla V_1(k_i)
\end{cases}
$$

$$(6.28)$$

are volume preserving for the highly oscillatory second-order system (6.25), where $(c_1, \cdots, c_s)^\mathsf{T}$, $(b_1, \cdots, b_s)^\mathsf{T}$ *and* $(a_{ij})_{s \times s}$ *are given in Definition 6.2.*

Remark 6.5.2 We remark that the above two corollaries are meaningful discoveries which are of great importance to Geometric Integration for second-order highly oscillatory problems.

6.5.2   Separable Partitioned Systems

The authors in [18] have proved that the set $\mathscr{S}$ contains all separable partitioned systems. For instance, we consider

$$
\begin{pmatrix} q \\ p \end{pmatrix}' = \begin{pmatrix} p \\ -\Omega q + \tilde{g}(q) \end{pmatrix} = \begin{pmatrix} 0 & I \\ -\Omega & 0 \end{pmatrix} \begin{pmatrix} q \\ p \end{pmatrix} + \begin{pmatrix} 0 \\ \tilde{g}(q) \end{pmatrix}, \tag{6.29}
$$

which is exactly the system (6.1) with

$$
K = \begin{pmatrix} 0 & I \\ -\Omega & 0 \end{pmatrix}, \quad g = \begin{pmatrix} 0 \\ \tilde{g}(q) \end{pmatrix}, \quad f = \begin{pmatrix} p \\ -\Omega q + \tilde{g}(q) \end{pmatrix}.
$$

It is easy to see that f and g both fall into $\mathscr{S}$ with the same $P = \mathrm{diag}(I, -I)$. For this special matrix K, it is clear that

$$
e^{xK} = \begin{pmatrix} \phi_0(x^2 \Omega) & x\phi_1(x^2 \Omega) \\ -x\Omega\phi_1(x^2 \Omega) & \phi_0(x^2 \Omega) \end{pmatrix} \quad \text{for } x \in \mathbb{R}, \tag{6.30}
$$

where

$$\phi_i(\Omega) := \sum_{l=0}^{\infty} \frac{(-1)^l \Omega^l}{(2l+i)!}$$

for $i = 0, 1$. Hence, the exponential integrator (6.3) has a special form, and then we present it by the following definition.

Definition 6.5 (See [34]) An s-stage ERKN integrator applied with stepsize h for solving (6.29) is defined by

$$\begin{cases} Q_i = \phi_0(c_i^2 V)q_n + hc_i\phi_1(c_i^2 V)p_n + h^2 \sum_{j=1}^{s} \bar{a}_{ij}(V)\tilde{g}(Q_j), \quad i = 1, \cdots, s, \\ q_{n+1} = \phi_0(V)q_n + h\phi_1(V)p_n + h^2 \sum_{i=1}^{s} \bar{b}_i(V)\tilde{g}(Q_i), \\ p_{n+1} = -h\Omega\phi_1(V)q_n + \phi_0(V)p_n + h \sum_{i=1}^{s} b_i(V)\tilde{g}(Q_i), \end{cases}$$

where c_i for $i = 1, \cdots, s$ are real constants, and $b_i(V)$, $\bar{b}_i(V)$ and $\bar{a}_{ij}(V)$ for $i, j = 1, \cdots, s$ are matrix-valued functions of $V \equiv h^2\Omega$.

ERKN integrators were first proposed in [34], which are oscillation preserving as stated in Chap. 1. Further efforts in connection with ERKN integrators have been made, including symmetric integrators (see [35]), symplectic integrators (see [17]), energy-preserving integrators (see [36]) and other kinds of integrators (see [37, 38]). However, the volume-preserving property of ERKN integrators has not received much attention in the literature. With the analysis given in this chapter, we obtain the following VP result of ERKN integrators.

Corollary 6.4 *Consider a type of s-stage ERKN integrator applied with stepsize h for (6.29)*

$$\begin{cases} Q_i = \phi_0(c_i^2 V)q_n + hc_i\phi_1(c_i^2 V)p_n + h^2 \sum_{j=1}^{s} a_{ij}(c_i - c_j)\phi_1((c_i - c_j)^2 V)\tilde{g}(Q_j), \\ \hspace{8cm} i = 1, \cdots, s, \\ q_{n+1} = \phi_0(V)q_n + h\phi_1(V)p_n + h^2 \sum_{i=1}^{s} b_i(1 - c_i)\phi_1((1 - c_i)^2 V)\tilde{g}(Q_i), \\ p_{n+1} = -h\Omega\phi_1(V)q_n + \phi_0(V)p_n + h \sum_{i=1}^{s} b_i\phi_0((1 - c_i)^2 V)\tilde{g}(Q_i), \end{cases}$$

$$(6.31)$$

where $(c_1, \cdots, c_s)^\mathsf{T}$, $(b_1, \cdots, b_s)^\mathsf{T}$ and $(a_{ij})_{s \times s}$ are given in Definition 6.2. Under the condition that $b_j \neq 0$ for $j = 1, \cdots, s$, all one-stage and two-stage (with (6.15)) ERKN integrators (6.31) (and any composition of these methods) are volume preserving for solving the separable partitioned system (6.29).

Proof According to Definition 6.2 and the result (6.30), we adapt the SSEI methods to the system (6.29) and then obtain (6.31). Hence, the volume-preserving result of (6.31) immediately follows from Theorem 6.4.   □

Remark 6.5.3 It is noted that this is an important result which shows the volume-preserving ERKN integrators for (6.29). Moreover, it can be observed from (6.31) that all one-stage ERKN integrators are explicit, which implies that explicit volume preserving ERKN integrators are obtained for the separable partitioned system (6.29).

Remark 6.5.4 If Ω is a symmetric and positive semi-definite matrix and $\tilde{g}(q) = -\nabla U(q)$, the system (6.29) is an oscillatory Hamiltonian system

$$\begin{pmatrix} q \\ p \end{pmatrix}' = \begin{pmatrix} 0 & I \\ -I & 0 \end{pmatrix} \nabla H(p, q)$$

with the Hamiltonian

$$H(p, q) = \frac{1}{2} p^\mathsf{T} p + \frac{1}{2} q^\mathsf{T} \Omega q + U(q). \tag{6.32}$$

It has been addressed in Sect. 6.5.1 that this vector field falls into the set $\mathscr{H}$. Thus Theorem 6.3 provides another way to prove the well-known fact that all symplectic ERKN integrators (6.31) are volume preserving for the oscillatory Hamiltonian system (6.32).

We next investigate the volume-preserving property of RKN methods for standard second-order ODEs. Consider the special case where $\Omega = 0$ for the above analysis and under this situation, ERKN integrators reduce to RKN methods. Therefore, we are now in a position to present the following volume-preserving property for RKN methods.

Corollary 6.5 *Consider the following s-stage RKN methods applied with stepsize h for the standard second-order ODE $q'' = \tilde{g}(q)$*

$$\begin{cases} Q_i = q_n + h c_i q_n' + h^2 \sum_{j=1}^{s} a_{ij}(c_i - c_j)\tilde{g}(Q_j), \quad i = 1, \cdots, s, \\ q_{n+1} = q_n + h q_n' + h^2 \sum_{i=1}^{s} b_i(1 - c_i)\tilde{g}(Q_i), \\ q_{n+1}' = q_n' + h \sum_{i=1}^{s} b_i \tilde{g}(Q_i) \end{cases} \tag{6.33}$$

with the coefficients $c = (c_1, \cdots, c_s)^\mathsf{T}$, $b = (b_1, \cdots, b_s)^\mathsf{T}$ and $A = (a_{ij})_{s \times s}$ of an s-stage RK method. If this RK method is symplectic and $b_j \neq 0$ for all $j = 1, \cdots, s$, then all one-stage and two-stage RKN methods (6.33) (and compositions thereof) are volume preserving for the standard second-order ODE $q'' = \tilde{g}(q)$.

Remark 6.5.5 The fact of this corollary can be derived in a different way. Hairer, Lubich and Wanner have proved in [13] that any symplectic RK method with at most two stages (and any composition of such methods) is volume preserving for separable divergence free vector fields. Rewriting the second-order ODE $q'' = \tilde{g}(q)$ as a first-order system and applying symplectic RK methods to it implies the result of Corollary 6.5. In other words, the analysis of volume-preserving ERKN integrators provides an alternative derivation of the volume-preserving RKN methods.

6.5.3   Other Applications

It has been shown in [18] that $\mathscr{F}^{(\infty)}$ contains the affine vector fields $f(y) = Ly + d$ such that $\left| I + \dfrac{h}{2}L \right| = \left| I - \dfrac{h}{2}L \right|$ for all $h > 0$. For solving the system in the affine vector fields, the exponential integrator (6.3) becomes

$$
\begin{cases}
k_i = e^{c_i hL} y_n + h \displaystyle\sum_{j=1}^{s} \bar{a}_{ij}(hL)d, \quad i = 1, 2, \cdots, s, \\
y_{n+1} = e^{hL} y_n + h \displaystyle\sum_{i=1}^{s} \bar{b}_i(hL)d.
\end{cases}
$$

In the light of Theorem 6.5, this SSEI method is volume preserving for the affine vector fields.

It was also noted in [18] that $\mathscr{F}^{(\infty)}$ contains the vector fields $f(y)$ such that $f'(y) = JS(y)$ with a skew-symmetric matrix J and the symmetric matrix $S(y)$. Assume that

$$
K = JM, \quad g'(y) = JS(y), \tag{6.34}
$$

where M is a symmetric matrix. The system (6.1) with the vector field (6.34) falls into $\mathscr{F}^{(\infty)}$. Thus all SSEI methods are volume preserving for the vector field (6.34).

6.6 Numerical Examples

The purpose of this section is to present a numerical study of our SSEI methods in comparison with other numerical methods in the literature. To this end, the methods chosen for comparison are as follows:

- SSRK1: the Gauss-Legendre method of order two whose coefficients are given as

$$
\begin{array}{c|c}
\dfrac{1}{2} & \dfrac{1}{2} \\
\hline
 & 1
\end{array}
$$

- SSEI1: the one-stage SSEI method of order two with the coefficients

$$
\begin{array}{c|c}
\dfrac{1}{2} & \dfrac{1}{2} \\
\hline
 & e^{\frac{1}{2}hK}
\end{array}
$$

- SSRK2: the Gauss-Legendre method of order four whose coefficients are given as

$$
\begin{array}{c|cc}
\dfrac{3-\sqrt{3}}{6} & \dfrac{1}{4} & \dfrac{3-2\sqrt{3}}{12} \\[2mm]
\dfrac{3+\sqrt{3}}{6} & \dfrac{3+2\sqrt{3}}{12} & \dfrac{1}{4} \\
\hline
 & \dfrac{1}{2} & \dfrac{1}{2}
\end{array}
$$

- SSEI2: the two-stage SSEI method of order four with the coefficients

$$
\begin{array}{c|cc}
\dfrac{3-\sqrt{3}}{6} & \dfrac{1}{4} & \dfrac{3-2\sqrt{3}}{12}e^{-\frac{\sqrt{3}}{3}hK} \\[2mm]
\dfrac{3+\sqrt{3}}{6} & \dfrac{3+2\sqrt{3}}{12}e^{\frac{\sqrt{3}}{3}hK} & \dfrac{1}{4} \\
\hline
 & \dfrac{1}{2}e^{\frac{3+\sqrt{3}}{6}hK} & \dfrac{1}{2}e^{\frac{3-\sqrt{3}}{6}hK}
\end{array}
$$

It is noted that all these methods are implicit in general and iterative solutions are needed. We use fixed-point iteration. We set 10^{-16} as the error tolerance and 100 as the maximum number of fixed-point iterations per time step.

Problem 6.1 As the first numerical example, we consider the Duffing equation

$$\begin{pmatrix} q \\ p \end{pmatrix}' = \begin{pmatrix} 0 & 1 \\ -\omega^2 - k^2 & 0 \end{pmatrix} \begin{pmatrix} q \\ p \end{pmatrix} + \begin{pmatrix} 0 \\ 2k^2 q^3 \end{pmatrix}, \quad \begin{pmatrix} q(0) \\ p(0) \end{pmatrix} = \begin{pmatrix} 0 \\ \omega \end{pmatrix}.$$

The exact solution of this system is $q(t) = sn(\omega t; k/\omega)$ with the Jacobi elliptic function sn. Since it is a Hamiltonian system, all the methods chosen for comparison are volume preserving. For this problem, we choose $k = 0.07$ and $\omega = 20$ and then solve it on the interval $[0, 100]$ with different stepsizes $h = 1/2, 1/10, 1/50, 1/200$. The numerical flows at the time points $\left\{ \frac{1}{2} i \right\}_{i=1,\cdots,200}$ of the four methods are given in Fig. 6.2. It can be observed from the numerical results that the integrators

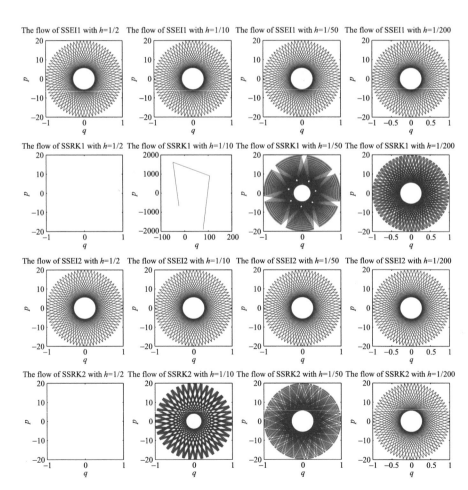

Fig. 6.2 Problem 6.1: The flows of different methods

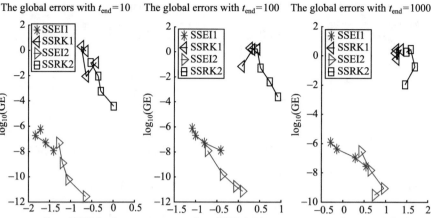

Fig. 6.3 Problem 6.1: The relative global errors

SSEI1 and SSEI2 perform better than Runge–Kutta methods since they present a uniform result for every different stepsizes. Finally, we integrate this problem on $[0, t_{end}]$ with the stepsizes $h = 0.1/2^i$ for $i = 1, \cdots, 4$. The relative global errors for different t_{end} are presented in Fig. 6.3. These results show again that exponential integrators have better accuracy than Runge–Kutta methods. It is noted that in Fig. 6.3, some methods do not show the correct convergence. The reason for this observation might be that we set 10^{-16} as the error tolerance and 100 as the maximum number of each fixed-point iteration, and implicit iterations converge incompletely for these methods.

Problem 6.2   Consider the following divergence free ODEs

$$
\begin{pmatrix} x \\ y \\ z \end{pmatrix}' = \begin{pmatrix} 0 & -\omega & 0 \\ \omega & 0 & -\omega \\ 0 & \omega & 0 \end{pmatrix} \begin{pmatrix} x \\ y \\ z \end{pmatrix} + \begin{pmatrix} \sin(x-z) \\ 0 \\ \sin(x-z) \end{pmatrix}.
$$

The choice of

$$
P = \begin{pmatrix} 0 & 0 & 1 \\ 0 & 1 & 0 \\ 1 & 0 & 0 \end{pmatrix},
$$

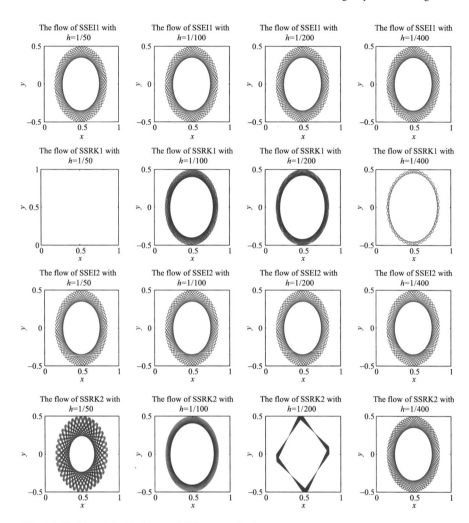

Fig. 6.4 Problem 6.2: The flows of different methods

ensures the vector field of this problem fall into $\mathscr{S}$. We consider $\omega = 100$ and the initial value $(x(0), y(0), z(0))^{\mathsf{T}} = (0.5, 0.5, 0.5)^{\mathsf{T}}$. This problem is first integrated on $[0, 100]$ with the stepsizes $h = 1/50, 1/100, 1/200, 1/400$ and the numerical flows x and y at the time points $\left\{\frac{1}{2}i\right\}_{i=1,\cdots,200}$ are shown in Fig. 6.4. Then the relative global errors for different t_{end} with the stepsizes $h = 0.1/2^i$ for $i = 2, \cdots, 5$ are given in Fig. 6.5. These results demonstrate clearly again that SSEI methods perform better than SSRK methods.

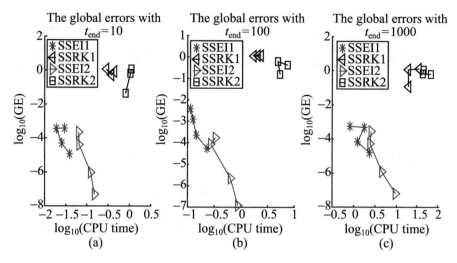

Fig. 6.5 Problem 6.2: The relative global errors

Problem 6.3 Consider the damped Helmholtz-Duffing oscillator (see [39])

$$q'' + 2\upsilon q' + Aq = -Bq^2 - \varepsilon q^3,$$

where q denotes the displacement of the system, A is the natural frequency, ε is a nonlinear system parameter, υ is the damping factor, and B is a system parameter independent of time. It is well known that the dynamical behaviour of eardrum oscillations, elasto-magnetic suspensions, thin laminated plates, graded beams, and other physical phenomena all fall into this category of equations. We choose the parameters

$$\upsilon = 0.01, \ A = 200, \ B = -0.5, \ \varepsilon = 1$$

and the initial values $q(0) = 1$ and $q'(0) = 15.199$. This problem is first integrated on $[0, 200]$ with the stepsizes $h = 1/2, 1/10, 1/50, 1/200$. We present the numerical flows q and $p = q'$ at the time points $\left\{\dfrac{1}{2}i\right\}_{i=1,\cdots,400}$ in Fig. 6.6. We then solve the problem with different $t_{\text{end}} = 10, 100, 1000$ and the stepsizes $h = 0.1/2^i$ for $i = 0, \cdots, 3$. The relative global errors are shown in Fig. 6.7. It follows again from the results that SSEI methods perform much better than SSRK methods.

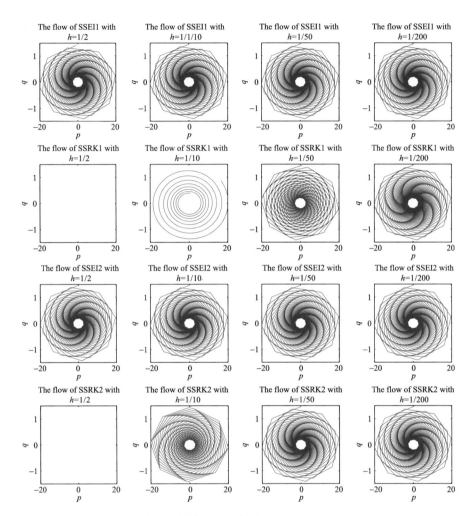

Fig. 6.6 Problem 6.3: The flows of different methods

Problem 6.4 This numerical experiment concerns the charged particle system with a constant magnetic field (see [33]). The system is given by (6.26) with the potential $U(x) = \dfrac{1}{100\sqrt{x_1^2 + x_2^2}}$ and the constant magnetic field $B = (0, 0, 10)^{\mathsf{T}}$. The initial values are chosen as $x(0) = (0.7, 1, 0.1)^{\mathsf{T}}$ and $x'(0) = (0.9, 0.5, 0.4)^{\mathsf{T}}$. We first integrate this system on $[0, 100]$ with the stepsizes $h = 1/2, 1/10, 1/50, 1/200$ and show the numerical flows x_2 and $v_2 = x_2'$ at the time points $\left\{ \dfrac{1}{2}i \right\}_{i=1,\cdots,200}$

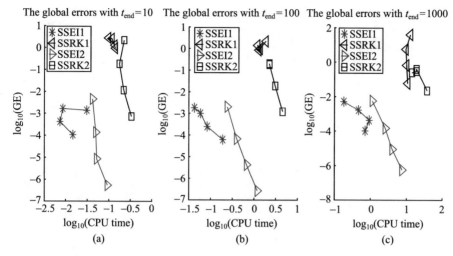

Fig. 6.7 Problem 6.3: The relative global errors

in Fig. 6.8. Then the problem is solved with $t_{end} = 10, 100, 1000$ and the stepsizes $h = 0.1/2^i$ for $i = 0, \cdots, 3$ and the relative global errors are shown in Fig. 6.9. The SSEI methods are also shown to be robust for this problem. Here, it is important to note that the SSEI1 method is explicit (see (6.28)) when applied to this problem, whereas the SSRK1 method is implicit and iterative solutions are required for solving this problem. This fact shows another significant advantage of our volume-preserving exponential integrators in comparison with volume-preserving RK methods.

Problem 6.5 We consider the following dynamical system for investigating fluid particle motion (see [25])

$$\dot{x}_1 = \frac{1}{2}(w_2 x_3 - w_3 x_2) + \frac{1}{2}\left[(5r^2 - 3)\frac{x_1}{1+\alpha} - 2x_1 \left(\frac{x_1^2}{1+\alpha} + \frac{\alpha x_1^2}{1+\alpha} - x_3^2 \right) \right],$$

$$\dot{x}_2 = \frac{1}{2}(w_3 x_1 - w_1 x_3) + \frac{1}{2}\left[(5r^2 - 3)\frac{\alpha x_2}{1+\alpha} - 2x_2 \left(\frac{x_1^2}{1+\alpha} + \frac{\alpha x_1^2}{1+\alpha} - x_3^2 \right) \right],$$

$$\dot{x}_3 = \frac{1}{2}(w_1 x_2 - w_2 x_1) + \frac{1}{2}\left[-(5r^2 - 3)x_3 - 2x_3 \left(\frac{x_1^2}{1+\alpha} + \frac{\alpha x_1^2}{1+\alpha} - x_3^2 \right) \right],$$

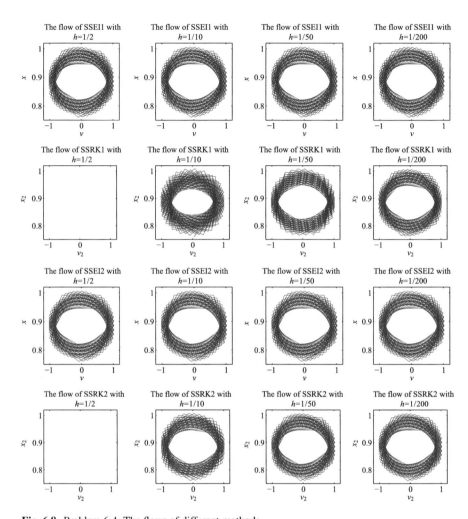

Fig. 6.8 Problem 6.4: The flows of different methods

with $\alpha = 1$, $(w_1, w_2, w_3) = (300, 500, 400)$, and the initial value $(x_1(0), x_2(0),$ $x_3(0))^\mathsf{T} = (-0.1689, 0, -0.0437)^\mathsf{T}$. We solve this problem on $[0, 1000]$ with the stepsizes $h = 1/50, 1/200, 1/500, 1/1000$. The numerical flows through the (x_1, x_3) plane at the time points $\left\{\dfrac{1}{50}i\right\}_{i=1,\cdots,1000}$ are plotted in Fig. 6.10. Then the problem is integrated with $t_{\mathrm{end}} = 10, 100, 1000$ and the stepsizes $h = 0.01/2^i$ for $i = 0, \cdots, 3$ and the relative global errors are shown in Fig. 6.11.

Counterexample We here present a counterexample to show that higher-order methods do not preserve the volume of the phase space for some vector fields. To

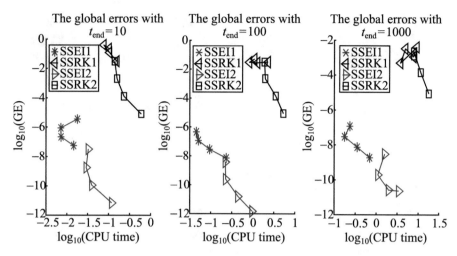

Fig. 6.9 Problem 6.4: The relative global errors

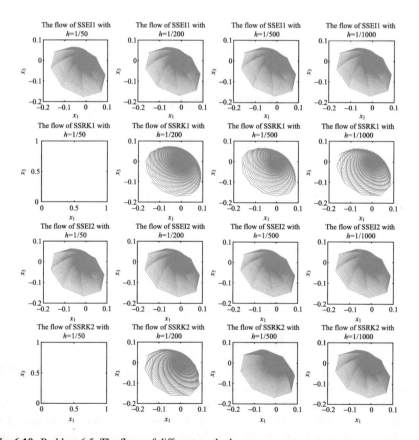

Fig. 6.10 Problem 6.5: The flows of different methods

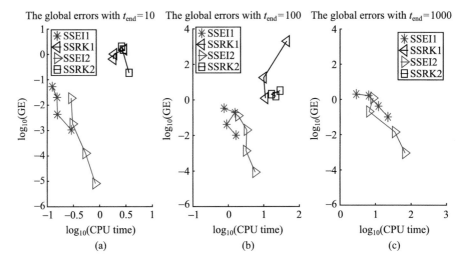

Fig. 6.11 Problem 6.5: The relative global errors

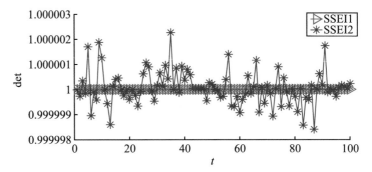

Fig. 6.12 The determinant of the derivative of the numerical flow as a function of time

this end, we consider the following dynamic system

$$\dot{x} = -y + \sin(z),$$
$$\dot{y} = -x + z + \cos(z),$$
$$\dot{z} = y + \cos(x) + \sin(y),$$

with the initial value $(x(0), y(0), z(0))^{\mathsf{T}} = (0, 0, 0)^{\mathsf{T}}$. This problem is solved on $[0, 10]$ with the stepsize $h = 0.1$. The determinant of the derivative of the numerical flow as a function of time is given in Fig. 6.12. It can be observed from the results that only the second-order method SSEI1 is volume-preserving, whereas the fourth-order method SSEI2 does not preserve the volume. It is noted here that for higher-order methods, the volume-preserving property is more likely to fail than low order methods, and one should pay attention to this point in the study of volume-preserving methods.

6.7 Conclusions

This chapter studied volume-preserving exponential integrators for dynamical systems. The necessary and sufficient volume-preserving condition for exponential integrators was derived and volume-preserving properties were discussed for four kinds of vector fields. It was shown that symplectic exponential integrators can be volume preserving for a much larger class of vector fields than just Hamiltonian systems. It should be noted that some interesting results on Geometric Integration were presented for second-order highly oscillatory problems and separable partitioned systems. In particular, an important result for Geometric Integration has been obtained that a type of adapted exponential integrator is volume preserving for the second-order highly oscillatory systems (6.22) and (6.25). Moreover, the volume-preserving property of ERKN/RKN methods was analysed for separable partitioned systems. Numerical experiments are implemented and the numerical results demonstrate the remarkable robustness and superiority of our volume-preserving exponential integrators in comparison with volume-preserving Runge–Kutta methods.

The material in this chapter is based on the work by Wang and Wu [40].

References

1. Brugnano, L., Frasca Caccia, G., Iavernaro, F.: Hamiltonian Boundary Value Methods (HBVMs) and their efficient implementation, mathematics in engineering, science and aerospace. MESA **5**, 343–411 (2014)
2. Celledoni, E., McLachlan, R.I., McLaren, D.I., et al.: Energy-preserving Runge–Kutta methods. M2AN Math. Model. Numer. Anal. **43**, 645–649 (2009)
3. Cohen, D., Hairer, E.: Linear energy-preserving integrators for Poisson systems. BIT Numer. Math. **51**, 91–101 (2011)
4. Hairer, E.: Energy-preserving variant of collocation methods. J. Numer. Anal. Ind. Appl. Math. **5**, 73–84 (2010)
5. Hairer, E., Lubich, C.: Long-time energy conservation of numerical methods for oscillatory differential equations. SIAM J. Numer. Anal. **38**, 414–441 (2000)
6. Hairer, E., Lubich, C.: Long-term analysis of the Störmer-Verlet method for Hamiltonian systems with a solution-dependent high frequency. Numer. Math. **134**, 119–138 (2016)
7. Li, Y.W., Wu, X.: Exponential integrators preserving first integrals or Lyapunov functions for conservative or dissipative systems. SIAM J. Sci. Comput. **38**, 1876–1895 (2016)
8. McLachlan, R.I., Quispel, G.R.W.: Discrete gradient methods have an energy conservation law. Discrete Contin. Dyn. Syst. **34**, 1099–1104 (2014)
9. Sanz-Serna, J.M.: Symplectic integrators for Hamiltonian problems: an overview. Acta Numer. **1**, 243–286 (1992)
10. Wang, B., Wu, X.: Functionally-fitted energy-preserving integrators for Poisson systems. J. Comput. Phys. **364**, 137–152 (2018)
11. Wang, B., Wu, X.: The formulation and analysis of energy-preserving schemes for solving high-dimensional nonlinear Klein-Gordon equations. IMA J. Numer. Anal. **39**, 2016–2044 (2019)
12. Feng, K., Qin, M.: Symplectic Geometric Algorithms for Hamiltonian Systems. Springer, Berlin (2010)

13. Hairer, E., Lubich, C., Wanner, G.: Geometric Numerical Integration: Structure-Preserving Algorithms for Ordinary Differential Equations, 2nd edn. Springer, Berlin (2006)
14. Wu, X., Liu, K., Shi, W.: Structure-Preserving Algorithms for Oscillatory Differential Equations II. Springer, Berlin (2015)
15. Wu, X., Wang, B., Mei, L.: Oscillation-preserving algorithms for efficiently solving highly oscillatory second-order ODEs. Numer. Algor. **86**, 693–727 (2021)
16. Wu, X., Wang, B.: Recent Developments in Structure-Preserving Algorithms for Oscillatory Differential Equations. Springer Nature Singapore Pte Ltd., Singapore (2018)
17. Wu, X., You, X., Wang, B.: Structure-preserving Algorithms for Oscillatory Differential Equations. Springer, Berlin (2013)
18. Bader, P., McLaren, D.I., Quispel, G.R.W., et al.: Volume preservation by Runge–Kutta methods. Appl. Numer. Math. **109**, 123–137 (2016)
19. Chartier, P., Murua, A.: Preserving first integrals and volume forms of additively split systems. IMA J. Numer. Anal. **27**, 381–405 (2007)
20. Feng, K., Shang, Z.J.: Volume-preserving algorithms for source-free dynamical systems. Numer. Math. **71**, 451–463 (1995)
21. He, Y., Sun, Y., Liu, J., et al.: Volume-preserving algorithms for charged particle dynamics. J. Comput. Phys. **281**, 135–147 (2015)
22. Iserles, A., Quispel, G.R.W., Tse, P.S.P.: B-series methods cannot be volume-preserving. BIT Numer. Math. **47**, 351–378 (2007)
23. McLachlan, R.I., Munthe-Kaas, H.Z., Quispel, G.R.W., et al.: Explicit volume preserving splitting methods for linear and quadratic divergence-free vector fields. Found. Comput. Math. **8**, 335–355 (2008)
24. Quispel, G.R.W.: Volume-preserving integrators. Phys. Lett. A **206**, 26–30 (1995)
25. Xue, H., Zanna, A.: Explicit volume-preserving splitting methods for polynomial divergence-free vector fields. BIT Numer. Math. **53**, 265–281 (2013)
26. McLachlan, R.I., Scovel, C.: A survey of open problems in symplectic integration. Fields Inst. Commun. **10**, 151–180 (1998)
27. Zanna, A.: Explicit volume-preserving splitting methods for divergence-free ODEs by tensor product basis decompositions. IMA J. Numer. Anal. **35**, 89–106 (2014)
28. Hochbruck, M., Ostermann, A.: Explicit exponential Runge–Kutta methods for semilinear parabolic problems. SIAM J. Numer. Anal. **43**, 1069–1090 (2005)
29. Hochbruck, M., Ostermann, A.: Exponential integrators. Acta Numer. **19**, 209–286 (2010)
30. Hochbruck, M., Ostermann, A., Schweitzer, J.: Exponential rosenbrock-type methods. SIAM J. Numer. Anal. **47**, 786–803 (2009)
31. Wang, B., Wu, X., Meng, F., et al.: Exponential Fourier collocation methods for solving first-order differential equations. J. Comput. Math. **35**, 711–736 (2017)
32. Mei, L., Wu, X.: Symplectic exponential Runge–Kutta methods for solving nonlinear Hamiltonian systems. J. Comput. Phys. **338**, 567–584 (2017)
33. Hairer, E., Lubich, C.: Symmetric multistep methods for charged-particle dynamics. SMAI J. Comput. Math. **3**, 205–218 (2017)
34. Wu, X., You, X., Shi, W., et al.: ERKN integrators for systems of oscillatory second order differential equations. Comput. Phys. Commun. **181**, 1873–1887 (2010)
35. Wang, B., Yang, H., Meng, F.: Sixth order symplectic and symmetric explicit ERKN schemes for solving multi-frequency oscillatory nonlinear Hamiltonian equations. Calcolo **54**, 117–140 (2017)
36. Wu, X., Wang, B., Shi, W.: Efficient energy preserving integrators for oscillatory Hamiltonian systems. J. Comput. Phys. **235**, 587–605 (2013)
37. Wang, B., Iserles, A., Wu, X.: Arbitrary-order trigonometric Fourier collocation methods for multi-frequency oscillatory systems. Found. Comput. Math. **6**, 151–181 (2016)

38. Wang, B., Wu, X., Meng, F.: Trigonometric collocation methods based on Lagrange basis polynomials for multi-frequency oscillatory second-order differential equations. J. Comput. Appl. Math. **313**, 185–201 (2017)
39. Elías-Zúñiga, A.: Analytical solution of the damped Helmholtz-Duffing equation. Appl. Math. Lett. **25**, 2349–2353 (2012)
40. Wang, B., Wu, X.: Volume-preserving exponential integrators and their applications. J. Comput. Phys. **396**, 867–887 (2019)

Chapter 7
Global Error Bounds of One-Stage Explicit ERKN Integrators for Semilinear Wave Equations

In this chapter, we analyse global error bounds for one-stage explicit extended Runge–Kutta–Nyström integrators for semilinear wave equations with periodic boundary conditions. We show optimal second-order convergence without requiring Lipschitz continuity and higher regularity of the exact solution.

7.1 Introduction

First of all, we denote by H^s the Sobolev space $H^s(\mathbb{T})$. In this chapter we pursue the error analysis of one-stage explicit extended Runge–Kutta–Nyström (ERKN) integrators for the semilinear wave equation with some integer $p \geqslant 2$

$$u_{tt} = u_{xx} + u^p, \quad u = u(x, t), \quad t \in [t_0, T]. \tag{7.1}$$

The initial values are given by $u(\cdot, t_0) \in H^{s+1}$ and $u_t(\cdot, t_0) \in H^s$ for $s \geqslant 0$. We consider here real-valued solutions to (7.1) with 2π-periodic boundary conditions in one space dimension ($x \in \mathbb{T} = \mathbb{R}/(2\pi\mathbb{Z})$). It is noted that the energy is finite in the special case $s = 0$.

Using a semidiscretisation in space, we can transform equation (7.1) into a system of second-order ordinary differential equations (ODEs) of the form

$$\ddot{y}(t) = My(t) + f(y(t)), \tag{7.2}$$

where the matrix M describes the discretised second spatial derivative and $f(y)$ denotes the polynomial nonlinearity. It is very important to note that the eigenvalues of the matrix M range from 0 to $\mathcal{O}(K)$, where $2K$ stands for the number of internal discretisation points in space (see, e.g. [1, 2]). This implies that the spatial semidiscretisation exhibits oscillations with a variety of frequencies, and the

© The Author(s), under exclusive license to Springer Nature Singapore Pte Ltd. 2021
X. Wu, B. Wang, *Geometric Integrators for Differential Equations with Highly Oscillatory Solutions*, https://doi.org/10.1007/978-981-16-0147-7_7

solution of (7.2) typically contains high-frequency oscillatory terms. Many effective integrators have been researched (see, e.g. [3–10], and the references therein) for (7.2). Gautschi-type methods have been well researched and analysed in [6, 11]. Exponential integrators have been widely developed and the reader is referred to [12–14] for instance. These methods have been applied to semilinear wave equations (see, e.g. [15–19]). As a standard form of trigonometric integrator (TI), ERKN integrators were formulated for highly oscillatory second-order differential equations in [20]. Further researches of these integrators are contained in [21–23].

As is known, the error analysis of TI for ODEs has been researched by many papers (see, e.g. [11–13, 24–27]). Unfortunately, however, this work is obviously insufficient because the nonlinearity is assumed to be Lipschitz continuous in all these publications. There is also much work about the error analysis of TI for PDEs (see, e.g. [28–31]). The author in [32] showed error bounds of TI for wave equations without requiring higher regularity of the exact solution, which was achieved by performing the error analysis in two stages. These two-stage arguments have also been used by many researchers such as in [33–37]. Recently, an error analysis has been presented for different schemes for quasilinear wave equations (see, e.g. [38–40]).

We note the fact that the error analysis of ERKN integrators has not been well researched yet in the literature for spatial semidiscretisations of (7.1) with initial values of finite energy. Thus, in this chapter, using the approach described in [32], we will analyse and present error bounds for one-stage explicit ERKN integrators when applied to a spectral semidiscretisation in space, requiring only that the exact solution is of finite energy. First, low-order error bounds will be considered in a higher-order Sobolev space, where the nonlinearity is, at least locally, Lipschitz continuous. From this low-order error bound, suitable regularity of the ERKN integrator will be obtained. Then higher-order error bounds will be shown in these spaces based on the regularity of the ERKN integrator. Optimal second-order convergence will be achieved without requiring Lipschitz continuity and higher regularity of the exact solution. Moreover, this approach to the error analysis is not restricted to spectral semidiscretisations in space.

7.2 Preliminaries

7.2.1 Spectral Semidiscretisation in Space

We consider the following trigonometric polynomial as an ansatz for the solution of the nonlinear wave equation (7.1)

$$u_{\mathcal{K}}(x, t) = \sum_{j \in \mathcal{K}} y_j(t) e^{ijx} \quad \text{with} \quad \mathcal{K} = \{-K, -K+1, \cdots, K-1\}, \quad (7.3)$$

where $y_j(t)$ for $j \in \mathcal{K}$ are the Fourier coefficients (see, e.g. [32, 41]). Inserting this ansatz into (7.1) and evaluating at the collocation points $x_k = \pi k/K$ with $k \in \mathcal{K}$, we obtain a system of second-order ODEs

$$\ddot{y}(t) = -\Omega^2 y(t) + f(y(t)), \tag{7.4}$$

where $y(t) = (y_j(t))_{j \in \mathcal{K}} \in \mathbb{C}^{\mathcal{K}}$ is the vector of Fourier coefficients, Ω is a nonnegative diagonal matrix $\Omega = \mathrm{diag}(\omega_j)_{j \in \mathcal{K}}$ with $\omega_j = |j|$, and the nonlinearity f is given by

$$f(y) = \underbrace{y * \cdots * y}_{p \text{ times}} \quad \text{with} \quad (y * z)_j = \sum_{k+l \equiv j \bmod 2K} y_k z_l, \quad j \in \mathcal{K}. \tag{7.5}$$

Here, '$*$' denotes the discrete convolution. The initial values $y(t_0)$ and $\dot{y}(t_0)$ for (7.4) are given respectively by

$$y_j(t_0) = \sum_{k \in \mathbb{Z}:\, k \equiv j \bmod 2K} u_k(t_0), \quad \dot{y}_j(t_0) = \sum_{k \in \mathbb{Z}:\, k \equiv j \bmod 2K} \dot{u}_k(t_0), \quad j \in \mathcal{K},$$
$$\tag{7.6}$$

where $u_k(t)$ and $\dot{u}_k(t)$ are the Fourier coefficients of $u(\cdot, t)$ and $u_t(\cdot, t)$, respectively. Once the initial values $u(\cdot, t_0)$ and $u_t(\cdot, t_0)$ are given in terms of their Fourier coefficients, we have the simpler expression:

$$y_j(t_0) = u_j(t_0), \quad \dot{y}_j(t_0) = \dot{u}_j(t_0), \quad j \in \mathcal{K}. \tag{7.7}$$

It is clear that the exact solution of the semidiscrete system (7.4) can be expressed by

$$\begin{pmatrix} y(t) \\ \dot{y}(t) \end{pmatrix} = R(t - t_0) \begin{pmatrix} y(t_0) \\ \dot{y}(t_0) \end{pmatrix} + \int_{t_0}^{t} \begin{pmatrix} \cos(h\Omega) & (t - \tau)\mathrm{sinc}(h\Omega) \\ -\Omega\sin(h\Omega) & \cos(h\Omega) \end{pmatrix} \begin{pmatrix} 0 \\ f(y(\tau)) \end{pmatrix} d\tau, \tag{7.8}$$

where $\mathrm{sinc}\, x = \sin x/x$ and

$$R(t) = \begin{pmatrix} \cos(t\Omega) & t\,\mathrm{sinc}(t\Omega) \\ -\Omega\sin(t\Omega) & \cos(t\Omega) \end{pmatrix}.$$

Throughout this chapter, we measure the error by the norm (see, e.g. [32, 41])

$$\|y\|_s := \left(\sum_{j \in \mathcal{K}} \langle j \rangle^{2s} |y_j|^2 \right)^{1/2} \quad \text{with} \quad \langle j \rangle = \max(1, |j|) \tag{7.9}$$

for $y \in \mathbb{C}^{\mathscr{K}}$, where $s \in \mathbb{R}$. This norm is (equivalent to) the Sobolev H^s-norm of the trigonometric polynomial $\sum_{j \in \mathscr{K}} y_j e^{ijx}$. Clearly, using this norm, we have $\|y\|_{s_1} \leqslant \|y\|_{s_2}$ if $s_1 \leqslant s_2$. The following result presented in [32] is needed in this chapter.

Proposition 7.1 (See [32]) *Assume that* $\sigma, \sigma' \in \mathbb{R}$ *with* $\sigma' \geqslant |\sigma|$ *and* $\sigma' \geqslant 1$. *If* $\|y\|_{\sigma'} \leqslant M$ *and* $\|z\|_{\sigma'} \leqslant M$, *then we have*

$$\|f(y) - f(z)\|_\sigma \leqslant C \|y - z\|_\sigma, \tag{7.10}$$

$$\|f(y)\|_{\sigma'} \leqslant C, \tag{7.11}$$

with a constant C depending only on M, $|\sigma|$, σ', and p.

7.2.2   ERKN Integrators

It is known that ERKN integrators are oscillation preserving for (7.4), as stated in Chap. 1. In this chapter, we consider one-stage explicit ERKN integrators which are formulated as follows.

Definition 7.1 (See [20]) A one-stage explicit ERKN integrator with stepsize h for solving (7.4) is defined by

$$\begin{cases} y^{(n+c_1)} = \phi_0(c_1^2 V) y^n + h c_1 \phi_1(c_1^2 V) \dot{y}^n, \\[2mm] y^{(n+1)} = \phi_0(V) y^n + h \phi_1(V) \dot{y}^n + h^2 \bar{b}_1(V) f(y^{(n+c_1)}), \\[2mm] \dot{y}^{(n+1)} = -h\Omega^2 \phi_1(V) y^n + \phi_0(V) \dot{y}^n + h b_1(V) f(y^{(n+c_1)}), \end{cases} \tag{7.12}$$

where c_1 is real constant, $b_1(V)$ and $\bar{b}_1(V)$ are matrix-valued functions of $V \equiv h^2 \Omega^2$, and $\phi_j(V) := \sum_{k=0}^{\infty} \dfrac{(-1)^k V^k}{(2k+j)!}$ for $j = 0, 1, \cdots$.

In particular, for $V = h^2 \Omega^2$, we have

$$\phi_0(V) = \cos(h\Omega), \qquad \phi_1(V) = \text{sinc}(h\Omega), \qquad \phi_2(V) = (h\Omega)^{-2}(I - \cos(h\Omega)).$$

In this chapter, we present five practical one-stage explicit ERKN integrators whose coefficients are displayed in Table 7.1. It can be seen from Table 7.1 that there are many different one-stage explicit ERKN integrators, and various methods with different properties can be constructed.

Table 7.1 Five one-stage explicit ERKN integrators

Methods	c_1	$\bar{b}_1(V)$	$b_1(V)$	Symmetric	Symplectic
ERKN1	1/2	$\phi_2(V)$	$\phi_0(V/4)$	Non	Non
ERKN2	1/2	$\phi_2(V)$	$\phi_1(V)$	Symmetric	Non
ERKN3	1/2	$1/2\,\phi_1(V/4)$	$\phi_0(V/4)$	Symmetric	Symplectic
ERKN4	1/2	$1/2\,\phi_1^2(V/4)$	$\phi_1(V/4)\phi_0(V/4)$	Symmetric	Non
ERKN5	1/2	$1/2\,\phi_1(V)\phi_1(V/4)$	$\phi_1(V)\phi_0(V/4)$	Symmetric	Non

7.3 Main Result

In order to present the error bounds, we need the following assumptions for the coefficients of the ERKN integrators. Similar assumptions on the filter functions of some trigonometric methods have been considered in [32].

Assumption 7.1 It is assumed that for a given $-1 \leqslant \beta \leqslant 1$, there exists a constant c such that

$$|\bar{b}_1(\xi^2)| \leqslant c\xi^\beta, \qquad \text{if} \qquad -1 \leqslant \beta \leqslant 0, \qquad (7.13)$$

$$|1/2\mathrm{sinc}^2(\xi/2) - \bar{b}_1(\xi^2)| \leqslant c\xi^\beta, \qquad \text{if} \qquad 0 < \beta \leqslant 1, \qquad (7.14)$$

$$|1 - b_1(\xi^2)| \leqslant c\xi^{(1+\beta)}, \qquad (7.15)$$

for all $\xi = h\omega_j$ with $j \in \mathcal{K}$ and $\omega_j \neq 0$. Furthermore, we assume that $c_1 = \dfrac{1}{2}$ for the ERKN integrators determined by (7.12).

It is easy to see that all the ERKN integrators displayed in Table 7.1 satisfy this assumption uniformly for $-1 \leqslant \beta \leqslant 1$ and $h > 0$. Under this assumption, we have the following property, which can be verified easily by the definition of the norm (7.9).

Proposition 7.2 *With the conditions of Assumption 7.1 it holds that*

$$\|y - b_1(V)y\|_{s-\beta} \leqslant ch^{(1+\beta)} \|y\|_{s+1}$$

for $s \in \mathbb{R}$. Moreover, we have

$$\left\|\bar{b}_1(V)y\right\|_{s-\beta} \leqslant ch^\beta \|y\|_s$$

for $-1 \leqslant \beta \leqslant 0$, and

$$\left\|\left(\frac{1}{2}\mathrm{sinc}^2\left(\frac{h\Omega}{2}\right) - \bar{b}_1(V)\right)y\right\|_{s-\beta} \leqslant ch^\beta \|y\|_s$$

for $0 < \beta \leqslant 1$.

The following theorem presents the main result of this chapter.

Theorem 7.1 *Let $c \geqslant 1$ and $s \geqslant 0$. Assume that the exact solution $(y(t), \dot{y}(t))$ of the spatial semidiscretisation (7.4) satisfies*

$$\|y(t)\|_{s+1} + \|\dot{y}(t)\|_s \leqslant M \qquad \text{for} \qquad 0 \leqslant t - t_0 \leqslant T. \tag{7.16}$$

Under Assumption 7.1 with the constant c for $\beta = 0$ and $\beta = \alpha \in [-1, 1]$, there exists $h_0 > 0$ such that for $0 < h \leqslant h_0$, the error bound for the numerical solution $(y^n, \dot{y}^n)$ obtained from the ERKN integrator (7.12) is given by

$$\left\| y(t_n) - y^n \right\|_{s+1-\alpha} + \left\| \dot{y}(t_n) - \dot{y}^n \right\|_{s-\alpha} \leqslant C h^{(1+\alpha)} \quad \text{for} \quad 0 \leqslant t_n - t_0 = nh \leqslant T,$$

where the constants C and h_0 depend only on M and s from (7.16), the power p, the final time T, and the constant c in Assumption 7.1.

Using the two-stage arguments described in [32, 33, 35–37], we divide the proof of Theorem 7.1 into two parts. We first show the proof of the lower-order error bounds in higher-order Sobolev spaces (i.e., $-1 \leqslant \alpha \leqslant 0$) in Sect. 7.4. We then present the proof of the higher-order error bounds in lower-order Sobolev spaces (i.e., $0 < \alpha \leqslant 1$) in Sect. 7.5.

Remark 7.1 We remark that the authors in [31] present an error analysis of ERKN integrators when applied to wave equations. The result is given by using the norm of a matrix and is proved by following [27, 42]. It is noted that the normal result and its proof, given in this chapter, are different from those in [31]. Moreover, Lipschitz continuity and higher regularity of the exact solution are not required in the analysis of this chapter, which is also different from [31].

Remark 7.2 One-stage ERKN integrators contain some trigonometric integrators of [32], and some ERKN integrators can be considered as trigonometric integrators of [32]. However, there is no inclusive relation for these two kinds of methods, which means that the analysis of [32] cannot be directly used for one-stage ERKN integrators. The analysis presented here essentially follows from [32] with some modifications arising from the ERKN discretisation.

7.4   The Lower-Order Error Bounds in Higher-Order Sobolev Spaces

Throughout the proof in this subsection, we assume that $0 < h \leqslant 1$ and use the norm $\||(y, \dot{y})\||_\sigma = (\|y\|_{\sigma+1}^2 + \|\dot{y}\|_\sigma^2)^{1/2}$ on $H^{\sigma+1} \times H^\sigma$ for $\sigma \in \mathbb{R}$.

7.4.1 Regularity Over One Time Step

We first show the preservation of regularity of (7.12) over one time step.

Lemma 7.1 *Let $s \geqslant 0$ and $-1 \leqslant \alpha \leqslant 0$. Suppose that Assumption 7.1 holds for $\beta = \alpha$ with a constant c and $|||(y^0, \dot{y}^0)|||_s \leqslant M$, then for the solution given by the ERKN integrator (7.12), we have $|||(y^1, \dot{y}^1)|||_s \leqslant C$, where C depends only on M, s, p, and c.*

Proof On noticing $\operatorname{sinc}(0) = 1 \leqslant h^{-1}$ and the bound $|\operatorname{sinc}(\xi)| \leqslant \xi^{-1}$ for $\xi > 0$, it follows from (7.12) that

$$\left\| y^{\frac{1}{2}} \right\|_{s+1} \leqslant \left\| y^0 \right\|_{s+1} + \left\| \dot{y}^0 \right\|_s \leqslant 2M, \tag{7.17}$$

which gives

$$\left\| f(y^{\frac{1}{2}}) \right\|_{s+1} \leqslant C, \tag{7.18}$$

by considering (7.11) with $\sigma' = s + 1$. On noticing the fact that $-1 \leqslant \alpha \leqslant 0$ and the bound (7.13) of $\bar{b}_1$, we have

$$\begin{aligned}
\left\| y^1 \right\|_{s+1} &\leqslant \left\| y^0 \right\|_{s+1} + \left\| \dot{y}^0 \right\|_s + h^{2+\alpha} \left\| f(y^{\frac{1}{2}}) \right\|_{s+1+\alpha} \\
&\leqslant \left\| y^0 \right\|_{s+1} + \left\| \dot{y}^0 \right\|_s + h^{2+\alpha} \left\| f(y^{\frac{1}{2}}) \right\|_{s+1}.
\end{aligned}$$

It follows from (7.18) that $\left\| y^1 \right\|_{s+1} \leqslant C$. Similarly, we obtain $\left\| \dot{y}^1 \right\|_s \leqslant C$, and then the proof is complete. $\qquad\square$

7.4.2 Local Error Bound

We now turn to the local error of the ERKN integrator (7.12).

Lemma 7.2 (Local Error in $H^{s+1-\alpha} \times H^{s-\alpha}$ for $-1 \leqslant \alpha \leqslant 0$) *With the conditions of Lemma 7.1, if $|||(y(\tau), \dot{y}(\tau))|||_s \leqslant M$ for $t_0 \leqslant \tau \leqslant t_1$, it holds that $|||(y(t_1), \dot{y}(t_1)) - (y^1, \dot{y}^1)|||_{s-\alpha} \leqslant Ch^{2+\alpha}$, where the constant C depends only on M, s, p, and c.*

Proof Throughout the proof, C stands for a generic constant depending only on M, s, p, and c.

(I) The local error of $y(t_1) - y^1$.

Using (7.8) and (7.12) we obtain

$$y(t_1) - y^1 = \int_{t_0}^{t_1} (t_1 - \tau)\mathrm{sinc}((t_1 - \tau)\Omega)f(y(\tau))\mathrm{d}\tau - h^2\bar{b}_1(V)f(y^{\frac{1}{2}}).$$

We note the fact that for $\xi > 0$ and $-1 \leqslant \alpha \leqslant 0$, $|\mathrm{sinc}(\xi)| \leqslant \xi^\alpha$, and $h^\alpha \geqslant 1$. By these results, (7.13) and (7.17), we have

$$\left\| y(t_1) - y^1 \right\|_{s+1-\alpha} \leqslant h^{2+\alpha} \sup_{t_0 \leqslant \tau \leqslant t_1} \| f(y(\tau)) \|_{s+1} + ch^{2+\alpha} \left\| f(y^{\frac{1}{2}}) \right\|_{s+1}.$$

It follows from (7.11) and (7.18) that $\| f(y(\tau)) \|_{s+1} \leqslant C$, which leads to

$$\left\| y(t_1) - y^1 \right\|_{s+1-\alpha} \leqslant Ch^{2+\alpha}.$$

(II) The local error of $\dot{y}(t_1) - \dot{y}^1$.

It follows from (7.8) and (7.12) that

$$\dot{y}(t_1) - \dot{y}^1 = \int_{t_0}^{t_1} [\cos((t_1 - \tau)\Omega) - I]f(y(\tau))\mathrm{d}\tau \tag{7.19}$$

$$+ \int_{t_0}^{t_1} f(y(\tau))\mathrm{d}\tau - hf\left(y\left(\frac{t_0 + t_1}{2}\right)\right) \tag{7.20}$$

$$+ hf\left(y\left(\frac{t_0 + t_1}{2}\right)\right) - hf(y^{\frac{1}{2}}) \tag{7.21}$$

$$+ h(I - b_1(V))f(y^{\frac{1}{2}}). \tag{7.22}$$

- Bound of (7.19). For $\xi > 0$ and $-1 \leqslant \alpha \leqslant 0$, it is easy to obtain that $|\cos(\xi) - 1| \leqslant 2\xi^{1+\alpha}$. On noticing (7.11) with $\sigma' = s + 1$, one arrives at

$$\left\| \int_{t_0}^{t_1} [\cos((t_1 - \tau)\Omega) - I]f(y(\tau))\mathrm{d}\tau \right\|_{s-\alpha} \leqslant 2h^{1+\alpha}\int_{t_0}^{t_1} C\mathrm{d}\tau \leqslant Ch^{2+\alpha}.$$

- Bound of (7.20). Since $1 \leqslant \xi^{1+\alpha} + \xi^\alpha$ for $\xi > 0$, we rewrite (7.20) as

$$\left\| \int_{t_0}^{t_1} f(y(\tau))\mathrm{d}\tau - hf\left(y\left(\frac{t_0 + t_1}{2}\right)\right) \right\|_{s-\alpha} \leqslant h^{1+\alpha}$$

$$\times \left\| \int_{t_0}^{t_1} f(y(\tau))\mathrm{d}\tau - hf\left(y\left(\frac{t_0 + t_1}{2}\right)\right) \right\|_{s+1}$$

$$+ h^\alpha \left\| \int_{t_0}^{t_1} f(y(\tau))\mathrm{d}\tau - hf\left(y\left(\frac{t_0 + t_1}{2}\right)\right) \right\|_s.$$

It then follows from (7.11) with $\sigma' = s + 1$ that

$$\left\| \int_{t_0}^{t_1} f(y(\tau))d\tau - hf\left(y\left(\frac{t_0 + t_1}{2} \right) \right) \right\|_{s+1} \leqslant \int_{t_0}^{t_1} C d\tau + Ch \leqslant Ch.$$

For an estimate in the norm $\|\cdot\|_s$, it is remarked that (7.20) is the quadrature error of the mid-point rule. With its first-order Peano kernel $K_1(\tau)$ and by the Peano kernel theorem, we obtain

$$\left\| \int_{t_0}^{t_1} f(y(\tau))d\tau - hf\left(y\left(\frac{t_0 + t_1}{2} \right) \right) \right\|_s$$

$$= h^2 \left\| \int_{t_0}^{t_1} K_1(\tau) \frac{d}{dt} f(y(t_0 + \tau h))d\tau \right\|_s \leqslant Ch^2,$$

where we have used (3.4a) in [32]. Thus, it is true that

$$\left\| \int_{t_0}^{t_1} f(y(\tau))d\tau - hf\left(y\left(\frac{t_0 + t_1}{2} \right) \right) \right\|_{s-\alpha} \leqslant Ch^{2+\alpha}. \tag{7.23}$$

- Bound of (7.21). Using (7.10) with $\sigma = s - \alpha$, we have

$$\left\| hf\left(y\left(\frac{t_0 + t_1}{2} \right) \right) - hf(y^{\frac{1}{2}}) \right\|_{s-\alpha} \leqslant Ch \left\| y\left(\frac{t_0 + t_1}{2} \right) - y^{\frac{1}{2}} \right\|_{s-\alpha}.$$

It follows from (7.8) and (7.12) that

$$y\left(\frac{t_0 + t_1}{2} \right) - y^{\frac{1}{2}} = \int_{t_0}^{\frac{t_0+t_1}{2}} \left(\frac{t_0 + t_1}{2} - \tau \right) \operatorname{sinc}\left(\left(\frac{t_0 + t_1}{2} - \tau \right) \Omega \right) f(y(\tau))d\tau. \tag{7.24}$$

In a similar way to the first part of this proof, we obtain

$$\left\| y\left(\frac{t_0 + t_1}{2} \right) - y^{\frac{1}{2}} \right\|_{s+1-\alpha} \leqslant h^{2+\alpha} \sup_{t_0 \leqslant \tau \leqslant \frac{t_0+t_1}{2}} \| f(y(\tau)) \|_{s+1} \leqslant Ch^{2+\alpha}.$$

Then, it is true that

$$\left\| y\left(\frac{t_0 + t_1}{2} \right) - y^{\frac{1}{2}} \right\|_{s-\alpha} \leqslant \left\| y\left(\frac{t_0 + t_1}{2} \right) - y^{\frac{1}{2}} \right\|_{s+1-\alpha} \leqslant Ch^{2+\alpha},$$

which leads to $\left\| hf\left(y\left(\frac{t_0 + t_1}{2} \right) \right) - hf(y^{\frac{1}{2}}) \right\|_{s-\alpha} \leqslant Ch^{3+\alpha}.$

- Bound of (7.22). According to (7.18) and the bound (7.15), we have

$$\left\| h(I - b_1(V)) f(y^{\frac{1}{2}}) \right\|_{s-\alpha} \leqslant Ch^{2+\alpha} \left\| f(y^{\frac{1}{2}}) \right\|_{s+1} \leqslant Ch^{2+\alpha}.$$

Clearly, all these estimates imply $\left\| \dot{y}(t_1) - \dot{y}^1 \right\|_{s-\alpha} \leqslant Ch^{2+\alpha}$.

The proof is complete.   □

7.4.3 Stability

In this subsection we analyse the stability of the ERKN integrator (7.12).

Lemma 7.3 (Stability in $H^{s+1-\alpha} \times H^{s-\alpha}$ for $-1 \leqslant \alpha \leqslant 0$) *Under the conditions of Lemma 7.1, if we consider the ERKN integrator (7.12) with different initial values $(y_0, \dot{y}_0)$ and $(z_0, \dot{z}_0)$ satisfying $|||(y_0, \dot{y}_0)|||_s \leqslant M$ and $|||(z_0, \dot{z}_0)|||_s \leqslant M$, then it holds that*

$$|||(y^1, \dot{y}^1) - (z^1, \dot{z}^1)|||_{s-\alpha} \leqslant (1 + Ch)|||(y^0, \dot{y}^0) - (z^0, \dot{z}^0)|||_{s-\alpha},$$

where the constant C depends only on M, s, p, and c.

Proof It follows from the result (3.8) in [32] and ERKN integrators (7.12) that

$$
\begin{aligned}
|||(y^1, \dot{y}^1) - (z^1, \dot{z}^1)|||_{s-\alpha} &\leqslant |||(y^0, \dot{y}^0) - (z^0, \dot{z}^0)|||_{s-\alpha} \\
&\quad + h|\dot{y}_0^0 - \dot{z}_0^0| \quad\quad (7.25) \\
&\quad + h^2 \left\| \bar{b}_1(V) \left(f(y^{\frac{1}{2}}) - f(z^{\frac{1}{2}}) \right) \right\|_{s+1-\alpha} \quad (7.26) \\
&\quad + h \left\| b_1(V) \left(f(y^{\frac{1}{2}}) - f(z^{\frac{1}{2}}) \right) \right\|_{s-\alpha}. \quad (7.27)
\end{aligned}
$$

- It is trivial for (7.25), that $h|\dot{y}_0^0 - \dot{z}_0^0| \leqslant h \left\| \dot{y}_0^0 - \dot{z}_0^0 \right\|_{s-\alpha}$.
- With regard to (7.26), combining the bound (7.13) of $\bar{b}_1$ and (7.10) with $\sigma = \sigma' = s + 1$ yields

$$h^2 \left\| \bar{b}_1(V) \left(f(y^{\frac{1}{2}}) - f(z^{\frac{1}{2}}) \right) \right\|_{s+1-\alpha} \leqslant Ch^{2+\alpha} \left\| y^{\frac{1}{2}} - z^{\frac{1}{2}} \right\|_{s+1}.$$

Using the formula for ERKN integrators (7.12) again, we confirm that $\left\| y^{\frac{1}{2}} - z^{\frac{1}{2}} \right\|_{s+1} \leqslant \left\| y^0 - z^0 \right\|_{s+1} + \left\| \dot{y}^0 - \dot{z}^0 \right\|_{s}$. This implies

$$h^2 \left\| \bar{b}_1(V) \left(f(y^{\frac{1}{2}}) - f(z^{\frac{1}{2}}) \right) \right\|_{s+1-\alpha} \leqslant Ch^{2+\alpha} \left\| y^0 - z^0 \right\|_{s+1} + Ch^{2+\alpha} \left\| \dot{y}^0 - \dot{z}^0 \right\|_{s}.$$

- Concerning (7.27), it follows from (7.15) that $|b_1(\xi)| \leq 1 + c\xi^{1+\alpha}$, and then we have

$$h \left\| b_1(V) \left(f(y^{\frac{1}{2}}) - f(z^{\frac{1}{2}}) \right) \right\|_{s-\alpha}$$

$$\leq h \left\| f(y^{\frac{1}{2}}) - f(z^{\frac{1}{2}}) \right\|_{s-\alpha} + ch^{2+\alpha} \left\| f(y^{\frac{1}{2}}) - f(z^{\frac{1}{2}}) \right\|_{s+1}$$

$$\leq Ch \left\| y^{\frac{1}{2}} - z^{\frac{1}{2}} \right\|_{s-\alpha} + Ch^{2+\alpha} \left\| y^{\frac{1}{2}} - z^{\frac{1}{2}} \right\|_{s+1}$$

$$\leq C(h + h^{2+\alpha}) \left\| y^0 - z^0 \right\|_{s+1} + C(h + h^{2+\alpha}) \left\| \dot{y}^0 - \dot{z}^0 \right\|_s.$$

The above estimates of (7.25)–(7.27) with $-1 \leq \alpha \leq 0$ complete the proof. □

7.4.4 Proof of Theorem 7.1 for $-1 \leq \alpha \leq 0$

We are now in a position to present the proof of Theorem 7.1 for $-1 \leq \alpha \leq 0$, based on the three lemmas stated above.

Proof

(I) We begin with the proof for the case where $\alpha = 0$. Let C_1 and C_2 be the constants of Lemmas 7.2 and 7.3 with $\alpha = 0$, respectively. It is noted that Lemma 7.3 is considered with $2M$ instead of M. Let $h_0 = M/(C_1 T e^{C_2 T})$ and we show by induction on n that for $h \leq h_0$

$$|||(y^n, \dot{y}^n) - (y(t_n), \dot{y}(t_n))|||_s \leq C_1 e^{C_2 nh} nh^2, \tag{7.28}$$

as long as $t_n - t_0 = nh \leq T$.

We first have $|||(y^0, \dot{y}^0) - (y(t_0), \dot{y}(t_0))|||_s = 0 \leq C_1$. We assume that the result (7.28) is true for $n = 0, \cdots, m - 1$. This implies that

$$|||(y^{m-1}, \dot{y}^{m-1}) - (y(t_{m-1}), \dot{y}(t_{m-1}))|||_s \leq C_1 e^{C_2(m-1)h} (m-1)h^2,$$

which gives

$$|||(y^{m-1}, \dot{y}^{m-1})|||_s \leq M + C_1 e^{C_2(m-1)h} (m-1)h^2 \leq M + C_1 e^{C_2 T} Th \leq 2M,$$

as long as $t_{m-1} - t_0 = (m-1)h \leq T$. Denoting by $\mathscr{E}$ one time step with the ERKN integrator (7.12), we obtain

$$|||(y^m, \dot{y}^m) - (y(t_m), \dot{y}(t_m))|||_s = |||\mathscr{E}(y^{m-1}, \dot{y}^{m-1}) - (y(t_m), \dot{y}(t_m))|||_s$$

$$\leq |||\mathscr{E}(y^{m-1}, \dot{y}^{m-1}) - \mathscr{E}(y(t_{m-1}), \dot{y}(t_{m-1}))|||_s \tag{7.29}$$

$$+ |||\mathscr{E}(y(t_{m-1}), \dot{y}(t_{m-1})) - (y(t_m), \dot{y}(t_m))|||_s. \tag{7.30}$$

In terms of Lemma 7.3, (7.29) admits the bound

$$|||\mathscr{E}(y^{m-1}, \dot{y}^{m-1}) - \mathscr{E}(y(t_{m-1}), \dot{y}(t_{m-1}))|||_s$$
$$\leqslant (1 + C_2 h)|||(y^{m-1}, \dot{y}^{m-1}) - (y(t_{m-1}), \dot{y}(t_{m-1}))|||_s$$
$$\leqslant (1 + C_2 h)C_1 e^{C_2(m-1)h}(m-1)h^2.$$

With regard to (7.30), it follows from Lemma 7.2 that $|||\mathscr{E}(y(t_{m-1}), \dot{y}(t_{m-1})) - (y(t_m), \dot{y}(t_m))|||_s \leqslant C_1 h^2$. We then obtain

$$|||(y^m, \dot{y}^m) - (y(t_m), \dot{y}(t_m))|||_s \leqslant (1 + C_2 h)C_1 e^{C_2(m-1)h}(m-1)h^2 + C_1 h^2.$$

Using Taylor expansions, we obtain that

$$(1 + C_2 h)C_1 e^{C_2(m-1)h}(m-1)h^2 + C_1 h^2 \leqslant C_1 e^{C_2 mh} m h^2.$$

Consequently, (7.28) holds, and hence

$$|||(y^n, \dot{y}^n) - (y(t_n), \dot{y}(t_n))|||_s \leqslant C_1 T e^{C_2 T} h \leqslant Ch,$$

which proves the statement of Theorem 7.1 for $\alpha = 0$.

(II) We next consider the case $-1 \leqslant \alpha < 0$. Let h_0 be as above and let C_1 and C_2 be as above but for the new α instead of $\alpha = 0$. We then prove, by induction on n, that

$$|||(y^n, \dot{y}^n) - (y(t_n), \dot{y}(t_n))|||_{s-\alpha} \leqslant C_1 e^{C_2 nh} n h^{2+\alpha}, \tag{7.31}$$

as long as $t_n - t_0 = nh \leqslant T$.

Obviously, this holds for $n = 0$. It follows from the proof stated above for the case $\alpha = 0$ that $|||(y^{n-1}, \dot{y}^{n-1})|||_s \leqslant 2M$, as long as $t_{n-1} - t_0 = (n-1)h \leqslant T$. This allows us to apply Lemmas 7.2 and 7.3 to (7.31), which gives

$$|||(y^n, \dot{y}^n) - (y(t_n), \dot{y}(t_n))|||_{s-\alpha} \leqslant |||\mathscr{E}(y^{n-1}, \dot{y}^{n-1})$$
$$-\mathscr{E}(y(t_{n-1}), \dot{y}(t_{n-1}))|||_{s-\alpha}$$
$$+|||\mathscr{E}(y(t_{n-1}), \dot{y}(t_{n-1})) - (y(t_n), \dot{y}(t_n))|||_{s-\alpha}$$
$$\leqslant (1 + C_2 h)C_1 e^{C_2(n-1)h}(n-1)h^{2+\alpha} + C_1 h^{2+\alpha} \leqslant C_1 e^{C_2 nh} n h^{2+\alpha}.$$

This confirms that (7.31) is true, and then we have

$$|||(y^n, \dot{y}^n) - (y(t_n), \dot{y}(t_n))|||_{s-\alpha} \leqslant C_1 T e^{C_2 T} h^{1+\alpha} \leqslant Ch^{1+\alpha}.$$

The proof is complete.   □

Remark 7.3 It follows from the above proof for $\alpha = 0$ that the numerical solutions are bounded in $H^{s+1} \times H^s$

$$|||(y^n, \dot{y}^n)|||_s \leqslant 2M \qquad \text{for} \qquad 0 \leqslant t_n - t_0 = nh \leqslant T. \tag{7.32}$$

This regularity of the numerical solution is essential for the proof of Theorem 7.1 for $0 < \alpha \leqslant 1$ in the next section.

7.5 Higher-Order Error Bounds in Lower-Order Sobolev Spaces

The following three lemmas are needed for the proof of Theorem 7.1 in lower-order Sobolev spaces.

Lemma 7.4 *Let $s \geqslant 0$ and $0 < \alpha \leqslant 1$. Suppose that Assumption 7.1 holds for $\beta = \alpha$ with constant c and $|||(y^0, \dot{y}^0)|||_s \leqslant M$. We have $|||(y^1, \dot{y}^1)|||_s \leqslant C$ with a constant C depending only on $M, s, p,$ and c.*

We omit the proof of Lemma 7.4 which is quite similar to that of Lemma 7.1.

Lemma 7.5 (Local Error in $H^{s+1-\alpha} \times H^{s-\alpha}$ for $0 < \alpha \leqslant 1$) *Under the conditions of Lemma 7.4, if $|||(y(\tau), \dot{y}(\tau))|||_s \leqslant M$ for $t_0 \leqslant \tau \leqslant t_1$, then it holds that $|||(y(t_1), \dot{y}(t_1)) - (y^1, \dot{y}^1)|||_{s-\alpha} \leqslant Ch^{2+\alpha}$, where the constant C depends only on $M, s, p,$ and c.*

Proof

(I) Local error of $y(t_1) - y^1$.

It follows from (7.8), (7.12) and

$$\int_{t_0}^{t_1} (t_1 - \tau)\mathrm{sinc}((t_1 - \tau)\Omega)\mathrm{d}\tau = \frac{1}{2}h^2\mathrm{sinc}^2\left(\frac{1}{2}h\Omega\right),$$

that

$$y(t_1) - y^1 = \int_{t_0}^{t_1} (t_1 - \tau)\mathrm{sinc}((t_1 - \tau)\Omega)$$

$$\left[f(y(\tau)) - f\left(y\left(\frac{t_0 + t_1}{2}\right)\right) \right]\mathrm{d}\tau \tag{7.33}$$

$$+ \frac{1}{2}h^2\mathrm{sinc}^2\left(\frac{1}{2}h\Omega\right)\left[f\left(y\left(\frac{t_0 + t_1}{2}\right)\right) - f(y^{\frac{1}{2}}) \right] \tag{7.34}$$

$$+ h^2\left[\frac{1}{2}\mathrm{sinc}^2\left(\frac{1}{2}h\Omega\right) - \bar{b}_1(V) \right] f(y^{\frac{1}{2}}). \tag{7.35}$$

- Bound of (7.33). For $\xi > 0$ and $0 < \alpha \leqslant 1$, it is clear that $|\mathrm{sinc}(\xi)| \leqslant \xi^{-1+\alpha}$. Using this result, the estimate (7.10) with $\sigma = s$, and the fact

$$\left\| y(\tau) - y\left(\frac{t_0 + t_1}{2}\right) \right\|_s \leqslant \int_{\frac{t_0+t_1}{2}}^{\tau} \|\dot{y}(t)\|_s \, dt \leqslant Ch,$$

we obtain

$$\left\| \int_{t_0}^{t_1} (t_1 - \tau)\mathrm{sinc}((t_1 - \tau)\Omega)\left[f(y(\tau)) - f\left(y\left(\frac{t_0 + t_1}{2}\right)\right)\right] d\tau \right\|_{s+1-\alpha}$$

$$\leqslant h^{-1+\alpha} \int_{t_0}^{t_1} |t_1 - \tau| \left\| f(y(\tau)) - f\left(y\left(\frac{t_0 + t_1}{2}\right)\right)\right\|_s d\tau$$

$$\leqslant Ch^{-1+\alpha} \int_{t_0}^{t_1} |t_1 - \tau| \left\| y(\tau) - y\left(\frac{t_0 + t_1}{2}\right)\right\|_s d\tau \leqslant Ch^{2+\alpha}.$$

- For (7.34), according to the fact that $|\mathrm{sinc}(\xi)|^2 \leqslant \dfrac{1 \cdot \xi}{\xi^2} = \xi^{-1}$ for $\xi > 0$ and the estimate (7.10) with $\sigma = s - \alpha$, we have

$$\left\| \frac{1}{2}h^2\mathrm{sinc}^2\left(\frac{1}{2}h\Omega\right)\left[f\left(y\left(\frac{t_0 + t_1}{2}\right)\right) - f(y^{\frac{1}{2}})\right]\right\|_{s+1-\alpha}$$

$$\leqslant Ch \left\| f\left(y\left(\frac{t_0 + t_1}{2}\right)\right) - f(y^{\frac{1}{2}})\right\|_{s-\alpha} \leqslant Ch \left\| y\left(\frac{t_0 + t_1}{2}\right) - y^{\frac{1}{2}}\right\|_{s-\alpha}.$$

Furthermore, the estimate (7.11) with $\sigma' = s + 1$ gives

$$\left\| y\left(\frac{t_0 + t_1}{2}\right) - y^{\frac{1}{2}}\right\|_{s-\alpha} \leqslant \left\| y(\frac{t_0 + t_1}{2}) - y^{\frac{1}{2}}\right\|_{s+2-\alpha}$$

$$= \left\| \int_{t_0}^{\frac{t_0+t_1}{2}} (\frac{t_0 + t_1}{2} - \tau)\mathrm{sinc}((\frac{t_0 + t_1}{2} - \tau)\Omega) f(y(\tau)) d\tau \right\|_{s+2-\alpha} \qquad (7.36)$$

$$\leqslant h^{-1+\alpha} \int_{t_0}^{\frac{t_0+t_1}{2}} \left| \frac{t_0 + t_1}{2} - \tau \right| \|f(y(\tau))\|_{s+1} \, d\tau \leqslant Ch^{1+\alpha}.$$

Thus, we obtain $\left\| \dfrac{1}{2}h^2\mathrm{sinc}^2\left(\dfrac{1}{2}h\Omega\right)\left[f\left(y\left(\dfrac{t_0 + t_1}{2}\right)\right) - f(y^{\frac{1}{2}})\right]\right\|_{s+1-\alpha} \leqslant Ch^{2+\alpha}.$

- With regard to (7.35), considering (7.14) and the estimate (7.11) with $\sigma' = s + 1$ yields

$$\left\| h^2 \left[\frac{1}{2} \mathrm{sinc}^2 \left(\frac{1}{2} h \Omega \right) - \bar{b}_1(V) \right] f(y^{\frac{1}{2}}) \right\|_{s+1-\alpha} \leqslant h^{2+\alpha} \left\| f(y^{\frac{1}{2}}) \right\|_{s+1} \leqslant C h^{2+\alpha}.$$

Finally, all the bounds of (7.33)–(7.35) imply

$$\left\| y(t_1) - y^1 \right\|_{s+1-\alpha} \leqslant C h^{2+\alpha}.$$

(II) Local error of $\dot{y}(t_1) - \dot{y}^1$.

Likewise, this error bound can be derived as the bound given in (II) of Lemma 7.2 with the first-order Peano kernel replaced by the second-order Peano kernel. □

Lemma 7.6 (Stability in $H^{s+1-\alpha} \times H^{s-\alpha}$ for $0 < \alpha \leqslant 1$) *With the conditions of Lemma 7.4, we consider different initial values $(y_0, \dot{y}_0)$ and $(z_0, \dot{z}_0)$ for the ERKN integrator (7.12). If $\max\{\|\|(y_0, \dot{y}_0)\|\|_s, \|\|(z_0, \dot{z}_0)\|\|_s\} \leqslant M$, then we have*

$$\|\|(y^1, \dot{y}^1) - (z^1, \dot{z}^1)\|\|_{s-\alpha} \leqslant (1 + Ch)\|\|(y^0, \dot{y}^0) - (z^0, \dot{z}^0)\|\|_{s-\alpha},$$

where the constant C depends only on M, s, p, and c.

Proof We begin with

$$\|\|(y^1, \dot{y}^1) - (z^1, \dot{z}^1)\|\|_{s-\alpha} \leqslant \|\|(y^0, \dot{y}^0) - (z^0, \dot{z}^0)\|\|_{s-\alpha} + h|\dot{y}_0^0 - \dot{z}_0^0|$$

$$+ h^2 \left\| \bar{b}_1(V) \left(f(y^{\frac{1}{2}}) - f(z^{\frac{1}{2}}) \right) \right\|_{s+1-\alpha} \quad (7.37)$$

$$+ h \left\| b_1(V) \left(f(y^{\frac{1}{2}}) - f(z^{\frac{1}{2}}) \right) \right\|_{s-\alpha}. \quad (7.38)$$

It is clear that $|\bar{b}_1(\xi)| \leqslant \dfrac{1}{2} + c\xi^\alpha$ due to (7.14). Hence, the bound of (7.37) is

$$h^2 \left\| \bar{b}_1(V) \left(f(y^{\frac{1}{2}}) - f(z^{\frac{1}{2}}) \right) \right\|_{s+1-\alpha} \leqslant \frac{1}{2} h^2 \left\| f(y^{\frac{1}{2}}) - f(z^{\frac{1}{2}}) \right\|_{s+1-\alpha}$$

$$+ \frac{1}{2} h^{2+\alpha} \left\| f(y^{\frac{1}{2}}) - f(z^{\frac{1}{2}}) \right\|_{s+1} \leqslant C h^2 (1/2 + h^\alpha) \left\| y^{\frac{1}{2}} - z^{\frac{1}{2}} \right\|_{s+1}$$

$$\leqslant C h^2 (1/2 + h^\alpha) \left\| y^0 - z^0 \right\|_{s+1} + C h^2 (1/2 + h^\alpha) \left\| \dot{y}^0 - \dot{z}^0 \right\|_s.$$

We turn to (7.38). Clearly, $|b_1(\xi)| \leqslant 1 + c\xi^{1+\alpha}$ due to (7.15), and then we obtain

$$
\begin{aligned}
h \left\| b_1(V)\left(f(y^{\frac{1}{2}}) - f(z^{\frac{1}{2}}) \right) \right\|_{s-\alpha} &\leqslant h \left\| f(y^{\frac{1}{2}}) - f(z^{\frac{1}{2}}) \right\|_{s-\alpha} \\
&\quad + ch^{2+\alpha} \left\| f(y^{\frac{1}{2}}) - f(z^{\frac{1}{2}}) \right\|_{s+1} \\
&\leqslant C(h + h^{2+\alpha}) \left\| y^0 - z^0 \right\|_{s+1} + C(h + h^{2+\alpha}) \left\| \dot{y}^0 - \dot{z}^0 \right\|_s .
\end{aligned}
$$

The proof is complete as a consequence of the above bounds.   □

Proof of Theorem 7.1 for $0 < \alpha \leqslant 1$.

Proof This proof is the same as that for $-1 \leqslant \alpha < 0$ given in Sect. 7.4. A key point used here is that the numerical solution is bounded in $H^{s+1} \times H^s$ on the basis of Remark 7.3.   □

Remark 7.4 We only consider one-stage ERKN integrators in the error analysis. The extension of the analysis to higher-stage ERKN integrators is not obvious since there are some technical difficulties which need to be overcome. This issue needs to be considered in future investigations.

7.6 Numerical Experiments

This section presents a numerical experiment to illustrate the error bounds of two one-stage explicit ERKN integrators.

We solve the problem (7.1) with $p = 2$, and use the spatial semidiscretisation with $K = 2^6$ and $K = 2^8$. Following [32], we choose the initial conditions for the coefficients $y_j(t_0)$ and $\dot{y}_j(t_0)$ on the complex unit circle and then scale them by $\langle j \rangle^{-1.51}$ and $\langle j \rangle^{-0.51}$, respectively. Here, it is important to note that these complex numbers are chosen such that the corresponding trigonometric polynomial (7.3) takes real values at the collocation points. Then, the corresponding initial values satisfy the condition (7.16) of Theorem 7.1 at time $t = t_0$ uniformly in K for $s = 0$. For the time discretisation, we choose ERKN3 and ERKN4 whose coefficients are displayed in Table 7.1. For comparison, we also consider a one-stage RKN method, which is obtained from these ERKN integrators by letting $M = 0$.

The problem is solved on the interval $[0, 10]$ with the stepsizes $h = 1/2^j$ for $j = 0, 1, \cdots, 10$. We measure the errors

$$
\text{erry} = \left\| y(t_n) - y^n \right\|_{1-\alpha}, \quad \text{errdy} = \left\| \dot{y}(t_n) - \dot{y}^n \right\|_{-\alpha}
$$

in different Sobolev norms $\alpha = 1, \dfrac{1}{2}, 0, -\dfrac{1}{2}, -1$. For the RKN method, it has been checked that the errors are too large for some big stepsizes. Therefore we use smaller stepsizes $h = 1/2^j$ for $j = 4, \cdots, 14$. We plot the logarithm of the errors against the logarithm of stepsizes for the results displayed in Figs. 7.1 and 7.2.

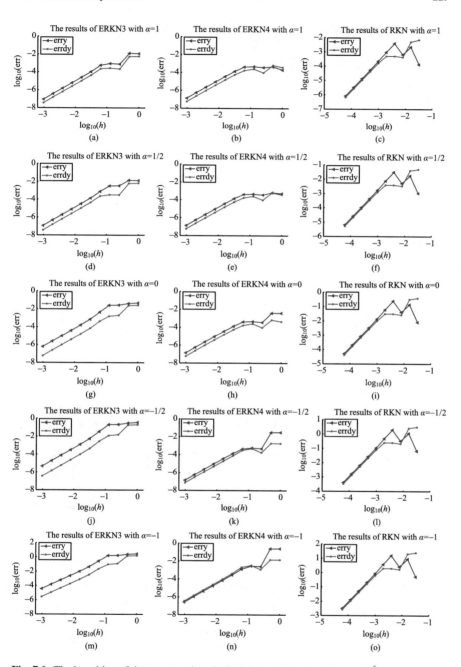

Fig. 7.1 The logarithm of the errors against the logarithm of stepsizes for $K = 2^6$

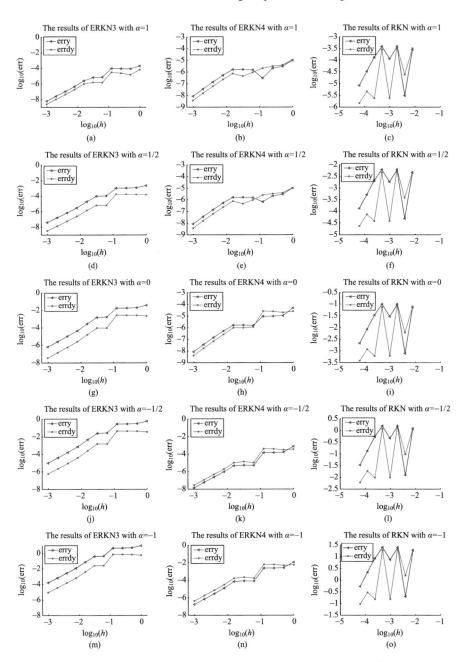

Fig. 7.2 The logarithm of the errors against the logarithm of stepsizes for $K = 2^8$

It follows from these results that the convergence order is not uniform for α, and when α goes from 1 to -1, the errors of ERKN integrators become large. This supports the result given in Theorem 7.1. Moreover, it can be observed from the computed results that ERKN integrators work much better for larger stepsizes, and they are more accurate for smaller stepsizes than the corresponding RKN method.

At the end of this section, we remark that the results for ERKN4 with a small stepsize are considered as the "exact" solutions of the underlying system for both values of K. We also note that a few errors for ERKN integrators for $K = 2^8$ are smaller than those for $K = 2^6$. This phenomenon may be caused by the choices of "exact" solutions for different K.

7.7 Concluding Remarks

In this chapter, we have analysed the error bounds of ERKN integrators when applied to spatial semidiscretisations of semilinear wave equations. Optimal second-order convergence has been obtained without requiring Lipschitz continuity and higher regularity of the exact solution. Moreover, the analysis is uniform in the spatial discretisation parameter. On the basis of this work, we are hopeful of obtaining an extension to two-stage ERKN integrators for semidiscrete semilinear wave equations. Another issue for future exploration is the error analysis of ERKN integrators in the case of quasi-linear wave equations.

The material in this chapter is based on the work by Wang and Wu [43].

References

1. Bueno-Orovio, A., Kay, D., Burrage, K.: Fourier spectral methods for fractional-in-space reaction-diffusion equations. BIT Numer. Math. **54**, 937–954 (2014)
2. Iserles, A.: A First Course in the Numerical Analysis of Differential Equations, 2nd edn. Cambridge University Press, Cambridge (2009)
3. Brugnano, L., Frasca Caccia, G., Iavernaro, F.: Energy conservation issues in the numerical solution of the semilinear wave equation. Appl. Math. Comput. **270**, 842–870 (2015)
4. Brugnano, L., Iavernaro, F.: Line Integral Methods for Conservative Problems. Chapman and Hall/CRC, Boca Raton (2016)
5. Butcher, J.C.: Numerical Methods for Ordinary Differential Equations, 2nd edn. Wiley, Chichester (2008)
6. Hairer, E., Lubich, C., Wanner, G.: Geometric Numerical Integration: Structure-preserving Algorithms for Ordinary Differential Equations, 2nd ed. Springer, Berlin (2006)
7. Hochbruck, M., Ostermann, A.: Exponential integrators. Acta Numer. **19**, 209–286 (2010)
8. Iserles, A.: On the global error of discretization methods for highly-oscillatory ordinary differential equations. BIT Numer. Math. **42**, 561–599 (2002)
9. Wu, X., You, X., Wang, B.: Structure-Preserving Algorithms for Oscillatory Differential Equations. Springer, Berlin (2013)
10. Wu, X., Liu, C., Mei, L.: An analytical expression of solutions to nonlinear wave equations in higher dimensions with Robin boundary conditions. J. Math. Anal. Appl. **426**, 1164–1173 (2015)

11. Hochbruck, M., Lubich, C.: A Gautschi-type method for oscillatory second-order differential equations. Numer. Math. **83**, 403–426 (1999)
12. Hochbruck, M., Ostermann, A.: Explicit exponential Runge–Kutta methods for semilinear parabolic problems. SIAM J. Numer. Anal. **43**, 1069–1090 (2005)
13. Hochbruck, M., Ostermann, A., Schweitzer, J.: Exponential rosenbrock-type methods. SIAM J. Numer. Anal. **47**, 786–803 (2009)
14. Li, Y.W., Wu, X.: Exponential integrators preserving first integrals or Lyapunov functions for conservative or dissipative systems. SIAM J. Sci. Comput. **38**, 1876–1895 (2016)
15. Bao, W., Dong, X.: Analysis and comparison of numerical methods for the Klein-Gordon equation in the nonrelativistic limit regime. Numer. Math. **120**, 189–229 (2012)
16. Cano, B.: Conservation of invariants by symmetric multistep cosine methods for second-order partial differential equations. BIT Numer. Math. **53**, 29–56 (2013)
17. Cano, B., Moreta, M.J.: Multistep cosine methods for second-order partial differential systems. IMA J. Numer. Anal. **30**, 431–461 (2010)
18. Cohen, D., Hairer, E., Lubich, C.: Conservation of energy, momentum and actions in numerical discretizations of nonlinear wave equations. Numer. Math. **110**, 113–143 (2008)
19. Gauckler, L., Weiss, D.: Metastable energy strata in numerical discretizations of weakly nonlinear wave equations. Disc. Contin. Dyn. Syst. **37**, 3721–3747 (2017)
20. Wu, X., You, X., Shi, W., et al.: ERKN integrators for systems of oscillatory second-order differential equations. Comput. Phys. Commun. **181**, 1873–1887 (2010)
21. Liu, C., Iserles, A., Wu, X.: Symmetric and arbitrarily high-order Birkhoff-Hermite time integrators and their long-time behaviour for solving nonlinear Klein-Gordon equations. J. Comput. Phys. **356**, 1–30 (2018)
22. Wang, B., Iserles, A., Wu, X.: Arbitrary-order trigonometric Fourier collocation methods for multi-frequency oscillatory systems. Found. Comput. Math. **16**, 151–181 (2016)
23. Wang, B., Yang, H., Meng, F.: Sixth order symplectic and symmetric explicit ERKN schemes for solving multi-frequency oscillatory nonlinear Hamiltonian equations. Calcolo **54**, 117–140 (2017)
24. García-Archilla, B., Sanz-Serna, J.M., Skeel, R.D.: Long-time-step methods for oscillatory differential equations. SIAM J. Sci. Comput. **20**, 930–963 (1999)
25. Grimm, V.: On error bounds for the Gautschi-type exponential integrator applied to oscillatory second-order differential equations. Numer. Math. **100**, 71–89 (2005)
26. Grimm, V., Hochbruck, M.: Error analysis of exponential integrators for oscillatory second order differential equations. J. Phys. A Math. Gen. **39**, 5495–5507 (2006)
27. Wang, B., Wu, X., Xia, J.: Error bounds for explicit ERKN integrators for systems of multifrequency oscillatory second-order differential equations. Appl. Numer. Math. **74**, 17–34 (2013)
28. Dong, X.: Stability and convergence of trigonometric integrator pseudospectral discretization or N-coupled nonlinear Klein-Gordon equations. Appl. Math. Comput. **232**, 752–765 (2014)
29. Faou, E.: Geometric Numerical Integration and Schrödinger Equations. EMS Zurich Lectures in Advanced Mathematics. European Mathematical Society, Zurich (2012)
30. Grimm, V.: On the Use of the Gautschi-type Exponential Integrator for Wave Equations. Numerical Mathematics and Advanced Applications, pp. 557–563. Springer, Berlin (2006)
31. Mei, L., Liu, C., Wu, X.: An essential extension of the finite-energy condition for extended Runge–Kutta–Nyström integrators when applied to nonlinear wave equations. Commun. Comput. Phys. **22**, 742–764 (2017)
32. Gauckler, L.: Error analysis of trigonometric integrators for semilinear wave equations. SIAM J. Numer. Anal. **53**, 1082–1106 (2015)
33. Gauckler, L.: Convergence of a split-step Hermite method for the Gross-Pitaevskii equation. IMA J. Numer. Anal. **31**, 396–415 (2011)
34. Holden, H., Lubich, C., Risebro, N.H.: Operator splitting for partial differential equations with Burgers nonlinearity. Math. Comput. **82**, 173–185 (2013)

35. Koch, O., Lubich, C.: Variational-splitting time integration of the multi-configuration time dependent Hartree-Fock equations in electron dynamics. IMA J. Numer. Anal. **31**, 379–395 (2011)
36. Lubich, C.: On splitting methods for Schrödinger-Poisson and cubic nonlinear Schrödinger equations. Math. Comput. **77**, 2141–2153 (2008)
37. Thalhammer, M.: Convergence analysis of high-order time-splitting pseudospectral methods for nonlinear Schrödinger equations. SIAM J. Numer. Anal. **50**, 3231–3258 (2012)
38. Gauckler, L., Lu, J., Marzuola, J.L., et al.: Trigonometric integrators for quasilinear wave equations. Math. Comput. **88**, 717–749 (2019)
39. Hochbruck, M., Pažur, T.: Error analysis of implicit Euler methods for quasilinear hyperbolic evolution equations. Numer. Math. **135**, 547–569 (2017)
40. Kovács, B., Lubich, C.: Stability and convergence of time discretizations of quasi-linear evolution equations of Kato type. Numer. Math. **138**, 365–388 (2018)
41. Hairer, E., Lubich, C.: Spectral semi-discretisations of weakly nonlinear wave equations over long times. Found. Comput. Math. **8**, 319–334 (2008)
42. Verwer, J.G., Sanz-Serna, J.M.: Convergence of method of lines approximations to partial differential equations. Computing **33**, 297–313 (1984)
43. Wang, B., Wu, X.: Global error bounds of one-stage extended RKN integrators for semilinear wave equations. Numer. Algor. **81**, 1203–1218 (2019)

Chapter 8
Linearly-Fitted Conservative (Dissipative) Schemes for Nonlinear Wave Equations

The discrete gradient method is a well-known scheme for the numerical integration of dynamic systems. Its extension to highly oscillatory Hamiltonian systems is called extended discrete gradient method. In this chapter, on the basis of the extended discrete gradient method, we present an efficient approach to devising a structure-preserving scheme for numerically solving conservative (dissipative) nonlinear wave equations. This scheme can preserve the energy exactly for conservative wave equations. With a minor improvement to the extended discrete gradient method, this scheme is applicable to dissipative wave equations, and can preserve the dissipation structure of the underlying dissipative wave equation.

8.1 Introduction

The idea of geometric integration has led to the rapid development of numerical integration. Numerical schemes that conserve geometric structure have been shown to possess excellent qualitative properties in studies of the long-time behaviour of dynamical systems. Such schemes are sometimes called geometric or structure-preserving integrators. The structure includes physical/geometric properties such as first integrals, symplecticity, oscillations, symmetries or reversing symmetries, phase-space volume, Lyapunov functions, and foliations. Since geometric integrators are excellent in preserving qualitative features of simulated differential equations, they have important applications in many fields, such as fluid dynamics, celestial mechanics, molecular dynamics, quantum physics, plasma physics, quantum mechanics, and meteorology. We refer the reader to [1–4] for surveys of this research. The consideration of qualitative properties in ordinary and partial differential equations is very important when designing numerical schemes. It is possible to devise relatively general frameworks for structure preservation for ordinary differential equations (ODEs). However, this seems somewhat much more difficult for partial differential equations (PDEs) because PDEs are a huge and diverse collection of problems and each equation under consideration normally requires a dedicated scheme (see, e.g. [5–9]). Fortunately, many attempts have been made to give a fairly general methodology to develop geometric schemes for PDEs. For example, in [10], by discretising the energy of the PDEs to get an ODE system, then applying the average vector field method to the resulting system, the authors proposed a systematic procedure to deal with evolutionary PDEs as far as conservation or dissipation of energy is concerned. Another example is the class of PDEs that can be formulated into multi-symplectic form to which, one can apply a scheme which preserves a discrete version of this form (see, e.g. [11] for a review of this approach). Many energy-preserving or multi-symplectic methods have been derived for Hamiltonian PDEs based on the multi-symplectic formulation (see, e.g. [12–17]), although further theoretical analysis for them is still needed.

It is noted that there has been an enormous advance during recent years in dealing with the system of highly oscillatory ODEs

$$\ddot{q} + Mq = f(q), \tag{8.1}$$

where $\|M\| \gg \max\left\{1, \left\|\frac{\partial f}{\partial q}\right\|\right\}$, which are frequently generated by spatial semidiscretisations of nonlinear wave equations. In the literature, some useful approaches for constructing Runge–Kutta–Nyström (RKN)-type integrators have been proposed (see, e.g. [18–23]). Taking full advantage of the special structure introduced by the linear term Mq, Wu et al. [23] formulated a standard form of extended RKN (ERKN) integrators for (8.1). ERKN integrators exhibit the correct qualitative behaviour much better than classical RKN methods due to the use of the special structure brought by the linear term Mq from (8.1). An essential feature is that this class of ERKN integrators is oscillation preserving (see Chap. 1). For further work on this topic, we refer the reader to [22, 24, 25]. If f is the negative gradient of a scalar function V, i.e., $f = -\nabla V$, where the operator ∇ is the standard gradient, then (8.1) is a multi-frequency highly oscillatory Hamiltonian system. Energy-preserving integrators, such as the AVF method, are an important approach for Hamiltonian systems. As is known, the AVF method is related to discrete gradient methods (see [26]). Combining the idea of the discrete gradient method with the ERKN integrator, the authors in [27] presented an extended discrete gradient formula for the highly oscillatory Hamiltonian system (8.1).

In this chapter, we will present an efficient approach for dealing with nonlinear wave equations following the line of [10]. We first approximate the function whose negative variational derivative is the right-hand side term of the underlying wave equation, and semidiscretise the conservative wave equations into a Hamiltonian system of ODEs, or the dissipative wave equations into a dissipative system of ODEs. We then apply the extended discrete gradient method to the resulting system of ODEs. This process yields a conservative scheme for conservative wave equations and a dissipative scheme for dissipative wave equations, and can be applied to a broad class of wave equations in a systematic and routine manner.

8.2 Preliminaries

In this chapter, we consider nonlinear wave equations of the form

$$\frac{\partial^2 u}{\partial t^2} = -\frac{\delta \mathscr{G}}{\delta u},$$
(8.2)

where

$$\mathscr{G}[u] = \int_\Omega G[u] dx, \quad \Omega \subseteq \mathbb{R}^d,$$
(8.3)

and $u : \mathbb{R}^d \times \mathbb{R} \to \mathbb{R}^m$, $dx = dx_1 \cdots dx_d$. The square brackets appearing in (8.3) denote that a function depends on u itself as well as the derivatives of u with respect to the independent variables $x = (x_1, \cdots, x_d)$ up to and including some degree v. The variational derivative $\frac{\delta \mathscr{G}}{\delta u}$ is an m-vector, which can be defined via the following relation

$$\frac{d}{d\varepsilon}\bigg|_{\varepsilon=0} \mathscr{G}[u + \varepsilon v] = \int_\Omega \frac{\delta \mathscr{G}}{\delta u} \cdot v dx,$$
(8.4)

for any sufficiently smooth m-vector of functions $v(x)$.

In what follows, we assume that the solution has sufficient regularity and the boundary conditions on Ω are chosen such that the boundary terms vanish when calculating integration by parts (for example, periodic boundary conditions).

We take $d = 1, m = 1$ as an example. This case gives

$$\mathscr{G}[u] = \int_\Omega G\left(u, \frac{\partial u}{\partial x}, \cdots, \frac{\partial^v u}{\partial x^v}\right) dx,$$

and

$$\frac{\delta \mathscr{G}}{\delta u} = \frac{\partial G}{\partial u} - \frac{\partial}{\partial x}\left(\frac{\partial G}{\partial u_x}\right) + \frac{\partial}{\partial x^2}\left(\frac{\partial G}{\partial u_{xx}}\right) + \cdots + (-1)^v \frac{\partial}{\partial x^v}\left(\frac{\partial G}{\partial u^{(v)}}\right).$$

In general, for any positive integers m, d, the variational derivatives can be calculated by applying the Euler operator to $G[u]$ (see, e.g. [28] for details).

It follows from our assumption that equations of the form (8.2) have in common with the energy conservation property

$$\frac{d}{dt}\mathscr{H}[u] = \frac{d}{dt}\int_\Omega \frac{1}{2}\left(\frac{\partial u}{\partial t}\right)^2 + G[u] dx = 0,$$
(8.5)

and we call them conservative systems. It is important to note that the wave equation (8.2) can be represented as a system of first-order PDEs

$$\begin{pmatrix} u_t \\ w_t \end{pmatrix} = \begin{pmatrix} 0 & I \\ -I & 0 \end{pmatrix} \begin{pmatrix} \dfrac{\delta \tilde{\mathscr{H}}}{\delta u} \\ \dfrac{\delta \tilde{\mathscr{H}}}{\delta w} \end{pmatrix}, \tag{8.6}$$

where

$$\tilde{\mathscr{H}}[u, w] = \int_{\Omega} \frac{1}{2} w^2 + G[u] dx, \tag{8.7}$$

and $w = u_t$ is an intermediate function. As

$$S = \begin{pmatrix} 0 & I \\ -I & 0 \end{pmatrix},$$

and S is skew-symmetric, with (8.6), the conservation property is rewritten as the modified energy conservation property

$$\frac{d}{dt} \tilde{\mathscr{H}}[u, w] = 0. \tag{8.8}$$

Discretising the energy functional $\tilde{\mathscr{H}}$ based on a consistent approximation $\bar{H} \Delta x$, the authors in [10] semidiscretised the conservative PDEs (8.6) into a Hamiltonian system of ODEs with 'skew-gradient' form

$$\dot{y} = S \nabla \bar{H}(y), \quad y = \begin{pmatrix} U \\ W \end{pmatrix}, \tag{8.9}$$

where U and W denote the discrete values of u and $w = u_t$ at the mesh points, respectively. Then applying the discrete gradient method (see the next section for details) to the semidiscretised system leads to the following scheme

$$\frac{y_{n+1} - y_n}{\Delta t} = S \bar{\nabla} \bar{H}(y_n, y_{n+1}),$$

for advancing the numerical solution y_n at time t_n to y_{n+1} at time $t_{n+1} = t_n + \Delta t$, where $\bar{\nabla} \bar{H}(y_n, y_{n+1})$ is the discrete gradient of $\bar{H}$. A distinct feature of this scheme is that it preserves the discretised energy exactly.

In this chapter, we here also consider wave equations with a damping term

$$\frac{\partial^2 u}{\partial t^2} + \alpha \frac{\partial u}{\partial t} = -\frac{\delta \mathscr{G}}{\delta u}, \tag{8.10}$$

where $\alpha > 0$ is a small positive constant. The term αu_t appearing in (8.10) represents a damping force proportional to the velocity u_t. Since these type of equations (8.10) have in common the energy dissipation property

$$\frac{d}{dt}\mathscr{H}[u] = -\alpha \int_\Omega \left(\frac{\partial u}{\partial t}\right)^2 dx < 0, \qquad (8.11)$$

we usually call them dissipative systems. In this case, $\mathscr{H}[u]$ is a Lyapunov function of (8.10). It is clear that wave equations (8.10) can be rewritten as

$$\begin{pmatrix} u_t \\ w_t \end{pmatrix} = D \begin{pmatrix} \dfrac{\delta\tilde{\mathscr{H}}}{\delta u} \\ \dfrac{\delta\tilde{\mathscr{H}}}{\delta w} \end{pmatrix}, \qquad (8.12)$$

where

$$D = \begin{pmatrix} 0 & I \\ -I & -\alpha I \end{pmatrix}.$$

It is easy to see that D is semi-negative definite. In a similar way to the conservative case, semidiscretising the dissipative PDEs (8.12) results in a dissipative system of ODEs as follows

$$\dot{y} = D\nabla\bar{H}(y), \quad y = \begin{pmatrix} U \\ W \end{pmatrix}, \qquad (8.13)$$

and then applying the discrete gradient method to system (8.13), we obtain the scheme

$$\frac{y_{n+1} - y_n}{\Delta t} = D\bar{\nabla}\bar{H}(y_n, y_{n+1}).$$

An advantage of this scheme is that it preserves the decay of the energy (see, e.g. [10]).

Remark 8.2.1 We here remark that in the case of the wave equations, in order to fit the framework in [10], we need to double the dimension of the systems, which should be avoided from the computational point of view. Moreover, the wave equations have their own structures, which cannot be fully taken account of, once they are transformed into the form (8.6) or (8.12). For instance, the nonlinear Klein–Gordon equation can be written in the form (8.2):

$$\frac{\partial^2 u}{\partial t^2} = -\frac{\delta\mathscr{G}}{\delta u}, \quad G[u] = \frac{u_x^2}{2} + \eta(u).$$

When spatial semidiscretisations are used, a linear term naturally arises due to the quadratic term $\dfrac{u_x^2}{2}$ in $G[u]$. The special structure brought by the linear term can be taken advantage of when an efficient numerical scheme is designed. The extended discrete gradient method is an example of such a scheme.

On noticing the fact stated above, in this chapter, instead of transforming the PDEs under consideration into the form (8.6), we deal directly with the original form (8.2). Furthermore, we apply the extended discrete gradient method instead of the traditional discrete gradient method to the semidiscretised system of ODEs.

Next, we discretise the functional $\mathscr{G}$ using a consistent approximation $\bar{G}\Delta x$. Before doing this, we quote the following lemma whose proof can be found in [10].

Lemma 8.1 *Let*

$$\mathscr{H}[u] = \int_{\Omega} H[u]\mathrm{d}x, \quad \Omega \subseteq \mathbb{R}^d, \tag{8.14}$$

and assume that $\bar{H}(U)\Delta x$ is any consistent (finite difference) approximation to $\mathscr{H}[u]$ (where $\Delta x = \Delta x_1 \cdots \Delta x_d$) with N degrees of freedom. Then in the finite-dimensional Hilbert space $\mathbb{R}^N$ with the Euclidean inner product, the variational derivative $\dfrac{\delta}{\delta U}(\bar{H}(U)\Delta x)$ is given by $\nabla \bar{H}(U)$.

Remark 8.2.2 We here remark that when approximating $\mathscr{H}[u]$ by a spectral discretisation, despite that the approximation is not of the form in Lemma 8.1, the lemma still works since such an approximation can be thought of as a finite difference approximation where the finite difference stencil has the same number of entries as the number of grid points on which it is defined.

In what follows, we let U represent the discrete values of u at the mesh grid points, in the multidimensional case after choosing an order. According to Lemma 8.1, the variational derivative $\dfrac{\delta \mathscr{G}}{\delta u}$ is approximated by $\nabla \bar{G}$. Hence, the wave equation (8.2) is semidiscretised into

$$\frac{\mathrm{d}^2 U}{\mathrm{d}t^2} = -\nabla \bar{G}(U), \tag{8.15}$$

and the wave equation (8.10) with damping term is semidiscretised into

$$\frac{\mathrm{d}^2 U}{\mathrm{d}t^2} + \alpha \frac{\mathrm{d}U}{\mathrm{d}t} = -\nabla \bar{G}(U). \tag{8.16}$$

Then apply the extended discrete gradient method to the systems (8.15) and (8.16), respectively. This process leads to a conservative scheme for conservative wave equations, and a dissipative scheme for dissipative wave equations. Remember that one of the benefits of using the extended discrete gradient method is that the scheme is linearly-fitted, which will be shown in the next section.

8.3 Extended Discrete Gradient Method

As is known, discrete gradient methods for ODEs were introduced by Gonzalez [29], for research on discrete gradient methods, and we refer the reader to [26, 30–34].

We take a look back at the definition of a discrete gradient. If $Q : \mathbb{R}^k \to \mathbb{R}$, the discrete gradient of Q is defined as follow.

Definition 8.1 Assume that Q is a differentiable function. Then $\overline{\nabla} Q$ is a discrete gradient of Q provided it is continuous and for $\forall\, u, v \in \mathbb{R}^k$, $u \neq v$, satisfies

$$\begin{cases} \overline{\nabla} Q(u, v) \cdot (u - v) = Q(u) - Q(v), \\ \overline{\nabla} Q(u, u) = \nabla Q(u). \end{cases} \tag{8.17}$$

We next consider continuous time systems of linear-gradient form:

$$\dot{y} = L(y) \nabla Q(y), \tag{8.18}$$

where $L(y)$ is a matrix-valued function, and is skew-symmetric for all y. We remark that any ODE system preserving Q can be written in this form. Then the corresponding discrete gradient method for (8.18) has the following form:

$$\frac{y_{n+1} - y_n}{h} = \overline{L}(y_n, y_{n+1}, h) \overline{\nabla} Q(y_n, y_{n+1}), \tag{8.19}$$

where $\overline{L}(y_n, y_{n+1}, h)$ is a skew-symmetric matrix, which approximates the original $L(y)$. Here, it is required that $\overline{L}(y, y, 0) = L(y)$ and $\overline{\nabla} Q(y, y) = \nabla Q(y)$ for the sake of consistency.

In the literature, there have been many possible choices of discrete gradients for a function Q (see, e.g. [29, 30, 35]). Among potential candidates is the version used in the average vector field (AVF) method due to the fact that the AVF method has some good features such as linear covariance, automatic preservation of linear symmetries, and reversibility with respect to linear reversing symmetries. Therefore, we next consider only the AVF method. The so-called average vector field is defined by

$$\overline{\nabla} Q(y_n, y_{n+1}) := \int_0^1 \nabla Q((1 - \tau) y_n + \tau y_{n+1}) d\tau. \tag{8.20}$$

As is known, a particular form of linear-gradient system (8.18) is the Hamiltonian system

$$\dot{y} = J^{-1}\nabla H(y), \tag{8.21}$$

with the Hamiltonian

$$H(p, q) = \frac{1}{2}p^{\mathsf{T}}p + \frac{1}{2}q^{\mathsf{T}}Mq + V(q), \tag{8.22}$$

where $y = (p^{\mathsf{T}}, q^{\mathsf{T}})^{\mathsf{T}}$, $q = (q_1, q_2, \cdots, q_N)^{\mathsf{T}}$, $p = (p_1, p_2, \cdots, p_N)^{\mathsf{T}}$, $M \in \mathbb{R}^{N \times N}$ is a symmetric and positive semi-definite matrix. J is the $2N \times 2N$ skew-symmetric matrix

$$J = \begin{pmatrix} 0 & I \\ -I & 0 \end{pmatrix}.$$

Applying the AVF method to the system (8.21) leads to the following scheme

$$\begin{cases} q_{n+1} = q_n + hp_n - \dfrac{h^2}{2}\left(\dfrac{1}{2}M(q_n + q_{n+1}) + \displaystyle\int_0^1 \nabla V((1-\tau)q_n + \tau q_{n+1})\mathrm{d}\tau\right), \\[2ex] p_{n+1} = p_n - h\left(\dfrac{1}{2}M(q_n + q_{n+1}) + \displaystyle\int_0^1 \nabla V((1-\tau)q_n + \tau q_{n+1})\mathrm{d}\tau\right). \end{cases} \tag{8.23}$$

Actually, (8.21) is exactly the following highly oscillatory Hamiltonian system of second-order ODEs

$$\begin{cases} \ddot{q}(t) + Mq(t) = f(q(t)), & t \in [t_0, T], \\ q(t_0) = q_0, & \dot{q}(t_0) = p_0, \end{cases} \tag{8.24}$$

where $f : \mathbb{R}^N \to \mathbb{R}^N$ is the negative gradient of $V(q)$ for some smooth function $V(q)$.

It is important to note from Sect. 8.2 that the semidiscretised systems of many conservative wave equations can be formulated in the form (8.24) if all the linear terms with respect to U_i for $i = 1, \cdots, N$ are attributed to MU.

We are now in a position to present the extended discrete gradient method for (8.24). Before doing that, we define the matrix-valued functions which originally appeared in [22]

$$\phi_l(M) := \sum_{k=0}^{\infty} \frac{(-1)^k M^k}{(2k+l)!}, \quad l = 0, 1, \cdots. \tag{8.25}$$

Then the following extended discrete gradient formula is derived for (8.24) (see [27])

$$
\begin{cases}
q_{n+1} = \phi_0(K)q_n + h\phi_1(K)p_n - h^2\phi_2(K)\overline{\nabla}V(q_n, q_{n+1}), \\
p_{n+1} = -hM\phi_1(K)q_n + \phi_0(K)p_n - h\phi_1(K)\overline{\nabla}V(q_n, q_{n+1}),
\end{cases}
\tag{8.26}
$$

where h is the stepsize, $K = h^2M$, $p_n = \dot{q}_n$, and $\overline{\nabla}V(q_n, q_{n+1})$ is the discrete gradient of $V(q)$.

Remark 8.3.1 The matrix-valued functions ϕ_i for $i = 0, 1, 2$ can be approximated by truncation of the Taylor expansion. Since the matrix-valued functions ϕ_i for $i = 0, 1, 2$ only need to be calculated once for every fixed stepsize h, it does not need much extra computational cost at each iteration step. In terms of the relationship among the ϕ_i for $i = 0, 1, 2$, other efficient algorithms for the computation of matrix-valued functions ϕ_i for $i = 0, 1, 2$ can be found in [36] and the references therein.

If we choose $\overline{\nabla}V(q_n, q_{n+1})$ in (8.26) to be the average vector field (8.20), then we obtain the extended AVF method as follows:

$$
\begin{cases}
q_{n+1} = \phi_0(K)q_n + h\phi_1(K)p_n \\
\qquad - h^2\phi_2(K)\displaystyle\int_0^1 \nabla V((1-\tau)q_n + \tau q_{n+1})d\tau, \\
p_{n+1} = -hM\phi_1(K)q_n \\
\qquad + \phi_0(K)p_n - h\phi_1(K)\displaystyle\int_0^1 \nabla V((1-\tau)q_n + \tau q_{n+1})d\tau.
\end{cases}
\tag{8.27}
$$

It is obvious that the extended discrete gradient method is linearly-fitted in the sense that in the particular case where $\nabla V(q) \equiv \nabla V_0$ is constant, the extended discrete gradient method gives the exact solution of the system (8.21) or (8.24).

Due to the fact that all the schemes under consideration are implicit, the iterative solution is required, in general. A simple and common choice would be fixed-point iteration. Fortunately, it has been shown in [27] that the convergence of fixed-point iteration for the extended discrete gradient method is independent of $\|M\|$. Unfortunately, however, the traditional discrete gradient method depends on $\|M\|$. This observation implies that, in general, a larger stepsize can be chosen for the extended discrete gradient scheme than that for the traditional discrete gradient method. The convergence rate of fixed-point iteration for the extended discrete gradient method is faster than that for the traditional discrete gradient method.

Remark 8.3.2 It is noted that the extended gradient methods can conserve the energy exactly when applied to the Hamiltonian system (8.15) (after reformulated into the form (8.24)). However, it cannot be applied directly to a system of the form (8.16) because of the existence of the damping term.

We now turn to the following damped system

$$\begin{cases} \ddot{q}(t) + \alpha \dot{q}(t) + Mq(t) = f(q(t)), & t \in [t_0, T], \\ q(t_0) = q_0, \quad \dot{q}(t_0) = p_0. \end{cases} \tag{8.28}$$

We next try to revise the extended gradient method so that the revision version can be applied to (8.28). To this end, we move the term $\alpha \dot{q}$ to the right-hand side of the system and consider formally $\tilde{f}(q) = f(q) - \alpha \dot{q}$ as the negative gradient of the potential $\tilde{V}(q) = V(q) + \dfrac{\alpha}{2} \dot{q}^2$. Accordingly, the system (8.28) is rewritten in the form

$$\begin{cases} \ddot{q}(t) + Mq(t) = \tilde{f}(q(t)), & t \in [t_0, T], \\ q(t_0) = q_0, \quad \dot{q}(t_0) = p_0. \end{cases} \tag{8.29}$$

As $p_n = \dot{q}_n$, applying the extended discrete gradient method to (8.29), we obtain

$$\begin{cases} q_{n+1} = \phi_0(K)q_n + h\phi_1(K)p_n - h^2\phi_2(K)\left(\overline{\nabla}V(q_n, q_{n+1}) + \alpha\dfrac{p_n + p_{n+1}}{2}\right), \\ p_{n+1} = -hM\phi_1(K)q_n + \phi_0(K)p_n - h\phi_1(K)\left(\overline{\nabla}V(q_n, q_{n+1}) + \alpha\dfrac{p_n + p_{n+1}}{2}\right). \end{cases} \tag{8.30}$$

The extended AVF method for (8.28) then is identical to

$$\begin{cases} q_{n+1} = \phi_0(K)q_n + h\phi_1(K)p_n \\ \qquad\quad - h^2\phi_2(K)\left(\displaystyle\int_0^1 \nabla V((1-\tau)q_n + \tau q_{n+1})\mathrm{d}\tau + \alpha\dfrac{p_n + p_{n+1}}{2}\right), \\ p_{n+1} = -hM\phi_1(K)q_n + \phi_0(K)p_n \\ \qquad\quad - h\phi_1(K)\left(\displaystyle\int_0^1 \nabla V((1-\tau)q_n + \tau q_{n+1})\mathrm{d}\tau + \alpha\dfrac{p_n + p_{n+1}}{2}\right). \end{cases} \tag{8.31}$$

Now all that remains is to prove that (8.30) preserves the dissipation or the decay of Lyapunov function $H(p, q)$. Before doing this, we present the following properties of matrix-valued functions which play an important role in the proof:

$$\phi_0^2(K) + K\phi_1^2(K) = I, \quad K\big(\phi_1^2(K) - \phi_0(K)\phi_2(K)\big) = I - \phi_0(K),$$
$$\phi_1^2(K) + K\phi_2^2(K) = 2\phi_2(K), \quad \phi_0(K) + K\phi_2(K) = I. \tag{8.32}$$

The properties of matrix-valued functions can be verified straightforwardly.

Theorem 8.1 *Let h be sufficiently small. Then the scheme* (8.30) *with the stepsize h preserves the dissipation or the decay of Lyapunov function $H(p, q)$ when applied to the damped system* (8.28), *i.e.,*

$$H(p_{n+1}, q_{n+1}) \leqslant H(p_n, q_n), \quad n = 0, 1, \cdots .$$

Proof Let

$$\overline{\nabla}\tilde{V}(q_n, q_{n+1}) = \overline{\nabla}V(q_n, q_{n+1}) + \alpha \frac{p_n + p_{n+1}}{2}.$$

We compute

$$H(p_{n+1}, q_{n+1}) = \frac{1}{2}p_{n+1}^{\mathsf{T}}p_{n+1} + \frac{1}{2}q_{n+1}^{\mathsf{T}}Mq_{n+1} + V(q_{n+1}). \tag{8.33}$$

According to the symmetry of K and commutativity of K and all the $\phi_l(K)$ and inserting (8.30) into (8.33), a tedious computation gives

$$
\begin{aligned}
&H(p_{n+1}, q_{n+1}) \\
&= \frac{1}{2}p_n^{\mathsf{T}}\big(\phi_0^2(K) + K\phi_1^2(K)\big)p_n + \frac{1}{2}q_n^{\mathsf{T}}M\big(\phi_0^2(K) + K\phi_1^2(K)\big)q_n \\
&\quad + q_n^{\mathsf{T}}K\big(\phi_1^2(K) - \phi_0(K)\phi_2(K)\big)\overline{\nabla}\tilde{V}(q_n, q_{n+1}) \\
&\quad - hp_n^{\mathsf{T}}\big(\phi_0(K)\phi_1(K) + K\phi_1(K)\phi_2(K)\big)\overline{\nabla}\tilde{V}(q_n, q_{n+1}) \\
&\quad + \frac{1}{2}h^2\overline{\nabla}\tilde{V}(q_n, q_{n+1})^{\mathsf{T}}\big(\phi_1(K)^2 + K\phi_2(K)^2\big)\overline{\nabla}\tilde{V}(q_n, q_{n+1}) + V(q_{n+1}).
\end{aligned}
\tag{8.34}
$$

Then substituting (8.32) into (8.34) yields

$$
\begin{aligned}
H(p_{n+1}, q_{n+1}) &= \frac{1}{2}p_n^{\mathsf{T}}p_n + \frac{1}{2}q_n^{\mathsf{T}}Mq_n + q_n^{\mathsf{T}}\big(I - \phi_0(K)\big)\overline{\nabla}\tilde{V}(q_n, q_{n+1}) \\
&\quad - hp_n^{\mathsf{T}}\phi_1(K)\overline{\nabla}\tilde{V}(q_n, q_{n+1}) \\
&\quad + h^2\overline{\nabla}\tilde{V}(q_n, q_{n+1})^{\mathsf{T}}\phi_2(K)\overline{\nabla}\tilde{V}(q_n, q_{n+1}) + V(q_{n+1}) \\
&= \frac{1}{2}p_n^{\mathsf{T}}p_n + \frac{1}{2}q_n^{\mathsf{T}}Mq_n + \Big(q_n - \big(\phi_0(K)q_n + h\phi_1(K)p_n \\
&\quad - h^2\phi_2(K)\overline{\nabla}\tilde{V}(q_n, q_{n+1})\big)\Big)^{\mathsf{T}}\overline{\nabla}\tilde{V}(q_n, q_{n+1}) + U(q_{n+1}) \\
&= \frac{1}{2}p_n^{\mathsf{T}}p_n + \frac{1}{2}q_n^{\mathsf{T}}Mq_n + (q_n - q_{n+1})^{\mathsf{T}}\overline{\nabla}\tilde{V}(q_n, q_{n+1}) + V(q_{n+1}) \\
&= \frac{1}{2}p_n^{\mathsf{T}}p_n + \frac{1}{2}q_n^{\mathsf{T}}Mq_n + \alpha(q_n - q_{n+1})^{\mathsf{T}}\frac{p_n + p_{n+1}}{2} + V(q_n). \quad (8.35)
\end{aligned}
$$

The last equality follows from the definition of discrete gradient. Therefore, we obtain

$$H(p_{n+1}, q_{n+1}) - H(p_n, q_n) = \alpha(q_n - q_{n+1})^{\mathsf{T}} \frac{p_n + p_{n+1}}{2}. \tag{8.36}$$

Solving the second equation of (8.30) for $\overline{\nabla}\tilde{V}(q_n, q_{n+1})$ and substituting it into the first equation of (8.30), we have

$$q_{n+1} = \phi_0(K)q_n + h\phi_1(K)p_n + \phi_1^{-1}(K)\phi_2(K)(hp_{n+1} + K\phi_1(K)q_n - \phi_0(K)hp_n). \tag{8.37}$$

Substituting (8.37) into (8.36) gives

$$H(p_{n+1}, q_{n+1}) - H(p_n, q_n)$$
$$= -\alpha\left(\Phi(K) + \phi_1^{-1}(K)\phi_2(K)(hp_{n+1} + K\phi_1(K)q_n - \phi_0(K)hp_n)\right)^{\mathsf{T}} \frac{p_n + p_{n+1}}{2}$$
$$= -2\alpha h\left(\frac{p_n + p_{n+1}}{2}\right)^{\mathsf{T}} \phi_1^{-1}(K)\phi_2(K)\frac{p_n + p_{n+1}}{2}, \tag{8.38}$$

where

$$\Phi(K) = (\phi_0(K) - I)q_n + h\phi_1(K)p_n.$$

According to the definition of matrix-valued functions and the hypothesis on h, it can be verified that $\phi_1^{-1}(K)\phi_2(K)$ is positive semi-definite. Hence, we obtain

$$H(p_{n+1}, q_{n+1}) - H(p_n, q_n) \leqslant 0.$$

Thus, the statement is proved.   □

Remark 8.3.3 When $M \to 0$, the extended discrete gradient method (8.30) reduces to the traditional discrete gradient method and (8.38) is identical to

$$\frac{H(p_{n+1}, q_{n+1}) - H(p_n, q_n)}{h} = -\alpha\left(\frac{p_n + p_{n+1}}{2}\right)^{\mathsf{T}} \frac{p_n + p_{n+1}}{2}.$$

Then the scheme preserves the dissipation property regardless of the magnitude of the stepsize h. This coincides with the fact that discrete gradient methods are unconditionally energy-diminishing methods for dissipative gradient system (see [37]).

We now summarise the extended discrete gradient scheme presented in this chapter (denoted by EAVF) as follows:

- First discretise the functional $\mathcal{G}[u] = \int_\Omega G[u]\mathrm{d}x$ using some consistent approximation $\bar{G}(U)\Delta x$ to yield a system of ODEs ((8.15) or (8.16)) as described in this chapter.
- Then the quadratic terms with respect to U in $\bar{G}(U)$ lead to the linear terms $\nabla \bar{G}(U)$ of the semidiscretised system of ODEs. By attributing all these linear terms to MU, we can rewrite the system of ODEs in a form to which the extended AVF method is applicable.
- Finally, apply the extended AVF method to the resulting system of ODEs.

Remark 8.3.4 It follows from (8.36) that

$$\frac{H(p_{n+1}, q_{n+1}) - H(p_n, q_n)}{h} = -\alpha\left(\frac{q_{n+1} - q_n}{h}\right)^\mathsf{T} \frac{p_n + p_{n+1}}{2}. \tag{8.39}$$

We can consider $\dfrac{q_{n+1} - q_n}{h}$ and $\dfrac{p_n + p_{n+1}}{2}$ as two different approximations of $p_n = \dot{q}_n$, and in this sense (8.39) is the discrete analogy of the dissipation property

$$\frac{\mathrm{d}}{\mathrm{d}t}\mathcal{H}[u] = -\alpha\int_\Omega \left(\frac{\partial u}{\partial t}\right)^2 \mathrm{d}x.$$

Remark 8.3.5 In the case where the damping coefficient α in (8.10) is space-dependent, i.e., $\alpha = \alpha(x)$ with the property $\alpha(x) > c$, where $c > 0$ is a positive constant, it can be verified that (8.10) is still dissipative. In order to design a corresponding linearly-fitted dissipative scheme, we only need to replace α in (8.30) by the diagonal matrix with diagonal entries being the discrete values of α at the mesh grid points. Then, in a similar way, the preservation of dissipation can be proved.

8.4 Numerical Experiments

In this section, we apply the scheme EAVF described in this chapter to conservative and dissipative wave equations to illustrate its efficiency in comparison with the scheme presented in [10]. The scheme given in [10] (denoted by AVF) for comparison is stated as follow:

- First discretise the energy functional $\mathcal{H}[u] = \int_\Omega \frac{1}{2}\left(\frac{\partial u}{\partial t}\right)^2 + G[u]\mathrm{d}x$ using some consistent approximation $\bar{H}(U)\Delta x$ to give a system of ODEs. We here choose $\bar{H}(U) = \sum_j \frac{1}{2}\dot{U}_j^2 + \bar{G}(U)$ when comparing the two schemes. In such a way, the resulting system of ODEs is the same as that obtained by the scheme EAVF.

• Then apply the traditional AVF method to the resulting system of ODEs.

In this section, five numerical experiments are described. All the computations and graphics are implemented in MATLAB 7 in IEEE double precision arithmetic.

8.4.1 Implementation Issues

1. Evaluation of the AVF method

We first consider the evaluation of the average vector field

$$\int_0^1 \nabla V((1-\tau)q_n + \tau q_{n+1})d\tau.$$

For the system of ODEs obtained by semidiscretising the underlying wave equation, the potential $V(q)$ is typically of the form $V(q) = \sum_{i=1}^N a_i W(q^i)$, where W is a scalar function, q^i is the i-th entry of q and $a = (a_1, \cdots, a_N)^\mathsf{T} \in \mathbb{R}^N$ is a constant vector (usually $a = (1, \cdots, 1)^\mathsf{T}$). As the following lemma claims, the average vector field can be evaluated exactly for this kind of special potential.

Lemma 8.2 Let $V(q) = \sum_{i=1}^N a_i W(q^i)$. We then have

$$\int_0^1 \nabla V((1-\tau)q_n + \tau q_{n+1})d\tau = \begin{pmatrix} a_1 \dfrac{W(q_{n+1}^1) - W(q_n^1)}{q_{n+1}^1 - q_n^1} \\ \vdots \\ a_i \dfrac{W(q_{n+1}^i) - W(q_n^i)}{q_{n+1}^i - q_n^i} \\ \vdots \\ a_N \dfrac{W(q_{n+1}^N) - W(q_n^N)}{q_{n+1}^N - q_n^N} \end{pmatrix}.$$

Proof The proof is straightforward and we omit it. □

2. Choice of starting approximations for fixed-point iteration

All the schemes considered in this chapter are implicit, so we need to solve a system of nonlinear algebraic equations iteratively. In this case, a good starting approximation will improve the efficiency of the iteration process. We here refer the reader to [1] for details on the choice of good starting approximations for implicit schemes.

With regard to conservative system, the unknowns q_{n+1} and p_{n+1} appearing in the two schemes AVF and EAVF are decoupled. This implies that we only need to solve q_{n+1} implicitly and p_{n+1} can be calculated explicitly. Unfortunately, however, for dissipative system, the unknowns q_{n+1} and p_{n+1} are no longer decoupled. Consequently, we have to solve them simultaneously by an implicit iteration process.

It can be observed from the AVF formula that simple starting approximations for q_{n+1} and p_{n+1} are $q_{n+1} = q_n$, $p_{n+1} = p_n$ or $q_{n+1} = q_n + hp_n$, $p_{n+1} = p_n$. However, they are not accurate enough. Moreover, it follows from Lemma 8.2 that the denominators in the evaluation of the integral are of the form $q_{n+1}^i - q_n^i$. Hence, it is not wise to choose the starting approximation $q_{n+1} = q_n$ or $q_{n+1} = q_n + hp_n$ if $p_n = 0$, since the implicit iteration will diverge immediately in this case. Consequently, it is very subtle and takes much effort to give a suitable starting approximation for the AVF method.

Fortunately, however, the scheme EAVF enlightens us to choose suitable starting approximations for q_{n+1} and p_{n+1}. In fact, we can choose

$$q_{n+1} = \phi_0(K)q_n + h\phi_1(K)p_n, \quad p_{n+1} = -hM\phi_1(K)q_n + \phi_0(K)p_n$$

as the starting approximations $q_{n+1}^{(0)}$ and $p_{n+1}^{(0)}$ which are accurate enough because the EAVF scheme integrates unperturbed systems exactly. Moreover, this process does not entail extra computational cost since they are exactly parts of the EAVF scheme.

In the numerical experiments, we choose the same starting approximations for both schemes, i.e.,

$$q_{n+1}^{(0)} = \phi_0(K)q_n + h\phi_1(K)p_n, \quad p_{n+1}^{(0)} = -hM\phi_1(K)q_n + \phi_0(K)p_n,$$

to compare their efficiency and convergence rate.

8.4.2   Conservative Wave Equations

Problem 8.1 We consider the sine-Gordon equation

$$\frac{\partial^2 u}{\partial t^2} = \frac{\partial^2 u}{\partial x^2} - \sin u, \quad t > 0.$$

Here, we only consider the so-called breather-solution [38]

$$u(x, t) = 4 \arctan\left(\frac{\sqrt{1 - \omega^2}}{\omega} \frac{\cos \omega t}{\cosh(x\sqrt{1 - \omega^2})}\right). \tag{8.40}$$

The initial conditions are

$$u(x, 0) = 4 \arctan \left(\frac{\sqrt{1 - \omega^2}}{\omega} \frac{1}{\cosh(x\sqrt{1 - \omega^2})} \right),$$

and

$$u_t(x, 0) = \frac{\partial}{\partial t} \left\{ 4 \arctan \left(\frac{\sqrt{1 - \omega^2}}{\omega} \frac{\cos \omega t}{\cosh(x\sqrt{1 - \omega^2})} \right) \right\}_{t=0}$$

where $\omega = 0.9$. This is a bump shaped solution which oscillates up and down on an infinite domain, with period $2\pi/\omega$. To exclude boundary effects, we use periodic boundary conditions with $L = 20$, i.e., we consider the sine-Gordon equation on the interval $[-20, 20]$.

It is clear that the sine-Gordon equation is of the type (8.2), where

$$\mathscr{G}[u] = \int_{-L}^{L} \frac{1}{2} \left(\frac{\partial u}{\partial x} \right)^2 + (1 - \cos u) \mathrm{d}x. \tag{8.41}$$

We denote $x_j = -L + j\Delta x$ for $j = 0, 1, \cdots, N$, where $\Delta x = \dfrac{2L}{N}$. Let

$$\bar{G}(U)\Delta x = \left(\sum_{j=0}^{N-1} \frac{1}{2(\Delta x)^2} (u_{j+1} - u_j)^2 + (1 - \cos(u_j)) \right) \Delta x$$

be the approximation of $\mathscr{G}[u]$. Then the resulting system of ODEs is given by

$$\frac{\mathrm{d}^2 U}{\mathrm{d}t^2} + MU = f(U), \tag{8.42}$$

where

$$U = (u_1, \cdots, u_N)^{\mathsf{T}},$$

$$M = \frac{1}{\Delta x^2} \begin{pmatrix} 2 & -1 & & & -1 \\ -1 & 2 & -1 & & \\ & \ddots & \ddots & \ddots & \\ & & -1 & 2 & -1 \\ -1 & & & -1 & 2 \end{pmatrix}, \tag{8.43}$$

$$f(U) = -\nabla V(U) = -\sin(U) = -\left(\sin u_1, \cdots, \sin u_N \right)^{\mathsf{T}},$$

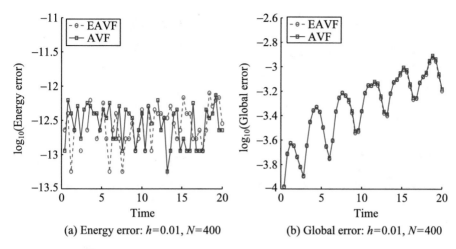

Fig. 8.1 The sine-Gordon equation with finite differences semidiscretisation: (**a**) The logarithm of energy error, and (**b**) the logarithm of global error vs time for AVF and EAVF methods

and

$$V(U) = 1 - \cos(u_1) + \cdots + 1 - \cos(u_N).$$

Note that we here use u_j to refer the value of u at x_j with a fixed time level.

First, system (8.42) is integrated on the interval $[0, 20]$ with $N = 400$ and $h = 0.01$. With regard to the fixed-point iteration at each time step, we set the error tolerance as 10^{-15}. In Fig. 8.1, we plot the logarithm of the energy errors and the logarithm of the global errors against time t, respectively. It can be observed from Fig. 8.1 that the errors of energy and the global errors are comparable for the two schemes. We show the numerical solution obtained by the EAVF scheme in Fig. 8.2a.

Furthermore, to illustrate the computational efficiency of the two schemes, we set the maximum iteration number as 1000 and the error tolerance as 10^{-15}. We apply the two schemes to the system on the interval $[0, 2]$ with the stepsize $h = 0.01$. We plot the total iteration number against the dimension of spatial discretisation N. The numerical results are shown in Fig. 8.2b.

Problem 8.2 We now consider the same sine-Gordon equation as that in Problem 8.1, but instead of using finite differences for the spatial discretisation in (8.41), we here use a spectral discretisation with a Fourier basis.

This means that the partial derivative $\dfrac{\partial u}{\partial x}$ can now be approximated by discrete Fourier transform (DFT), which is denoted by $\mathscr{F}_N^{-1} D_N \mathscr{F}_N U$, where the DFT matrix $\mathscr{F}_N$ has entries $[\mathscr{F}_N]_{j,k} = \mathrm{e}^{-jki2\pi/N}$, $[\mathscr{F}_N^{-1}]_{j,k} = \dfrac{1}{N}\mathrm{e}^{jki2\pi/N}$, and D_N

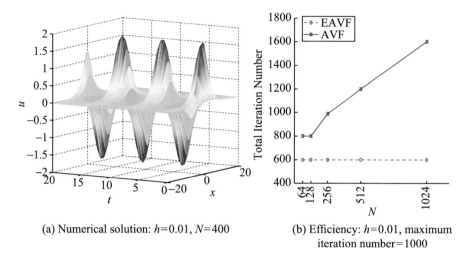

(a) Numerical solution: $h=0.01$, $N=400$

(b) Efficiency: $h=0.01$, maximum iteration number$=1000$

Fig. 8.2 The sine-Gordon equation with finite differences semidiscretisation: (**a**) Numerical solution obtained by EAVF method, and (**b**) efficiency: the total iteration number against the dimension of spatial discretisation

is a diagonal matrix whose diagonal entries are given by (see, e.g. [10])

$$
\mathrm{diag}(D_N) = \begin{cases} \dfrac{\pi \mathrm{i}}{L}\left[0, 1, 2, \cdots, \dfrac{N-1}{2}, -\dfrac{N-1}{2}, \cdots, -2, -1\right], \text{ for } N \text{ is odd} \\[2mm] \dfrac{\pi \mathrm{i}}{L}\left[0, 1, 2, \cdots, \dfrac{N}{2} - 1, 0, -\dfrac{N}{2} + 1, \cdots, -2, -1\right], \text{ for } N \text{ is even.} \end{cases}
$$

Following the notation in Problem 8.1, (8.41) can be approximated by

$$
\bar{G}(U)\Delta x = \left(\sum_{j=0}^{N-1} \frac{1}{2}[\mathscr{F}_N^{-1} D_N \mathscr{F}_N U]_j^2 + (1 - \cos(u_j)) \right) \Delta x.
$$

Accordingly, the resulting system of ODEs is of the form

$$
\frac{\mathrm{d}^2 U}{\mathrm{d}t^2} + MU = f(U), \tag{8.44}
$$

where U and $f(U)$ are the same as those in Problem 8.1. The matrix M now becomes

$$
M = \left(\mathscr{F}_N^{-1} D_N \mathscr{F}_N \right)^{\mathsf{T}} \left(\mathscr{F}_N^{-1} D_N \mathscr{F}_N \right). \tag{8.45}
$$

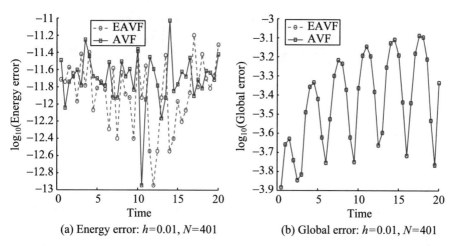

Fig. 8.3 The sine-Gordon equation with spectral semidiscretisation: (**a**) The logarithm of energy error, and (**b**) the logarithm of global error vs time for AVF and EAVF methods

We integrate system (8.44) on the interval $[0, 20]$ with $N = 401$ and $h = 0.01$. We first set the error tolerance as 10^{-15} for the fixed-point iteration. We show the logarithm of the energy errors and the logarithm of the global errors against time t in Fig. 8.3, respectively. We plot the numerical solution obtained by the EAVF method in Fig. 8.4a.

We then set the maximum iteration number as 1000 and the error tolerance as 10^{-15}. We apply the two methods to system (8.44) on the interval $[0, 2]$ with the stepsize $h = 0.01$. Figure 8.4b indicates the numerical results for the total iteration number against the dimension of spatial discretisation, N.

It can be observed from the numerical results that the errors of energy and the global errors are comparable for the two methods. However, the total number of iterations for the AVF method grows very fast with the increase of N. Fortunately, the total number of iterations for the EAVF method remains almost the same as N grows, which is much less than that of the AVF method. Here, it is important to note that when N is large, the iteration of the AVF method does not converge for some time steps within the maximum iteration number.

Problem 8.3 Consider the nonlinear 2D wave equation

$$\frac{\partial^2 u}{\partial t^2} = \Delta u - \frac{\partial V(u)}{\partial u}, \quad (x, y) \in [-1, 1] \times [-1, 1], \quad t > 0,$$

where $V(u) = \dfrac{u^4}{4}$. This equation is subject to periodic boundary conditions, and the initial conditions are given by

$$u(x, y, 0) = \text{sech}(10x)\text{sech}(10y).$$

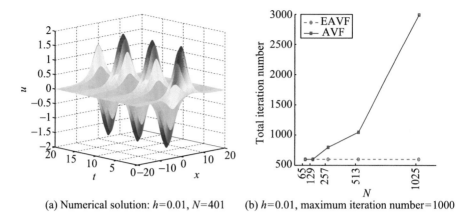

(a) Numerical solution: h=0.01, N=401   (b) h=0.01, maximum iteration number=1000

Fig. 8.4 The sine-Gordon equation with spectral semidiscretisation: (**a**) Numerical solution by EAVF method, and (**b**) efficiency: the total iteration number against the dimension of spatial discretisation

Obviously, this equation is of the type (8.2) with

$$\mathscr{G}[u] = \int_{-1}^{1} \int_{-1}^{1} \frac{1}{2} \left(\left(\frac{\partial u}{\partial x} \right)^2 + \left(\frac{\partial u}{\partial y} \right)^2 \right) + V(u) \mathrm{d}x. \tag{8.46}$$

Following [10], we use the spectral elements method to semidiscretise the wave equation. For the sake of self-containment in this chapter, we restate the setup given in [10]. $\mathscr{G}[u]$ is discretised in space with a tensor product Lagrange quadrature formula based on $p + 1$ Gauss–Lobatto–Legendre (GLL) quadrature nodes in each space direction. We then obtain

$$\bar{G}(U) = \frac{1}{2} \sum_{j_1=0}^{p} \sum_{j_2=0}^{p} w_{j_1} w_{j_2}$$

$$\cdot \left(\left(\sum_{k=0}^{p} d_{j_1,k} u_{k,j_2} \right)^2 + \left(\sum_{m=0}^{p} d_{j_2,m} u_{j_1,m} \right)^2 + \frac{1}{2} u_{j_1,j_2}^4 \right),$$

where $d_{j_1,k} = \dfrac{\mathrm{d} l_k(x)}{\mathrm{d}x}$, $l_k(x)$ is the k-th Lagrange basis function based on the GLL quadrature nodes $x_0, \cdots, x_p$, and $w_0, \cdots, w_p$ are the corresponding quadrature weights. The numerical approximation of u is

$$u_p(x, y, t) = \sum_{k=0}^{p} \sum_{m=0}^{p} u_{k,m}(t) l_k(x) l_m(y)$$

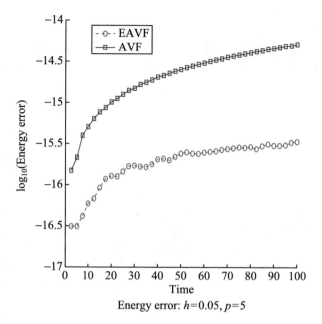

Fig. 8.5 2D wave equation with spectral elements semidiscretisation: The logarithm of energy error vs time for AVF and EAVF methods

with the property $u_p(x_{j_1}, y_{j_2}, t) = u_{j_1, j_2}(t)$. Here we remark that the quadratic terms with respect to u_{j_1, j_2} for $j_1, j_2 = 0, \cdots, p$ in $\bar{G}(U)$ lead to the linear terms in the semidiscretised system of ODEs, which will be attributed to MU. Moreover, it can be verified that M is a symmetric positive semi-definite matrix.

The system of ODEs is integrated on the interval $[0, 100]$ with $p = 5$ and $h = 0.05$. Likewise, the error tolerance of the fixed-point iteration is set as 10^{-15} in this numerical experiment. The energy errors are indicated in Fig. 8.5. Some snapshots of the numerical solution obtained by EAVF method are shown in Fig. 8.6.

8.4.3 Dissipative Wave Equations

We next consider dissipative nonlinear wave equations, including 1D and 2D, respectively.

Problem 8.4 We first consider the dissipative sine-Gordon equation

$$\frac{\partial^2 u}{\partial t^2} + \alpha \frac{\partial u}{\partial t} = \frac{\partial^2 u}{\partial x^2} - \sin u, \quad t > 0,$$

where $\alpha = 0.1$.

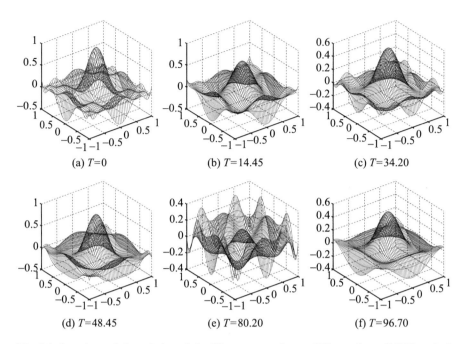

Fig. 8.6 Snapshots of the solution of the 2D wave equation at different times. EAVF method with the time stepsize $h = 0.05$. Space discretisation with six Gauss Lobatto nodes in each spatial direction. Numerical solution interpolated on an equidistant grid of 26 nodes in each spatial direction

With the same setting as that in Problems 8.1 and 8.2, we show the energy curves and the efficiency curves for finite differences and spectral semidiscretisation, respectively. The results are presented in Figs. 8.7 and 8.8. We remark that once the iteration number is too large, we do not plot the points in these figures. It can be seen from the numerical results that the two schemes preserve the decay of energy, and again EAVF method is much more efficient than AVF method regarding the computational efficiency.

Problem 8.5 Finally, we consider the nonlinear 2D dissipative wave equation

$$\frac{\partial^2 u}{\partial t^2} + \alpha \frac{\partial u}{\partial t} = \Delta u - \frac{\partial V(u)}{\partial u}, \quad (x, y) \in [-1, 1] \times [-1, 1], \quad t > 0$$

where $\alpha = 0.1$.

We integrate this equation on the interval $[0, 100]$ with $p = 5$ and $h = 0.05$. We present the energy and some snapshots of the numerical solution obtained by EAVF method in Figs. 8.9 and 8.10, respectively.

It can be observed from Figs. 8.10 and 8.6 that the shapes of the solutions of the conservative and dissipative 2D wave equations are similar at the same time step.

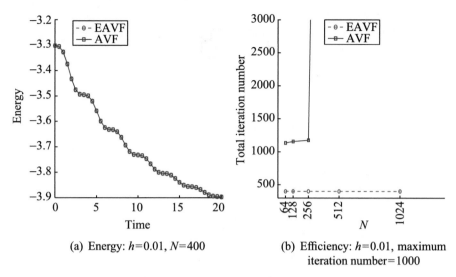

(a) Energy: $h=0.01$, $N=400$

(b) Efficiency: $h=0.01$, maximum iteration number$=1000$

Fig. 8.7 Dissipative sine-Gordon equation with finite differences semidiscretisation: (**a**) Energy vs time, and (**b**) the total iteration number against the dimension of spatial discretisation for AVF and EAVF methods

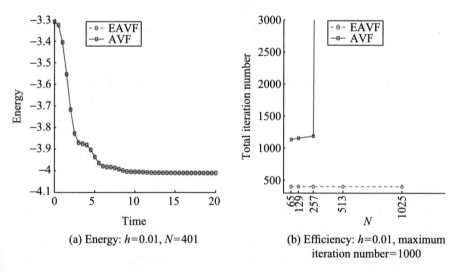

(a) Energy: $h=0.01$, $N=401$

(b) Efficiency: $h=0.01$, maximum iteration number$=1000$

Fig. 8.8 Dissipative sine-Gordon equation with spectral semidiscretisation: (**a**) Energy vs time, and (**b**) the total iteration number against the dimension of spatial discretisation for AVF and EAVF methods

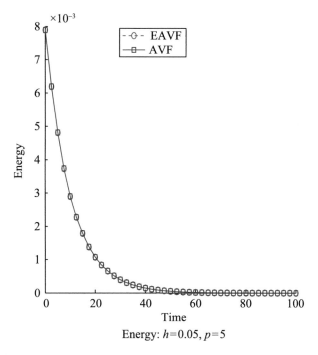

Energy: $h=0.05, p=5$

Fig. 8.9 2D dissipative wave equation with spectral elements semidiscretisation: Energy vs time for AVF and EAVF methods

Because of the damping term, the magnitude of the solution diminishes as time increases. This is consistent with the fact that the energy decays in the dissipative case.

8.5   Conclusions

In this chapter, we presented and analysed an interesting approach to designing conservative (dissipative) schemes for nonlinear conservative (dissipative) wave equations. Following the framework in [10], but with some modifications, we dealt directly with the original form of the underling wave equation rather than transforming it into a system of first-order PDEs, and discretised the functional $\mathscr{G}[u]$ instead of the energy functional $\mathscr{H}[u]$. Using this approach leads to a system of second-order ODEs in time. Under this framework, we apply the extended AVF method instead of the traditional AVF method to the system of second-order ODEs. This procedure presents an interesting linearly-fitted conservative (dissipative) scheme for nonlinear conservative (dissipative) wave equations. An outstanding benefit of this procedure is that, the implicit iteration involved in this scheme, which incorporates with the extended AVF method, converges much faster

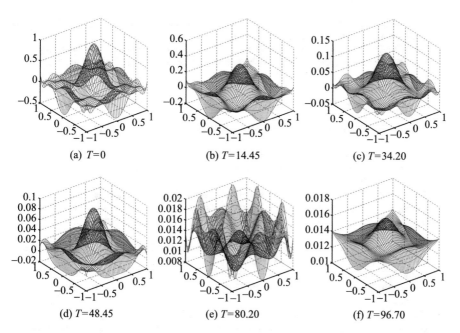

Fig. 8.10 Snapshots of the solution of the 2D dissipative wave equation at different times. EAVF method with time stepsize $h = 0.05$. Space discretisation with six Gauss Lobatto nodes in each space direction. Numerical solution interpolated on an equidistant grid of 26 nodes in each space direction

than those of the traditional AVF method. This implies that the procedure presented in this chapter is much more efficient. Moreover, the numerical results also illustrate this point.

The material in this chapter is based on the work by Liu et al. [39].

References

1. Hairer, E., Lubich, C., Wanner, G.: Geometric Numerical Integration: Structure-Preserving Algorithms, 2nd edn. Springer-Verlag, Berlin (2006)
2. McLachlan, R.I., Quispel, G.R.W.: Splitting methods. Acta Numer. **11**, 341–434 (2002)
3. Sanz-Serna, J.M.: Symplectic integrators for Hamiltonian problems: An overview. Acta Numer. **1**, 243–286 (1992)
4. McLachlan, R.I., Quispel, G.R.W.: Geometric integrators for ODEs. J. Phys. A **39**, 5251–5285 (2006)
5. Chang, Q., Jia, E., Sun, W.: Difference schemes for solving the generalized nonlinear Schrödinger equation. J. Comput. Phys. **148**, 397–415 (1999)
6. Chen, J.B., Qin, M.Z.: Multi-symplectic Fourier pseudospectral method for the nonlinear Schrödinger equation. Electron. Trans. Numer. Anal. **12**, 193–204 (2001)
7. Cai, J.X., Wang, Y.S., Liang, H.: Local energy-preserving and momentum-preserving algorithms for coupled nonlinear Schrödinger system. J. Comput. Phys. **239**, 30–50 (2013)

8. Cai, J.X., Wang, Y.S.: Local structure-preserving algorithms for the "good" Boussinesq equation. J. Comput. Phys. **239**, 72–89 (2013)
9. Yan, J.L., Zhang, X.Y.: New energy-preserving schemes using Hamiltonian Boundary Value and Fourier pseudospectral methods for the numerical solution of the "good" Boussinesq equation. Comput. Phys. Commun. **201**, 33–42 (2016)
10. Celledoni, E., Grimm, V., McLachlan, R.I., et al.: Preserving energy resp. dissipation in numerical PDEs using the 'Average Vector Field' method. J. Comput. Phys. **231**, 6770–6789 (2012)
11. Bridges, T.J., Reich, S.: Numerical methods for Hamiltonian PDEs. J. Phys. A **39**, 5287–5320 (2006)
12. Chabassier, J., Joly, P.: Energy preserving schemes for nonlinear Hamiltonian systems of wave equations: Application to the vibrating piano string. Comput. Methods Appl. Mech. Eng. **199**, 2779–2795 (2010)
13. Chen, Y.M., Song, S.H., Zhu, H.J.: The multi-symplectic Fourier pseudospectral method for solving two-dimensional Hamiltonian PDEs. J. Comput. Appl. Math. **236**, 1354–1369 (2011)
14. Shi, W., Wu, X., Xia, J.: Explicit multi-symplectic extended leap-frog methods for Hamiltonian wave equations. J. Comput. Phys. **231**, 7671–7694 (2012)
15. Wang, Y.S., Hong, J.L.: Multi-symplectic algorithms for Hamiltonian partial differential equations. Commun. Appl. Math. Comput. **27**, 163–230 (2013)
16. Gong, Y.Z., Cai, J.X., Wang, Y.S.: Some new structure-preserving algorithms for general multi-symplectic formulations of Hamiltonian PDEs. J. Comput. Phys. **279**, 80–102 (2014)
17. Li, Y.W., Wu, X.: General local energy-preserving integrators for solving multi-symplectic Hamiltonian PDEs. J. Comput. Phys. **301**, 141–166 (2015)
18. González, A.B., Martín, P., Farto, J.M.: A new family of Runge–Kutta type methods for the numerical integration of perturbed oscillators. Numer. Math. **82**, 635–646 (1999)
19. García-Archilla, B.J., Sanz-Serna, M., Skeel, R.D.: Long-time-step methods for oscillatory differential equations. SIAM J. Sci. Comput. **20**, 930–963 (1999)
20. Hochbruck, M., Lubich, C.: A Gautschi-type method for oscillatory second-order differential equations. Numer. Math. **83**, 403–426 (1999)
21. Hairer, E., Lubich, C.: Long-time energy conservation of numerical methods for oscillatory differential equations. SIAM J. Numer. Anal. **38**, 414–441 (2000)
22. Wu, X., You, X., Xia, J.: Order conditions for ARKN methods solving oscillatory systems. Comput. Phys. Commun. **180**, 2250–2257 (2009)
23. Wu, X., You, X., Shi, W., et al.: ERKN integrators for systems of oscillatory second order differential equations. Comput. Phys. Commun. **181**, 1873–1887 (2010)
24. Yang, H.L., Wu, X.Y.: Trigonometrically-fitted ARKN methods for perturbed oscillators. Appl. Numer. Math. **58**, 1375–1395 (2008)
25. Wu, X., You, X., Li, J.: Note on derivation of order conditions for ARKN methods for perturbed oscillators. Comput. Phys. Commun. **180**, 1545–1549 (2009)
26. McLachlan, R.I., Quispel, G.R.W., Robidoux, N.: Geometric integration using discrete gradients. Philos. Trans. R. Soc. Lond. A **357**, 1021–1045 (1999)
27. Liu, K., Shi, W., Wu, X.: An extended discrete gradient formula for oscillatory Hamiltonian systems. J. Phys. A Math. Theor. **46**, 165203 (2013)
28. Dahlby, M., Owren, B.A.: A general framework for deriving integral preserving numerical methods for PDEs. SIAM J. Sci. Comput. **33**, 2318–2340 (2011)
29. Gonzalez, O.: Time integration and discrete Hamiltonian systems. J. Nonlinear Sci. **6**, 449–467 (1996)
30. Itoh, T., Abe, K.: Hamiltonian conserving discrete canonical equations based on variational difference quotients. J. Comput. Phys. **77**, 85–102 (1988)
31. Quispel, G.R.W., Capel, H.W.: Solving ODEs numerically while preserving a first integral. Phys. Lett. A **218**, 223–228 (1996)
32. Quispel, G.R.W., Turner, G.S.: Discrete gradient methods for solving ODEs numerically while preserving a first integral. J. Phys. A Math. Gen. **29**, L341–L349 (1996)

33. Quispel, G.R.W., McLaren, D.I.: A new class of energy-preserving numerical integration methods. J. Phys. A Math. Theor. **41**, 045206 (2008)
34. Cieśliński, J.L., Ratkiewicz, B.: Energy-preserving numerical schemes of high accuracy for one-dimensional Hamiltonian systems. J. Phys. A Math. Theor. **44**, 155206 (2011)
35. Harten, A., Lax, P.D., van Leer, B.: On upstream differencing and Godunov-type schemes for hyperbolic conservation laws. SIAM Rev. **25**, 35–61 (1983)
36. Higham, N.J., Smith, M.I.: Computing the matrix cosine. Numer. Algorithms **34**, 13–26 (2003)
37. Hairer, E., Lubich, C.: Energy-diminishing integration of gradient systems. IMA J. Numer. Anal. **34**, 452–461 (2014)
38. Reich, S.: Multi-symplectic Runge–Kutta collocation methods for Hamilton wave equations. J. Comput. Phys. **157**, 473–499 (2000)
39. Liu, K., Wu, X., Shi, W.: A linearly-fitted conservative (dissipative) scheme for efficiently solving nonlinear wave equations. J. Comput. Math. **35**, 780–800 (2017)

Chapter 9
Energy-Preserving Schemes for High-Dimensional Nonlinear KG Equations

The main theme of this chapter is the analysis of energy-preserving schemes for solving high-dimensional nonlinear Klein–Gordon equations. An energy-preserving scheme is presented based on the discrete gradient method and the Duhamel Principle. The local error, global convergence, and nonlinear stability of the scheme are analysed in detail.

9.1 Introduction

The nonlinear Klein–Gordon (KG) equation is one of the important models in quantum mechanics and mathematical physics. It is well known that a key feature of the KG equation is energy preservation. The main purpose of this chapter is to formulate and analyse energy-preserving schemes for the high-dimensional nonlinear KG equation

$$
\begin{cases}
\dfrac{\partial u}{\partial t} = v, & u(x, 0) = g_1(x), \\[2mm]
\dfrac{\partial v}{\partial t} = \omega^2 \Delta u - G'(u), & v(x, 0) = g_2(x),
\end{cases}
\tag{9.1}
$$

where

$$
\Delta = \sum_{j=1}^{d} \frac{\partial^2}{\partial x_j^2},
$$

ω is a real parameter, the real-valued function $u(x, t)$, representing the wave displacement at position x and time t, is defined in $(x, t) \in \Omega \times [0, T]$ with $\Omega := (0, X_1) \times \cdots \times (0, X_d) \subset \mathbb{R}^d$, and $G(u)$ is a smooth potential energy function

X. Wu, B. Wang, *Geometric Integrators for Differential Equations with Highly Oscillatory Solutions*, https://doi.org/10.1007/978-981-16-0147-7_9

with $G(u) \geqslant 0$. We consider (9.1) under periodic boundary conditions

$$u(x, t)|_{\partial \Omega \cap \{x_j = 0\}} = u(x, t)|_{\partial \Omega \cap \{x_j = X_j\}}, \quad j = 1, 2, \cdots, d. \tag{9.2}$$

As is known, the high-dimensional nonlinear KG equation is a Hamiltonian partial differential equation and the Hamiltonian of system (9.1) is

$$\mathscr{H}[u, v](t) = \int_{\Omega} \left[\frac{1}{2} |v(x, t)|^2 + \frac{1}{2} \omega^2 |\nabla u(x, t)|^2 + G\big(u(x, t)\big) \right] dx, \tag{9.3}$$

where $dx = dx_1 dx_2 \cdots dx_d$.

The KG equation (9.1) can be rewritten as the following infinite-dimensional Hamiltonian system

$$z_t = J \frac{\delta \mathscr{H}}{\delta z}, \tag{9.4}$$

where

$$J = \begin{pmatrix} 0 & -1 \\ -1 & 0 \end{pmatrix}, \quad z = \begin{pmatrix} u \\ v \end{pmatrix},$$

and

$$\frac{\delta \mathscr{H}}{\delta z} = \left(\frac{\delta \mathscr{H}}{\delta u}, \frac{\delta \mathscr{H}}{\delta v} \right)^{\mathsf{T}}$$

is the functional derivative of $\mathscr{H}$.

The nonlinear KG equation, which is also termed Schrödinger's relativistic wave equation, arises frequently in various fields of scientific applications such as solid state physics, fluid dynamics, nonlinear optics, quantum field theory and relativistic quantum mechanics. It is used to model many different nonlinear phenomena, including the propagation of dislocations in crystals and the behaviour of elementary particles. Several numerical schemes have been developed to solve KG equations (see, e.g. [1–7]). In recent decades, geometric algorithms have been received much attention and they have been shown to be important in studying the long-time behaviour of dynamical systems. Various structure-preserving algorithms have been designed to preserve as much as possible the physical/geometric properties of the underlying systems. For research papers related to this topic, we refer the reader to [8–17] and references therein. As for a good theoretical foundation of structure-preserving algorithms for ordinary differential equations (ODEs), we refer the reader to [18]. Surveys of structure-preserving algorithms for oscillatory differential equations are given in [19–21]. By extending ideas and tools related to structure-preserving algorithms of ODEs, various integrators have been proposed and investigated for specific or general classes of partial differential equations

(PDEs). Typical categories include multi-symplectic methods, semidiscretisation by means of the method of lines, and energy-preserving schemes.

In the literature, many multi-symplectic schemes have been proposed for multi-symplectic Hamiltonian PDEs. By assigning a distinct symplectic operator for each unbounded space direction and time, the multi-symplectic structure for a class of PDEs generalises the classical symplectic structure of Hamiltonian ODEs. Multi-symplectic integrators precisely conserve a discrete space-time symplectic structure of Hamiltonian PDEs and we refer to [22–28] for related work. With regard to the method of lines, the spatial derivatives are usually approximated by finite differences or by discrete Fourier transform and the resulting system is then integrated in time by a suitable ODE integrator. This approach has become one of the central topics in PDEs, and we refer the readers to [29–33] for examples on this subject.

Among the most widely used properties is energy preservation. The study of energy-preserving schemes for Hamiltonian PDEs has a long history, which dates back to an old paper of [34] where a discrete energy conservation law for the 5-point finite difference approximation was derived. A historical survey of energy preserving methods for PDEs and their applications has been made in [5]. Some recent relevant work can be found in [27, 35–40]. Energy-conserving methods have also been the subject of many investigations for ODEs in the last two decades. Various effective energy-preserving methods have been proposed and investigated, such as discrete gradient (DG) methods (see, e.g. [41–44]), time finite elements (see, e.g. [45, 46]), the average vector field (AVF) method (see, e.g. [47–49]), Hamiltonian boundary value methods (HBVMs) (see, e.g. [50, 51]), and the adapted average vector field (AAVF) method (see, e.g. [52, 53]). All the energy-preserving methods mentioned above were originally designed for ODEs and have also been considered in the PDE setting. In [54], the authors discussed numerical methods for PDEs that are based on the discrete gradient method. A systematic introduction to finite element methods for efficient numerical solution of PDEs is presented in [55]. The AVF method for discretising Hamiltonian PDEs was studied in [35], and HBVMs related to the numerical solution of the semilinear wave equation were researched in [56]. The authors in [57] analysed the AAVF method for one-dimensional Hamiltonian wave equations.

We note that an important aspect in the numerical simulation of Hamiltonian systems is the approximate conservation of the total energy over long times. DG methods have also been applied to PDEs in the form of the AVF method in [35] and in a somewhat more general setting, the discrete variational derivative method in [54, 58]. On the other hand, the authors in [59] established an operator-variation-of-constants formula for wave equations, and based on that, many efficient numerical methods have been designed and analysed (see, e.g. [6, 33, 57, 60, 61]). In this chapter we focus on the formulation and analyses of energy-preserving schemes for high-dimensional nonlinear KG equations based on DG methods and the Duhamel Principle. We will formulate the energy-preserving scheme and analyse its errors, nonlinear stability, convergence and implementation issues. Numerical results will be presented to show the remarkable superiority of the energy-preserving scheme in comparison with well-known energy-preserving methods in the literature.

9.2 Formulation of Energy-Preserving Schemes

The main aim of this section is to formulate energy-preserving schemes for high-dimensional nonlinear KG equations. To this end, we first formulate Eq. (9.1) as an abstract Hamiltonian system of ODEs on an infinity-dimensional Hilbert space $L^2(\Omega)$.

It will be convenient to define the linear differential operator $\mathscr{A}$ by

$$(\mathscr{A}u)(x,t) = -\omega^2 \Delta u(x,t). \tag{9.5}$$

Then, for the periodic boundary condition (9.2), the domain of this operator is

$$
\begin{aligned}
D(\mathscr{A}) = \{u \in H^1(\Omega) : \, &u(x,t)|_{\partial\Omega\cap\{x_j=0\}} = u(x,t)|_{\partial\Omega\cap\{x_j=X_j\}}, \\
&\nabla u(x,t)|_{\partial\Omega\cap\{x_j=0\}} = \nabla u(x,t)|_{\partial\Omega\cap\{x_j=X_j\}}, \\
&j = 1, 2, \cdots, d\}.
\end{aligned}
\tag{9.6}
$$

In terms of self-adjoint operator theories, we introduce the following semi-group generated by $\mathscr{A}$:

$$\phi_j(\mathscr{A}) := \sum_{k=0}^{\infty} \frac{(-1)^k \mathscr{A}^k}{(2k+j)!}, \quad j = 0, 1, 2, \cdots. \tag{9.7}$$

With regard to the definitions and properties of these operator-argument functions for different boundary conditions, which have been studied in detail, we refer the readers to [61]. We next summarise the following two useful properties which are needed in this chapter.

Proposition 9.1 (See [61]) *All the operator-argument functions ϕ_j for $j \in \mathbb{N}$ are symmetric operators with respect to the inner product of the space $L^2(\Omega)$:*

$$(p,q) = \int_{\Omega} p(x)\overline{q(x)}\mathrm{d}x. \tag{9.8}$$

The norm of the function in $L^2(\Omega)$ can be characterized in the frequency space by

$$\|q\|^2 = (q,q) = \int_{\Omega} |q(x)|^2 \mathrm{d}x.$$

Proposition 9.2 (See [61]) *All the operator-argument functions defined by (9.7) are bounded as follows:*

$$\|\phi_j(t\mathscr{A})\|_* \leqslant \gamma_j, \quad j \in \mathbb{N}, \quad t \geqslant 0, \tag{9.9}$$

where $\| \cdot \|_*$ *is the Sobolev norm, and* γ_j *are the bounds of the functions* $\phi_j(x)$ *with* $x \geqslant 0$. *It follows from* (9.9) *that*

$$\|\phi_j(t\mathscr{A})\|_* \leqslant 1, \quad j = 0, 1, \qquad \|\phi_2(t\mathscr{A})\|_* \leqslant \frac{1}{2}.$$

It is clear that the operator $\mathscr{A}$ defined by (9.5) is a positive semi-definite operator, i.e., $\forall\, u(x, t) \in D(\mathscr{A})$

$$\left(\mathscr{A} u(x, t), u(x, t)\right) = \int_\Omega \mathscr{A} u(x, t) \cdot u(x, t) \mathrm{d}x = \omega^2 \int_\Omega |\nabla u(x, t)|^2 \mathrm{d}x \geqslant 0.$$
(9.10)

The exact energy (9.3) can be presented in the following form:

$$\mathscr{H}[u, v](t) \equiv \frac{1}{2}\left(v(x, t), v(x, t)\right) + \frac{1}{2}\left(\mathscr{A} u(x, t), u(x, t)\right) + \int_\Omega G\big(u(x, t)\big)\mathrm{d}x$$
$$= \mathscr{H}[u, v](0).$$
(9.11)

We now define $u(t)$ as the function that maps x to $u(x, t)$:

$$u(t) = [x \mapsto u(x, t)].$$

We then formulate the original system (9.1) as follows:

$$\begin{cases} u'(t) = v(t), & u(0) = g_1(x), \\ v'(t) = -\mathscr{A}u(t) - G'(u(t)), & v(0) = g_2(x). \end{cases}$$
(9.12)

Using the Duhamel Principle leads to the following operator-variation-of-constants formula for the nonlinear KG equation (9.1):

$$\begin{cases} u(t) = \phi_0(t^2\mathscr{A})u(0) + t\phi_1(t^2\mathscr{A})v(0) - \displaystyle\int_0^t (t - \zeta)\phi_1\big((t - \zeta)^2\mathscr{A}\big)G'\big(u(\zeta)\big)\mathrm{d}\zeta, \\[2ex] v(t) = -t\mathscr{A}\phi_1(t^2\mathscr{A})u(0) + \phi_0(t^2\mathscr{A})v(0) - \displaystyle\int_0^t \phi_0\big((t - \zeta)^2\mathscr{A}\big)G'\big(u(\zeta)\big)\mathrm{d}\zeta, \end{cases}$$
(9.13)

where $\forall t \in [0, T]$.

Remark 9.1 For the nonlinear KG equation (9.12), the formula (9.13) is a pair of nonlinear integral equations which reflect the changes of the solution and its derivative with time t. This pattern will assist in the design of structure-preserving schemes for solving (9.12).

On the one hand, it follows from (9.13) that

$$
\begin{cases}
u(t_{n+1}) = \phi_0(\mathscr{V})u(t_n) + h\phi_1(\mathscr{V})v(t_n) \\
\qquad\qquad - h^2 \displaystyle\int_0^1 (1-\zeta)\phi_1\big((1-\zeta)^2\mathscr{V}\big)G'\big(u(t_n+h\zeta)\big)d\zeta, \\
v(t_{n+1}) = -h\mathscr{A}\phi_1(\mathscr{V})u(t_n) + \phi_0(\mathscr{V})v(t_n) \\
\qquad\qquad - h \displaystyle\int_0^1 \phi_0\big((1-\zeta)^2\mathscr{V}\big)G'\big(u(t_n+h\zeta)\big)d\zeta,
\end{cases}
\tag{9.14}
$$

where h is the time stepsize, $t_n = nh$ and $\mathscr{V} = h^2\mathscr{A}$.

On the other hand, we recall the definition of a DG method and define the following discrete gradient of function $G(u)$

$$
\begin{cases}
\overline{\nabla}G(u,\hat{u}) \cdot (\hat{u}-u) = G(\hat{u}) - G(u), \\
\overline{\nabla}G(u,u) = G'(u).
\end{cases}
\tag{9.15}
$$

Here, by replacing $G'\big(u(t_n+h\zeta)\big)$ with $\overline{\nabla}G(u_n, u_{n+1})$, the integrals appearing in (9.14) can be approximated by:

$$
\int_0^1 (1-\zeta)\phi_1\big((1-\zeta)^2\mathscr{V}\big)G'\big(u(t_n+h\zeta)\big)d\zeta
$$

$$
\approx \int_0^1 (1-\zeta)\phi_1\big((1-\zeta)^2\mathscr{V}\big)d\zeta\,\overline{\nabla}G(u_n, u_{n+1}) = \phi_2(\mathscr{V})\overline{\nabla}G(u_n, u_{n+1}),
$$

$$
\int_0^1 \phi_0\big((1-\zeta)^2\mathscr{V}\big)G'\big(u(t_n+h\zeta)\big)d\zeta \approx \int_0^1 \phi_0\big((1-\zeta)^2\mathscr{V}\big)d\zeta\,\overline{\nabla}G(u_n, u_{n+1})
$$

$$
= \phi_1(\mathscr{V})\overline{\nabla}G(u_n, u_{n+1}),
$$

where we have used the following results (see [57]):

$$
\int_0^1 (1-\zeta)\phi_1\big((1-\zeta)^2\mathscr{V}\big)d\zeta = \phi_2(\mathscr{V}), \quad \int_0^1 \phi_0\big((1-\zeta)^2\mathscr{V}\big)d\zeta = \phi_1(\mathscr{V}).
$$

Obviously, the above analysis leads to a continuous function $u_n := u_n(x) \approx u(x, t_n)$. With this function, we then define the numerical scheme for the high-dimensional nonlinear KG equation (9.12) as follows.

Definition 9.1 The discrete gradient scheme for solving the high-dimensional nonlinear KG equation (9.12) is defined by

$$
\begin{cases}
u_{n+1} = \phi_0(\mathscr{V})u_n + h\phi_1(\mathscr{V})v_n - h^2\phi_2(\mathscr{V})\overline{\nabla}G(u_n, u_{n+1}), \\
v_{n+1} = -h\mathscr{A}\phi_1(\mathscr{V})u_n + \phi_0(\mathscr{V})v_n - h\phi_1(\mathscr{V})\overline{\nabla}G(u_n, u_{n+1}),
\end{cases}
\tag{9.16}
$$

which is called the KGDG scheme, where $\overline{\nabla}G(u_n, u_{n+1})$ is determined by (9.15).

Remark 9.2 Here, we remark that this KGDG scheme is relevant to the operator-variation-of-constants formula. There have been some other numerical integrators for KG equations, which are also relevant to the formula (see, e.g. [6, 60]). It is important to note that they are derived for different purposes. Those schemes proposed in [6, 60] are formulated to be of arbitrarily high order, and the KGDG scheme is to preserve the continuous energy (9.3) exactly.

Theorem 9.1 *The KGDG scheme* (9.16) *exactly preserves the Hamiltonian* (9.11), *i.e.,*

$$
\mathscr{H}[u_{n+1}, v_{n+1}] = \mathscr{H}[u_n, v_n], \quad n = 0, 1, \cdots .
\tag{9.17}
$$

Proof We insert (9.16) into (9.11) and calculate carefully with the Proposition 9.1. This yields

$$
\begin{aligned}
\mathscr{H}[u_{n+1}, v_{n+1}] = \frac{1}{2}\Big(& \big(\phi_0^2(\mathscr{V}) \\
& + \mathscr{V}\phi_1^2(\mathscr{V})\big)v_n, v_n\Big) + \frac{1}{2}\Big(\mathscr{A}\big(\phi_0^2(\mathscr{V}) + \mathscr{V}\phi_1^2(\mathscr{V})\big)u_n, u_n\Big) \\
& + \Big(\mathscr{V}\big(\phi_1^2(\mathscr{V}) - \phi_0(\mathscr{V})\phi_2(\mathscr{V})\big)u_n, \overline{\nabla}G(u_n, u_{n+1})\Big) \\
& - \Big(h\big(\phi_0(\mathscr{V})\phi_1(\mathscr{V}) + \mathscr{V}\phi_1(\mathscr{V})\phi_2(\mathscr{V})\big)v_n, \overline{\nabla}G(u_n, u_{n+1})\Big) \\
& + \frac{1}{2}\Big(h^2\big(\phi_1^2(\mathscr{V}) + \mathscr{V}\phi_2^2(\mathscr{V})\big)\overline{\nabla}G(u_n, u_{n+1}), \overline{\nabla}G(u_n, u_{n+1})\Big) \\
& + \int_\Omega G(u_{n+1})dx.
\end{aligned}
\tag{9.18}
$$

It follows from the following results on the operator-argument functions (see [61])

$$
\begin{cases}
\phi_0^2(\mathscr{V}) + \mathscr{V}\phi_1^2(\mathscr{V}) = I, \\
\phi_1^2(\mathscr{V}) - \phi_0(\mathscr{V})\phi_2(\mathscr{V}) = \phi_2(\mathscr{V}), \\
\phi_0(\mathscr{V})\phi_1(\mathscr{V}) + \mathscr{V}\phi_1(\mathscr{V})\phi_2(\mathscr{V}) = \phi_1(\mathscr{V}), \\
\frac{1}{2}\big(\phi_1^2(\mathscr{V}) + \mathscr{V}\phi_2^2(\mathscr{V})\big) = \phi_2(\mathscr{V}),
\end{cases}
$$

that (9.18) can be simplified as

$$\mathcal{H}[u_{n+1}, v_{n+1}] = \frac{1}{2}(v_n, v_n) + \frac{1}{2}(\mathscr{A}u_n, u_n) + \left(\mathscr{V}\phi_2(\mathscr{V})u_n, \overline{\nabla}G(u_n, u_{n+1})\right)$$

$$- \left(h\phi_1(\mathscr{V})v_n, \overline{\nabla}G(u_n, u_{n+1})\right) + \left(h^2\phi_2(\mathscr{V})\overline{\nabla}G(u_n, u_{n+1}), \overline{\nabla}G(u_n, u_{n+1})\right)$$

$$+ \int_\Omega G(u_{n+1})dx. \tag{9.19}$$

In what follows, we consider two cases. In the case where $u_{n+1} - u_n = 0$, the first equation of (9.16) gives

$$0 = u_{n+1} - u_n = \left(\phi_0(\mathscr{V}) - I\right)u_n + h\phi_1(\mathscr{V})v_n - h^2\phi_2(\mathscr{V})\overline{\nabla}G(u_n, u_{n+1})$$

$$= -\mathscr{V}\phi_2(\mathscr{V})u_n + h\phi_1(\mathscr{V})v_n - h^2\phi_2(\mathscr{V})\overline{\nabla}G(u_n, u_{n+1}),$$

where we have used the following result presented in [61]

$$\phi_0(\mathscr{V}) - I = -\mathscr{V}\phi_2(\mathscr{V}).$$

Taking $\overline{\nabla}G(u_n, u_n) = \nabla G(u_n)$ into account, (9.19) becomes

$$\mathcal{H}[u_{n+1}, v_{n+1}] = \mathcal{H}[u_n, v_{n+1}]$$

$$= \frac{1}{2}(v_n, v_n) + \frac{1}{2}(\mathscr{A}u_n, u_n) + \left(\mathscr{V}\phi_2(\mathscr{V})u_n, G'(u_n)\right)$$

$$- \left(h\phi_1(\mathscr{V})v_n, G'(u_n)\right) + \left(h^2\phi_2(\mathscr{V})G'(u_n), G'(u_n)\right) + \int_\Omega G(u_n)dx$$

$$= \frac{1}{2}(v_n, v_n) + \frac{1}{2}(\mathscr{A}u_n, u_n) + \int_\Omega G(u_n)dx$$

$$+ \left(\mathscr{V}\phi_2(\mathscr{V})u_n - h\phi_1(\mathscr{V})v_n + h^2\phi_2(\mathscr{V})G'(u_n), \nabla G(u_n)\right)$$

$$= \mathcal{H}[u_n, v_n] + \left(0, G'(u_n)\right) = \mathcal{H}[u_n, v_n].$$

In the case where $u_{n+1} - u_n \neq 0$, we have

$$G(u_{n+1}) - G(u_n) = \overline{\nabla}G(u_n, u_{n+1}) \cdot (u_{n+1} - u_n),$$

and we obtain

$$\int_\Omega G(u_{n+1})dx = \int_\Omega G(u_n)dx + \int_\Omega \Big(G(u_{n+1}) - G(u_n)\Big)dx$$

$$= \int_\Omega G(u_n)dx + \Big((u_{n+1} - u_n)I, \overline{\nabla}G(u_n, u_{n+1})\Big). \quad (9.20)$$

It follows from the first equation of (9.16) that

$$u_{n+1} - u_n = -\mathscr{V}\phi_2(\mathscr{V})u_n + h\phi_1(\mathscr{V})v_n - h^2\phi_2(\mathscr{V})\overline{\nabla}G(u_n, u_{n+1}).$$

Accordingly, (9.20) can be rewritten as

$$\int_\Omega G(u_{n+1})dx = \int_\Omega G(u_n)dx - \Big(\mathscr{V}\phi_2(\mathscr{V})u_n, \overline{\nabla}G(u_n, u_{n+1})\Big)$$

$$+ \Big(h\phi_1(\mathscr{V})v_n, \overline{\nabla}G(u_n, u_{n+1})\Big) - \Big(h^2\phi_2(\mathscr{V})\overline{\nabla}G(u_n, u_{n+1}), \overline{\nabla}G(u_n, u_{n+1})\Big).$$

$$(9.21)$$

We insert (9.21) into (9.19) and obtain

$$\mathscr{H}[u_{n+1}, v_{n+1}]$$

$$= \frac{1}{2}(v_n, v_n) + \frac{1}{2}(\mathscr{A}u_n, u_n) + \Big(\mathscr{V}\phi_2(\mathscr{V})u_n, \overline{\nabla}G(u_n, u_{n+1})\Big)$$

$$- \Big(h\phi_1(\mathscr{V})v_n, \overline{\nabla}G(u_n, u_{n+1})\Big)$$

$$+ \Big(h^2\phi_2(\mathscr{V})\overline{\nabla}G(u_n, u_{n+1}), \overline{\nabla}G(u_n, u_{n+1})\Big) + \int_\Omega G(u_n)dx$$

$$- \Big(\mathscr{V}\phi_2(\mathscr{V})u_n, \overline{\nabla}G(u_n, u_{n+1})\Big)$$

$$+ \Big(h\phi_1(\mathscr{V})v_n, \overline{\nabla}G(u_n, u_{n+1})\Big) - \Big(h^2\phi_2(\mathscr{V})\overline{\nabla}G(u_n, u_{n+1}), \overline{\nabla}G(u_n, u_{n+1})\Big)$$

$$= \frac{1}{2}(v_n, v_n) + \frac{1}{2}(\mathscr{A}u_n, u_n) + \int_\Omega G(u_n)dx = \mathscr{H}[u_n, v_n].$$

The proof is finished. ∎

Remark 9.3 It is noted that there exist many possible choices of discrete gradients for a function (see e.g. [18, 44]). Among typical discrete gradients is the well-known AVF method defined by

$$\overline{\nabla}_{\text{AVF}}G(u_n, u_{n+1}) = \int_0^1 G'((1-\tau)u_n + \tau u_{n+1})d\tau.$$

With this special choice, the KGDG scheme (9.16) reduces to the AAVF method which has been studied in [57, 61] for nonlinear wave equations. Moreover, when $\omega = 0$, the energy-preserving scheme (9.16) reduces to the AVF method for the semidiscretised Hamiltonian PDEs considered in [35]. In other words, the KGDG scheme (9.16) is an essential extension of AAVF methods and AVF methods from Hamiltonian ODEs to Hamiltonian PDEs. Furthermore, it is remarked that existing analyses of errors, nonlinear stability and convergence of AAVF methods and AVF methods for PDEs are insufficient. Therefore, a primary mission of this work is to analyse the errors, nonlinear stability and convergence of KGDG methods for high-dimensional nonlinear KG equations.

Remark 9.4 Here we remark that an extended DG method for Hamiltonian ODEs was researched in [62] and applied to conservative (dissipative) nonlinear wave PDEs in [58]. However, this method was presented in a scheme for ODEs and applied only to semidiscrete PDEs; i.e., spatial derivatives are discretised in advance. Thus, it cannot preserve the continuous energy of the PDEs exactly. Moreover, the errors, nonlinear stability and convergence of this method were not analysed in [58]. It is remarked that the scheme (9.16) reduces to the extended DG method when applied to second-order oscillatory ODEs considered in [62]. Furthermore, the method for nonlinear KG equations is based on the operator-variation-of-constants formula and depends on the differential operator $\mathcal{V}$. This means that the scheme does not require the PDEs to be discretised in space and avoids the semidiscretisation of the spacial derivative. Moreover, the scheme can preserve the continuous energy of the PDEs exactly. The scheme (9.16) is more suitable and competitive since different efficient ways to approximate the operator in the literature can be chosen in a flexible approach, according to different situations and requirements.

Theorem 9.2 *The KGDG scheme* (9.16) *is symmetric with respect to the time variable.*

Proof Exchanging $u_{n+1} \leftrightarrow u_n$, $v_{n+1} \leftrightarrow v_n$ and replacing h by $-h$ in (9.16) gives this result straightforwardly. We here skip the details.   □

9.3   Error Analysis

This section will be devoted to local error bounds for the energy-preserving scheme (9.16) under the following assumption.

Assumption 9.1 We assume that $f(u) = -G'(u) : D(\mathscr{A}) \to \mathbb{R}$ is sufficiently often Fréchet differentiable in a strip along the exact solution and is sufficiently smooth with respect to the time. There exists a real number k such that

$$\|f(w_1) - f(w_2)\| \leqslant k\|w_1 - w_2\|$$

for all $w_1, w_1 \in L^2(\Omega)$.

Theorem 9.3 *Suppose that the KG equation (9.1) possesses uniformly bounded and sufficiently smooth solutions with respect to the time and $f(u)$ satisfies Assumption 9.1. Assume that $f_t'' \in L^\infty(0, T; L^2(\Omega))$. Under the local assumptions of $u_n = u(t_n)$, $v_n = v(t_n)$, if the sufficiently small time stepsize h satisfies $0 < h \leqslant \sqrt{\dfrac{2}{k}}$, then the local error bounds of (9.16) are given by*

$$\|u(t_{n+1}) - u_{n+1}\| \leqslant Ch^3 \quad and \quad \|v(t_{n+1}) - v_{n+1}\| \leqslant Ch^3. \tag{9.22}$$

Proof Inserting the exact solution of (9.12) into the approximation (9.16) yields

$$\begin{cases} u(t_{n+1}) = \phi_0(\mathcal{V})u(t_n) + h\phi_1(\mathcal{V})v(t_n) - h^2\phi_2(\mathcal{V})\overline{\nabla}G\big(u(t_n), u(t_{n+1})\big) + \delta_{n+1}, \\ v(t_{n+1}) = -h\mathscr{A}\phi_1(\mathcal{V})u(t_n) + \phi_0(\mathcal{V})v(t_n) - h\phi_1(\mathcal{V})\overline{\nabla}G\big(u(t_n), u(t_{n+1})\big) + \delta_{n+1}', \end{cases} \tag{9.23}$$

where δ_{n+1} and δ_{n+1}' denote the discrepancies.

Using (9.23) and (9.14), we obtain

$$\begin{cases} \delta_{n+1} = h^2 \displaystyle\int_0^1 (1-\zeta)\phi_1\big((1-\zeta)^2\mathcal{V}\big)f\big(u(t_n + h\zeta)\big)d\zeta + h^2\phi_2(\mathcal{V})\overline{\nabla}G\big(u(t_n), u(t_{n+1})\big), \\ \delta_{n+1}' = h \displaystyle\int_0^1 \phi_0\big((1-\zeta)^2\mathcal{V}\big)f\big(u(t_n + h\zeta)\big)d\zeta + h\phi_1(\mathcal{V})\overline{\nabla}G\big(u(t_n), u(t_{n+1})\big). \end{cases} \tag{9.24}$$

Expressing f and G of the formula (9.24) by the Taylor series expansion at $u(t_n)$ gives

$$\delta_{n+1} = h^2\phi_2(\mathcal{V})f\big(u(t_n)\big) + \mathcal{O}(h^3) - h^2\phi_2(\mathcal{V})f\big(u(t_n)\big) + \mathcal{O}(h^3) = \mathcal{O}(h^3),$$

and

$$\begin{aligned} \delta_{n+1}' =& h\phi_1(\mathcal{V})f\big(u(t_n)\big) + h^2\phi_2(\mathcal{V})\frac{\partial f\big(u(t)\big)}{\partial t}\Big|_{t=t_n} + \mathcal{O}(h^3) \\ & - h\phi_1(\mathcal{V})f\big(u(t_n)\big) - \frac{1}{2}h\phi_1(\mathcal{V})f'\big(u(t_n)\big)\big(u(t_{n+1}) - u(t_n)\big) + \mathcal{O}(h^3) \\ =& h^2\phi_2(\mathcal{V})\frac{\partial f\big(u(t)\big)}{\partial t}\Big|_{t=t_n} - \frac{1}{2}h\phi_1(\mathcal{V})f'\big(u(t_n)\big)\big(u(t_{n+1}) - u(t_n)\big) + \mathcal{O}(h^3) \\ =& h^2\phi_2(\mathcal{V})\frac{\partial f\big(u(t)\big)}{\partial t}\Big|_{t=t_n} - \frac{1}{2}h\phi_1(\mathcal{V})f'\big(u(t_n)\big)\Big(h\frac{\partial u(t)}{\partial t}\Big|_{t=t_n} + \mathcal{O}(h^2)\Big) + \mathcal{O}(h^3) \\ =& h^2\Big(\phi_2(\mathcal{V}) - \frac{1}{2}\phi_1(\mathcal{V})\Big)\frac{\partial f\big(u(t)\big)}{\partial t}\Big|_{t=t_n} + \mathcal{O}(h^3) = \mathcal{O}(h^3). \end{aligned}$$

For convenience, define errors

$$e_n^u = u(t_n) - u_n, \qquad e_n^v = v(t_n) - v_n.$$

Then, subtracting (9.16) from (9.23) yields

$$
\begin{cases}
e_{n+1}^u = h^2 \phi_2(\mathscr{V}) \overline{\nabla} G(u_n, u_{n+1}) - h^2 \phi_2(\mathscr{V}) \overline{\nabla} G\big(u(t_n), u(t_{n+1})\big) + \delta_{n+1}, \\
e_{n+1}^v = h \phi_1(\mathscr{V}) \overline{\nabla} G(u_n, u_{n+1}) - h \phi_1(\mathscr{V}) \overline{\nabla} G\big(u(t_n), u(t_{n+1})\big) + \delta_{n+1}',
\end{cases}
$$

which leads to

$$
\begin{aligned}
\|e_{n+1}^u\| &\leqslant h^2 \|\phi_2(\mathscr{V})\| \frac{1}{2} k \|e_{n+1}^u\| + \|\delta_{n+1}\| \leqslant \frac{1}{4} h^2 k \|e_{n+1}^u\| + \|\delta_{n+1}\|, \\
\|e_{n+1}^v\| &\leqslant h \|\phi_1(\mathscr{V})\| \frac{1}{2} k \|e_{n+1}^u\| + \|\delta_{n+1}'\| \leqslant \frac{1}{2} hk \|e_{n+1}^u\| + \|\delta_{n+1}'\|.
\end{aligned}
\tag{9.25}
$$

Under the condition that the time stepsize h satisfies $h \leqslant \sqrt{\dfrac{2}{k}}$, the first inequality in (9.25) implies that

$$\|e_{n+1}^u\| \leqslant 2\|\delta_{n+1}\| \leqslant Ch^3.$$

With this result and the second line of (9.25), we deduce that $\|e_{n+1}^v\| \leqslant Ch^3$. The proof is complete.   □

9.4 Analysis of the Nonlinear Stability

In this section, we study the nonlinear stability of the energy-preserving scheme (9.16). To accomplish this purpose, we consider the following perturbed problem of (9.12)

$$
\begin{cases}
\tilde{u}'(t) = \tilde{v}(t), & \tilde{u}(0) = g_1(x) + \tilde{g}_1(x), \\
\tilde{v}'(t) = -\mathscr{A}\tilde{u}(t) - G'(\tilde{u}(t)), & \tilde{v}(0) = g_2(x) + \tilde{g}_2(x),
\end{cases}
\tag{9.26}
$$

where $\tilde{g}_1(x)$ and $\tilde{g}_2(x)$ are perturbation functions. Let

$$\hat{u}(t) = \tilde{u}(t) - u(t), \qquad \hat{v}(t) = \tilde{v}(t) - v(t).$$

It can be obtained by subtracting (9.12) from (9.26) that

$$
\begin{cases}
\hat{u}'(t) = \hat{v}(t), & \hat{u}(0) = \tilde{g}_1(x), \\
\hat{v}'(t) = -\mathscr{A}\hat{u}(t) - G'(\tilde{u}(t)) + G'(u(t)), & \hat{v}(0) = \tilde{g}_2(x).
\end{cases}
\tag{9.27}
$$

We then apply the approximation (9.16) respectively to (9.12) and (9.26), and obtain two numerical schemes, which give an approximation of (9.27) as follows

$$
\begin{cases}
\hat{u}_{n+1} = \phi_0(\mathscr{V})\hat{u}_n + h\phi_1(\mathscr{V})\hat{v}_n + h^2\phi_2(\mathscr{V})\tilde{\mathscr{I}}, \\
\hat{v}_{n+1} = -h\mathscr{A}\phi_1(\mathscr{V})\hat{u}_n + \phi_0(\mathscr{V})\hat{v}_n + h\phi_1(\mathscr{V})\tilde{\mathscr{I}},
\end{cases}
\tag{9.28}
$$

where

$$
\tilde{\mathscr{I}} = \overline{\nabla}G(u_n, u_{n+1}) - \overline{\nabla}G(\tilde{u}_n, \tilde{u}_{n+1}).
\tag{9.29}
$$

Clearly, contrary to the traditional manner, until now we have mostly considered the situation where the PDE is discretised in time while remaining continuous in space. To achieve practical numerical schemes, it remains to deal with the differential operator $\mathscr{A}$ in an appropriate way. A straightforward approach to the treatment of the space derivatives is simply to discretise them in the Hamiltonian. In this situation, the operator $\mathscr{A}$ is approximated by a suitable differentiation matrix. For ODEs it is common to devise relatively general frameworks for structure preservation. Hence, we approximate the operator $\mathscr{A}$ by a symmetric and positive semi-definite differentiation matrix. In such a way we can derive corresponding Hamiltonian ODEs, and commence rigorous nonlinear stability and convergence analysis. Fortunately, there have been many publications which proposed various effective ways to deal with the spatial derivatives (see, e.g. [30, 63, 64]), and it is not difficult to find the symmetric and positive semi-definite differentiation matrix. Using the space discretisation of a Hamiltonian PDE, we can obtain a system of Hamiltonian ODEs to which a geometric time integrator may be applied.

The following theorem presents the nonlinear stability of the energy-preserving approximation (9.16) when the operator $\mathscr{A}$ is approximated by a suitable differentiation matrix.

Theorem 9.4 *Let the conditions of Theorem 9.3 hold and assume that the operator $\mathscr{A}$ is approximated by a symmetric and positive semi-definite differentiation matrix A. If the time stepsize h satisfies $0 < h \leqslant \sqrt{\dfrac{2}{k}}$, then*

$$
\|\hat{u}_n\| \leqslant \exp(\hat{C}T)\left(\|\tilde{g}_1\| + \sqrt{\|\sqrt{A}\tilde{g}_1\|^2 + \|\tilde{g}_2\|^2}\right),
$$
$$
\|\hat{v}_n\| \leqslant \exp(\hat{C}T)\left(\|\tilde{g}_1\| + \sqrt{\|\sqrt{A}\tilde{g}_1\|^2 + \|\tilde{g}_2\|^2}\right),
\tag{9.30}
$$

where $k > 0$ is a Lipschitz constant, and $\hat{C}$ and $\sqrt{A}$ are defined in the proof.

Proof First of all, the matrix A can be expressed as

$$A = PD^2 P^\mathsf{T} = \sqrt{A}^2,$$

where P is an orthogonal matrix, D is a positive semi-definite diagonal matrix and $\sqrt{A} = PDP^\mathsf{T}$. Consequently, (9.28) becomes

$$
\begin{cases}
\hat{u}_{n+1} = \phi_0(K)\hat{u}_n + h\phi_1(K)\hat{v}_n + h^2\phi_2(K)\tilde{\mathscr{I}}, \\
\hat{v}_{n+1} = -hA\phi_1(K)\hat{u}_n + \phi_0(K)\hat{v}_n + h\phi_1(K)\tilde{\mathscr{I}},
\end{cases}
\tag{9.31}
$$

with $K = h^2 A$. It follows from the first formula of (9.31) that

$$
\left\| \hat{u}_{n+1} \right\| \leqslant \left\| \hat{u}_n \right\| + h \left\| \hat{v}_n \right\| + \frac{1}{2} h^2 \left\| \tilde{\mathscr{I}} \right\|.
\tag{9.32}
$$

Rewriting (9.31) as the following compact form:

$$
\begin{pmatrix} \sqrt{A}\hat{u}_{n+1} \\ \hat{v}_{n+1} \end{pmatrix} = \begin{pmatrix} \phi_0(K) & h\sqrt{A}\phi_1(K) \\ -h\sqrt{A}\phi_1(K) & \phi_0(K) \end{pmatrix} \begin{pmatrix} \sqrt{A}\hat{u}_n \\ \hat{v}_n \end{pmatrix}
$$
$$
+ h \begin{pmatrix} h\sqrt{A}\phi_2(K) \\ & \phi_1(K) \end{pmatrix} \begin{pmatrix} \tilde{\mathscr{I}} \\ \tilde{\mathscr{I}} \end{pmatrix},
$$

we then obtain

$$
\sqrt{ \| \sqrt{A}\hat{u}_{n+1} \|^2 + \| \hat{v}_{n+1} \|^2 } \leqslant \sqrt{ \| \sqrt{A}\hat{u}_n \|^2 + \| \hat{v}_n \|^2 } + h\tilde{\alpha} \left\| \tilde{\mathscr{I}} \right\|.
$$

Here $\tilde{\alpha}$ is the uniformly bound of $\left\| \begin{pmatrix} h\sqrt{A}\phi_2(K) \\ & \phi_1(K) \end{pmatrix} \right\|$ and we have used the fact (see [6]) that

$$
\left\| \begin{pmatrix} \phi_0(K) & h\sqrt{A}\phi_1(K) \\ -h\sqrt{A}\phi_1(K) & \phi_0(K) \end{pmatrix} \right\| = 1.
$$

Summing up the above results yields

$$
\| \hat{u}_{n+1} \| + \sqrt{ \| \sqrt{A}\hat{u}_{n+1} \|^2 + \| \hat{v}_{n+1} \|^2 }
$$
$$
\leqslant \| \hat{u}_n \| + \sqrt{ \| \sqrt{A}\hat{u}_n \|^2 + \| \hat{v}_n \|^2 } + h\| \hat{v}_n \| + h \left(\tilde{\alpha} + \frac{h}{2} \right) \left\| \tilde{\mathscr{I}} \right\|.
\tag{9.33}
$$

According to the definition (9.29) of $\tilde{\mathscr{I}}$, the following result is obtained:

$$\left\| \tilde{\mathscr{I}} \right\| \leqslant k \int_0^1 \left\| \left((1-\tau)\hat{u}_n + \tau \hat{u}_{n+1} \right) \right\| d\tau \leqslant \frac{1}{2} k \left(\left\| \hat{u}_n \right\| + \left\| \hat{u}_{n+1} \right\| \right). \qquad (9.34)$$

Inserting (9.34) into (9.32) yields

$$\left\| \hat{u}_{n+1} \right\| \leqslant \left\| \hat{u}_n \right\| + h \left\| \hat{v}_n \right\| + \frac{1}{4} h^2 k \left(\left\| \hat{u}_n \right\| + \left\| \hat{u}_{n+1} \right\| \right),$$

and then we have

$$\left(1 - \frac{1}{4} h^2 k \right) \left\| \hat{u}_{n+1} \right\| \leqslant \left(1 + \frac{1}{4} h^2 k \right) \left\| \hat{u}_n \right\| + h \left\| \hat{v}_n \right\|.$$

Under the condition that $h \leqslant \sqrt{\dfrac{2}{k}}$, we obtain

$$\left\| \hat{u}_{n+1} \right\| \leqslant 3 \left\| \hat{u}_n \right\| + 2h \left\| \hat{v}_n \right\|.$$

Combining this result with (9.33)–(9.34) gives

$$\left\| \hat{u}_{n+1} \right\| + \sqrt{ \left\| \sqrt{A}\hat{u}_{n+1} \right\|^2 + \left\| \hat{v}_{n+1} \right\|^2 }$$

$$\leqslant \left\| \hat{u}_n \right\| + \sqrt{ \left\| \sqrt{A}\hat{u}_n \right\|^2 + \left\| \hat{v}_n \right\|^2 } + h \left\| \hat{v}_n \right\| + \frac{h}{2} \left(\tilde{\alpha} + \frac{h}{2} \right)$$

$$k \left(\left\| \hat{u}_n \right\| + 3 \left\| \hat{u}_n \right\| + 2h \left\| \hat{v}_n \right\| \right)$$

$$\leqslant \left\| \hat{u}_n \right\| + \sqrt{ \left\| \sqrt{A}\hat{u}_n \right\|^2 + \left\| \hat{v}_n \right\|^2 } + \left(h + h^2 \left(\tilde{\alpha} + \frac{h}{2} \right) k \right)$$

$$\left\| \hat{v}_n \right\| + 2h \left(\tilde{\alpha} + \frac{h}{2} \right) k \left\| \hat{u}_n \right\|$$

$$\leqslant \left\| \hat{u}_n \right\| + \sqrt{ \left\| \sqrt{A}\hat{u}_n \right\|^2 + \left\| \hat{v}_n \right\|^2 } + h\hat{C} \left(\left\| \hat{u}_n \right\| + \sqrt{ \left\| \sqrt{A}\hat{u}_n \right\|^2 + \left\| \hat{v}_n \right\|^2 } \right),$$

where $\hat{C} = \max \left(1 + \tilde{\alpha} k h + \dfrac{k h^2}{2}, 2\tilde{\alpha} k + \dfrac{k h}{2} \right)$. Then, it follows from mathematical induction that

$$\left\| \hat{u}_{n+1} \right\| + \sqrt{ \left\| \sqrt{A}\hat{u}_{n+1} \right\|^2 + \left\| \hat{v}_{n+1} \right\|^2 } \leqslant (1 + h\hat{C})^n \left(\left\| \hat{u}_0 \right\| + \sqrt{ \left\| \sqrt{A}\hat{u}_0 \right\|^2 + \left\| \hat{v}_0 \right\|^2 } \right)$$

$$\leqslant \exp\left(\hat{C} T \right) \left(\left\| \tilde{g}_1 \right\| + \sqrt{ \left\| \sqrt{A}\tilde{g}_1 \right\|^2 + \left\| \tilde{g}_2 \right\|^2 } \right).$$

This proves this theorem.   □

9.5   Convergence

This section is concerned with the convergence of the fully discrete approximation. Using some suitable spatial discretisation strategies, we discretise the original continuous system (9.12) as follows:

$$\begin{cases} U'(t) = V(t), & U(0) = g_1(x), \\ V'(t) = -AU(t) - \nabla \widetilde{G}(U(t)) + \hat{\delta}(\Delta x), & V(0) = g_2(x), \end{cases} \tag{9.35}$$

where U, $V \in \mathbb{R}^M$ are vectors, A is a symmetric and positive semi-definite differentiation matrix, Δx is the spatial stepsize for the space discretisation, $\widetilde{G}(U) = \sum_{j=1}^{M} G(U_j)$, and $\hat{\delta}(\Delta x)$ is the truncation error introduced by the approximation of spatial differential operator $\mathscr{A}$ through the matrix A.

Applying the numerical approximation (9.16) to (9.35) and ignoring $\hat{\delta}(\Delta x)$ yields

$$\begin{cases} U_{n+1} = \phi_0(K)U_n + h\phi_1(K)V_n - h^2\phi_2(K)\overline{\nabla}\widetilde{G}(U_n, U_{n+1}), \\ V_{n+1} = -hA\phi_1(K)U_n + \phi_0(K)V_n - h\phi_1(K)\overline{\nabla}\widetilde{G}(U_n, U_{n+1}). \end{cases} \tag{9.36}$$

We then have the following convergence result of (9.36).

Theorem 9.5 *Under the conditions of Theorem 9.4, there exists a constant C such that*

$$\begin{cases} \|U(t_n) - U_n\| \leqslant CT \exp(\hat{C}T)(h^2 + \|\hat{\delta}(\Delta x)\|), \\ \|V(t_n) - V_n\| \leqslant CT \exp(\hat{C}T)(h^2 + \|\hat{\delta}(\Delta x)\|), \end{cases} \tag{9.37}$$

where C is a constant independent of n, h and Δx.

Proof Inserting the exact solution of (9.35) into the numerical approximation (9.16), we obtain

$$\begin{cases} U(t_{n+1}) = \phi_0(K)U(t_n) + h\phi_1(K)V(t_n) \\ \qquad\qquad + h^2\phi_2(K)\int_0^1 f\big((1-\tau)U(t_n) + \tau U(t_{n+1})\big)d\tau + \hat{\delta}_{n+1}, \\ V(t_{n+1}) = -hA\phi_1(K)U(t_n) + \phi_0(K)V(t_n) \\ \qquad\qquad + h\phi_1(K)\int_0^1 f\big((1-\tau)U(t_n) + \tau U(t_{n+1})\big)d\tau + \hat{\delta}'_{n+1}, \end{cases} \tag{9.38}$$

where $\hat{\delta}_{n+1}$ and $\hat{\delta}'_{n+1}$ are the discrepancies. In terms of these formulae and the operator-variation-of-constants formula (9.13) of (9.35), we have

$$
\begin{cases}
\hat{\delta}_{n+1} = h^2 \displaystyle\int_0^1 (1-\zeta)\phi_1\big((1-\zeta)^2 K\big) f\big(U(t_n + h\zeta)\big)\mathrm{d}\zeta + h^2\phi_2(K)\overline{\nabla}\widetilde{G}(U_n, U_{n+1}) \\
\qquad + h^2 \displaystyle\int_0^1 (1-z)\phi_1\big((1-z)^2 K\big)\hat{\delta}(\Delta x)\mathrm{d}z, \\[2mm]
\hat{\delta}'_{n+1} = h \displaystyle\int_0^1 \phi_0\big((1-\zeta)^2 K\big) f\big(U(t_n + h\zeta)\big)\mathrm{d}\zeta + h\phi_1(K)\overline{\nabla}\widetilde{G}(U_n, U_{n+1}) \\
\qquad + h \displaystyle\int_0^1 \phi_0\big((1-z)^2 K\big)\hat{\delta}(\Delta x)\mathrm{d}z.
\end{cases}
\tag{9.39}
$$

Repeating the similar steps which we did in Sect. 9.3, we deduce the following results associated with these discrepancies

$$
\left\|\hat{\delta}_{n+1}\right\| \leqslant C_1 h^3 + h^2 \left\|\int_0^1 (1-z)\phi_1\big((1-z)^2 K\big)\hat{\delta}(\Delta x)\mathrm{d}z\right\|,
$$

$$
\left\|\hat{\delta}'_{n+1}\right\| \leqslant C_2 h^3 + h \left\|\int_0^1 \phi_0\big((1-z)^2 K\big)\hat{\delta}(\Delta x)\mathrm{d}z\right\|.
$$

Thus, we have

$$
\|\hat{\delta}_{n+1}\| \leqslant C_1 h^3 + \frac{1}{2}h^2\|\hat{\delta}(\Delta x)\|, \quad \|\hat{\delta}'_{n+1}\| \leqslant C_2 h^3 + h\|\hat{\delta}(\Delta x)\|.
\tag{9.40}
$$

We now denote

$$
e_n^U = U(t_n) - U_n, \qquad e_n^V = V(t_n) - V_n.
$$

Then, subtracting (9.36) from (9.38) results in

$$
\begin{cases}
e_{n+1}^U = \phi_0(K)e_n^U + h\phi_1(K)e_n^V + h^2\phi_2(K)\hat{\mathscr{I}} + \hat{\delta}_{n+1}, \\
e_{n+1}^V = -hA\phi_1(K)e_n^U + \phi_0(K)e_n^V + h\phi_1(K)\hat{\mathscr{I}} + \hat{\delta}'_{n+1},
\end{cases}
\tag{9.41}
$$

where

$$
\hat{\mathscr{I}} = \overline{\nabla}\widetilde{G}(U_n, U_{n+1}) - \overline{\nabla}\widetilde{G}\big(U(t_n), U(t_{n+1})\big)
\tag{9.42}
$$

with the initial conditions $e_0^U = \mathbf{0}$ and $e_0^V = \mathbf{0}$. We rewrite (9.41) as follows:

$$
\begin{pmatrix} \sqrt{A}e_{n+1}^U \\ e_{n+1}^V \end{pmatrix} = \begin{pmatrix} \phi_0(K) & h\sqrt{A}\phi_1(K) \\ -h\sqrt{A}\phi_1(K) & \phi_0(K) \end{pmatrix} \begin{pmatrix} \sqrt{A}e_n^U \\ e_n^V \end{pmatrix}
$$
$$
+ h \begin{pmatrix} h\sqrt{A}\phi_2(K) \\ & \phi_1(K) \end{pmatrix} \begin{pmatrix} \hat{\mathscr{I}} \\ \hat{\mathscr{I}} \end{pmatrix} + \begin{pmatrix} \sqrt{A}\hat{\delta}_{n+1} \\ \hat{\delta}'_{n+1} \end{pmatrix},
$$

which implies

$$
\sqrt{\|\sqrt{A}e_{n+1}^U\|^2 + \|e_{n+1}^V\|^2} \leqslant \sqrt{\|\sqrt{A}e_n^U\|^2 + \|e_n^V\|^2} + h\tilde{\alpha}\left\|\hat{\mathscr{I}}\right\|
$$
$$
+ \sqrt{\|\sqrt{A}\hat{\delta}_{n+1}\|^2 + \|\hat{\delta}'_{n+1}\|^2}.
$$

According to (9.42), we obtain

$$
\left\|\hat{\mathscr{I}}\right\| \leqslant k \int_0^1 \left\|(1-\tau)e_n^U + \tau e_{n+1}^U\right\| \, d\tau \leqslant \frac{1}{2}k\left(\left\|e_n^U\right\| + \left\|e_{n+1}^U\right\|\right).
$$

It then follows from the first equality of (9.41) that

$$
\left\|e_{n+1}^U\right\| \leqslant \left\|e_n^U\right\| + h\left\|e_n^V\right\| + \frac{1}{2}h^2\left\|\hat{\mathscr{I}}\right\| + \left\|\hat{\delta}_{n+1}\right\|
$$
$$
\leqslant \left\|e_n^U\right\| + h\left\|e_n^V\right\| + \frac{1}{4}h^2 k\left(\left\|e_n^U\right\| + \left\|e_{n+1}^U\right\|\right) + \left\|\hat{\delta}_{n+1}\right\|.
$$

Using the condition $h \leqslant \sqrt{\dfrac{2}{k}}$, we obtain

$$
\left\|e_{n+1}^U\right\| \leqslant 3\left\|e_n^U\right\| + 2h\left\|e_n^V\right\| + 2\left\|\hat{\delta}_{n+1}\right\|,
$$

and

$$
\left\|\hat{\mathscr{I}}\right\| \leqslant k\left(2\left\|e_n^U\right\| + h\left\|e_n^V\right\| + \left\|\hat{\delta}_{n+1}\right\|\right).
$$

Now summing up the above results gives

$$
\left\|e_{n+1}^U\right\| + \sqrt{\|\sqrt{A}e_{n+1}^U\|^2 + \|e_{n+1}^V\|^2} \leqslant \left\|e_n^U\right\| + \sqrt{\|\sqrt{A}e_n^U\|^2 + \|e_n^V\|^2}
$$
$$
+ h\left\|e_n^V\right\| + h\left(\tilde{\alpha} + \frac{1}{2}h\right)\left\|\hat{\mathscr{I}}\right\| + \left\|\hat{\delta}_{n+1}\right\| + \sqrt{\|\sqrt{A}\hat{\delta}_{n+1}\|^2 + \|\hat{\delta}'_{n+1}\|^2}
$$

$$\leqslant \left\| e_n^U \right\| + \sqrt{\left\| \sqrt{A} e_n^U \right\|^2 + \left\| e_n^V \right\|^2} + \sqrt{\left\| \sqrt{A} \hat{\delta}_{n+1} \right\|^2 + \left\| \hat{\delta}'_{n+1} \right\|^2}$$

$$+ h \left\| e_n^V \right\| + h \left(\tilde{\alpha} + \frac{1}{2} h \right) k \left(2 \left\| e_n^U \right\| + h \left\| e_n^V \right\| + \left\| \hat{\delta}_{n+1} \right\| \right) + \left\| \hat{\delta}_{n+1} \right\|$$

$$\leqslant (1 + h\hat{C}) \left(\left\| e_n^U \right\| + \sqrt{\left\| \sqrt{A} e_n^U \right\|^2 + \left\| e_n^V \right\|^2} \right)$$

$$+ \sqrt{\left\| \sqrt{A} \hat{\delta}_{n+1} \right\|^2 + \left\| \hat{\delta}'_{n+1} \right\|^2} + \left(1 + h \left(\tilde{\alpha} + \frac{1}{2} h \right) k \right) \left\| \hat{\delta}_{n+1} \right\| .$$

On noting that the truncation errors (9.40), there exists a constant C such that

$$\sqrt{\left\| \sqrt{A} \hat{\delta}_{n+1} \right\|^2 + \left\| \hat{\delta}'_{n+1} \right\|^2} + \left(1 + h \left(\tilde{\alpha} + \frac{1}{2} h \right) k \right) \left\| \hat{\delta}_{n+1} \right\| \leqslant Ch \left(h^2 + \left\| \hat{\delta}(\Delta x) \right\| \right).$$

It then follows form the Gronwall's inequality that

$$\left\| e_{n+1}^U \right\| + \sqrt{\left\| \sqrt{A} e_{n+1}^U \right\|^2 + \left\| e_{n+1}^V \right\|^2}$$

$$\leqslant \exp(nh\hat{C}) \left(\left\| e_0^U \right\| + \sqrt{\left\| \sqrt{A} e_0^U \right\|^2 + \left\| e_0^V \right\|^2} + Cnh \left(h^2 + \left\| \hat{\delta}(\Delta x) \right\| \right) \right).$$

This confirms the theorem. $\square$

9.6 Implementation Issues of KGDG Scheme

This section focuses on the implementation issues of the energy-preserving KGDG scheme (9.16). Obviously, (9.16) itself falls well short of being a practical scheme unless the $\overline{\nabla} G(u_n, u_{n+1})$ can be approximated. Fortunately, however, due to the special structure of the function $G'(u)$ appearing in the KG equation (9.1), the $\overline{\nabla} G(u_n, u_{n+1})$ can be calculated as follows:

$$\begin{cases} \overline{\nabla} G(u_n, u_{n+1}) = \dfrac{1}{(u_{n+1} - u_n)} G(u_{n+1}) - G(u_n), & \text{if } u_{n+1} - u_n \neq 0, \\ \overline{\nabla} G(u_n, u_n) = G'(u_n), & \text{if } u_{n+1} - u_n = 0. \end{cases}$$

$$(9.43)$$

Thus, we are now in a position to present the following practical energy-preserving scheme for solving the high-dimensional nonlinear KG equation. We call it the KGDG scheme again.

Definition 9.2 A practical KGDG scheme for the high-dimensional nonlinear KG equation (9.1) is defined by

$$
\begin{cases}
u_{n+1} = \phi_0(\mathcal{V})u_n + h\phi_1(\mathcal{V})v_n - h^2\phi_2(\mathcal{V})\dfrac{1}{u_{n+1} - u_n}\Big(G\,(u_{n+1}) - G(u_n)\Big), \\[4mm]
v_{n+1} = -h\mathscr{A}\phi_1(\mathcal{V})u_n + \phi_0(\mathcal{V})v_n - h\phi_1(\mathcal{V})\dfrac{1}{u_{n+1} - u_n}\Big(G\,(u_{n+1}) - G(u_n)\Big),
\end{cases}
\tag{9.44}
$$

provided $u_{n+1} - u_n \neq 0$, whereas if $u_{n+1} - u_n = 0$, the corresponding scheme reduces to

$$
\begin{cases}
u_{n+1} = \phi_0(\mathcal{V})u_n + h\phi_1(\mathcal{V})v_n - h^2\phi_2(\mathcal{V})G'(u_n), \\[2mm]
v_{n+1} = -h\mathscr{A}\phi_1(\mathcal{V})u_n + \phi_0(\mathcal{V})v_n - h\phi_1(\mathcal{V})G'(u_n).
\end{cases}
\tag{9.45}
$$

As stated above, a notable feature of the analysis is that the underlying KG equation is discretised in time while remaining continuous in space.

Contrarily, in practice, one usually discretises the spatial derivative first, for instance, by finite differences methods or spectral methods. This implies that the differential operator $\mathscr{A}$ is replaced by a symmetric and semi-definite matrix A, namely, the high-dimensional nonlinear KG equation (9.1) is approximated by the following system of ODEs

$$
\begin{cases}
q'(t) = p(t), & q(0) = g_1(x), \\
p'(t) = -Aq(t) - \nabla\widetilde{G}(q(t)), & p(0) = g_2(x),
\end{cases}
\tag{9.46}
$$

where $q(t),\, p(t) \in \mathbb{R}^M$ and

$$
\widetilde{G}\big(q(t)\big) = \sum_{j=1}^{M} G\big(q_j(t)\big).
$$

This Hamiltonian system of ODEs has a corresponding energy conservation law of the form:

$$
\tilde{H}(p,q) = \frac{\Delta x}{2}p^{\mathsf{T}}p + \frac{\Delta x}{2}q^{\mathsf{T}}Aq + \Delta x\widetilde{G}\big(q(t)\big) = \cdots = \tilde{H}(p(0), q(0)),
\tag{9.47}
$$

where Δx is the spatial stepsize. This energy $\tilde{H}(p, q)$ is termed semidiscrete energy, which can be thought of as an approximate energy of the original nonlinear KG equation (9.1). We then have the following practical energy-preserving scheme for the nonlinear KG equation (9.1), which is termed semidiscrete KGDG (SKGDG) scheme.

Definition 9.3 An energy-preserving SKGDG scheme for (9.46) is defined by

$$\begin{cases} q_{n+1} = \phi_0(K)q_n + h\phi_1(K)p_n - h^2\phi_2(K)\overline{\nabla}_S G(q_n, q_{n+1}), \\ p_{n+1} = -hA\phi_1(K)q_n + \phi_0(K)p_n - h\phi_1(K)\overline{\nabla}_S G(q_n, q_{n+1}), \end{cases} \tag{9.48}$$

where $K = h^2 A$, $\overline{\nabla}_S G(q_n, q_{n+1})$ is determined as follows:

$$\begin{cases} \overline{\nabla}_S G(q_n, q_{n+1}) = \big(G\,(q_{n+1}) - G(q_n)\big)/(q_{n+1} - q_n), & \text{if } q_{n+1} - q_n \neq 0, \\ \overline{\nabla}_S G(q_n, q_n) = \nabla\tilde{G}(u_n), & \text{if } q_{n+1} - q_n = 0, \end{cases} \tag{9.49}$$

and $\cdot/\cdot$ denotes elementwise division of vectors. We here remark that for a vector $q = (q_1, q_2, \cdots, q_M)^{\mathsf{T}}$, $G(q)$ is defined by

$$G(q) = \big(G(q_1), G(q_2), \cdots, G(q_M)\big)^{\mathsf{T}}. \tag{9.50}$$

Remark 9.5 Actually, the SKGDG scheme (9.48) can be straightforwardly obtained by replacing $\mathscr{V}$, u_n, $\big(G(u_{n+1}) - G(u_n)\big)/(u_{n+1} - u_n)$, and $G'(u_n)$ in the KGDG scheme (9.44)–(9.45) with K, q_n, $\overline{\nabla}_S G(q_n, q_{n+1})$, and $\overline{\nabla}_S G(q_n, q_n)$, respectively. Remember that $\mathscr{V} = h^2\mathscr{A}$ and $K = h^2 A$ here. This provides new insight into the design of geometric numerical integration for solving high-dimensional nonlinear KG equations.

Theorem 9.6 *The SKGDG scheme* (9.48) *exactly preserves the semidiscrete energy* $\tilde{H}$, *i.e., we have*

$$\tilde{H}(p_{n+1}, q_{n+1}) = \tilde{H}(p_n, q_n), \quad n = 0, 1, \cdots.$$

Proof The proof is divided into two cases: $q_{n+1} - q_n \neq 0$ and $q_{n+1} - q_n = 0$.

Case (**i**): $q_{n+1} - q_n \neq 0$.

In this case, we first need to show that

$$(q_{n+1} - q_n)^{\mathsf{T}}\overline{\nabla}_S G(q_n, q_{n+1}) = \tilde{G}\,(q_{n+1}) - \tilde{G}(q_n).$$

In fact, it follows from (9.48) and (9.50) that

$$(q_{n+1} - q_n)^{\mathsf{T}}\overline{\nabla}_S G(q_n, q_{n+1}) = (q_{n+1} - q_n)^{\mathsf{T}}$$
$$\Big((G(q_{n+1}) - G(q_n))/(q_{n+1} - q_n)\Big) \tag{9.51}$$
$$= \sum_{j=1}^{M}\big(G(q_{n+1,j}) - G(q_{n,j})\big) = \tilde{G}\,(q_{n+1}) - \tilde{G}(q_n),$$

where $q_{n,j}$ denotes the j-th component of the vector q_n.

We insert the SKGDG scheme (9.48) into the semidiscrete energy $\tilde{H}$ (9.47) with some careful calculations, and obtain

$$
\begin{aligned}
\tilde{H}(p_{n+1}, q_{n+1}) = {} & \frac{\Delta x}{2} p_n^\mathsf{T} p_n + \frac{\Delta x}{2} q_n^\mathsf{T} A q_n + \Delta x \tilde{G}(q_n) + \Delta x \left(\tilde{G}(q_{n+1}) - \tilde{G}(q_n) \right) \\
& + \Delta x q_n^\mathsf{T} K \phi_2(K) \overline{\nabla}_S G(q_n, q_{n+1}) \\
& - \Delta x h p_n^\mathsf{T} K \phi_1(K) \overline{\nabla}_S G(q_n, q_{n+1}) \\
& + \Delta x h^2 \overline{\nabla}_S G(q_n, q_{n+1})^\mathsf{T} \phi_2(K) \overline{\nabla}_S G(q_n, q_{n+1}). \qquad (9.52)
\end{aligned}
$$

Then, using the first formula of (9.48) and (9.51) yields

$$
\begin{aligned}
\tilde{G}(q_{n+1}) - \tilde{G}(q_n) = {} & (q_{n+1} - q_n)^\mathsf{T} \overline{\nabla}_S G(q_n, q_{n+1}) = -q_n^\mathsf{T} K \phi_2(K) \overline{\nabla}_S G(q_n, q_{n+1}) \\
& + h p_n^\mathsf{T} K \phi_1(K) \overline{\nabla}_S G(q_n, q_{n+1}) - h^2 \overline{\nabla}_S G(q_n, q_{n+1})^\mathsf{T} \phi_2(K) \overline{\nabla}_S G(q_n, q_{n+1}).
\end{aligned}
$$

Inserting this result into (9.52), we immediately obtain

$$
\tilde{H}(p_{n+1}, q_{n+1}) = \frac{\Delta x}{2} p_n^\mathsf{T} p_n + \frac{\Delta x}{2} q_n^\mathsf{T} A q_n + \Delta x \tilde{G}(q_n) = \tilde{H}(p_n, q_n).
$$

Case (**ii**): $q_{n+1} - q_n = \mathbf{0}$.

In this case, the conclusion can be deduced straightforwardly from considering (9.48) and (9.52).

The proof is complete.   □

Remark 9.6 It can be observed that the schemes (9.44) and (9.48) are formulated in a completely closed form since the $\overline{\nabla} G(u_n, u_{n+1})$ is evaluated exactly. This makes the schemes more practical and robust, which can be thought of as an efficient and straightforward approach to the implementation of the energy-preserving scheme (9.16) for nonlinear KG equations.

Remark 9.7 It is very important to note that the schemes (9.44) and (9.48) for nonlinear KG equations are different from the AVF method and AAVF method for Hamiltonian ODEs since both the AVF method and the AAVF method are dependent on the evaluation of the integrals appearing in their formulae. As time integration methods for nonlinear KG equations, both the AVF method and the AAVF method cannot exactly preserve the energy in practical computation, in general. Moreover, for the scheme (9.48), the difference between the continuous energy and the discrete energy is dependent only on the spatial discretisation and independent of the time integration, since the integral appearing in the formula has been calculated without error, which is completely different from the case for Hamiltonian ODEs.

Remark 9.8 In the spirit of Geometric Integration, the scheme (9.48) is a genuine energy-preserving formula from both theoretical and computation aspects because (9.48) exactly preserves the discrete energy without error. This advantage is significant for the study of structure-preserving algorithms for nonlinear Hamiltonian PDEs.

Remark 9.9 As is known, one of the important applications of numerical ODEs is to solve PDEs efficiently. The traditional and popular way is to discretise the spatial derivative first. Then numerical methods of ODEs are applied to the underlying semidiscretised ODEs. Sometimes this approach will result in a gap between PDEs and ODEs in view of numerical analysis. In particular, the analysis of global errors, since the true solution of the underlying nonlinear PDEs, or the true solution of the semidiscretised ODEs may not be available. Fortunately, however, on the basis of the so-called operator-variation-of-constants formula and the matrix-variation-of-constants formula, we can get an insight into the true solutions of both the underlying PDEs and the semidiscretised ODEs. Hence, we can deal with the important issues of numerical analysis for structure-preserving schemes when applied to nonlinear KG equations.

In what follows, we analyse the convergence of fixed-point iteration for (9.16), because the scheme (9.16) is implicit and an iterative procedure is required.

Theorem 9.7 *It is assumed that the DG* $\overline{\nabla}G(\cdot, y)$ *satisfies the Lipschitz condition with respect to y and the Lipschitz constant is L. If the time stepsize h satisfies*

$$0 < h \leqslant \tilde{h} \leqslant \sqrt{\frac{2}{L}}, \text{ then the fixed-point iteration for (9.16) is convergent.}$$

Proof Let

$$\Psi : y \to \phi_0(\mathcal{V})u_n + h\phi_1(\mathcal{V})v_n - h^2\phi_2(\mathcal{V})\overline{\nabla}G(u_n, y).$$

Using Proposition 9.2, we obtain

$$\|\Psi(y_1) - \Psi(y_2)\| \leqslant \frac{1}{2}h^2 L \|y_1 - y_2\|,$$

which proves the result by the contraction mapping theorem.   □

Remark 9.10 It should be pointed out that the convergence of the fixed-point iteration for the KGDG scheme is independent of $\mathscr{A}$. Unfortunately, however, the convergence of other methods such as the AVF method and HBVMs depends on $\mathscr{A}$. This allows us to use a large time stepsize and a simple iteration method with a small iteration number for the KGDG scheme. This point will be demonstrated clearly by the numerical experiments presented in next section.

9.7 Numerical Experiments

The purpose of this section is to present the numerical results of applying the energy-preserving KGDG scheme to several nonlinear KG equations in comparison with the well-known leap-frog scheme and two classes of energy-preserving methods, AVF methods and HBVMs.

We choose the following solvers for comparison:

- KGDG: the energy-preserving KGDG scheme of order two presented in this chapter;
- AVF: the energy-preserving AVF method of order two proposed in [35];
- HBVM11: the HBVM(1,1) of order two derived in [50, 56];
- LFS: the well-known leap-frog scheme;
- HBVM22: the HBVM(2,2) of order four given in [50, 56].

It is remarked that the first four methods are all of order two while HBVM22 is of order four. The main aim of choosing the fourth-order method is to show the superiority of the KGDG scheme of order two in comparison with higher-order methods in the literature. It is also noted that KGDG, AVF, HBVM11 and HBVM22 are all implicit, and iterative solutions are required for them. In order to demonstrate the advantage of the KGDG scheme, i.e., the KGDG scheme can perform well even though a simple iteration method is employed with a small iteration number, we use standard fixed-point iteration in the practical computations. We set 10^{-15} as the error tolerance, and 10 as the maximum number of iterations for all the experiments. Of course, it is possible to change standard fixed-point iteration to the quasi-Newton or Newton iteration. In this chapter we do not consider this issue further for brevity.

9.7.1 One-Dimensional Problems

We start from three one-dimensional nonlinear KG equations and approximate the operator $\mathscr{A}$ by the Fourier spectral collocation (FSC) (see [30, 64]) as follows:

$$\mathscr{A} \approx A = \frac{1}{(\pi/L)^2}\left(a_{kj}\right)_{M \times M} \text{ with } a_{kj} = \begin{cases} \dfrac{(-1)^{k+j}}{2}\sin^{-2}\left(\dfrac{(k-j)\pi}{M}\right), & k \neq j, \\[2mm] \dfrac{M^2}{12} + \dfrac{1}{6}, & k = j. \end{cases}$$
$$(9.53)$$

Problem 9.1 Consider the following sine-Gordon equation (see, e.g. [56])

$$\frac{\partial^2 u}{\partial t^2} = \frac{\partial^2 u}{\partial x^2} - \sin u, \qquad x \in [-20, 20], \qquad t \in [0, T].$$

We here consider soliton-like solutions, defined by the initial conditions:

$$u(x, 0) \equiv 0, \quad u_t(x, 0) = 4/\gamma \operatorname{sech}\left(x/\gamma\right), \quad \gamma \geq 0.$$

The solution of Problem 9.1 depends on the value of γ, and is given by

$$u(x, t) = 4 \arctan \left(\psi(t, \gamma) \operatorname{sech}(x/\gamma)\right)$$

where

$$\psi(t, \gamma) = \begin{cases} \sinh\left(\dfrac{\sqrt{1-\gamma^2}t}{\gamma}\right) / \sqrt{1-\gamma^2}, & 0 < \gamma < 1, \\ t, & \gamma = 1, \\ \sin\left(\dfrac{\sqrt{\gamma^2-1}t}{\gamma}\right) / \sqrt{\gamma^2-1}, & \gamma > 1. \end{cases}$$

The exact solutions with $\gamma = 0.99$, 1 and 1.01 are shown in Fig. 9.1. As is known, the case $0 < \gamma < 1$ is named kink-antikink and the case $\gamma > 1$ is termed breather. The case $\gamma = 1$ is named double-pole which separates the two different types of dynamics.

After the Fourier spectral collocation semidiscretisation of operator $\mathscr{A}$ by (9.53) with $M = 200$, we integrate the semidiscrete system with $h = 0.1$ and $T = 1000$. The errors of the semidiscrete energy conservation for $\gamma = 0.99$, 1, 1.01 are shown in Fig. 9.2a. We here use the semidiscrete energy (9.47) at the initial values as the "exact" energy of this problem and show the energy errors for each numerical method. Similar situations are encountered in the next four problems.

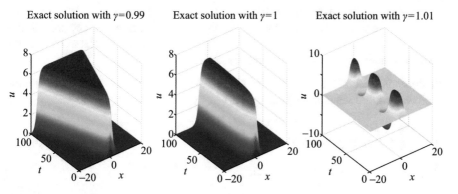

Exact solution with $\gamma=0.99$ Exact solution with $\gamma=1$ Exact solution with $\gamma=1.01$

Fig. 9.1 Exact solutions of Problem 9.1

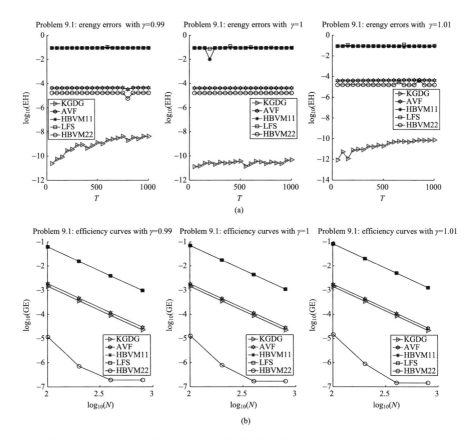

Fig. 9.2 Energy errors and efficiency curves for Problem 9.1

Problem 9.1 is solved on $[0, 10]$ with different stepsizes $h = 0.1/2^j$ for $j = 0, 1, 2, 3$. We use the log-log plots of the global errors. The efficiency curves (the global error versus $N = T/h$) are shown in Fig. 9.2b. It can be observed that the KGDG scheme shows remarkable numerical behaviour.

Problem 9.2 Consider the dimensionless relativistic KG equation with highly oscillatory solutions in time (see, e.g. [1])

$$\begin{cases} \varepsilon^2 \dfrac{\partial^2 u}{\partial t^2} - \dfrac{\partial^2 u}{\partial x^2} + \dfrac{1}{\varepsilon^2} u + f(u) = 0, & -L \leqslant x \leqslant L, \ 0 \leqslant t \leqslant T, \quad u(-L, t) = u(L, t), \\ u(x, 0) = \phi(x), \quad u_t(x, 0) = \dfrac{1}{\varepsilon^2} \gamma(x), \end{cases}$$

$$(9.54)$$

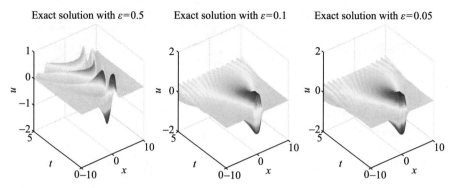

Exact solution with $\varepsilon=0.5$ Exact solution with $\varepsilon=0.1$ Exact solution with $\varepsilon=0.05$

Fig. 9.3 The 'exact' solutions of Problem 9.2

where ε is a dimensionless parameter $\varepsilon > 0$. Similarly to [1], we choose

$$f(u) = 4u^3, \quad \phi(x) = \frac{2}{\exp(x^2) + \exp(-x^2)}, \quad \gamma(x) = 0,$$

in (9.54).

Let $u(x, t)$ be the 'exact' solution of this problem, which is obtained numerically by using the fourth-order method HBVM22 with small time step $h = 1/1000$. The results with $\varepsilon = 0.5, 0.1$, and 0.05 are presented in Fig. 9.3.

After the Fourier spectral collocation semidiscretisation of operator $\mathscr{A}$ with $L = 10$, $M = 100$ for $\varepsilon = 0.5$ and $M = 400$ for $\varepsilon = 0.1$ and 0.05, the problem is solved with $T = 1000$, $h = 0.05$ for $\varepsilon = 0.5$, $T = 200$, $h = 0.004$ for $\varepsilon = 0.1$ and $T = 100$, $h = 0.002$ for $\varepsilon = 0.05$. The errors of the semidiscrete energy conservation are shown in Fig. 9.4a. We then choose $T = 10$ and $h = 0.04/2^j$ for $j = 0, 1, 2, 3$ with $\varepsilon = 0.2$ and for $j = 2, 3, 4, 5$ with $\varepsilon = 0.1$ and 0.05. The efficiency curves are displayed in Fig. 9.4b. Clearly, the KGDG scheme shows remarkable numerical behaviour. Moreover, it can be observed from Fig. 9.4b that the KGDG method is allowed to take larger time stepsizes than the other methods for this highly oscillatory system.

Problem 9.3 Consider the nonlinear KG equation (see, e.g. [31, 65])

$$\begin{cases} \dfrac{\partial^2 u}{\partial t^2} - a^2 \dfrac{\partial^2 u}{\partial x^2} = bu^3 - au, & -20 \leqslant x \leqslant 20, \ 0 \leqslant t \leqslant T, \quad u(-20, t) = u(20, t), \\[2mm] u(x, 0) = \sqrt{\dfrac{2a}{b}} \, \mathrm{sech}(\lambda x), \quad u_t(x, 0) = c\lambda \sqrt{\dfrac{2a}{b}} \, \mathrm{sech}(\lambda x) \tanh(\lambda x) \end{cases}$$

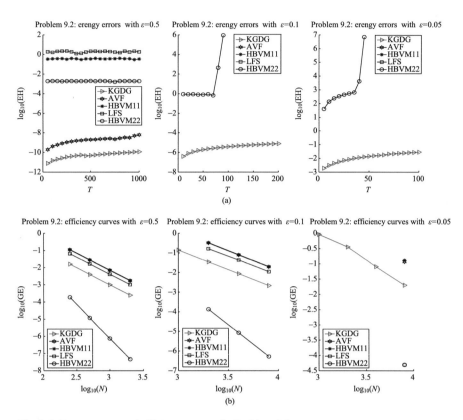

Fig. 9.4 Energy errors and efficiency curves for Problem 9.2

with $\lambda = \sqrt{\dfrac{a}{a^2 - c^2}}$ and $a, b, a^2 - c^2 > 0$. The exact solution of Problem 9.3 is

$$u(x, t) = \sqrt{\frac{2a}{b}} \mathrm{sech}(\lambda(x - ct)),$$

as shown in Fig. 9.5a. In this numerical experiment, we choose the parameters $a = 0.3, b = 1$ and $c = 0.25$. After the Fourier spectral collocation semi-discretisation of operator $\mathscr{A}$ with $M = 200$, we integrate the semidiscrete system with $T = 1000$ and $h = 0.2$. The errors of the semidiscrete energy conservation are presented in Fig. 9.5b. The efficiency curves with $T = 10$ and $h = 0.05/2^j$ for $j = 0, 1, 2, 3$ are shown in Fig. 9.5c. Again, the KGDG scheme gives good numerical behaviour.

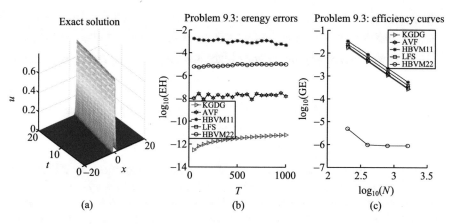

Fig. 9.5 Exact solutions and energy errors and efficiency curves for Problem 9.3

9.7.2 Two-Dimensional Problems

We are now concerned with high-dimensional problems. We next show the remarkable efficiency of the KGDG scheme for high-dimensional nonlinear KG equations by considering two two-dimensional problems.

Problem 9.4 Consider the following two-dimensional nonlinear wave equation (see, e.g. [35])

$$\frac{\partial^2 u(x, y, t)}{\partial t^2} = \Delta u(x, y, t) - u^3(x, y, t), \quad (x, y) \in [-1, 1] \times [-1, 1], \quad t > 0$$

with periodic boundary conditions. The initial conditions are

$$u(x, y, 0) = \text{sech}(10x)\text{sech}(10y), \quad u_t(x, y, 0) = 0.$$

Here, we use spectral elements method (see, e.g. [35]) to semidiscretise the wave equation. The space is discretised with a tensor product Lagrange quadrature formula using $p + 1$ Gauss–Lobatto–Legendre (GLL) quadrature nodes in each space direction. This leads to (see [35] for more details)

$$\tilde{H}(U) = \frac{1}{2} \sum_{j_1=0}^{p} \sum_{j_2=0}^{p} w_{j_1} w_{j_2} \left(\left(\sum_{k=0}^{p} d_{j_1,k} u_{k,j_2} \right)^2 + \left(\sum_{m=0}^{p} d_{j_2,m} u_{j_1,m} \right)^2 + \frac{1}{2} u_{j_1,j_2}^4 \right),$$

where $l_k(x)$ is the k-th Lagrange basis function based on the GLL quadrature nodes $x_0, \cdots, x_p$ with the corresponding quadrature weights $w_0, \cdots, w_p$, and $d_{j_1,k} = \dfrac{\mathrm{d}l_k(x)}{\mathrm{d}x}$. The numerical approximation of the solution can be expressed in the form

$$u_p(x, y, t) = \sum_{k=0}^{p} \sum_{m=0}^{p} u_{k,m}(t) l_k(x) l_m(y)$$

where $u_p(x_{j_1}, y_{j_2}, t) = u_{j_1, j_2}(t)$.

In Fig. 9.6, we show some snapshots of the numerical solutions by KGDG and AVF with a small stepsize $h = 0.005$. In this experiment, the spatial derivatives are discretised with six Gauss Lobatto nodes in each spatial direction and numerical solutions interpolate on an equidistant grid of 26 nodes in each spatial direction.

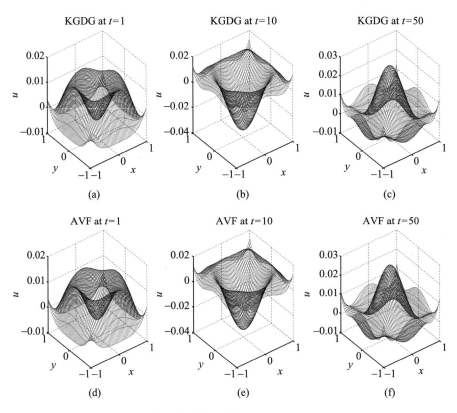

Fig. 9.6 Snapshots of the solution of Problem 9.4

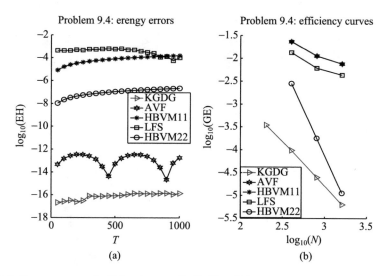

Fig. 9.7 Energy errors and efficiency curves for Problem 9.4

Then the semidiscretised system is integrated with $p = 5$, $h = \dfrac{1}{5}$ and $T = 1000$, and the results of energy conservation are presented in Fig. 9.7a. Finally, we integrate the problem on the interval $[0, 100]$ with $h = \dfrac{1}{2^i}$ for $i = 0, 1, 2, 3$. The global errors are shown in Fig. 9.7b. It can be observed from Fig. 9.7b that the KGDG scheme is the most efficient among the underlying methods.

Problem 9.5 Consider the following two-dimensional sine-Gordon equation:

$$\begin{cases} u_{tt} - (u_{xx} + u_{yy}) = -\sin u, \ (x, y) \in [-1, 1] \times [-1, 1], \ t > 0, \\[2mm] u(x, y, 0) = 4 \arctan\left(\exp\left(\dfrac{4 - \sqrt{(x + 3)^2 + (y + 3)^2}}{0.436}\right)\right), \\[2mm] u_t(x, y, 0) = \dfrac{4.13}{\cosh\left(\exp\left((4 - \sqrt{(x + 3)^2 + (y + 3)^2})/0.436\right)\right)}, \end{cases}$$

with periodic boundary conditions.

Likewise, we use the spectral elements method to semidiscrete the two-dimensional sine-Gordon equation, and space is discretised with a tensor product Lagrange quadrature formula based on 6 GLL quadrature nodes in each spatial direction. We first solve the system with $h = \dfrac{1}{5}$ and $T = 1000$. The results of

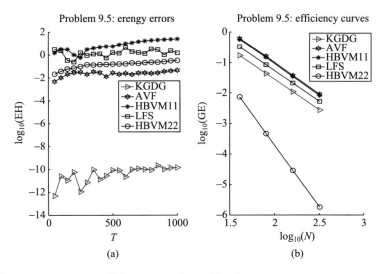

Fig. 9.8 Energy errors and efficiency curves for Problem 9.5

energy conservation are shown in Fig. 9.8a. We then integrate the problem on $[0, 100]$ with $h = \dfrac{1}{2^i}$ for $i = 1, 2, 3, 4$. The global errors are presented in Fig. 9.8b. It can be seen again that the KGDG scheme is the most efficient among these methods.

9.8   Concluding Remarks

Since the concept of energy preservation has far reaching consequences in the physical sciences, many energy-preserving methods have been proposed for ODEs and PDEs. In this chapter, using the blend of the discrete gradient method and the operator-variation-of-constants formula, we have presented a systematic and unified approach to the discretisation of high-dimensional nonlinear KG equations, so that the semidiscrete energy can be preserved precisely. The resulting energy-preserving scheme was analysed in detail for local truncation error, nonlinear stability, convergence and implementations. Moreover, the remarkable efficiency of the energy-preserving scheme was demonstrated by the numerical experiments in comparison with some existing numerical methods in the literature. In this chapter, we also make an effort to bridge the gap between numerical ODEs and PDEs in numerical analysis based on the operator-variation-of-constants formula for nonlinear KG equations. This formula provides insight into the solution to the underlying nonlinear KG equation, which is needed for the numerical analysis. Both the nonlinear stability analysis and convergence analysis are essential in the study of Geometric Numerical Integration for Hamiltonian PDEs. We have addressed this point in this chapter.

Last but definitely not least, compared with symplectic methods, the prominent advantage of energy-preserving methods is that they can preserve the energy of the underlying Hamiltonian system exactly. However, we believe that further exploration is needed. Future research should explore the numerical behaviour of energy-preserving methods in other aspects, such as the long-time numerical conservation of momentum and actions of Hamiltonian PDEs. The key technique for the analysis is modulated Fourier expansion, and we are hopeful of obtaining some interesting results on this important subject. In Chap. 14, we will commence this potentially interesting study.

The material in this chapter is based on the work by Wang and Wu [66].

References

1. Bank R, Graham R L, Stoer J, et al. Hight Order Difference Method for Time Dependent PDE. Berlin, Heidelberg: Springer-Verlag, 2008.
2. Dehghan M, Shokri A. Numerical solution of the nonlinear Klein-Gordon equation using radial basis functions. J. Comput. Appl. Math., 2009, 230: 400–410.
3. Duncan D B. Symplectic finite difference approximations of the nonlinear Klein-Gordon Equation. SIAM J. Numer. Anal., 1997, 34: 1742–1760.
4. Lakestani M, Dehghan M. Collocation and finite difference-collocation methods for the solution of nonlinear Klein-Gordon equation. Comput. Phys. Commun., 2010, 181: 392–1401.
5. Li S, Vu-Quoc L. Finite difference calculus invariant structure of a class of algorithms for the nonlinear Klein-Gordon equation. SIAM J. Numer. Anal., 1995, 32: 1839–1875.
6. Liu C, Wu X. Arbitrarily high-order time-stepping schemes based on the operator spectrum theory for high-dimensional nonlinear Klein-Gordon equations. J. Comput. Phys., 2017, 340: 243–275.
7. Shakeri F, Dehghan M. Numerical solution of the Klein-Gordon equation via He's variational iteration method. Nonl. Dyn., 2008, 51: 89–97.
8. Hairer E, Lubich C. Long-time energy conservation of numerical methods for oscillatory differential equations. SIAM J. Numer. Anal., 2000, 38: 414–441.
9. Hochbruck M, Ostermann A. Exponential integrators. Acta Numer., 2010, 19: 209–286.
10. Mei L, Wu X. Symplectic exponential Runge-Kutta methods for solving nonlinear Hamiltonian systems. J. Comput. Phys., 2017, 338, 567–584.
11. Sanz-Serna J M. Symplectic integrators for Hamiltonian problems: An overview. Acta Numer., 1992, 1: 243–286.
12. Wang B, Iserles A, Wu X. Arbitrary-order trigonometric Fourier collocation methods for multi-frequency oscillatory systems. Found. Comput. Math., 2016, 16: 151–181.
13. Wang B, Wu X, Meng F. Trigonometric collocation methods based on Lagrange basis polynomials for multi-frequency oscillatory second-order differential equations. J. Comput. Appl. Math., 2017, 313: 185–201.
14. Wang B, Yang H, Meng F. Sixth order symplectic and symmetric explicit ERKN schemes for solving multi-frequency oscillatory nonlinear Hamiltonian equations. Calcolo, 2017, 54: 117–140.
15. Wang B, Meng F, Fang Y. Efficient implementation of RKN-type Fourier collocation methods for second-order differential equations. Appl. Numer. Math., 2017, 119: 164–178.
16. Wang B, Wu X, Meng F, et al. Exponential Fourier collocation methods for solving first-order differential equations. J. Comput. Math., 2017, 35: 711–736.
17. Wu X, Wang B, Xia J. Explicit symplectic multidimensional exponential fitting modified Runge-Kutta-Nyström methods. BIT Numer. Math., 2012, 52: 773–795.

18. Hairer E, Lubich C, Wanner G. Geometric Numerical Integration: Structure-Preserving Algorithms for Ordinary Differential Equations. 2nd ed. Berlin, Heidelberg: Springer-Verlag, 2006.
19. Wu X, Wang B. Recent Developments in Structure-Preserving Algorithms for Oscillatory Differential Equations. Singapore: Springer Nature Singapore Pte Ltd., 2018.
20. Wu X, Liu K, Shi W. Structure-Preserving Algorithms for Oscillatory Differential Equations II. Heidelberg: Springer-Verlag, 2015.
21. Wu X, You X, Wang B. Structure-Preserving Algorithms for Oscillatory Differential Equations. Berlin, Heidelberg: Springer-Verlag, 2013.
22. Bridges T J. Multi-symplectic structures and wave propagation. Math. Proc. Cambridge Philos. Soc., 1997, 121, 147–190.
23. Bridges T J, Reich S. Numerical methods for Hamiltonian PDEs. J. Phys. A: Math. Gen., 2006, 39: 5287–5320.
24. Feng K, Qin M. The Symplectic Methods for the Computation of Hamiltonian Equations//Numerical Methods for Partial Differential Equations. Berlin, Heidelberg: Springer, 2006: 1–37.
25. Frank J, Moore B E, Reich S. Linear PDEs and numerical methods that preserve a multi symplectic conservation law. SIAM J. Sci. Comput., 2006, 28: 260–277.
26. Hu W, Deng Z, Han S, et al. Generalized multi-symplectic integrators or a class of Hamiltonian nonlinear wave PDEs. J. Comput. Phys., 2013, 235: 394–406.
27. Li Y W, Wu X. General local energy-preserving integrators for solving multi symplectic Hamiltonian PDEs. J. Comput. Phys., 2015, 301: 141–166.
28. Shi W, Wu X, Xia J. Explicit multi-symplectic extended leap-frog methods for Hamiltonian wave equations. J. Comput. Phys., 2012, 231: 7671–7694.
29. Bratsos, A G. The solution of the two-dimensional sine-Gordon equation using the method of lines. J. Comput. Appl. Math., 2007, 206: 251–277.
30. Hesthaven J S, Gottlieb S, Gottlieb D. Spectral Methods for Time Dependent Problems//Cambridge Monographs on Applied and Computational Mathematics. Cambridge: Cambridge University Press, 2007.
31. Liu C, Shi W, Wu X. An efficient high-order explicit scheme for solving Hamiltonian nonlinear wave equations. Appl. Math. Comput., 2014, 246: 696–710.
32. Schiesser W. The Numerical Methods of Lines: Integration of Partial Differential Equation. San Diego: Academic Press, 1991.
33. Wu X, Liu C, Mei L. A new framework for solving partial differential equations using semi-analytical explicit RK(N)-type integrators. J. Comput. Appl. Math., 2016, 301: 74–90.
34. Courant R, Friedrichs K, Lewy H. Über die partiellen differenzengleichungen der mathematischen physik. (German) Math. Ann., 1928, 100: 32–74.
35. Celledoni E, Grimm V, Mclachlan R I, et al. Preserving energy resp. dissipation in numerical PDEs using the "Average Vector Field" method. J. Comput. Phys., 2012, 231: 6770–6789.
36. Cohen D, Hairer E, Lubich C. Conservation of energy, momentum and actions in numerical discretizations of non-linear wave equations. Numer. Math., 2008, 110: 113–143.
37. Matsuo T. New conservative schemes with discrete variational derivatives for nonlinear wave equations. J. Comput. Appl. Math., 2007, 203: 32–56.
38. Matsuo T, Yamaguchi H. An energy-conserving Galerkin scheme for a class of nonlinear dispersive equations. J. Comput. Phys., 2009, 228: 4346–4358.
39. Mei L, Liu C, Wu X. An essential extension of the finite-energy condition for extended Runge-Kutta-Nyström integrators when applied to nonlinear wave equations. Commun. Comput. Phys., 2017, 22: 742–764.
40. Mclachlan R I, Quispel G R W. Discrete gradient methods have an energy conservation law. Discrete Contin. Dyn. Syst., 2014, 34: 1099–1104.
41. Gonzalez O. Time integration and discrete Hamiltonian systems. J. Nonlinear Sci., 1996, 6: 449–467.
42. Li Y W, Wu X. Exponential integrators preserving first integrals or Lyapunov functions for conservative or dissipative systems. SIAM J. Sci. Comput., 2016, 38: 1876–1895.

43. Mclachlan R I, Quispel G R W. Discrete gradient methods have an energy conservation law. Discrete Contin. Dyn. Syst., 2014, 34: 1099–1104.
44. Mclachlan R I, Quispel G R W, Robidoux N. Geometric integration using discrete gradient. Philos. Trans. R. Soc. Lond. A, 1999, 357: 1021–1045.
45. Betsch P, Steinmann P. Inherently energy conserving time finite elements for classical mechanics. J. Comput. Phys., 2000, 160: 88–116.
46. Betsch P, Steinmann P. Conservation properties of a time FE method. Part I: Time-stepping schemes for N-body problems. Int. J. Numer. Methods Eng., 2000, 49: 599–638.
47. Celledoni E, Mclachlan R I, Mclaren D I, et al. Energy-preserving Runge-Kutta methods. M2AN Math. Model. Numer. Anal., 2009, 43: 645–649.
48. Celledoni E, Owren B, Sun Y. The minimal stage, energy preserving Runge-Kutta method for polynomial Hamiltonian systems is the averaged vector field method. Math. Comput., 2014, 83: 1689–1700.
49. Quispel G R W, Mclaren D I. A new class of energy-preserving numerical integration methods. J. Phys. A, 2008, 41: 045206.
50. Brugnano L, Frasca Caccia G, Iavernaro F. Hamiltonian Boundary Value Methods (HBVMs) and their efficient implementation. Math. Eng. Sci. Aero. MESA, 2014, 5: 343–411.
51. Brugnano L, Iavernaro F, Trigiante D. Analysis of Hamiltonian Boundary Value Methods (HBVMs): A class of energy-preserving Runge-Kutta methods for the numerical solution of polynomial Hamiltonian systems. Commun. Nonl. Sci. Numer. Simul., 2015, 20: 650–667.
52. Wang B, Wu X. A new high precision energy-preserving integrator for system of oscillatory second-order differential equations. Phys. Lett. A, 2012, 376: 1185–1190.
53. Wu X, Wang B, Shi W. Efficient energy preserving integrators for oscillatory Hamiltonian systems. J. Comput. Phys., 2013, 235: 587–605.
54. Dahlby M, Owren B. A general framework for deriving integral preserving numerical methods for PDEs. SIAM J. Sci. Comput., 2010, 33: 2318–2340.
55. Šolin P. Partial Differential Equations and the Finite Element Method. Pure and Applied Mathematics. New York: Wiley-Interscience, 2006.
56. Brugnano L, Frasca Caccia G, Iavernaro F. Energy conservation issues in the numerical solution of the semilinear wave equation. Appl. Math. Comput., 2015, 270: 842–870.
57. Liu C, Wu X. An energy-preserving and symmetric scheme for nonlinear Hamiltonian wave equations. J. Math. Anal. Appl., 2016, 440: 167–182.
58. Liu K, Wu X, Shi W. A linearly-fitted conservative (dissipative) scheme for efficiently solving conservative (dissipative) nonlinear wave PDEs. J. Comput. Math., 2017, 35: 780–800.
59. Wu X, Mei L, Liu C. An analytical expression of solutions to nonlinear wave equations in higher dimensions with Robin boundary conditions. J. Math. Anal. Appl., 2015, 426: 1164–1173.
60. Liu C, Iserles A, Wu X. Symmetric and arbitrarily high-order Birkhoff Hermite time integrators and their long-time behavior for solving nonlinear Klein-Gordon equations. J. Comput. Phys., 2018, 356: 1–30.
61. Liu C, Wu X. The boundness of the operator-valued functions for multidimensional nonlinear wave equations with applications. Appl. Math. Lett., 2017, 74: 60–67.
62. Liu C, Shi W, Wu X. An extended discrete gradient formula for oscillatory Hamiltonian systems. J. Phys. A-Math. Theor., 2013, 46: 165–203.
63. Bank R, Graham R L, Stoer J, et al. Hight Order Difference Method for Time Dependent PDE. Berlin, Heidelberg: Springer-Verlag, 2008.
64. Shen J, Tang T, Wang L L. Spectral Methods: Algorithms, Analysis, Applications. Berlin: Springer, 2011.
65. Schiesser W E, Griffiths G W. A Compendium of Partial Differential Equation Models: Method of Lines Analysis with Matlab. Cambridge, New York: Cambridge University Press, 2009.
66. Wang B, Wu X. The formulation and analysis of energy-preserving schemes for solving high-dimensional nonlinear Klein-Gordon equations. IMA J. Numer. Anal., 2019, 39: 2016–2044.

Chapter 10
High-Order Symmetric Hermite–Birkhoff Time Integrators for Semilinear KG Equations

The computation of the Klein–Gordon equation featuring a nonlinear potential function is of great importance in a wide range of application areas in science and engineering. It represents major challenges because of the nonlinear potential. The main aim of this chapter is to present symmetric and arbitrarily high-order time-stepping integrators and analyse their stability, convergence and long-time behaviour for the semilinear Klein–Gordon equation. To achieve this, under the assumption of periodic boundary conditions, an abstract ordinary differential equation (ODE) and its operator-variation-of-constants formula are formulated on a suitable function space based on operator spectrum theory. By applying a two-point Hermite–Birkhoff interpolation to the nonlinear integrals that appear in the operator-variation-of-constants formula, as a result, a suitable spatial discretisation leads to the fully discrete scheme, which needs only a weak temporal smoothness assumption.

10.1 Introduction

It is well known that the nonlinear wave equation plays a prominent role in a wide range of applications in engineering and science, including nonlinear optics, solid state physics and quantum field theory [1]. Most importantly, the *Klein–Gordon (KG) equation,* a relativistic counterpart of the Schrödinger equation, is used to model diverse nonlinear phenomena, such as the propagation of dislocations in crystals and the behaviour of elementary particles and of Josephson junctions (see Chap. 2 in [2] for details). Numerical computations play an important role in the study of nonlinear waves. We here restrict ourselves to the one-dimensional case, although all ideas, algorithms and analysis described in this chapter can be easily extended to the solution of semilinear KG equations in a moderate number of space dimensions.

© The Author(s), under exclusive license to Springer Nature Singapore Pte Ltd. 2021 299
X. Wu, B. Wang, *Geometric Integrators for Differential Equations with Highly Oscillatory Solutions*, https://doi.org/10.1007/978-981-16-0147-7_10

We now consider the following semilinear KG equation in a single space variable:

$$\begin{cases} u_{tt} - a^2 \Delta u = f(u), & t_0 < t \leqslant T, \quad x \in \Omega, \\ u(x, t_0) = \varphi_1(x), & u_t(x, t_0) = \varphi_2(x), \; x \in \bar{\Omega}, \end{cases} \tag{10.1}$$

where $u(x, t)$ represents the wave displacement at position x and time t, and the nonlinear function $f(u)$ is the negative derivative of a potential energy $V(u) \geqslant 0$. Here it is assumed that the initial value problem (10.1) is equipped with the periodic boundary conditions on the domain $\Omega = (-\pi, \pi)$,

$$u(x, t) = u(x + 2\pi, t), \qquad x \in (-\pi, \pi], \tag{10.2}$$

where 2π is the fundamental period with respect to x. The semilinear KG equation (10.1) is used to model many different nonlinear phenomena, including the propagation of dislocations in crystals and the behaviour of elementary particles and of Josephson junctions (see Chap. 8.2 in [2] for details). In general, it has also been the subject of detailed investigation in studies of solitons and in nonlinear science. In the literature, there are various choices of the potential $f(u)$. Among typical examples is the best known sine-Gordon equation

$$u_{tt} - a^2 \Delta u + \sin(u) = 0,$$

and it also appears with polynomial $f(u)$, and other nonlinear functions. Another point is that, if $u(\cdot, t) \in H^1(\Omega)$ and $u_t(\cdot, t) \in L^2(\Omega)$, the energy conservation is a key feature of the KG equation (10.1) with periodic boundary condition (10.2), that is

$$E(t) = \frac{1}{2} \int_\Omega \left(u_t^2 + a^2 |\nabla u|^2 + 2V(u) \right) dx \equiv E(t_0). \tag{10.3}$$

This is an essential property in the theory of solitons. Therefore, it is also very important to test the effectiveness of a numerical method for (10.1) for the preservation of the corresponding discrete energy.

The KG equation has received much attention in both its numerical and analytical aspects. With regard to analytical issues, the initial value problem (10.1) was investigated by many authors (see, e.g. [3–7]). In particular, for the defocusing case, $V(u) \geqslant 0, u \in \mathbb{R}$, the global existence of solutions was established in [3], and for the focusing case, $V(u) \leqslant 0, u \in \mathbb{R}$, possible finite time blow-up was investigated. In numerical analysis, various solution procedures have been proposed and studied including classical finite difference methods such as explicit, semi-implicit, compact finite difference and symplectic conservative discretisations [8–12]. Other effective numerical methods, such as the finite element method and the spectral method were also studied in [13–16]. Although various numerical methods for the semilinear KG equation have been derived and investigated in the literature, their accuracy is limited, and little attention has been paid to the special structure brought by spatial discretisations.

It is known that recent interest in exponential integrators for semilinear parabolic problems has led to the development of numerical schemes (see, e.g. [17–21]). Motivated by this and based on the operator spectrum theory (see, e.g. [22]), we first formulate the nonlinear KG equation (10.1)–(10.2) as an abstract second-order ordinary differential equation. Then, the operator-variation-of-constants formula (also is termed the *Duhamel Principle*) for the abstract equation is introduced, which is in fact an implicit expression of the solution of the semilinear KG equation (see [23]). In a similar way to the useful approach to dealing with the semiclassical Schrödinger equation in [24], we forego the standard steps, of first semidiscretising and then dealing with the semidiscretisation, in a totally different approach which greatly reduces the requirement of the smoothness with respect to time. Employing the operator-variation-of-constants formula, we interpolate the nonlinear integrators by two-point Hermite interpolation, and then a class of symmetric and arbitrarily high-order time integration formulae is derived and analysed. In fact, the space semidiscretisation is deferred to the very last moment, and this helps us take a subtle but powerful advantage of dealing with the undiscretised operator Δ and incorporate the special structure brought by spatial discretisations into the underlying numerical integrator.

10.2   The Symmetric and High-Order Hermite–Birkhoff Time Integration Formula

In this section, using operator theory (see, e.g. [22]), we firstly formulate the nonlinear problem (10.1)–(10.2) as an abstract ordinary differential equation on the Hilbert space $L^2(\Omega)$. Then, the operator-variation-of-constants formula for the abstract equation is presented, which is in fact an implicit expression of the solution for the system (see, e.g. [23, 25]). Keeping the eventual discretisation in mind and applying Hermite–Birkhoff interpolation to the operator-variation-of-constants formula, we will present a class of symmetric and arbitrarily high-order time integrators in a suitable infinite-dimensional function space.

10.2.1   The Operator-Variation-of-Constants Formula

In this subsection, we start with recalling the abstract second-order ordinary differential equation and its operator-variation-of-constants formula (see [23]) before considering the design of the numerical integrators,.

To formulate an abstract formulation for the problem (10.1)–(10.2), we first consider the differential operator $\mathscr{A}$ defined by

$$(\mathscr{A}v)(x) = -a^2 v_{xx}(x),$$

where $\mathscr{A}$ is a linear, unbounded positive semi-definite operator, whose domain is

$$D(\mathscr{A}) := \left\{ v \in H^1(\Omega) : v(x) = v(x + 2\pi) \right\}.$$

Clearly, the operator $\mathscr{A}$ has a complete system of orthogonal eigenfunctions $\left\{ e^{ikx} : k \in \mathbb{Z} \right\}$. The linear span of all these eigenfunctions

$$X := \mathrm{lin}\left\{ e^{ikx} : k \in \mathbb{Z} \right\} \tag{10.4}$$

is dense in the Hilbert space $L^2(\Omega)$. Thus, we obtain the orthonormal basis of eigenvectors of the operator $\mathscr{A}$ with the corresponding eigenvalues $a^2 k^2$ for $k \in \mathbb{Z}$.

We next introduce the functions as follows:

$$\phi_j(x) := \sum_{k=0}^{\infty} \frac{(-1)^k x^k}{(2k+j)!}, \qquad j \in \mathbb{N} \quad \text{for} \quad \forall x \geqslant 0. \tag{10.5}$$

It is easy to see that the functions ϕ_j for $j = 0, 1, 2, \cdots$ are bounded for any $x \geqslant 0$. For instance,

$$\phi_0(x) = \cos(\sqrt{x}), \quad \phi_1(x) = \mathrm{sinc}(\sqrt{x}),$$

and it is obvious that $|\phi_j(x)| \leqslant 1$ for $j = 0, 1$ and $\forall x \geqslant 0$. These functions (10.5) can induce the bounded operators

$$\phi_j(t\mathscr{A}) : L^2(\Omega) \to L^2(\Omega)$$

for $j \in \mathbb{N}$ and $t_0 \leqslant t \leqslant T$:

$$\phi_j(t\mathscr{A})v(x) = \sum_{k=-\infty}^{\infty} \hat{v}_k \phi_j(ta^2 k^2) e^{ikx} \quad \text{for} \quad v(x) = \sum_{k=-\infty}^{\infty} \hat{v}_k e^{ikx}. \tag{10.6}$$

The boundedness follows from the definition of the operator norm that

$$\|\phi_j(t\mathscr{A})\|_*^2 = \sup_{\|v\| \neq 0} \frac{\|\phi_j(t\mathscr{A})v\|^2}{\|v\|^2} \leqslant \sup_{t_0 \leqslant t \leqslant T} |\phi_j(ta^2 k^2)|^2 \leqslant \gamma_j^2, \tag{10.7}$$

where $\| \cdot \|_*$ is the Sobolev norm $\| \cdot \|_{L^2(\Omega) \leftarrow L^2(\Omega)}$, and γ_j for $j \in \mathbb{N}$ are the bounds of the functions $|\phi_j(x)|$ for $j \in \mathbb{N}$ and $x \geqslant 0$.

In what follows, we define $u(t)$ as the function that maps x to $u(x, t)$:

$$u(t) := [x \mapsto u(x, t)].$$

The system (10.1)–(10.2) can be formulated as an abstract second-order ordinary differential equation

$$\begin{cases} u''(t) + \mathscr{A}u(t) = f\big(u(t)\big), & t_0 < t \leqslant T, \\ u(t_0) = \varphi_1(x), & u'(t_0) = \varphi_2(x), \end{cases} \tag{10.8}$$

on the closed subspace

$$\mathscr{X} := \left\{ u(x, \cdot) \in X \mid u(x, \cdot) \text{ satisfies the corresponding boundary conditions} \right\}$$

$$\subseteq L^2(\Omega). \tag{10.9}$$

The next theorem characterizes the solution of the abstract second-order ordinary differential equation (10.8) (see [23]).

Theorem 10.1 *The solution of* (10.8) *and its derivative satisfy the following operator-variation-of-constants formula*

$$\begin{cases} u(t) = \phi_0\big((t - t_0)^2 \mathscr{A}\big)u(t_0) + (t - t_0)\phi_1\big((t - t_0)^2 \mathscr{A}\big)u'(t_0) \\ \qquad + \displaystyle\int_{t_0}^t (t - \zeta)\phi_1\big((t - \zeta)^2 \mathscr{A}\big) f\big(u(\zeta)\big)\mathrm{d}\zeta, \\ u'(t) = -(t - t_0)\mathscr{A}\phi_1\big((t - t_0)^2 \mathscr{A}\big)u(t_0) + \phi_0\big((t - t_0)^2 \mathscr{A}\big)u'(t_0) \\ \qquad + \displaystyle\int_{t_0}^t \phi_0\big((t - \zeta)^2 \mathscr{A}\big) f\big(u(\zeta)\big)\mathrm{d}\zeta, \end{cases} \tag{10.10}$$

for $t_0 \leqslant t \leqslant T$, *where both* $\phi_0\big((t - t_0)^2 \mathscr{A}\big)$ *and* $\phi_1\big((t - t_0)^2 \mathscr{A}\big)$ *are bounded operators.*

10.2.2   The Formulation of the Time Integrators

According to the operator-variation-of-constants formula (10.12) and the two-point Hermite interpolation, we develop a class of arbitrarily high-order and symmetric time integration formulae. We start with a few useful preliminaries.

Lemma 10.1 *The bounded functions* $\phi_j(\mathscr{A})$, $j \in \mathbb{N}$ *of the operator* $\mathscr{A}$ *reduced by* (10.5) *satisfy*

$$\int_0^1 (1-z)\phi_1\big((1-z)^2\mathscr{A}\big)z^j\,\mathrm{d}z = j!\phi_{j+2}(\mathscr{A}), \qquad j \in \mathbb{N},$$

$$\int_0^1 \phi_0\big((1-z)^2\mathscr{A}\big)z^j\,\mathrm{d}z = j!\phi_{j+1}(\mathscr{A}), \qquad j \in \mathbb{N}. \tag{10.11}$$

Proof The first formula can be proved as follows

$$\int_0^1 (1-z)\phi_1\big((1-z)^2\mathscr{A}\big)z^j\,\mathrm{d}z = \sum_{k=0}^{\infty} \frac{(-1)^k \int_0^1 (1-z)^{2k+1} z^j\,\mathrm{d}z}{(2k+1)!}\mathscr{A}^k$$

$$= \sum_{k=0}^{\infty} \frac{(-1)^k j!}{(2k+j+2)!}\mathscr{A}^k = j!\phi_{j+2}(\mathscr{A}).$$

Likewise, we can obtain the second formula. □

Corollary 10.1 *For every* $m, n \in \mathbb{N}$ *the operators* (10.6) *satisfy*

$$\int_0^1 (1-z)^{m+1}\phi_1\big((1-z)^2\mathscr{A}\big)z^n\,\mathrm{d}z = \sum_{i=0}^{m} C_m^i(-1)^{m-i}(m+n-i)!\phi_{m+n-i+2}(\mathscr{A}),$$

$$\int_0^1 (1-z)^m \phi_0\big((1-z)^2\mathscr{A}\big)z^n\,\mathrm{d}z = \sum_{i=0}^{m} C_m^i(-1)^{m-i}(m+n-i)!\phi_{m+n-i+1}(\mathscr{A}),$$

where $C_m^i = \binom{m}{i}$ *is the binomial symbol.*

Proof We only prove the first formula

$$\int_0^1 (1-z)^{m+1}\phi_1\big((1-z)^2\mathscr{A}\big)z^n\,\mathrm{d}z$$

$$= \sum_{i=0}^{m} C_m^i(-1)^{m-i}\int_0^1 (1-z)\phi_1\big((1-z)^2\mathscr{A}\big)z^{m+n-i}\,\mathrm{d}z$$

$$= \sum_{i=0}^{m} C_m^i(-1)^{m-i}(m+n-i)!\phi_{m+n-i+2}(\mathscr{A}).$$

Likewise, the second formula can be obtained. □

It follows from Theorem 10.1 that the solution of (10.8) and its derivative at a time point $t_{n+1} = t_n + \Delta t$, $n \in \mathbb{N}$ are

$$
\begin{cases}
u(t_{n+1}) = \phi_0(\mathcal{V})u(t_n) + \Delta t \phi_1(\mathcal{V})u'(t_n) + \Delta t^2 \displaystyle\int_0^1 (1-z)\phi_1\big((1-z)^2\mathcal{V}\big)\tilde{f}(z)\mathrm{d}z, \\[2ex]
u'(t_{n+1}) = -\Delta t \mathcal{A}\phi_1(\mathcal{V})u(t_n) + \phi_0(\mathcal{V})u'(t_n) + \Delta t \displaystyle\int_0^1 \phi_0\big((1-z)^2\mathcal{V}\big)\tilde{f}(z)\mathrm{d}z,
\end{cases}
$$

$$(10.12)$$

where $\mathcal{V} = \Delta t^2 \mathcal{A}$ and $\tilde{f}(z) = f\big(u(t_n + z\Delta t)\big)$. Clearly, in order to obtain the time integration formula from (10.12), we need to consider efficient integrators for approximating the nonlinear integrals

$$
\begin{aligned}
I_1 &:= \int_0^1 (1-z)\phi_1\big((1-z)^2\mathcal{V}\big)\tilde{f}(z)\mathrm{d}z, \\[1ex]
I_2 &:= \int_0^1 \phi_0\big((1-z)^2\mathcal{V}\big)\tilde{f}(z)\mathrm{d}z.
\end{aligned}
$$

$$(10.13)$$

Usually, the potential function $f(u)$ is nonlinear, and only the endpoints' information can be used directly when we deal with the two nonlinear integrals in (10.13) and design numerical methods. Accordingly, we are particularly concerned with fitting function values and derivatives at the two boundary points of the finite interval $[0, 1]$. This motivates us to interpolate $\tilde{f}(z)$ by a two-point Hermite interpolation $p_r(z)$ of degree $2r + 1$ (see, e.g. [26, 27]).

Lemma 10.2 *Assume that* $\tilde{f} \in C^{2r+2}([0, 1])$. *Then there exists a Hermite interpolating polynomial* $p_r(z)$ *of degree* $2r + 1$

$$
p_r(z) = \sum_{j=0}^{r} \Big[\beta_{r,j}(z)\tilde{f}^{(j)}(0) + (-1)^j \beta_{r,j}(1-z)\tilde{f}^{(j)}(1) \Big],
$$

$$(10.14)$$

satisfying the interpolation conditions

$$
p_r^{(j)}(0) = \tilde{f}^{(j)}(0), \quad p_r^{(j)}(1) = \tilde{f}^{(j)}(1), \qquad j = 0, 1, 2, \cdots, r,
$$

where

$$
\beta_{r,j}(z) = \frac{z^j}{j!}(1-z)^{r+1} \sum_{s=0}^{r-j} C_{r+s}^s z^s,
$$

$$(10.15)$$

and the error on [0, 1] *is*

$$R_r = \tilde{f}(z) - p_r(z) = (-1)^{r+1} z^{r+1} (1-z)^{r+1} \frac{\tilde{f}^{(2r+2)}(\xi)}{(2r+2)!}, \quad \xi \in (0, 1). \quad (10.16)$$

Replacing $\tilde{f}(z)$ in (10.13) by the Hermite interpolation $p_r(z)$ where $\tilde{f}(z) = f\big(u(t_n + z\Delta t)\big)$ and $\tilde{f}^{(j)}(z) = \Delta t^j f_t^{(j)}\big(u(t_n + z\Delta t)\big)$ yields

$$\tilde{I}_1^r = \sum_{j=0}^{r} \Delta t^j \Big[I_1[\beta_{r,j}(z)] f_t^{(j)}\big(u(t_n)\big) + (-1)^j I_1[\beta_{r,j}(1-z)] f_t^{(j)}\big(u(t_{n+1})\big) \Big],$$

$$\tilde{I}_2^r = \sum_{j=0}^{r} \Delta t^j \Big[I_2[\beta_{r,j}(z)] f_t^{(j)}\big(u(t_n)\big) + (-1)^j I_2[\beta_{r,j}(1-z)] f_t^{(j)}\big(u(t_{n+1})\big) \Big],$$

$$(10.17)$$

where $f_t^{(j)}\big(u(t)\big)$ denotes the j-th derivative of $f\big(u(t)\big)$ with respect to t. In terms of the Hermite–Birkhoff quadrature formula (see, e.g. [28–30]), we will determine the coefficients $I_1[\beta_j(z)]$, $I_2[\beta_j(z)]$, $I_1[\beta_j(1-z)]$ and $I_2[\beta_j(1-z)]$. These coefficients are given by

$$I_1[\beta_{r,j}(z)] := \int_0^1 (1-z)\phi_1\big((1-z)^2 \mathcal{V}\big)\beta_{r,j}(z)\,dz$$

$$= \sum_{s=0}^{r-j} \sum_{i=0}^{r+1} (-1)^{r-i+1} C_{r+s}^s C_{r+1}^i \frac{(r+s+j-i+1)!}{j!} \phi_{r+s+j-i+3}(\mathcal{V}), \quad (10.18)$$

$$I_2[\beta_{r,j}(z)] := \int_0^1 \phi_0\big((1-z)^2 \mathcal{V}\big)\beta_{r,j}(z)\,dz$$

$$= \sum_{s=0}^{r-j} \sum_{i=0}^{r+1} (-1)^{r-i+1} C_{r+s}^s C_{r+1}^i \frac{(r+s+j-i+1)!}{j!} \phi_{r+s+j-i+2}(\mathcal{V}), \quad (10.19)$$

$$I_1[\beta_{r,j}(1-z)] := \int_0^1 (1-z)\phi_1\big((1-z)^2 \mathcal{V}\big)\beta_{r,j}(1-z)\,dz$$

$$= \sum_{s=0}^{r-j} \sum_{i=0}^{r+j} (-1)^{s+j-i} C_{r+s}^s C_{s+j}^i \frac{(r+s+j-i+1)!}{j!} \phi_{r+s+j-i+3}(\mathcal{V}), \quad (10.20)$$

$$I_2[\beta_{r,j}(1-z)] := \int_0^1 \phi_0\big((1-z)^2 \mathcal{V}\big)\beta_{r,j}(1-z)\,dz$$

$$= \sum_{s=0}^{r-j} \sum_{i=0}^{r+j} (-1)^{s+j-i} C_{r+s}^s C_{s+j}^i \frac{(r+s+j-i+1)!}{j!} \phi_{r+s+j-i+2}(\mathcal{V}). \quad (10.21)$$

From the definitions stated above, it is evident that the coefficients are bounded for any $j = 0, 1, \cdots, r$,

$$\|I_1[\beta_{r,j}(z)]\|_* \leqslant \max_{0 \leqslant z \leqslant 1} |\beta_{r,j}(z)| \leqslant 1 \quad \text{and} \quad \|I_1[\beta_{r,j}(1-z)]\|_*$$

$$\leqslant \max_{0 \leqslant z \leqslant 1} |\beta_{r,j}(1-z)| \leqslant 1,$$

$$\|I_2[\beta_{r,j}(z)]\|_* \leqslant \max_{0 \leqslant z \leqslant 1} |\beta_{r,j}(z)| \leqslant 1 \quad \text{and} \quad \|I_2[\beta_{r,j}(1-z)]\|_*$$

$$\leqslant \max_{0 \leqslant z \leqslant 1} |\beta_{r,j}(1-z)| \leqslant 1.$$

Suppose that the following approximations have been given

$$u^n \approx u(t_n) \quad \text{and} \quad \mu^n \approx u'(t_n).$$

On the basis of the above analysis and the formula (10.12), we present the following time integration formula for the abstract ODE (10.8).

Definition 10.1 The Hermite–Birkhoff (HB) time integration formula for solving the abstract ODE (10.8) is defined by

$$
\begin{cases}
u^{n+1} = \phi_0(\mathcal{V})u^n + \Delta t \phi_1(\mathcal{V})\mu^n \\
\qquad + \sum_{j=0}^{r} \Delta t^{j+2} \left\{ I_1[\beta_{r,j}(z)] f_t^{(j)}(u^n) + (-1)^j I_1[\beta_{r,j}(1-z)] f_t^{(j)}(u^{n+1}) \right\}, \\
\mu^{n+1} = - \Delta t \mathscr{A} \phi_1(\mathcal{V})u^n + \phi_0(\mathcal{V})\mu^n \\
\qquad + \sum_{j=0}^{r} \Delta t^{j+1} \left\{ I_2[\beta_{r,j}(z)] f_t^{(j)}(u^n) + (-1)^j I_2[\beta_{r,j}(1-z)] f_t^{(j)}(u^{n+1}) \right\},
\end{cases}
$$

$$(10.22)$$

where $I_1[\beta_{r,j}(z)]$, $I_2[\beta_{r,j}(z)]$, $I_1[\beta_{r,j}(1-z)]$ and $I_2[\beta_{r,j}(1-z)]$ have been defined by (10.18)–(10.21), respectively.

Remark 10.1 The HB time integration formula (10.22) is derived by using a two-point Hermite interpolation to approximate the nonlinear function $\tilde{f}(z)$ appearing in the nonlinear integrals (10.13). Here, the high order derivatives $\dfrac{d^m \tilde{f}(z)}{dz^m}$ will be used. Fortunately, the high order derivative $u^{(m)}(t_n + z\Delta t)$ can be calculated from lower order derivative via the abstract equation (10.8), namely,

$$\frac{d^m}{dz^m} u(t_n + z\Delta t) = \frac{d^{m-2}}{dz^{m-2}}\left(- \mathscr{A}u(t_n + z\Delta t) + f\left(u(t_n + z\Delta t)\right)\right)\Delta t^2, \quad m \geqslant 2.$$

Hence, the high order derivatives $\dfrac{d^m \tilde{f}(z)}{dz^m}$ satisfy the following recursive relationship

$$\tilde{f}'(z) = f'\big(u(t_n + z\Delta t)\big)u'(t_n + z\Delta t)\Delta t,$$

$$\frac{d^m \tilde{f}(z)}{dz^m} = \frac{d^{m-2}}{dz^{m-2}}\left\{ f''\big(u(t_n + z\Delta t)\big)\big(u'(t_n + z\Delta t)\big)^2 \right.$$

$$\left. + f'\big(u(t_n + z\Delta t)\big)\big(-\mathscr{A}u(t_n + z\Delta t) + f\big(u(t_n + z\Delta t)\big)\big)\right\}\Delta t^2, \quad m \geqslant 2.$$

This means that the high order derivatives $u^{(m)}(\cdot)$ for $m \geqslant 2$ will not be affected in the HB time integration formula (10.22).

Concerning the local error bounds of the formula (10.22), we have the following theorem.

Theorem 10.2 *Assume that* $f\big(u(\cdot, t)\big) \in C^{2r+2}([t_0, T])$ *and* $f_t^{(2r+2)}\big(u(x, \cdot)\big) \in L^2(\Omega)$. *Under the local assumptions of* $u^n = u(t_n)$, $\mu^n = u'(t_n)$, *the local error bounds of the HB time integration formula* (10.22) *are*

$$\|u(t_{n+1}) - u^{n+1}\| \leqslant C_1\Delta t^{2r+4} \quad and \quad \|u'(t_{n+1}) - \mu^{n+1}\| \leqslant C_2\Delta t^{2r+3},$$
$$(10.23)$$

where the constants C_1 *and* C_2 *are given by*

$$C_1 = \frac{(r+2)!(r+1)!}{(2r+2)!(2r+4)!} \max_{t_0 \leqslant t \leqslant T} \left\| f_t^{(2r+2)}\big(u(t)\big)\right\|$$

and

$$C_2 = \frac{[(r+1)!]^2}{(2r+2)!(2r+3)!} \max_{t_0 \leqslant t \leqslant T} \left\| f_t^{(2r+2)}\big(u(t)\big)\right\|.$$

Proof Using (10.12) and (10.22), we obtain

$$u(t_{n+1}) - u^{n+1} = \Delta t^2 \int_0^1 (1-z)\phi_1\big((1-z)^2\mathscr{V}\big)\Big[f\big(u(t_n + z\Delta t)\big) - p_r(z)\Big]dz,$$
$$(10.24)$$

and

$$u'(t_{n+1}) - \mu^{n+1} = \Delta t \int_0^1 \phi_0\big((1-z)^2\mathscr{V}\big)\Big[f\big(u(t_n + z\Delta t)\big) - p_r(z)\Big]dz.$$
$$(10.25)$$

As $\tilde{f}^{(j)}(z) = \Delta t^j f_t^{(j)}\big(u(t_n + z\Delta t)\big)$, it follows from Lemma 10.2 that

$$f\big(u(t_n + z\Delta t)\big) - p_r(z) = \Delta t^{2r+2}(-1)^{r+1}z^{r+1}(1-z)^{r+1}\frac{f_t^{(2r+2)}\big(u(t_n + \xi^n \Delta t)\big)}{(2r+2)!}.$$

(10.26)

Then inserting (10.26) into (10.24) and (10.25) yields

$$\|u(t_{n+1}) - u^{n+1}\| \leqslant \Delta t^{2r+4}\frac{\big\| f_t^{(2r+2)}\big(u(t_n + \xi^n \Delta t)\big)\big\|}{(2r+2)!}\int_0^1 (1-z)^{r+2}z^{r+1}dz$$

$$\leqslant C_1\Delta t^{2r+4},$$

and

$$\|u'(t_{n+1}) - \mu^{n+1}\| \leqslant \Delta t^{2r+3}\frac{\big\| f_t^{(2r+2)}\big(u(t_n + \xi^n \Delta t)\big)\big\|}{(2r+2)!}\int_0^1 (1-z)^{r+1}z^{r+1}dz$$

$$\leqslant C_2\Delta t^{2r+3}.$$

The statement of this theorem is proved. $\square$

Since the KG equation (10.1) is time symmetric and a most welcome feature of (10.22) is that it preserves time symmetry, in what follows, we show the symmetry of the formula (10.22). As a first step, we introduce some useful properties of the operator-valued functions $\phi_0(\mathscr{A})$, $\phi_1(\mathscr{A})$ and the coefficients defined by (10.18)–(10.21) in the following two lemmas.

Lemma 10.3 *The bounded operators $\phi_0(\mathscr{A})$ and $\phi_1(\mathscr{A})$ defined by (10.6) satisfy*

$$\phi_0^2(\mathscr{A}) + \mathscr{A}\phi_1^2(\mathscr{A}) = I,$$

(10.27)

where $\mathscr{A}$ is an arbitrary positive semi-definite operator or matrix.

Lemma 10.4 *The coefficients $I_1[\beta_{r,j}(z)]$, $I_2[\beta_{r,j}(z)]$, $I_1[\beta_{r,j}(1-z)]$ and $I_2[\beta_{r,j}(1-z)]$ from (10.22) satisfy*

$$\phi_0(\mathscr{V})I_1[\beta_{r,j}(z)] - \phi_1(\mathscr{V})I_2[\beta_{r,j}(z)] = -I_1[\beta_{r,j}(1-z)],$$

$$\mathscr{V}\phi_1(\mathscr{V})I_1[\beta_{r,j}(z)] + \phi_0(\mathscr{V})I_2[\beta_{r,j}(z)] = I_0[\beta_{r,j}(1-z)],$$

(10.28)

where $\beta_{r,j}(z)$ for $j = 0, 1, \cdots, r$ are defined by (10.15) and $\mathscr{V} = \Delta t^2 \mathscr{A}$ with $\mathscr{A}$, an arbitrary positive semi-definite operator or matrix.

Proof According to the definitions of $I_1[\beta_{r,j}(z)]$ and $I_2[\beta_{r,j}(z)]$, we have

$$\phi_0(\mathscr{V})I_1[\beta_{r,j}(z)] - \phi_1(\mathscr{V})I_2[\beta_{r,j}(z)]$$

$$= \int_0^1 \left[(1-z)\phi_0(\mathscr{V})\phi_1\big((1-z)^2\mathscr{V}\big) - \phi_1(\mathscr{V})\phi_0\big((1-z)^2\mathscr{V}\big)\right]\beta_{r,j}(z)\mathrm{d}z$$

$$= \int_0^1 \left[z\phi_0(\mathscr{V})\phi_1(z^2\mathscr{V}) - \phi_1(\mathscr{V})\phi_0(z^2\mathscr{V})\right]\beta_{r,j}(1-z)\mathrm{d}z$$

$$= -\int_0^1 (1-z)\phi_1\big((1-z)^2\mathscr{V}\big)\beta_j(1-z)\mathrm{d}z = -I_1[\beta_{r,j}(1-z)],$$

and

$$\mathscr{V}\phi_1(\mathscr{V})I_1[\beta_j(z)] + \phi_0(\mathscr{V})I_2[\beta_j(z)]$$

$$= \int_0^1 \left((1-z)\mathscr{V}\phi_1(\mathscr{V})\phi_1\big((1-z)^2\mathscr{V}\big) + \phi_0(\mathscr{V})\phi_0\big((1-z)^2\mathscr{V}\big)\right)\beta_j(z)\mathrm{d}z$$

$$= \int_0^1 \left(z\mathscr{V}\phi_1(\mathscr{V})\phi_1(z^2\mathscr{V}) + \phi_0(\mathscr{V})\phi_0(z^2\mathscr{V})\right)\beta_j(1-z)\mathrm{d}z$$

$$= \int_0^1 \phi_0\big((1-z)^2\mathscr{V}\big)\beta_j(1-z)\mathrm{d}z = I_0[\beta_j(1-z)]. \tag{10.29}$$

Hence, the theorem is proved. □

We note that Hairer et al. [31] have pointed out that symmetric methods have excellent long-time behaviour when solving reversible differential systems. Therefore, it is an important aspect of the design and analysis of symmetric integrators in numerical PDEs. We are now in a position to prove the time symmetry of (10.22).

Theorem 10.3 *The HB time integration formula* (10.22) *is symmetric with respect to the time variable.*

Proof Exchanging $u^{n+1} \leftrightarrow u^n$, $\mu^{n+1} \leftrightarrow \mu^n$ and replacing Δt by $-\Delta t$ in formula (10.22), we obtain

$$u^n = \phi_0(\mathscr{V})u^{n+1} - \Delta t\phi_1(\mathscr{V})\mu^{n+1}$$
$$+ \sum_{j=0}^r \Delta t^{j+2} \left\{(-1)^j I_1[\beta_{r,j}(z)]f_t^{(j)}(u^{n+1}) + I_1[\beta_{r,j}(1-z)]f_t^{(j)}(u^n)\right\},$$

$$\tag{10.30}$$

$$\mu^n = \Delta t \mathscr{A} \phi_1(\mathscr{V}) u^{n+1} + \phi_0(\mathscr{V}) \mu^{n+1}$$

$$- \sum_{j=0}^{r} \Delta t^{j+1} \left\{ (-1)^j I_2[\beta_{r,j}(z)] f_t^{(j)}(u^{n+1}) + I_2[\beta_{r,j}(1-z)] f_t^{(j)}(u^n) \right\}.$$

$$(10.31)$$

It follows from the calculation $\phi_0(\mathscr{V}) \times (10.30) + \Delta t \phi_1(\mathscr{V}) \times (10.31)$ that

$$u^{n+1} = \phi_0(\mathscr{V}) u^n + \Delta t \phi_1(\mathscr{V}) \mu^n$$

$$- \sum_{j=0}^{r} \Delta t^{j+2} \left\{ (-1)^j \left[\phi_0(\mathscr{V}) I_1[\beta_{r,j}(z)] - \phi_1(\mathscr{V}) I_2[\beta_{r,j}(z)] \right] f_t^{(j)}(u^{n+1}) \right.$$

$$\left. + \left[\phi_0(\mathscr{V}) I_1[\beta_{r,j}(1-z)] - \phi_1(\mathscr{V}) I_2[\beta_{r,j}(1-z)] \right] f_t^{(j)}(u^n) \right\}. \quad (10.32)$$

Likewise, the calculation $-\Delta t \mathscr{A} \phi_1(\mathscr{V}) \times (10.30) + \phi_0(\mathscr{V}) \times (10.31)$ results in

$$\mu^{n+1} = -\Delta t \mathscr{A} \phi_1(\mathscr{V}) u^n + \phi_0(\mathscr{V}) \mu^n$$

$$+ \sum_{j=0}^{r} \Delta t^{j+1} \left\{ (-1)^j \left[\mathscr{V} \phi_1(\mathscr{V}) I_1[\beta_{r,j}(z)] + \phi_0(\mathscr{V}) I_2[\beta_{r,j}(z)] \right] f_t^{(j)}(u^{n+1}) \right.$$

$$\left. + \left[\mathscr{V} \phi_1(\mathscr{V}) I_1[\beta_{r,j}(1-z)] + \phi_0(\mathscr{V}) I_2[\beta_{r,j}(1-z)] \right] f_t^{(j)}(u^n) \right\}.$$

$$(10.33)$$

Then applying Lemma 10.4 to (10.32) and (10.33) yields the statement of the theorem. $\qquad\qquad\qquad\qquad\qquad\qquad\qquad\qquad\qquad\qquad\qquad\qquad\qquad\qquad$ □

10.3   Stability of the Fully Discrete Scheme

This section will show the stability of the fully discrete scheme after the differential operator $\mathscr{A}$ is replaced by a suitable matrix A. Throughout this section $\| \cdot \|$ represents both the vector 2-norm and the matrix 2-norm (the spectral norm).

Under the assumption of the following *finite-energy condition* (see, e.g. [32–34])

$$\frac{1}{2} \| u'(t) \|^2 + \frac{\kappa^2}{2} u(t)^\mathsf{T} A u(t) \leqslant \frac{K^2}{2}, \quad (10.34)$$

where K is a constant, global error bounds of the Gaustchi-type method were proved to be independent on $\|A\|$. Consequently, the Gaustchi-type time integrator of order two coupled with suitable spatial discretisation is an excellent choice to solve nonlinear wave equations. Moreover, it is a most important result that the long-time energy conservation for numerical methods can be achieved with the finite-energy condition (see, e.g. [33]). We here also suppose that the exact solution of the nonlinear system (10.8) after suitable spatial discretisation satisfies the finite-energy condition (10.34).

Assume that the perturbed problem of (10.8) is

$$\begin{cases} v''(t) + \mathscr{A}v(t) = f\big(v(t)\big), & t \in [t_0, T], \\ v(t_0) = \varphi_1(x) + \tilde{\varphi}_1(x), & v'(t_0) = \varphi_2(x) + \tilde{\varphi}_2(x), \end{cases} \tag{10.35}$$

where $\tilde{\varphi}_1$ and $\tilde{\varphi}_2$ are perturbation functions. Let

$$\eta(t) = v(t) - u(t).$$

Subtracting (10.8) from (10.35) yields

$$\begin{cases} \eta''(t) + \mathscr{A}\eta(t) = f\big(v(t)\big) - f\big(u(t)\big), & t \in [t_0, T], \\ \eta(t_0) = \tilde{\varphi}_1(x), & \eta'(t_0) = \tilde{\varphi}_2(x). \end{cases} \tag{10.36}$$

We approximate the operator $\mathscr{A}$ by a symmetric positive semi-definite differentiation matrix A on an M-dimensional space since this assists in structure preservation. This implies that there exists an orthogonal matrix P and a diagonal matrix Λ with non-negative diagonal such that

$$A = P\Lambda P^{\mathsf{T}}.$$

Then $A = D^2$, where $D = P\Lambda^{\frac{1}{2}}P^{\mathsf{T}}$. Accordingly, the bounded operators $\phi_j(t^2\mathscr{A})$ are replaced by the matrix functions $\phi_j(t^2A)$. Likewise, we also have

$$\|\phi_j(t^2A)\| = \sqrt{\lambda_{\max}\left(\phi_j^2(t^2A)\right)} \leqslant \gamma_j, \qquad j \in \mathbb{N}. \tag{10.37}$$

Moreover, it is clear that

$$\|D\alpha\|^2 = \alpha^{\mathsf{T}}A\alpha, \qquad \forall \alpha \in \mathbb{R}^M,$$

because A is a symmetric positive semi-definite matrix.

We next analyse the stability for the HB time integrators (10.22). We assume that

$$\eta^n \approx \eta(t_n), \quad \zeta^n \approx \eta'(t_n) \qquad \text{and} \qquad v^n \approx v(t_n), \quad w^n \approx v'(t_n).$$

Applying HB time integration to (10.36) yields

$$
\begin{cases}
\eta^{n+1} = \phi_0(V)\eta^n + \Delta t \phi_1(V)\zeta^n + \sum_{j=0}^{r} \Delta t^{j+2} \Big\{ I_1[\beta_{r,j}(z)][f_t^{(j)}(v^n) \\
\qquad - f_t^{(j)}(u^n)] + (-1)^j I_1[\beta_{r,j}(1-z)][f_t^{(j)}(v^{n+1}) - f_t^{(j)}(u^{n+1})] \Big\}, \\[2mm]
\zeta^{n+1} = -\Delta t A \phi_1(V)\eta^n + \phi_0(V)\zeta^n + \sum_{j=0}^{r} \Delta t^{j+1} \Big\{ I_2[\beta_{r,j}(z)][f_t^{(j)}(v^n) \\
\qquad - f_t^{(j)}(u^n)] + (-1)^j I_2[\beta_{r,j}(1-z)][f_t^{(j)}(v^{n+1}) - f_t^{(j)}(u^{n+1})] \Big\},
\end{cases}
$$

$$(10.38)$$

where $V = \Delta t^2 A$, $I_1[\beta_{r,j}(z)]$, $I_2[\beta_{r,j}(z)]$, $I_1[\beta_{r,j}(1-z)]$ and $I_2[\beta_{r,j}(1-z)]$ are defined by (10.18)–(10.21), respectively. Similarly, we obtain

$$
\|I_1[\beta_{r,j}(z)]\| \leqslant \max_{0 \leqslant z \leqslant 1} |\beta_{r,j}(z)| \leqslant 1 \quad \text{and} \quad \|I_1[\beta_{r,j}(1-z)]\| \leqslant \max_{0 \leqslant z \leqslant 1} |\beta_{r,j}(1-z)| \leqslant 1,
$$

$$
\|I_2[\beta_{r,j}(z)]\| \leqslant \max_{0 \leqslant z \leqslant 1} |\beta_{r,j}(z)| \leqslant 1 \quad \text{and} \quad \|I_2[\beta_{r,j}(1-z)]\| \leqslant \max_{0 \leqslant z \leqslant 1} |\beta_{r,j}(1-z)| \leqslant 1.
$$

The schemes (10.38) can be rewritten in a compact form:

$$
\begin{bmatrix} D\eta^{n+1} \\ \zeta^{n+1} \end{bmatrix} = \Psi(V) \begin{bmatrix} D\eta^n \\ \zeta^n \end{bmatrix} + \sum_{j=0}^{r} \Delta t^{j+1} \int_0^1 \Psi_j(\beta(z), V) dz \begin{bmatrix} 0 \\ f_t^{(j)}(v^n) - f_t^{(j)}(u^n) \end{bmatrix}
$$
$$
+ \sum_{j=0}^{r} (-1)^j \Delta t^{j+1} \int_0^1 \Psi_j(\beta(1-z), V) dz \begin{bmatrix} 0 \\ f_t^{(j)}(v^{n+1}) - f_t^{(j)}(u^{n+1}) \end{bmatrix},
$$

$$(10.39)$$

where

$$
\Psi(V) = \begin{bmatrix} \phi_0(V) & \Delta t D \phi_1(V) \\ -\Delta t D \phi_1(V) & \phi_0(V) \end{bmatrix}
$$

$$(10.40)$$

and

$$
\Psi_j(\beta(z), V) = \beta_{r,j}(z) \begin{bmatrix} \phi_0((1-z)^2 V) & \Delta t (1-z) D \phi_1((1-z)^2 V) \\ -\Delta t (1-z) D \phi_1((1-z)^2 V) & \phi_0((1-z)^2 V) \end{bmatrix}.
$$

$$(10.41)$$

Before dealing with stability analysis, we should investigate the spectral norm of matrices $\Psi(V)$ and $\Psi_j(\beta(z), V)$ for $j = 0, 1, \cdots, r$.

Lemma 10.5 *Suppose that A is a symmetric positive semi-definite matrix and that $V = \Delta t^2 A$. Then the spectral norms of matrices $\Psi(V)$ and $\Psi_j(\beta(z), V)$ satisfy*

$$\|\Psi(V)\| = 1 \quad \text{and} \quad \|\Psi_j(\beta(z), V)\| = |\beta_{r,j}(z)| \leqslant 1,$$
$$z \in [0, 1], \quad j = 0, 1, \cdots, r. \tag{10.42}$$

Proof It is trivial to verify the results based on Lemma 10.3, formulae (10.40) and (10.41) and the definition of the matrix 2-norm. The reader is referred to [23] for details. □

10.3.1 Linear Stability Analysis

We begin with the stability analysis of HB time integrators for the linear problem, i.e. $f(u) = u$. In this case, we have

$$f_t^{(2k)}\big(u(t)\big) = (I - \mathscr{A})^k u(t) \quad \text{and} \quad f_t^{(2k+1)}\big(u(t)\big) = (I - \mathscr{A})^k u'(t), \quad k \in \mathbb{N}. \tag{10.43}$$

Lemma 10.6 *Suppose that A is a symmetric matrix. Then*

$$\|(I - A)^k\| \leqslant \big[1 + \rho(A)\big]^k, \qquad k \in \mathbb{N},$$

where $\rho(A)$ is the spectral radius of A.

Proof It is immediately from the definition of the spectral norm that

$$\|(I - A)^k\| = \sqrt{\lambda_{\max}\big((I - A)^{2k}\big)} \leqslant \big(1 + \max_{1 \leqslant j \leqslant M} |\lambda_j|\big)^k = \big[1 + \rho(A)\big]^k,$$

where λ_j for $j = 1, 2, \cdots, M$ are the eigenvalues of A. □

Theorem 10.4 *Assume that the operator $\mathscr{A}$ is approximated by a symmetric positive semi-definite differentiation matrix A and let the finite energy condition (10.34) be satisfied. If the sufficiently small time stepsize Δt satisfies $\Delta t^2(1 + \rho(A)) \leqslant 1$ with $\Delta t \leqslant [4(r + 1)]^{-1}$, then we have the following stability results:*

$$\|\eta^n\| \leqslant \exp\big(2(2r + 3)T\big)\Big(\|\tilde{\varphi}_1\| + \sqrt{\tilde{\varphi}_1^{\mathsf{T}} A \tilde{\varphi}_1 + \|\tilde{\varphi}_2\|^2}\Big),$$

$$\|\zeta^n\| \leqslant \exp\big(2(2r + 3)T\big)\Big(\|\tilde{\varphi}_1\| + \sqrt{\tilde{\varphi}_1^{\mathsf{T}} A \tilde{\varphi}_1 + \|\tilde{\varphi}_2\|^2}\Big),$$

where $\tilde{\varphi}_l = \left(\tilde{\varphi}_l(x_0), \tilde{\varphi}_l(x_1), \cdots, \tilde{\varphi}_l(x_{M-1})\right)^{\mathsf{T}}$, while $\tilde{\varphi}_l(x_i)$ for $l = 1, 2$ are the values of the perturbation functions $\tilde{\varphi}_l$ for $l = 1, 2$, at the grid points $\{x_i\}_{i=0}^{M-1}$.

Proof Using the first formula in (10.38) and (10.39), we obtain

$$\|\eta^{n+1}\| \leqslant \|\eta^n\| + \Delta t \|\zeta^n\| + \sum_{j=0}^{r} \Delta t^{j+2} \left(\|f_t^{(j)}(v^n) - f_t^{(j)}(u^n)\| + \|f_t^{(j)}(v^{n+1})\right.$$

$$\left. - f_t^{(j)}(u^{n+1})\|\right),$$

and

$$\sqrt{(\eta^{n+1})^{\mathsf{T}} A \eta^{n+1} + \|\zeta^{n+1}\|^2} \leqslant \sqrt{(\eta^n)^{\mathsf{T}} A \eta^n + \|\zeta^n\|^2}$$

$$+ \sum_{j=0}^{r} \Delta t^{j+1} \left(\|f_t^{(j)}(v^n) - f_t^{(j)}(u^n)\| + \|f_t^{(j)}(v^{n+1}) - f_t^{(j)}(u^{n+1})\|\right).$$

Then summing up the above and using (10.43), we have

$$\|\eta^{n+1}\| + \sqrt{(\eta^{n+1})^{\mathsf{T}} A \eta^{n+1} + \|\zeta^{n+1}\|^2} \leqslant \|\eta^n\| + \sqrt{(\eta^n)^{\mathsf{T}} A \eta^n + \|\zeta^n\|^2} + \Delta t \|\zeta^n\|$$

$$+ \Delta t (1 + \Delta t) \sum_{j=0}^{r} \Delta t^j \left\|(I - A)^{\lceil \frac{j}{2} \rceil}\right\| \left(\|\eta^n\| + \|\zeta^n\| + \|\eta^{n+1}\| + \|\zeta^{n+1}\|\right).$$

$$(10.44)$$

Applying Lemma 10.6 to inequality (10.44) leads to

$$\|\eta^{n+1}\| + \sqrt{(\eta^{n+1})^{\mathsf{T}} A \eta^{n+1} + \|\zeta^{n+1}\|^2} \leqslant \|\eta^n\| + \sqrt{(\eta^n)^{\mathsf{T}} A \eta^n + \|\zeta^n\|^2} + \Delta t \|\zeta^n\|$$

$$+ \Delta t (1 + \Delta t) \sum_{j=0}^{r} \Delta t^j \left(1 + \rho(A)\right)^{\lceil \frac{j}{2} \rceil} \left(\|\eta^n\| + \|\zeta^n\| + \|\eta^{n+1}\| + \|\zeta^{n+1}\|\right).$$

Under the assumption that the stepsize satisfies $\Delta t^2 \left(1 + \rho(A)\right) \leqslant 1$, we obtain

$$\|\eta^{n+1}\| + \sqrt{(\eta^{n+1})^{\mathsf{T}} A \eta^{n+1} + \|\zeta^{n+1}\|^2}$$

$$\leqslant \left[1 + \frac{\Delta t (2r + 3)}{1 - 2\Delta t (r + 1)}\right] \left(\|\eta^n\| + \sqrt{(\eta^n)^{\mathsf{T}} A \eta^n + \|\zeta^n\|^2}\right).$$

As $\Delta t \leqslant [4(r+1)]^{-1}$, we have

$$\|\eta^{n+1}\| + \sqrt{(\eta^{n+1})^{\mathsf{T}} A \eta^{n+1} + \|\zeta^{n+1}\|^2} \leqslant \Big[1 + 2(2r+3)\Delta t\Big]$$

$$\Big(\|\eta^n\| + \sqrt{(\eta^n)^{\mathsf{T}} A \eta^n + \|\zeta^n\|^2}\Big).$$

Then an inductive argument yields the following result

$$\|\eta^{n+1}\| + \sqrt{(\eta^{n+1})^{\mathsf{T}} A \eta^{n+1} + \|\zeta^{n+1}\|^2} \leqslant \exp\big(2(2r+3)T\big)$$

$$\Big(\|\tilde{\varphi}_1\| + \sqrt{\tilde{\varphi}_1^{\mathsf{T}} A \tilde{\varphi}_1 + \|\tilde{\varphi}_2\|^2}\Big).$$

Therefore, we have

$$
\begin{aligned}
\|\eta^n\| &\leqslant \exp\big(2(4r+5)T\big)\Big(\|\tilde{\varphi}_1\| + \sqrt{\|D\tilde{\varphi}_1\|^2 + \|\tilde{\varphi}_2\|^2}\Big), \\
\|\zeta^n\| &\leqslant \exp\big(2(4r+5)T\big)\Big(\|\tilde{\varphi}_1\| + \sqrt{\|D\tilde{\varphi}_1\|^2 + \|\tilde{\varphi}_2\|^2}\Big),
\end{aligned}
\tag{10.45}
$$

and linear stability is proved.   □

10.3.2 Nonlinear Stability Analysis

We will further analyse in this subsection the stability of HB time integrators for nonlinear problems. The analysis relies upon some assumptions.

Assumption 10.1 It is assumed that both (10.8) and (10.35) possess sufficiently smooth solutions and $f : D(\mathscr{A}) \to \mathbb{R}$ is sufficiently Fréchet differentiable in a strip along the exact solution.

As is known, it follows from Chap. 3 in [35] that

$$f_t^{(k)}\big(u(t)\big) = \sum_{\tilde{t} \in \mathrm{SENT}_{k+2}^f} \alpha(\tilde{t}) \mathscr{F}(\tilde{t})\big(u(t), u'(t)\big),
\tag{10.46}$$

where $\mathrm{SENT}^f = \{\tau_2\} \cup \big\{\tilde{t} = [\tilde{t}_1, \cdots, \tilde{t}_m]_2 : \tilde{t}_i \in \mathrm{SENT}\big\}$ and SENT is the set of special extended Nyström trees defined in [35], $\alpha(\tilde{t})$ is the number of possible monotonic labellings of an extended Nyström tree $\tilde{t}$, and $\mathscr{F}(\tilde{t})\big(u, u'\big)$ is the corresponding elementary differential.

Assumption 10.2 We assume that $d^k f(u)/du^k : D(\mathscr{A}) \to \mathbb{R}$ for $k = 0, 1, 2, \cdots, r$ are locally Lipschitz continuous in a strip along the exact solution u. Hence, there exist real numbers $L(R, \rho(A)^{\lfloor \frac{k}{2} \rfloor})$ such that

$$\| \mathscr{F}(\tilde{t})\big(v(t), v'(t)\big) - \mathscr{F}(\tilde{t})\big(w(t), w'(t)\big) \|$$

$$\leqslant L(R, \rho(A)^{\lfloor \frac{k}{2} \rfloor})\Big(\|v(t) - w(t)\| + \|v'(t) - w'(t)\|\Big), \quad \forall \tilde{t} \in \mathrm{SENT}^f_{k+2}$$

for all $t \in [t_0, T]$ and $\max\Big(\|v - u(t)\|, \|w - u(t)\|, \|v' - u'(t)\|, \|w' - u'(t)\|\Big) \leqslant R$.
The next theorem shows the statement on nonlinear stability.

Theorem 10.5 *With Assumptions 10.1 and 10.2, suppose that the sufficiently small time stepsize satisfies*

$$\Delta t^2 L(R, \rho(A)) \leqslant 1 \quad and \quad \Delta t \sum_{j=0}^{r} \sum_{\tilde{t} \in \mathrm{SENT}^f_{j+2}} \alpha(\tilde{t}) \leqslant \frac{1}{4}.$$

Then, if the operator $\mathscr{A}$ is approximated by a symmetric positive semi-definite matrix A, we have the following stability results,

$$\|\eta^n\| \leqslant \exp\left(2T\left(1 + 2\sum_{j=0}^{r} \sum_{\tilde{t} \in \mathrm{SENT}^f_{j+2}} \alpha(\tilde{t})\right)\right)\left(\|\tilde{\varphi}_1\| + \sqrt{\tilde{\varphi}_1^\mathsf{T} A \tilde{\varphi}_1 + \|\tilde{\varphi}_2\|^2}\right),$$

$$\|\zeta^n\| \leqslant \exp\left(2T\left(1 + 2\sum_{j=0}^{r} \sum_{\tilde{t} \in \mathrm{SENT}^f_{j+2}} \alpha(\tilde{t})\right)\right)\left(\|\tilde{\varphi}_1\| + \sqrt{\tilde{\varphi}_1^\mathsf{T} A \tilde{\varphi}_1 + \|\tilde{\varphi}_2\|^2}\right),$$

where $\tilde{\varphi}_l = \big(\tilde{\varphi}_l(x_0), \tilde{\varphi}_l(x_1), \cdots, \tilde{\varphi}_l(x_{M-1})\big)^\mathsf{T}$ and $\tilde{\varphi}_l(x_i)$ for $l = 1, 2$ are the values of the perturbation functions $\tilde{\varphi}_l$ for $l = 1, 2$, at the spatial grid points $\{x_i\}_{i=0}^{M-1}$.

Proof Using the first formula in (10.38) and (10.39), we obtain

$$\|\eta^{n+1}\| \leqslant \|\eta^n\| + \Delta t \|\zeta^n\| + \sum_{j=0}^{r} \Delta t^{j+2}\Big[\|f_t^{(j)}(v^n) - f_t^{(j)}(u^n)\|$$

$$+ \|f_t^{(j)}(v^{n+1}) - f_t^{(j)}(u^{n+1})\|\Big],$$

$$\sqrt{(\eta^{n+1})^\mathsf{T} A \eta^{n+1} + \|\zeta^{n+1}\|^2} \leqslant \sqrt{(\eta^n)^\mathsf{T} A \eta^n + \|\zeta^n\|^2}$$

$$+ \sum_{j=0}^{r} \Delta t^{j+1}\Big[\|f_t^{(j)}(v^n) - f_t^{(j)}(u^n)\| + \|f_t^{(j)}(v^{n+1}) - f_t^{(j)}(u^{n+1})\|\Big].$$

$$(10.47)$$

Summing up (10.47) and inserting (10.46) into the right-hand side, we have

$$
\|\eta^{n+1}\| + \sqrt{(\eta^{n+1})^{\mathsf{T}} A \eta^{n+1} + \|\zeta^{n+1}\|^2} \leqslant \|\eta^n\| + \sqrt{(\eta^n)^{\mathsf{T}} A \eta^n + \|\zeta^n\|^2} + \Delta t \|\zeta^n\|
$$

$$
+ \Delta t (1 + \Delta t) \sum_{j=0}^{r} \sum_{\tilde{t} \in \mathrm{SENT}^f_{j+2}} \alpha(\tilde{t}) \Delta t^j \Big[\big\| \mathscr{F}(\tilde{t})(v^n, w^n) - \mathscr{F}(\tilde{t})(u^n, \mu^n) \big\|
$$

$$
+ \big\| \mathscr{F}(\tilde{t})(v^{n+1}, w^{n+1}) - \mathscr{F}(\tilde{t})(u^{n+1}, \mu^{n+1}) \big\| \Big].
$$

$$(10.48)$$

On the other hand, the use of Assumption 10.2 on the right-hand side of (10.48) gives

$$
\|\eta^{n+1}\| + \sqrt{(\eta^{n+1})^{\mathsf{T}} A \eta^{n+1} + \|\zeta^{n+1}\|^2}
$$

$$
\leqslant \|\eta^n\| + \sqrt{(\eta^n)^{\mathsf{T}} A \eta^n + \|\zeta^n\|^2} + \Delta t \|\zeta^n\|
$$

$$
+ \Delta t (1 + \Delta t) \sum_{j=0}^{r} \sum_{\tilde{t} \in \mathrm{SENT}^f_{j+2}} \alpha(\tilde{t}) \Delta t^j L(R, \rho(A)^{\lfloor \frac{j}{2} \rfloor})
$$

$$
\big(\|\eta^n\| + \|\zeta^n\| + \|\eta^{n+1}\| + \|\zeta^{n+1}\| \big).
$$

$$(10.49)$$

As Δt satisfies $\Delta t^2 L(R, \rho(A)) \leqslant 1$, the inequality (10.49) results in

$$
\|\eta^{n+1}\| + \sqrt{(\eta^{n+1})^{\mathsf{T}} A \eta^{n+1} + \|\zeta^{n+1}\|^2}
$$

$$
\leqslant \left\{ 1 + \frac{\Delta t \Big[1 + 2 \sum_{j=0}^{r} \sum_{\tilde{t} \in \mathrm{SENT}^f_{j+2}} \alpha(\tilde{t}) \Big]}{1 - 2 \Delta t \sum_{j=0}^{r} \sum_{\tilde{t} \in \mathrm{SENT}^f_{j+2}} \alpha(\tilde{t})} \right\} \left(\|\eta^n\| + \sqrt{(\eta^n)^{\mathsf{T}} A \eta^n + \|\zeta^n\|^2} \right).
$$

Furthermore, as $\Delta t \sum_{j=0}^{r} \sum_{\tilde{t} \in \mathrm{SENT}^f_{j+2}} \alpha(\tilde{t}) \leqslant \dfrac{1}{4}$, we obtain

$$
\|\eta^{n+1}\| + \sqrt{(\eta^{n+1})^{\mathsf{T}} A \eta^{n+1} + \|\zeta^{n+1}\|^2}
$$

$$
\leqslant \left[1 + 2 \Delta t \Big(1 + 2 \sum_{j=0}^{r} \sum_{\tilde{t} \in \mathrm{SENT}^f_{j+2}} \alpha(\tilde{t}) \Big) \right] \left(\|\eta^n\| + \sqrt{(\eta^n)^{\mathsf{T}} A \eta^n + \|\zeta^n\|^2} \right).
$$

Then an argument by induction gives the following result

$$\|\eta^{n+1}\| + \sqrt{(\eta^{n+1})^{\mathsf{T}} A \eta^{n+1} + \|\zeta^{n+1}\|^2}$$

$$\leqslant \exp\left(2T\left(1 + 2\sum_{j=0}^{r}\sum_{\tilde{\iota}\in\mathrm{SENT}_{j+2}^{f}}\alpha(\tilde{\iota})\right)\right)\left(\|\tilde{\varphi}_1\| + \sqrt{\tilde{\varphi}_1^{\mathsf{T}} A \tilde{\varphi}_1 + \|\tilde{\varphi}_2\|^2}\right).$$

Consequently, we obtain

$$\|\eta^{n}\| \leqslant \exp\left(2T\left(1 + 2\sum_{j=0}^{r}\sum_{\tilde{\iota}\in\mathrm{SENT}_{j+2}^{f}}\alpha(\tilde{\iota})\right)\right)\left(\|\tilde{\varphi}_1\| + \sqrt{\tilde{\varphi}_1^{\mathsf{T}} A \tilde{\varphi}_1 + \|\tilde{\varphi}_2\|^2}\right),$$

$$\|\zeta^{n}\| \leqslant \exp\left(2T\left(1 + 2\sum_{j=0}^{r}\sum_{\tilde{\iota}\in\mathrm{SENT}_{j+2}^{f}}\alpha(\tilde{\iota})\right)\right)\left(\|\tilde{\varphi}_1\| + \sqrt{\tilde{\varphi}_1^{\mathsf{T}} A \tilde{\varphi}_1 + \|\tilde{\varphi}_2\|^2}\right),$$

This shows the stability of HB time integrators for nonlinear problems. $\square$

10.4 Convergence of the Fully Discrete Scheme

10.4.1 Consistency

Under suitable assumptions on smoothness and the spatial discretisation, it is not difficult to obtain a spatial semidiscrete scheme which is consistent with the original system (10.1) or (10.8). We only require that the truncation error $\delta(\Delta x)$ in (10.58) satisfies $\delta(\Delta x) \to 0$ as $\Delta x \to 0$. In what follows, we analyse the consistency of the fully discrete scheme (10.62) or (10.63). To this end, we first analyse the truncation error of the fully discrete scheme (10.62) or (10.63).

Inserting the exact solution $U(t) = \big(u(x_0, t), u(x_1, t), \cdots, u(x_{M-1}, t)\big)^{\mathsf{T}}$ into the fully discrete scheme (10.62), we obtain

$$\begin{cases} \mathscr{T}^{n} = U(t_{n+1}) - \phi_0(V)U(t_n) - \Delta t \phi_1(V)U'(t_n) \\[2mm] \quad - \sum_{j=0}^{r}\Delta t^{j+2}\Big\{I_1[\beta_{r,j}(z)]f_t^{(j)}\big(U(t_n)\big) + (-1)^j I_1[\beta_{r,j}(1-z)]f_t^{(j)}\big(U(t_{n+1})\big)\Big\}, \\[2mm] \Gamma^{n} = U'(t_{n+1}) + \Delta t A \phi_1(V)U(t_n) - \phi_0(V)U'(t_n) \\[2mm] \quad - \sum_{j=0}^{r}\Delta t^{j+1}\Big\{I_2[\beta_{r,j}(z)]f_t^{(j)}\big(U(t_n)\big) + (-1)^j I_2[\beta_{r,j}(1-z)]f_t^{(j)}\big(U(t_{n+1})\big)\Big\}, \end{cases}$$

$$(10.50)$$

where $\mathcal{T}^n$ and Γ^n are the truncation errors of the fully discrete scheme (10.62) at time t_n. Applying the variation-of-constants formula to (10.58) and comparing the result with (10.50) leads to

$$
\begin{cases}
\mathcal{T}^n = \Delta t^2 \int_0^1 (1-z)\phi_1\big((1-z)^2 V\big) f\big(U(t_n + z\Delta t)\big)dz\| \\[2mm]
\quad - \sum_{j=0}^{r} \Delta t^{j+2}\Big\{ I_1[\beta_{r,j}(z)]f_t^{(j)}\big(U(t_n)\big) + (-1)^j I_1[\beta_{r,j}(1-z)]f_t^{(j)}\big(U(t_{n+1})\big)\Big\} \\[2mm]
\quad + \Delta t^2 \int_0^1 (1-z)\phi_1\big((1-z)^2 V\big)\delta(\Delta x)dz, \\[4mm]
\Gamma^n = \Delta t \int_0^1 \phi_0\big((1-z)^2 V\big) f\big(U(t_n + z\Delta t)\big)dz - \sum_{j=0}^{r} \Delta t^{j+1}\Big\{ I_2[\beta_{r,j}(z)]f_t^{(j)}\big(U(t_n)\big) \\[2mm]
\quad + (-1)^j I_2[\beta_{r,j}(1-z)]f_t^{(j)}\big(U(t_{n+1})\big)\Big\} + \Delta t \int_0^1 \phi_0\big((1-z)^2 V\big)\delta(\Delta x)dz.
\end{cases}
\tag{10.51}
$$

Using the Hermite–Birkhoff interpolation polynomial (see Lemma 10.2) to approximate the nonlinear function $f\big(U(t_n + z\Delta t)\big)$ appearing in (10.51), we have

$$
\mathcal{T}^n = (-1)^{r+1}\Delta t^{2r+4} \int_0^1 (1-z)^{r+2}\phi_1\big((1-z)^2 V\big)z^{r+1}dz \frac{f_t^{(2r+2)}\big(U(t_n + \xi^n \Delta t)\big)}{(2r+2)!}
$$

$$
+ \Delta t^2 \int_0^1 (1-z)\phi_1\big((1-z)^2 V\big)\delta(\Delta x)dz
\tag{10.52}
$$

and

$$
\Gamma^n = (-1)^{r+1}\Delta t^{2r+3} \int_0^1 (1-z)^{r+1}\phi_0\big((1-z)^2 V\big)z^{r+1}dz \frac{f_t^{(2r+2)}\big(U(t_n + \xi^n \Delta t)\big)}{(2r+2)!}
$$

$$
+ \Delta t \int_0^1 \phi_0\big((1-z)^2 V\big)\delta(\Delta x)dz.
\tag{10.53}
$$

We next prove consistency, based on the truncation error analysis.

Theorem 10.6 *Suppose that the exact solution $u(x,t)$ of the original continuous equations (10.1) or (10.8) is sufficiently smooth such that $f\big(u(\cdot,t)\big) \in C^{2r+2}([t_0, T])$ and $f_t^{(2r+2)}\big(u(x,\cdot)\big) \in L^2(\Omega)$. Then, the fully discrete scheme (10.62) or (10.63) is consistent over a finite time interval $t \in [t_0, T]$, i.e.,*

$$
\|\mathcal{T}^n\| \to 0 \quad and \quad \|\Gamma^n\| \to 0 \quad as \quad \Delta t, \Delta x \to 0.
\tag{10.54}
$$

Proof Taking the l_2-norm on both sides of the truncation errors (10.52) and (10.53) leads to

$$\|\mathcal{T}^n\| \leqslant \tilde{C}_1 \Delta t^{2r+4} + \frac{\Delta t^2}{2}\|\delta(\Delta x)\| \quad \text{and} \quad \|\Gamma^n\| \leqslant \tilde{C}_2 \Delta t^{2r+3} + \Delta t\|\delta(\Delta x)\|, \tag{10.55}$$

where the constants $\tilde{C}_1$ and $\tilde{C}_2$ are given by

$$\tilde{C}_1 = \frac{(r+2)!(r+1)!}{(2r+2)!(2r+4)!} \max_{t_0 \leqslant t \leqslant T} \left\| f_t^{(2r+2)}(u(\cdot,t)) \right\| \tag{10.56}$$

$$\tilde{C}_2 = \frac{[(r+1)!]^2}{(2r+4)!(2r+3)!} \max_{t_0 \leqslant t \leqslant T} \left\| f_t^{(2r+2)}(u(\cdot,t)) \right\|. \tag{10.57}$$

It can be confirmed from (10.55) and (10.56) that the constants $\tilde{C}_1$ and $\tilde{C}_2$ are only dependent on the bounds for derivatives of the exact solution $U(t)$ over a finite time interval $t \in [t_0, T]$. Then, the consistency of the fully discrete scheme (10.62) or (10.63) directly follows from the fact that

$$\|\mathcal{T}^n\| \to 0 \quad \text{and} \quad \|\Gamma^n\| \to 0 \quad \text{as} \quad \Delta t, \Delta x \to 0.$$

The proof of this theorem is complete.   □

10.4.2   Convergence

Our next objective is to analyse convergence for the fully discrete schemes. It is well known that convergence of classical methods for linear partial differential equations is governed by the Lax equivalence theorem: convergence is equivalent to consistency plus stability [36]. The HB time integrators are consistent (see Theorem 10.6), and the stability of the fully discrete scheme for linear problems has been proved in Sect. 10.3.1. Consequently, the convergence of the HB time integrators for linear problems can be obtained by applying the Lax equivalence theorem. Unfortunately, however, the Lax equivalence theorem might be less useful for nonlinear problems.

In what follows, we analyse the convergence of the fully discrete scheme for nonlinear problems. Based on some suitable assumptions on smoothness and spatial discretisation strategies, the original continuous system (10.1) or (10.8) can be discretised as:

$$\begin{cases} U''(t) + AU(t) = f(U(t)) + \delta(\Delta x), & t \in [t_0, T], \\ U(t_0) = \varphi_1, \quad U'(t_0) = \varphi_2, \end{cases} \tag{10.58}$$

where $U(t) = \big(u(x_0, t), u(x_1, t), \cdots, u(x_{M-1}, t)\big)^{\mathsf{T}}$, A is a positive semi-definite differentiation matrix and $\varphi_l = \big(\varphi_l(x_0), \varphi_l(x_1), \cdots, \varphi_l(x_{M-1})\big)^{\mathsf{T}}$ for $l = 1, 2$. Let $\delta(\Delta x) = \big(\delta_0(\Delta x), \delta_1(\Delta x), \cdots, \delta_{M-1}(\Delta x), \big)^{\mathsf{T}}$ be the truncation error brought by approximating the spatial differential operator $\mathscr{A}$ with a positive semi-definite matrix A, and the truncation error $\delta(\Delta x)$ satisfies $\delta_j(\Delta x) \to 0$ as $\Delta x \to 0$ for $j = 0, 1, \cdots, M - 1$. For instance, if we replace the spatial derivative by the classical forth-order finite difference method (see, e.g. [37, 38]), the truncation error $\delta(\Delta x)$ is $\|\delta(\Delta x)\| = O(\Delta x^4)$.

Applying the time integration formula (10.22) to (10.58) leads to

$$
\begin{cases}
U(t_{n+1}) = \phi_0(V)U(t_n) + \Delta t \phi_1(V)U'(t_n) + \sum_{j=0}^{r} \Delta t^{j+2} \Big\{ I_1[\beta_{r,j}(z)] f_t^{(j)}\big(U(t_n)\big) \\
\qquad\qquad + (-1)^j I_1[\beta_{r,j}(1-z)] f_t^{(j)}\big(U(t_{n+1})\big) \Big\} + R^n, \\[2mm]
U'(t_{n+1}) = - \Delta t A \phi_1(V)U(t_n) + \phi_0(V)U'(t_n) + \sum_{j=0}^{r} \Delta t^{j+1} \Big\{ I_2[\beta_{r,j}(z)] f_t^{(j)}\big(U(t_n)\big) \\
\qquad\qquad + (-1)^j I_2[\beta_{r,j}(1-z)] f_t^{(j)}\big(U(t_{n+1})\big) \Big\} + r^n,
\end{cases}
$$
(10.59)

where $R^n = \big(R_0^n, \cdots, R_{M-1}^n\big)^{\mathsf{T}}$ and $r^n = \big(r_0^n, \cdots, r_{M-1}^n\big)^{\mathsf{T}}$ are truncation errors, and

$$
\begin{aligned}
R_j^n =\ & (-1)^{r+1} \Delta t^{2r+4} \frac{f_t^{(2r+2)}\big(u(x_j, t_n + \xi^n \Delta t)\big)}{(2r+2)!} \int_0^1 (1-z)^{r+2}\phi_1\big((1-z)^2 V\big) z^{r+1} \mathrm{d}z \\
& + \Delta t^2 \int_0^1 (1-z)\phi_1\big((1-z)^2 V\big)\delta_j(\Delta x)\mathrm{d}z,
\end{aligned}
$$

and

$$
\begin{aligned}
r_j^n =\ & (-1)^{r+1} \Delta t^{2r+3} \frac{f_t^{(2r+2)}\big(u(x_j, t_n + \xi^n \Delta t)\big)}{(2r+2)!} \int_0^1 (1-z)^{r+1}\phi_0\big((1-z)^2 V\big) z^{r+1} \mathrm{d}z \\
& + \Delta t \int_0^1 \phi_0\big((1-z)^2 V\big)\delta_j(\Delta x)\mathrm{d}z
\end{aligned}
$$

respectively. Under suitable assumptions of smoothness, the errors R_j^n and r_j^n satisfy

$$
|R_j^n| \leqslant \frac{(r+2)!(r+1)!}{(2r+2)!(2r+4)!} \max_{t_0 \leqslant t \leqslant T} \max_{x \in \bar{\Omega}} |f_t^{(2r+2)}\big(u(x, t)\big)|\Delta t^{2r+4} + \frac{\Delta t^2}{2}|\delta_j(\Delta x)|,
$$
(10.60)

and

$$|r_j^n| \leqslant \frac{[(r+1)!]^2}{(2r+4)!(2r+3)!} \max_{t_0 \leqslant t \leqslant T} \max_{x \in \bar{\Omega}} |f_t^{(2r+2)}(u(x,t))| \Delta t^{2r+3} + \Delta t |\delta_j(\Delta x)|.$$

$$(10.61)$$

Disregarding the small terms R^n and r^n in (10.59) and letting $u_j^n \approx u(x_j, t_n)$, $\mu_j^n \approx u_t(x_j, t_n)$, the following fully discrete scheme follows

$$
\begin{cases}
u^{n+1} = \phi_0(V)u^n + \Delta t \phi_1(V)\mu^n \\
\qquad + \sum_{j=0}^{r} \Delta t^{j+2} \left\{ I_1[\beta_{r,j}(z)] f_t^{(j)}(u^n) + (-1)^j I_1[\beta_{r,j}(1-z)] f_t^{(j)}(u^{n+1}) \right\}, \\
\mu^{n+1} = -\Delta t A \phi_1(V)u^n + \phi_0(V)\mu^n \\
\qquad + \sum_{j=0}^{r} \Delta t^{j+1} \left\{ I_2[\beta_{r,j}(z)] f_t^{(j)}(u^n) + (-1)^j I_2[\beta_{r,j}(1-z)] f_t^{(j)}(u^{n+1}) \right\}.
\end{cases}
$$

$$(10.62)$$

In terms of the notation in (10.46), we rewrite the fully discrete scheme (10.62) in the following form

$$
\begin{cases}
u^{n+1} = \phi_0(V)u^n + \Delta t \phi_1(V)\mu^n + \sum_{j=0}^{r} \sum_{\tilde{i} \in \text{SENT}_{j+2}^f} \alpha(\tilde{i}) \Delta t^{j+2} \\
\qquad \times \left[I_1[\beta_{r,j}(z)] \mathcal{F}(\tilde{i})(u^n, \mu^n) + (-1)^j I_1[\beta_{r,j}(1-z)] \mathcal{F}(\tilde{i})(u^{n+1}, \mu^{n+1}) \right], \\
\mu^{n+1} = -\Delta t A \phi_1(V)u^n + \phi_0(V)\mu^n + \sum_{j=0}^{r} \sum_{\tilde{i} \in \text{SENT}_{j+2}^f} \alpha(\tilde{i}) \Delta t^{j+1} \\
\qquad \times \left[I_2[\beta_{r,j}(z)] \mathcal{F}(\tilde{i})(u^n, \mu^n) + (-1)^j I_2[\beta_{r,j}(1-z)] \mathcal{F}(\tilde{i})(u^{n+1}, \mu^{n+1}) \right],
\end{cases}
$$

$$(10.63)$$

where $I_1[\beta_{r,j}(z)]$, $I_2[\beta_{r,j}(z)]$, $I_1[\beta_{r,j}(1-z)]$ and $I_2[\beta_{r,j}(1-z)]$ have been defined by (10.18)–(10.21).

We next consider from first principles the convergence of the fully discrete scheme (10.63) for nonlinear problems. We let $e_j^n = u(x_j, t_n) - u_j^n$ and $\omega_j^n = u_t(x_j, t_n) - \mu_j^n$ for $j = 0, 1, \cdots, M-1$, i.e., $e^n = U(t_n) - u^n$ and $\omega^n = U'(t_n) - \mu^n$. Subtracting (10.63) from (10.59), and inserting exact initial conditions, we get a recurrence relation for the errors,

$$
\begin{cases}
e^{n+1} = \phi_0(V)e^n + \Delta t \phi_1(V)\omega^n \\
\quad + \displaystyle\sum_{j=0}^{r} \sum_{\tilde{\imath}\in\text{SENT}^f_{j+2}} \alpha(\tilde{\imath})\Delta t^{j+2} \left\{ I_1[\beta_{r,j}(z)] \left[\mathscr{F}(\tilde{\imath})\big(U(t_n), U'(t_n)\big) - \mathscr{F}(\tilde{\imath})\big(u^n, \mu^n\big) \right] \right. \\
\qquad \left. + (-1)^j I_1[\beta_{r,j}(1-z)] \left[\mathscr{F}(\tilde{\imath})\big(U(t_{n+1}), U'(t_{n+1})\big) - \mathscr{F}(\tilde{\imath})\big(u^{n+1}, \mu^{n+1}\big) \right] \right\} + R^n, \\
\omega^{n+1} = -\Delta t A\phi_1(V)e^n + \phi_0(V)\omega^n \\
\quad + \displaystyle\sum_{j=0}^{r} \sum_{\tilde{\imath}\in\text{SENT}^f_{j+2}} \alpha(\tilde{\imath})\Delta t^{j+1} \left\{ I_2[\beta_{r,j}(z)] \left[\mathscr{F}(\tilde{\imath})\big(U(t_n), U'(t_n)\big) - \mathscr{F}(\tilde{\imath})\big(u^n, \mu^n\big) \right] \right. \\
\qquad \left. + (-1)^j I_2[\beta_{r,j}(1-z)] \left[\mathscr{F}(\tilde{\imath})\big(U(t_{n+1}), U'(t_{n+1})\big) - \mathscr{F}(\tilde{\imath})\big(u^{n+1}, \mu^{n+1}\big) \right] \right\} + r^n,
\end{cases}
$$
$$\tag{10.64}$$

with the initial conditions $e^0 = 0$, $\omega^0 = 0$.

For the convergence analysis, we quote the Gronwall's inequality (see, e.g. [39]), which plays an important role in the analysis.

Lemma 10.7 *Let λ be positive, a_k, b_k, $k \in \mathbb{N}$, be nonnegative and assume further that*

$$
a_k \leqslant (1 + \lambda\Delta t)a_{k-1} + \Delta t b_k, \qquad k \in \mathbb{N}.
$$

Then

$$
a_k \leqslant \exp(\lambda k \Delta t)\left(a_0 + \Delta t \sum_{m=1}^{k} b_m \right), \qquad k \in \mathbb{N}.
$$

Theorem 10.7 *With Assumptions 10.1 and 10.2, suppose that $u(x,t)$ satisfies suitable smoothness assumptions. If the time stepsize Δt satisfies*

$$
\Delta t^2 L\big(R, \rho(A)\big) \leqslant 1 \qquad and \qquad \Delta t \sum_{j=0}^{r} \sum_{\tilde{\imath}\in\text{SENT}^f_{j+2}} \alpha(\tilde{\imath}) \leqslant \frac{1}{4},
$$

then there exists a constant C such that

$$\|e^n\| \leqslant CT \exp\left(2T\left(1 + 2\sum_{j=0}^{r}\sum_{\tilde{i}\in\text{SENT}_{j+2}^f}\alpha(\tilde{i})\right)\right)(\Delta t^{2r+2} + \|\delta(\Delta x)\|),$$

$$\|\omega^n\| \leqslant CT \exp\left(2T\left(1 + 2\sum_{j=0}^{r}\sum_{\tilde{i}\in\text{SENT}_{j+2}^f}\alpha(\tilde{i})\right)\right)(\Delta t^{2r+2} + \|\delta(\Delta x)\|).$$

Proof The error system (10.64) can be rewritten in a compact form

$$\mathscr{F}(\tilde{t})_n \equiv \mathscr{F}(\tilde{t})\big(U(t_n), U'(t_n)\big) - \mathscr{F}(\tilde{t})\big(u^n, \mu^n\big)$$

and

$$\mathscr{F}(\tilde{t})_{n+1} \equiv \mathscr{F}(\tilde{t})\big(U(t_{n+1}), U'(t_{n+1})\big) - \mathscr{F}(\tilde{t})\big(u^{n+1}, \mu^{n+1}\big),$$

$$\begin{bmatrix} De^{n+1} \\ \omega^{n+1} \end{bmatrix} = \Omega(V)\begin{bmatrix} De^n \\ \omega^n \end{bmatrix} + \sum_{j=0}^{r}\sum_{\tilde{i}\in\text{SENT}_{j+2}^f}\Delta t^{j+1}\int_0^1 \Omega_j\big(\beta(z), V\big)dz\begin{bmatrix} 0 \\ \mathscr{F}(\tilde{t})_n \end{bmatrix}$$

$$+ \sum_{j=0}^{r}\sum_{\tilde{i}\in\text{SENT}_{j+2}^f}(-1)^j\Delta t^{j+1}\int_0^1 \Omega_j\big(\beta(1-z), V\big)dz\begin{bmatrix} 0 \\ \mathscr{F}(\tilde{t})_{n+1} \end{bmatrix} + \begin{bmatrix} DR^n \\ r^n \end{bmatrix},$$

$$(10.65)$$

where $\Omega(V)$ and $\Omega\big(\beta(z), V\big)$ were defined in (10.40) and (10.41), respectively. On the one hand, taking the l_2-norm on both sides of the first formula in (10.64) and (10.65) and summing up the outcomes, we have

$$\|e^{n+1}\| + \sqrt{(e^{n+1})^{\mathsf{T}}Ae^{n+1} + \|\omega^{n+1}\|^2} \leqslant \|e^n\| + \sqrt{(e^n)^{\mathsf{T}}Ae^n + \|\omega^n\|^2} + \Delta t\|\omega^n\|$$

$$+ \Delta t(1 + \Delta t)\sum_{j=0}^{r}\sum_{\tilde{i}\in\text{SENT}_{j+2}^f}\Delta t^j\left[\left\|\mathscr{F}(\tilde{t})\big(U(t_n), U'(t_n)\big) - \mathscr{F}(\tilde{t})\big(u^n, \mu^n\big)\right\|\right.$$

$$\left. + \left\|\mathscr{F}(\tilde{t})\big(U(t_{n+1}), U'(t_{n+1})\big) - \mathscr{F}(\tilde{t})\big(u^{n+1}, \mu^{n+1}\big)\right\|\right]$$

$$+ \|R^n\| + \sqrt{\|DR^n\|^2 + \|r^n\|^2}.$$

$$(10.66)$$

On the other hand, applying Assumption 10.2 to the right-hand side of (10.66) results in

$$\|e^{n+1}\| + \sqrt{(e^{n+1})^\mathsf{T} A e^{n+1} + \|\omega^{n+1}\|^2} \leqslant \|e^n\| + \sqrt{(e^n)^\mathsf{T} A e^n + \|\omega^n\|^2} + \Delta t \|\omega^n\|$$

$$+ \Delta t (1 + \Delta t) \sum_{j=0}^{r} \sum_{\tilde{\iota} \in \mathrm{SENT}_{j+2}^f} \alpha(\tilde{\iota}) \Delta t^j L\big(R, \rho(A)^{\lfloor \frac{j}{2} \rfloor}\big) \Big(\|e^n\| + \|\omega^n\| + \|e^{n+1}\|$$

$$+ \|\omega^{n+1}\| \Big) + \|R^n\| + \sqrt{\|DR^n\|^2 + \|r^n\|^2}.$$

$$(10.67)$$

As $\Delta t^2 L\big(R, \rho(A)\big) \leqslant 1$, the inequality (10.67) leads to

$$\|e^{n+1}\| + \sqrt{(e^{n+1})^\mathsf{T} A e^{n+1} + \|\omega^{n+1}\|^2}$$

$$\leqslant \left\{ 1 + \frac{\Delta t \left[1 + 2 \sum_{j=0}^{r} \sum_{\tilde{\iota} \in \mathrm{SENT}_{j+2}^f} \alpha(\tilde{\iota}) \right]}{1 - 2\Delta t \sum_{j=0}^{r} \sum_{\tilde{\iota} \in \mathrm{SENT}_{j+2}^f} \alpha(\tilde{\iota})} \right\} \left(\|e^n\| + \sqrt{(e^n)^\mathsf{T} A e^n + \|\omega^n\|^2} \right)$$

$$+ \frac{1}{1 - 2\Delta t \sum_{j=0}^{r} \sum_{\tilde{\iota} \in \mathrm{SENT}_{j+2}^f} \alpha(\tilde{\iota})} \left(\|R^n\| + \sqrt{\|DR^n\|^2 + \|r^n\|^2} \right). \qquad (10.68)$$

If the time stepsize Δt also satisfies $\Delta t \sum_{j=0}^{r} \sum_{\tilde{\iota} \in \mathrm{SENT}_{j+2}^f} \alpha(\tilde{\iota}) \leqslant \dfrac{1}{4}$, then the inequality (10.68) results in

$$\|e^{n+1}\| + \sqrt{(e^{n+1})^\mathsf{T} A e^{n+1} + \|\omega^{n+1}\|^2}$$

$$\leqslant \left\{ 1 + 2\Delta t \left[1 + 2 \sum_{j=0}^{r} \sum_{\tilde{\iota} \in \mathrm{SENT}_{j+2}^f} \alpha(\tilde{\iota}) \right] \right\} \left(\|e^n\| + \sqrt{(e^n)^\mathsf{T} A e^n + \|\omega^n\|^2} \right)$$

$$+ 2 \left(\|R^n\| + \sqrt{\|DR^n\|^2 + \|r^n\|^2} \right).$$

$$(10.69)$$

Note that R_j^n and r_j^n satisfy (10.60) and (10.61), respectively. Hence, there exists a constant C such that

$$\|R^n\| + \sqrt{\|DR^n\|^2 + \|r^n\|^2} \leqslant C\Delta t\left(\Delta t^{2r+2} + \|\delta(\Delta x)\|\right).$$

Applying the Gronwall's inequality (Lemma 10.7) to (10.69) yields

$$\|e^n\| + \sqrt{(e^n)^\mathsf{T} A e^n + \|\omega^n\|^2} \leqslant \exp\left(2n\Delta t\left(1 + 2\sum_{j=0}^{r}\sum_{\tilde{i}\in\mathrm{SENT}_{j+2}^f}\alpha(\tilde{i})\right)\right)$$

$$\times\left[\|e^0\| + \sqrt{(e^0)^\mathsf{T} A e^0 + \|\omega^0\|^2} + Cn\Delta t\left(\Delta t^{2r+2} + \|\delta(\Delta x)\|\right)\right].$$

Therefore, we obtain

$$\|e^n\| \leqslant CT\exp\left(2T\left(1 + 2\sum_{j=0}^{r}\sum_{\tilde{i}\in\mathrm{SENT}_{j+2}^f}\alpha(\tilde{i})\right)\right)\left(\Delta t^{2r+2} + \|\delta(\Delta x)\|\right),$$

$$\|\omega^n\| \leqslant CT\exp\left(2T\left(1 + 2\sum_{j=0}^{r}\sum_{\tilde{i}\in\mathrm{SENT}_{j+2}^f}\alpha(\tilde{i})\right)\right)\left(\Delta t^{2r+2} + \|\delta(\Delta x)\|\right).$$

$$(10.70)$$

Then the proof of this theorem is complete.   □

Obviously, it follows from the analysis of Theorem 10.7 that the precision of the derived HB time integrators can be of order $(2r + 2)$ in time, provided the exact solution $u(x, t)$ of the semilinear KG equations (10.1) satisfies $u(\cdot, t) \in C^{2r+2}([t_0, T])$. Unfortunately, however, existing numerical schemes, such as the finite difference method and the finite element method, have only limited accuracy for solving the semilinear KG equations (10.1). Here, in order to design high-order numerical methods, higher smoothness assumptions of the underlying problem are required. For instance, assume that $u(x, t)$ has appropriately continuous derivatives with respect to the temporal variable, and we use the following fourth-order finite difference approximation

$$\frac{\partial^2 u(x_j, t_n)}{\partial t^2} = \frac{-u(x_j, t_{n+2}) + 16u(x_j, t_{n+1}) - 30u(x_j, t_n) + 16u(x_j, t_{n-1}) - u(x_j, t_{n-2})}{\Delta t^4}$$

$$-\frac{\Delta t^4}{90}\frac{\partial^6 u(x_j, \hat{\xi}^n)}{\partial t^6}.$$

This implies that the approximation needs the solution to satisfy $u(\cdot, t) \in C^6([t_0, T])$ at last. However, under the same smoothness assumption of $u(\cdot, t) \in C^6([t_0, T])$, we can obtain a sixth-order HB time integrator by Theorem 10.7. In particular, as an important example, if the exact solution satisfies $u(\cdot, t) \in C^4([t_0, T])$, the well-known *leap-frog scheme* or the *Störmer–Verlet formula* is of order two in time. Fortunately, the derived HB time integrator with $r = 1$ can achieve fourth-order convergence. This is definitely a major improvement.

Moreover, under the smoothness assumption of $u(\cdot, t) \in C^2([t_0, T])$, and taking $r = 0$ in the time integration formula (10.22), we obtain an interesting scheme as follows:

$$
\begin{cases}
u^{n+1} = \phi_0(\mathscr{V})u^n + \Delta t \phi_1(\mathscr{V})\mu^n \\
\qquad + \Delta t^2 \left\{ I_1[\beta_{0,0}(z)]f(u^n) + (-1)^j I_1[\beta_{0,0}(1-z)]f(u^{n+1}) \right\}, \\
\mu^{n+1} = -\Delta t \mathscr{A} \phi_1(\mathscr{V})u^n + \phi_0(\mathscr{V})\mu^n \\
\qquad + \Delta t \left\{ I_2[\beta_{0,0}(z)]f(u^n) + (-1)^j I_2[\beta_{0,0}(1-z)]f(u^{n+1}) \right\}.
\end{cases}
\tag{10.71}
$$

The scheme (10.71) is of order two.

Remark 10.2 Compared with the well-known *Störmer–Verlet method*, the second-order HB time integrator needs a much weaker smoothness assumption, whereas the interesting second-order scheme (10.71) exhibits excellent numerical behaviour. This remarkable superiority will be shown in the numerical experiments.

10.5 Spatial Discretisation

As stated above, the symmetric and arbitrarily high-order time integration formula (10.22) has been presented in operatorial terms in an infinite-dimensional function space $\mathscr{X}$. In order to render them into proper numerical algorithms, we need to replace the differential operator $\mathscr{A}$ with an suitable differentiation matrix A. Keeping the stability and convergence analysis in mind, we approximate the differential operator $\mathscr{A}$ by a positive semi-definite matrix A. Fortunately, there exists a great body of research investigating the replacement of spatial derivatives of nonlinear system (10.1) with periodic boundary conditions (10.2), and it is not difficult to find positive semi-definite differentiation matrices in this setting. Here, we mainly consider two types of spatial discretisations: Symmetric finite difference and Fourier spectral collocation discretisations.

1. *Symmetric Finite Difference (SFD)* (see, e.g. [37])

As is known, finite difference methods are achieved when approximating a function by local polynomial interpolation. Its derivatives are then approximated by differentiating this local polynomial, where 'local' refers to the use of nearby grid points to approximate the function or its derivative at a given point. In general, a finite difference approximation is of moderate order. For instance, we approximate the operator $\mathscr{A}$ by the following differentiation matrix

$$
A_{\text{sfd}} = \frac{a^2}{12\Delta x^2}
\begin{bmatrix}
30 & -16 & 1 & & & & 1 & -16 \\
-16 & 30 & -16 & 1 & & & & 1 \\
1 & -16 & 30 & -16 & 1 & & & \\
& \ddots & \ddots & \ddots & \ddots & \ddots & & \\
& & 1 & -16 & 30 & -16 & 1 & \\
1 & & & 1 & -16 & 30 & -16 \\
-16 & 1 & & & 1 & -16 & 30
\end{bmatrix}_{M \times M} .
$$

The approximation is of order four and the differentiation matrix A_{sfd} is clearly positive semi-definite.

2. *Fourier Spectral Collocation (FSC)* (see, e.g. [40, 41])

A distinctive feature of Spectral methods is their global nature, and the computation at any given point depends not only on the information at neighbouring points, but on the entire domain. The topic of spectral methods is very wide, and various methods and sub-methods have been proposed and are actively used. The Fourier spectral collocation method is our method of choice, which can be presented as a limit of local finite difference approximations of increasing order of accuracy (see [40]). We concentrate on differentiation being performed in the physical space. The key point here is to interpolate the solution at the nodal values using a trigonometric polynomial. The entries of the second-derivative Fourier differentiation matrix $A_{\text{fsc}} = (a_{kj})_{M \times M}$ are given by

$$
a_{kj} =
\begin{cases}
\dfrac{(-1)^{k+j}}{2} a^2 \sin^{-2}\left(\dfrac{(k-j)\pi}{M}\right), & k \neq j, \\[2ex]
a^2\left(\dfrac{M^2}{12} + \dfrac{1}{6}\right), & k = j.
\end{cases}
\tag{10.72}
$$

It is known that the main appeal of spectral methods is that they exhibit spectral convergence to $\mathscr{A}$: the error decays for C^∞ functions faster than $O(M^{-\alpha})\ \forall\, \alpha > 0$ for sufficiently large M. Another advantage is that the differentiation matrix A_{fsc} is positive semi-definite.

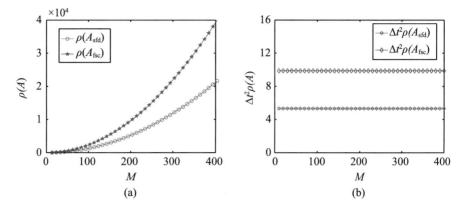

Fig. 10.1 Plots of $\rho(A)$ and $\Delta t^2 \rho(A)$ for the differentiation matrices A_{sfd} and A_{fsc} for $M = 10i$ and $i = 1, 2, \cdots, 40$

Figure 10.1 illustrates the size of the spectral radius for the differentiation matrices A_{sfd} and A_{fsc}. In Fig. 10.1a, we show the spectral radius of A_{sfd} and A_{fsc} as a function of M. If we take the time stepsize $\Delta t = \dfrac{2\pi}{M}, M = 10i$ for $i = 1, 2, \cdots, 40$, Fig. 10.1b shows that $\Delta t^2 \rho(A)$ is a constant. i.e. that $\rho(A) = O(\Delta t^{-2})$. Therefore, a small Δt $\left(\Delta t \leqslant \dfrac{2\pi}{M} \right)$ can be chosen to guarantee stability and convergence and obtain effective numerical methods.

We have already noted that energy conservation (10.3) is a crucial property of semilinear KG equations (10.1)–(10.2). As we approximate the operator $\mathscr{A}$ by a positive semi-definite differentiation matrix A, there is a corresponding discrete energy conservation law, which can be characterized by the following form:

$$\tilde{E}(t) = \frac{\Delta x}{2} \|u'(t)\|^2 + \frac{\Delta x}{2} \|Du(t)\|^2 + \Delta x \sum_{j=0}^{M-1} V\big(u_j(t)\big) \equiv \tilde{E}(t_0), \qquad (10.73)$$

where the norm $\| \cdot \|$ is the standard vector 2-norm and $\Delta x = 2\pi/M$ is the spatial grid size. Actually, this energy can be thought of as an approximate energy (a semidiscrete energy) of the original continuous system. Consequently, discussing numerical experiments, we will also test the effectiveness of HB time integrators in preserving (10.73).

We are now concerned with how accurately the discrete energy conservation law (10.73) is preserved by the HB time integrators. We first rewrite the semidiscrete system of the nonlinear KG equation (10.1)–(10.2) in the form

$$\begin{bmatrix} u(t) \\ u'(t) \end{bmatrix}' = \begin{bmatrix} u'(t) \\ -Au(t) + f\big(u(t)\big) \end{bmatrix}. \qquad (10.74)$$

On the one hand, we note that if we define

$$H\big(u(t), u'(t)\big) = \frac{1}{2}u'(t)^\mathsf{T} u'(t) + \frac{1}{2}u(t)^\mathsf{T} Au(t) + \tilde{V}\big(u(t)\big), \tag{10.75}$$

where $\tilde{V}\big(u(t)\big) = \sum_{j=0}^{M-1} V\big(u_j(t)\big)$, the discrete energy conservation law (10.73) is identical to

$$\Delta x\, H\big(u(t), u'(t)\big) \equiv \Delta x\, H\big(u(t_0), u'(t_0)\big). \tag{10.76}$$

By letting $y(t) = \big[u(t)^\mathsf{T}, u'(t)^\mathsf{T}\big]^\mathsf{T}$, where $u(t)$ and $u'(t)$ are the exact solution of (10.74) and its derivative, respectively, the system (10.74) can be further expressed as:

$$y'(t) = J^{-1}\nabla H\big(y(t)\big) \quad \text{with} \quad J = \begin{bmatrix} 0 & -I_{M\times M} \\ I_{M\times M} & 0 \end{bmatrix}. \tag{10.77}$$

On the other hand, if the numerical solutions u^{n+1} and μ^{n+1} are regarded as functions of Δt, and by denoting $z(t_n + \xi\,\Delta t) = \big[u^n(\xi\,\Delta t)^\mathsf{T}, \mu^n(\xi\,\Delta t)^\mathsf{T}\big]^\mathsf{T}$, it can be observed that the solutions of the HB time integration formula (10.22) satisfy

$$z'(t_n + \xi\,\Delta t) = \begin{bmatrix} \mu^n(\xi\,\Delta t) \\ \Upsilon^n(\xi\,\Delta t, u) \end{bmatrix}, \tag{10.78}$$

where

$$\Upsilon^n(\xi\,\Delta t, u) \equiv -Au^n(\xi\,\Delta t)$$

$$+ \sum_{j=0}^{r} \sum_{\tilde{i}\in\mathrm{SENT}_{j+2}^f} \alpha(\tilde{i})\Delta t^j \Big[\beta_{r,j}(\xi)\mathscr{F}\big(u^n, \mu^n\big) + (-1)^j\beta_{r,j}(1-\xi)\mathscr{F}\big(u^{n+1}, \mu^{n+1}\big)\Big],$$

$\xi \in [0, 1]$, and $z(t_n + \xi\,\Delta t)$ satisfies:

$$z(t_n + \xi\,\Delta t)\big|_{\xi=0} = \begin{bmatrix} u^n \\ \mu^n \end{bmatrix} \quad \text{and} \quad z(t_n + \xi\,\Delta t)\big|_{\xi=1} = \begin{bmatrix} u^{n+1} \\ \mu^{n+1} \end{bmatrix}, \quad 0 \leqslant n \leqslant N.$$

Theorem 10.8 *Let u^n and μ^n be the solutions of the HB time integration formula (10.22). Then the discrete energy defined in (10.75) satisfies*

$$\max_{0\leqslant n\leqslant N} \big|H(u^n, u'^n) - H(u^0, u'^0)\big| = \mathcal{O}(\Delta t^{2r+2}), \tag{10.79}$$

and this implies the order of preservation of the discrete energy is $2r + 2$.

Proof Using (10.75) and (10.78), we obtain

$$H\big(z(t_{n+1})\big) - H\big(z(t_n)\big) = \Delta t \int_0^1 \nabla H\big(z(t_n + \xi \Delta t)\big)^{\mathsf{T}} z'(t_n + \xi \Delta t) d\xi$$

$$= \Delta t \int_0^1 \left[\big(Au^n(\xi \Delta t) - f\big(u^n(\xi \Delta t)\big)\big)^{\mathsf{T}}, \mu^n(\xi \Delta t)^{\mathsf{T}} \right] \left[\begin{matrix} \mu^n(\xi \Delta t) \\ \Upsilon^n(\xi \Delta t, u) \end{matrix} \right] d\xi$$

$$= (-1)^{r+1} \Delta t^{2r+3} \left(\int_0^1 \mu^n(\xi \Delta t) \xi^{r+1}(1-\xi)^{r+1} d\xi \right)^{\mathsf{T}} \frac{f_t^{(2r+2)}\big(u^n(\theta^n \Delta t)\big)}{(2r+2)!},$$

$$\theta^n \in [0, 1].$$

This leads to

$$\big| H\big(z(t_{n+1})\big) - H\big(z(t_n)\big) \big| = \mathscr{O}(\Delta t^{2r+3}).$$

It then follows from

$$\big| H\big(z(t_n)\big) - H\big(z(t_0)\big) \big| \leqslant \sum_{j=0}^{n-1} \big| H\big(z(t_{j+1})\big) - H\big(z(t_j)\big) \big| = n\mathscr{O}(\Delta t^{2r+3}),$$

that

$$\max_{0 \leqslant n \leqslant N} \big| H\big(z(t_n)\big) - H\big(z(t_0)\big) \big| = \mathscr{O}(\Delta t^{2r+2}).$$

The proof of this is complete.   □

10.6   Waveform Relaxation and Its Convergence

The previous sections derived and analysed the fully discrete scheme for (10.1)–(10.2) and presented its properties. However, the scheme (10.63) is implicit in general and iteration cannot be avoided in practical computation. In this section we introduce a *waveform relaxation method* as a suitable iterative procedure. The waveform relaxation method has been investigated by many authors (see, e.g. [42–46]).

For simplicity, in terms of the notation in (10.46), we first rewrite the fully discrete scheme (10.63),

$$
\left\{
\begin{aligned}
u^{n+1} &= \phi_0(V)u^n + \Delta t \phi_1(V)\mu^n + \sum_{j=0}^{r} \sum_{\tilde{\imath}\in\mathrm{SENT}_{j+2}^f} \alpha(\tilde{\imath})\Delta t^{j+2} \\
&\quad \times \Big[I_1[\beta_j(z)]\mathscr{F}(\tilde{\imath})(u^n,\mu^n) + (-1)^j I_1[\beta_j(1-z)]\mathscr{F}(\tilde{\imath})(u^{n+1},\mu^{n+1}) \Big], \\
\mu^{n+1} &= -\Delta t A\phi_1(V)u^n + \phi_0(V)\mu^n + \sum_{j=0}^{r} \sum_{\tilde{\imath}\in\mathrm{SENT}_{j+2}^f} \alpha(\tilde{\imath})\Delta t^{j+1} \\
&\quad \times \Big[I_2[\beta_j(z)]\mathscr{F}(\tilde{\imath})(u^n,\mu^n) + (-1)^j I_2[\beta_j(1-z)]\mathscr{F}(\tilde{\imath})(u^{n+1},\mu^{n+1}) \Big],
\end{aligned}
\right.
$$

where $I_1[\beta_j(z)]$, $I_2[\beta_j(z)]$, $I_1[\beta_j(1-z)]$ and $I_2[\beta_j(1-z)]$ have been defined in (10.19)–(10.18). We then define the waveform relaxation method as follows:

$$
\left\{
\begin{aligned}
u_{[0]}^{n+1} &= \phi_0(V)u^n + \Delta t\phi_1(V)\mu^n, \\
\mu_{[0]}^{n+1} &= -\Delta t A\phi_1(V)u^n + \phi_0(V)\mu^n,
\end{aligned}
\right.
\tag{10.80}
$$

and subsequently iterate

$$
\left\{
\begin{aligned}
u_{[m+1]}^{n+1} &= u_{[0]}^{n+1} + \sum_{j=0}^{r} \sum_{\tilde{\imath}\in\mathrm{SENT}_{j+2}^f} \alpha(\tilde{\imath})\Delta t^{j+2}\Big\{ I_1[\beta_{r,j}(z)]\mathscr{F}(\tilde{\imath})(u^n,\mu^n) \\
&\quad + (-1)^j I_1[\beta_{r,j}(1-z)]\mathscr{F}(\tilde{\imath})(u_{[m]}^{n+1},\mu_{[m]}^{n+1}) \Big\}, \\
\mu_{[m+1]}^{n+1} &= \mu_{[0]}^{n+1} + \sum_{j=0}^{r} \sum_{\tilde{\imath}\in\mathrm{SENT}_{j+2}^f} \alpha(\tilde{\imath})\Delta t^{j+1}\Big\{ I_2[\beta_{r,j}(z)]\mathscr{F}(\tilde{\imath})(u^n,\mu^n) \\
&\quad + (-1)^j I_2[\beta_{r,j}(1-z)]\mathscr{F}(\tilde{\imath})(u_{[m]}^{n+1},\mu_{[m]}^{n+1}) \Big\}
\end{aligned}
\right.
\tag{10.81}
$$

for $m = 0, 1, \cdots$.

In what follows, we analyse the convergence of the algorithm (10.80)–(10.81).

Theorem 10.9 *Suppose that f satisfies Assumptions* 10.1 *and* 10.2. *Under the conditions*

$$\Delta t^2 L(R, \rho(A)) \leqslant 1 \qquad and \qquad \Delta t(1 + \Delta t) \sum_{j=0}^{r} \sum_{\tilde{\imath} \in \text{SENT}^f_{j+2}} \alpha(\tilde{\imath}) < 1,$$

the iterative procedure determined by (10.80)–(10.81) *is convergent.*

Proof According to Assumption 10.2 and (10.81), the following inequalities are true:

$$
\begin{cases}
\| u^{n+1}_{[m+1]} - u^{n+1}_{[m]} \| \\[4pt]
\leqslant \Delta t^2 \sum_{j=0}^{r} \sum_{\tilde{\imath} \in \text{SENT}^f_{j+2}} \alpha(\tilde{\imath}) \Delta t^j L(R, \rho(A)^{\lfloor \frac{j}{2} \rfloor}) \Big(\| u^{n+1}_{[m]} - u^{n+1}_{[m-1]} \| + \| \mu^{n+1}_{[m]} - \mu^{n+1}_{[m-1]} \| \Big), \\[10pt]
\| \mu^{n+1}_{[m+1]} - \mu^{n+1}_{[m]} \| \\[4pt]
\leqslant \Delta t \sum_{j=0}^{r} \sum_{\tilde{\imath} \in \text{SENT}^f_{j+2}} \alpha(\tilde{\imath}) \Delta t^j L(R, \rho(A)^{\lfloor \frac{j}{2} \rfloor}) \Big(\| u^{n+1}_{[m]} - u^{n+1}_{[m-1]} \| + \| \mu^{n+1}_{[m]} - \mu^{n+1}_{[m-1]} \| \Big).
\end{cases}
$$

$$(10.82)$$

Summing up (10.82) and noting that $\Delta t^2 L(R, \rho(A)) \leqslant 1$, we obtain

$$
\| u^{n+1}_{[m+1]} - u^{n+1}_{[m]} \| + \| \mu^{n+1}_{[m+1]} - \mu^{n+1}_{[m]} \|
$$

$$
\leqslant \Delta t(1 + \Delta t) \sum_{j=0}^{r} \sum_{\tilde{\imath} \in \text{SENT}^f_{j+2}} \alpha(\tilde{\imath}) \Big(\| u^{n+1}_{[m]} - u^{n+1}_{[m-1]} \| + \| \mu^{n+1}_{[m]} - \mu^{n+1}_{[m-1]} \| \Big).
$$

An argument by induction then gives

$$
\| u^{n+1}_{[m+1]} - u^{n+1}_{[m]} \| + \| \mu^{n+1}_{[m+1]} - \mu^{n+1}_{[m]} \|
$$

$$
\leqslant \left[\Delta t(1 + \Delta t) \sum_{j=0}^{r} \sum_{\tilde{\imath} \in \text{SENT}^f_{j+2}} \alpha(\tilde{\imath}) \right]^m \Big(\| u^{n+1}_{[1]} - u^{n+1}_{[0]} \| + \| \mu^{n+1}_{[1]} - \mu^{n+1}_{[0]} \| \Big).
$$

The condition $\Delta t(1 + \Delta t) \sum_{j=0}^{r} \sum_{\tilde{\imath} \in \text{SENT}^f_{j+2}} \alpha(\tilde{\imath}) < 1$ results in

$$\lim_{m \to +\infty} \Big(\| u^{n+1}_{[m+1]} - u^{n+1}_{[m]} \| + \| \mu^{n+1}_{[m+1]} - \mu^{n+1}_{[m]} \| \Big) = 0. \tag{10.83}$$

Therefore, the iterative procedure (10.80)–(10.81) is convergent.   □

10.7 Numerical Experiments

For the demonstration of the properties and performance of the HB time integrator, in this section, we derive three practical time integration formulae and use them to illustrate the solution of two semilinear wave equations.

The choice of $r = 0$ in (10.15) yields the first example of a symmetric time-stepping integrator for (10.1)–(10.2):

$$\beta_{0,0}(z) = (1 - z), \tag{10.84}$$

and the corresponding time integration formula, determined by (10.84) and (10.18)–(10.21), is defined by HB0.

As the second example, we take $r = 1$ in (10.15)

$$\beta_{1,0}(z) = (1 - z)^2(1 + 2z), \quad \beta_{1,1}(z) = z(1 - z)^2. \tag{10.85}$$

The time integration formula determined by (10.85) and (10.18)–(10.21) is denoted by HB1.

Letting $r = 2$ in (10.15) gives the third example:

$$\beta_{2,0}(z) = (1 - z)^3(1 + 3z + 6z^2), \quad \beta_{2,1}(z) = z(1 - z)^3(1 + 3z),$$

$$\beta_{2,2}(z) = \frac{1}{2}z^2(1 - z)^3. \tag{10.86}$$

The corresponding time integration formula determined by (10.86) and (10.18)–(10.21) as HB2.

In order to compare different algorithms, we briefly describe a number of standard finite difference schemes and method-of-lines schemes for the semilinear KG equation (see, e.g. [9, 10, 39]).

1. *Standard Finite Difference Schemes*
   Let u_j^n be the approximation of $u(x_j, t_n)$ for $j = 0, 1, \cdots, M - 1$ and $n = 0, 1, \cdots, N$. We also introduce the standard central difference operators

$$\delta_t^2 u_j^n = \frac{u_j^{n+1} - 2u_j^n + u_j^{n-1}}{\Delta t^2} \quad \text{and} \quad \delta_x^2 u_j^n = \frac{u_{j+1}^n - 2u_j^n + u_{j-1}^n}{\Delta x^2}.$$

We here consider three frequently used *finite difference schemes* to discretise the semilinear KG equation:

- An explicit finite difference scheme Expt-FD

$$\delta_t^2 u_j^n - a^2 \delta_x^2 u_j^n = f(u_j^n);$$

- Semi-implicit finite difference scheme **Simpt-FD**

$$\delta_t^2 u_j^n - \frac{a^2}{2}\left(\delta_x^2 u_j^{n+1} + \delta_x^2 u_j^{n-1}\right) = f(u_j^n);$$

- Compact finite difference scheme **Compt-FD**

$$\left(I + \frac{\Delta x^2}{12}\delta_x^2\right)\delta_t^2 u_j^n - \frac{a^2}{2}\left(\delta_x^2 u_j^{n+1} + \delta_x^2 u_j^{n-1}\right) = \left(I + \frac{\Delta x^2}{12}\delta_x^2\right)f(u_j^n).$$

2. *Method-of-lines Schemes*

The method-of-lines approach to the approximation of (10.1)–(10.2) is composed of two stages: space and time discretisations. We first approximate the spatial differential operator $\mathscr{A}$ to obtain a semidiscrete scheme of the form

$$u''(t) + Au(t) = f\big(u(t)\big),$$

where A is a symmetric positive semi-definite matrix. We then use an ODE solver to deal with the semidiscrete scheme. Here, the time integrators we select for comparison are

- **Gauss2s4**: the two-stage Gauss method of order four from [31];
- **Gauss3s6**: the three-stage fourth-order Gauss method in [31];
- **RKN3s4**: the three-stage Runge–Kutta–Nyström (RKN) method of order four from [31];
- **IRKN2s4**: the two-stage implicit symplectic RKN method of order four derived in [47];
- **IRKN3s6**: the three-stage implicit symplectic RKN method of order six derived in [47];
- **SV**: classical Störmer–Verlet formula [31].

For the time integrators **HB0**, **HB1** and **HB2** derived in this chapter, we use the tolerance 10^{-15} and choose $m = 2$ in the waveform relaxation algorithm (10.80)–(10.81), which implies that just one iteration is needed at each step. Consequently, these two integrators can be implemented at lower cost. Here, it should be noted that when the error of a method under consideration is very large for some Δt, we do not plot the corresponding points in efficiency curves. Moreover, in order to compute the convergence order, we denote

$$\text{EU}(\Delta x, \Delta t) = \max_{0 \leqslant n \leqslant N} \sqrt{\Delta x \sum_{i=0}^{M-1}(U_i^n - u_i^n)^2}$$

and

$$\mathrm{EH}(\Delta x, \Delta t) = \max_{0 \leqslant n \leqslant N} \left| H(u^n, u'^n) - H(u^0, u'^0) \right|.$$

The computational order of the method is calculated with the following formulae:

$$\log_2 \left(\frac{\mathrm{EU}(\Delta x, \Delta t)}{\mathrm{EU}(\Delta x, \Delta t/2)} \right) \quad \text{and} \quad \log_2 \left(\frac{\mathrm{EH}(\Delta x, \Delta t)}{\mathrm{EH}(\Delta x, \Delta t/2)} \right).$$

Problem 10.1 We consider the semilinear KG equation

$$\frac{\partial^2 u(x,t)}{\partial t^2} - a^2 \frac{\partial^2 u(x,t)}{\partial x^2} + au(x,t) - bu^3(x,t) = 0,$$

in the region $(x,t) \in [-20, 20] \times [0, T]$ with the initial conditions

$$u(x,0) = \sqrt{\frac{2a}{b}} \operatorname{sech}(\lambda x), \qquad u_t(x,0) = c\lambda \sqrt{\frac{2a}{b}} \operatorname{sech}(\lambda x) \tanh(\lambda x),$$

where $\lambda = \sqrt{a/(a^2 - c^2)}$ and $a, b, a^2 - c^2 > 0$. The exact solution of Problem 10.1 is

$$u(x,t) = \sqrt{\frac{2a}{b}} \operatorname{sech}(\lambda(x - ct)).$$

The real parameter $\sqrt{2a/b}$ represents the amplitude of a soliton which travels with velocity c. The potential function is $V(u) = au^2/2 - bu^4/4$. The problem can be found in [23]. We consider the parameters $a = 0.3, b = 1$ and $c = 0.25$ which are similar to those in [23].

In Figs. 10.2 and 10.3, we integrate the Problem 10.1 on the region $(x,t) \in [-20, 20] \times [0, 10]$ by using the time integrator **HB2**, coupled with the fourth-order symmetric finite difference (**SFD**) and Fourier spectral collocation (**FSC**). The graphs of errors are shown in Figs. 10.2 and 10.3 with the time stepsize $\Delta t = 0.01$ and different values of M. The numerical results demonstrate that the accuracy of the spatial discretisation is consistent with the theory presented in this chapter. It is evident that the Fourier spectral collocation method is the best choice to discretise the spatial variable.

Table 10.1 provides the computational results with $M = 800$. The data demonstrate that the temporal convergence orders of **HB0**, **HB1** and **HB2** are second, fourth and sixth, respectively. The results show that the temporal accuracy is completely consistent with the theory presented in Theorem 10.7.

To compare the integrators presented in this chapter with classical finite difference and method-of-lines schemes, we integrate the problem in the region $(x,t) \in$

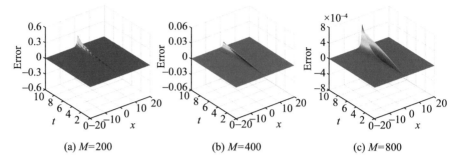

(a) M=200   (b) M=400   (c) M=800

Fig. 10.2 The errors for Problem 10.1 obtained by combining the time integrator **HB2** with the fourth-order finite difference spatial discretisation for $\Delta t = 0.01$ with $M = 200, 400$ and 800

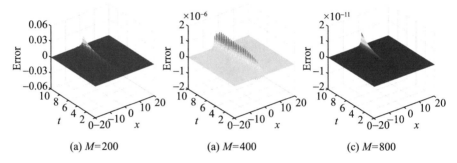

(a) M=200   (a) M=400   (c) M=800

Fig. 10.3 The errors for Problem 10.1 obtained by combining the time integrator **HB2** with Fourier spectral collocation method for $\Delta t = 0.01$ with $M = 200, 400$ and 800

Table 10.1 Numerical convergence in time with different Δt, fixed $M = 800$ and up to $T = 10$

Δt	HB0		HB1		HB2	
	EU$(\Delta x, \Delta t)$	Order	EU$(\Delta x, \Delta t)$	Order	EU$(\Delta x, \Delta t)$	Order
0.8	1.7941×10^{-1}	*	2.0990×10^{-3}	*	1.2742×10^{-4}	*
0.4	4.3627×10^{-2}	2.0400	2.6159×10^{-4}	3.0043	3.1967×10^{-6}	5.3168
0.2	1.0065×10^{-2}	2.1158	1.8650×10^{-5}	3.8101	5.3224×10^{-8}	5.9084
0.1	2.4614×10^{-3}	2.0318	1.2006×10^{-6}	3.9573	8.5113×10^{-10}	5.9666
0.05	6.1189×10^{-4}	2.0081	7.5579×10^{-8}	3.9896	1.2662×10^{-11}	6.0708

$[-20, 20] \times [0, 10]$ with different time stepsizes Δt, and the number of spatial nodal values is M. The numerical results are shown in Fig. 10.4. We compare the integrators presented in this chapter with the standard finite difference schemes with stepsizes $\Delta t = 0.01 \times 2^{3-j}$ for $j = 0, 1, 2, 3$ and $M = 1000$ for the finite difference schemes **Expt-FD**, **Simpt-FD** and **Compt-FD** and $M = 800$ for **HB0-FSC**, **HB1-FSC** and **HB2-FSC**. The logarithms of the global errors GE $= \|u(t_n) - u^n\|_\infty$ are plotted in Fig. 10.4a.

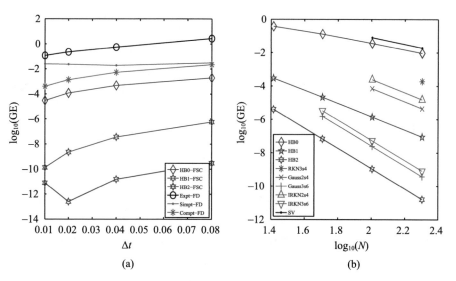

Fig. 10.4 The efficiency curves for Problem 10.1: (**a**) Comparison with standard finite difference schemes, (**b**) Comparison with method-of-lines schemes

Compared with the method-of-lines schemes, we discretise the spatial derivative using Fourier spectral collocation method with fixed $M = 800$ and integrate the KG equation with $\Delta t = 0.2/2^j$ for $j = 0, 1, 2, 3$. The efficiency curves (accuracy versus the computational cost measured by the number of function evaluations required by each method) are shown in Fig. 10.4b.

In conclusion, the numerical results in Fig. 10.4 demonstrate that the time integrators **HB0**, **HB1** and **HB2** derived in this chapter, combined with Fourier spectral collocation, have much better accuracy and are more efficient than those occurring in the literature.

The numerical results in Fig. 10.5 present the error of the semidiscrete energy conservation law as a function of the time-step calculated by $\tilde{E}(t)$, where EH $= |\tilde{E}(t) - \tilde{E}(t_0)|$. It can be observed form Fig. 10.5 that the error of **HB0** is $\approx 10^{-4}$, for **HB1** it is $\approx 10^{-11}$, while that of **HB2** is $\approx 10^{-13}$. Moreover, the convergence orders of the preservation of the discrete energy by the **HB** time integrators are computed which are listed in Table 10.2. The numerical results show that the accuracy of discrete energy preservation by **HB0** is of order two, by **HB1** of order four and by **HB2** of order six.

Problem 10.2 We consider the sine-Gordon equation

$$\frac{\partial^2 u}{\partial t^2}(x, t) - \frac{\partial^2 u}{\partial x^2}(x, t) + \sin(u(x, t)) = 0$$

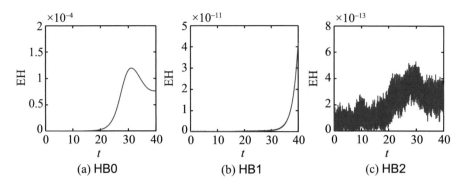

Fig. 10.5 Discrete energy conservation by HB0, HB1 and HB2 with the spatial discretisation by Fourier spectral collocation with $M = 800$ up to $T = 40$, using $\Delta t = 0.02$

Table 10.2 Numerical precision of the preservation of the semidiscrete energy up to $T = 40$ with various Δt and fixed $M = 200$

Δt	HB0		HB1		HB2	
	$\text{EH}(\Delta x, \Delta t)$	Order	$\text{EH}(\Delta x, \Delta t)$	Order	$\text{EH}(\Delta x, \Delta t)$	Order
0.16	1.8989×10^{-3}	*	5.2428×10^{-6}	*	1.1049×10^{-8}	*
0.08	4.7939×10^{-4}	1.9859	3.2875×10^{-7}	3.9952	1.7178×10^{-10}	6.0072
0.04	1.2005×10^{-4}	1.9976	2.0564×10^{-8}	3.9988	2.6613×10^{-12}	6.0123
0.02	3.0026×10^{-5}	1.9994	1.2860×10^{-9}	3.9992	5.0293×10^{-14}	5.7256
0.01	7.5071×10^{-6}	1.9999	8.0451×10^{-11}	3.9986	—	—

in the region $-20 \leqslant x \leqslant 20, 0 \leqslant t \leqslant T$, subject to the initial conditions

$$u(x, 0) = 0, \qquad u_t(x, 0) = 4 \operatorname{sech}(x/\sqrt{1 + c^2})/\sqrt{1 + c^2}.$$

The exact solution of Problem 10.2 is

$$u(x, t) = 4 \arctan\left(c^{-1} \sin(ct/\sqrt{1 + c^2}) \operatorname{sech}(x/\sqrt{1 + c^2})\right).$$

This problem is known as *breather solution* of the sine-Gordon equation and represents a pulse-type structure of a soliton. The parameter c is the velocity and we choose $c = 0.5$. The potential function is $V(u) = 1 - \cos(u)$. Problem 10.2 is integrated by HB2, coupled either with the fourth-order symmetric finite difference SFD or Fourier spectral collocation FSC. The error graphs are shown in Figs. 10.6 and 10.7 with $\Delta t = 0.01$ and several values of M. They demonstrate how the accuracy of the spatial discretisation varies with M, and also indicate that the Fourier spectral collocation FSC is decisively superior to the fourth-order symmetric finite difference SFD.

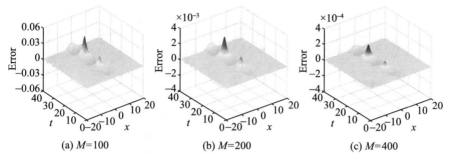

Fig. 10.6 The error for the sine-Gordon equation, blending the time integrator HB2 with fourth-order finite difference spatial discretisation for $\Delta t = 0.01$ and $M = 100, 200, 400$

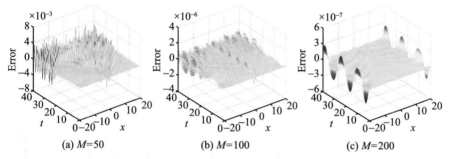

Fig. 10.7 The errors blending the time integrator HB2 with Fourier spectral method for $\Delta t = 0.01$ and $M = 50, 100, 200$

Table 10.3 Numerical convergence in time with different Δt, fixed $M = 200$ and up to $T = 40$

Δt	HB0		HB1		HB2	
	EU$(\Delta x, \Delta t)$	Order	EU$(\Delta x, \Delta t)$	Order	EU$(\Delta x, \Delta t)$	Order
0.8	21.85583982	*	2.762406385	*	2.2240×10^{-1}	*
0.4	5.486416853	1.9941	1.2514×10^{-1}	4.4643	3.6983×10^{-3}	5.9101
0.2	1.176387217	2.2215	7.3631×10^{-3}	4.0871	4.8228×10^{-5}	6.2609
0.1	2.8235×10^{-1}	2.0588	4.5373×10^{-4}	4.0204	7.3285×10^{-7}	6.0402
0.05	6.9931×10^{-2}	2.0135	2.8264×10^{-5}	4.0048	–	–

The computational results in Table 10.3 demonstrate that the temporal convergence orders of HB0, HB1 and HB2 are of two, four and six, respectively. The results again verify the convergence accuracy in time is consistent with the theory in Theorem 10.7.

The efficiency curves are shown in Fig. 10.8. In order to compare the integrators with a standard finite difference scheme, in Fig. 10.8a we integrate the problem for $\Delta t = 0.04, 0.03, 0.02, 0.01$. We use $M = 1000$ for the finite difference scheme Expt-FD, Simpt-FD and Compt-FD, and $M = 200$ for the HB0-FSC, HB1-FSC and HB2-FSC.

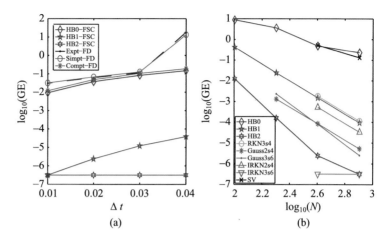

Fig. 10.8 Efficiency curves for Problem 10.2: (**a**) Comparison with standard finite difference schemes, (**b**) Comparison with method-of-lines schemes

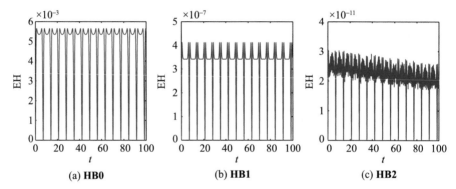

Fig. 10.9 Energy conservation by HB0, HB1 and HB2, both blended with FSC, using $M = 200$, $\Delta t = 0.02$ and $T = 100$

In Fig. 10.8b we compare the integrators presented in this chapter with method-of-lines schemes. The problem is integrated over the time interval [0, 40] with fixed $M = 200$ and time stepsizes $\Delta t = 0.4/2^j$ for $j = 0, 1, 2, 3$. It can be observed that the time integrators HB0, HB1 and HB2, coupled with Fourier spectral collocation, are more efficient than other chosen methods.

The numerical results in Fig. 10.9 represent the error of the semidiscrete energy conservation law. It can be seen that the error does not grow with time. The errors obtained by HB0, HB1 and HB2 reach magnitudes of $\approx 10^{-3}$, $\approx 10^{-7}$ and $\approx 10^{-11}$, respectively. The precisions of the preservation of the discrete energy by the HB time integrators are listed in Table 10.4. It is shown that the accuracy of the discrete energy preservation by HB0 is of order two, by HB1 is of order four and by HB2 is of order six.

Table 10.4 Numerical precision of the preservation of the semidiscrete energy up to $T = 100$ with different Δt and fixed $M = 200$

Δt	HB0		HB1		HB2	
	$EH(\Delta x, \Delta t)$	Order	$EH(\Delta x, \Delta t)$	Order	$EH(\Delta x, \Delta t)$	Order
0.16	3.6368×10^{-1}	*	1.7078×10^{-3}	*	8.0039×10^{-6}	*
0.08	9.0444×10^{-2}	2.0076	1.0578×10^{-4}	4.0130	1.2237×10^{-7}	6.0314
0.04	2.2580×10^{-2}	2.0020	6.5969×10^{-6}	4.0031	1.9184×10^{-9}	5.9952
0.02	5.6432×10^{-3}	2.0005	4.1208×10^{-7}	4.0008	3.0326×10^{-11}	5.9832
0.01	1.4107×10^{-3}	2.0001	2.5774×10^{-8}	3.9989	–	–

Below is an example of a high-dimensional problem.

Problem 10.3 We consider the 2D sine-Gordon equation (see, e.g. [23, 48–50]):

$$u_{tt} - (u_{xx} + u_{yy}) = -\sin(u), \quad t > 0, \tag{10.87}$$

in the spatial region $\Omega = [-14, 14] \times [-14, 14]$, with the initial conditions

$$u(x, y, 0) = 4 \arctan\left(\exp\left(3 - \sqrt{x^2 + y^2}\right)\right), \quad u_t(x, y, 0) = 0, \tag{10.88}$$

and the homogeneous Neumann boundary conditions

$$u_x(\pm 14, y, t) = u_y(x, \pm 14, t) = 0. \tag{10.89}$$

The exact solution of this problem is a phenomenon called a circular ring soliton (see, e.g. [48, 50]), and different initial conditions will result in different numerical phenomena. We here use the time integrators HB0, HB1 and HB2 coupled with the *discrete Fast Cosine Transformation* (see, e.g. [51, 52]) to simulate the particular circular ring solitons. In Figs. 10.10 and 10.11, we show the simulation results and the corresponding contour plots at the time points $t = 0, 2, 4, 6, 8$ and 10 with spatial stepsizes $\Delta x = \Delta y = 0.07$ and the time stepsize $\Delta t = 0.01$. The CPU time required to reach $t = 10$ is 1191.445350 s.

Likewise, to verify the theoretical results in Theorem 10.7, we fixed the spatial stepsizes as $\Delta x = \Delta y = 0.07$ and integrate the Problem 10.3 by the time integrators HB0, HB1 and HB2 with various time stepsizes. The data listed in Tables 10.5 and 10.6 demonstrate that the convergence order of the time integrators HB0, HB1 and HB2 are of order two, four and six, respectively. The results again verify the correctness of the theory presented in Theorem 10.7.

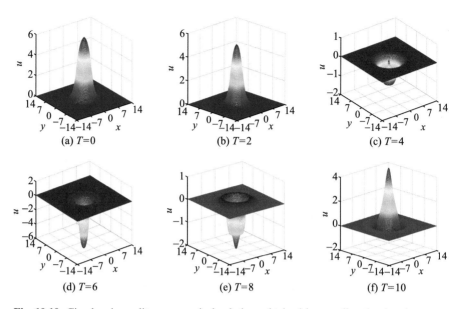

Fig. 10.10 Circular ring solitons: numerical solutions obtained by coupling the time integrator HB2 with the *discrete Fast Cosine Transformation* at the time points $t = 0, 2, 4, 6, 8$ and 10

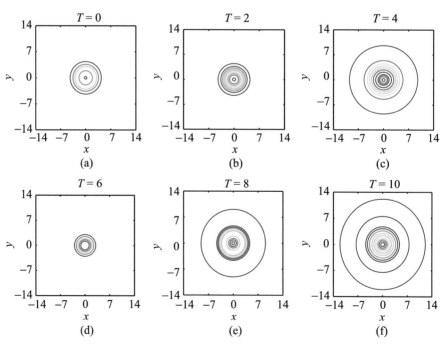

Fig. 10.11 Circular ring solitons: contours of the numerical solutions at the time points $t = 0, 2, 4, 6, 8$ and 10

Table 10.5 Numerical convergence "$u(x, y, t)$" in time with various Δt at time $T = 10$

$\Delta t_0 = 0.5$	HB0		HB1		HB2	
	EU($\Delta x, \Delta y, \Delta t$)	Order	EU($\Delta x, \Delta y, \Delta t$)	Order	EU($\Delta x, \Delta y, \Delta t$)	Order
$\dfrac{\Delta t_0}{2}$	7.9911×10^{-3}	$*$	7.9016×10^{-7}	$*$	4.4097×10^{-7}	$*$
$\dfrac{\Delta t_0}{4}$	1.9992×10^{-3}	1.9989	5.2878×10^{-7}	3.9014	9.4394×10^{-9}	5.5458
$\dfrac{\Delta t_0}{8}$	4.9990×10^{-4}	1.9997	3.3850×10^{-8}	3.9655	1.5975×10^{-10}	5.8848
$\dfrac{\Delta t_0}{16}$	1.2498×10^{-4}	1.9999	2.1293×10^{-9}	3.9907	2.5899×10^{-12}	5.9468

Table 10.6 Numerical convergence "$u_t(x, y, t)$" in time with different Δt at time $T = 10$

$\Delta t_0 = 0.5$	HB0		HB1		HB2	
	EU($\Delta x, \Delta y, \Delta t$)	Order	EU($\Delta x, \Delta y, \Delta t$)	Order	EU($\Delta x, \Delta y, \Delta t$)	Order
$\dfrac{\Delta t_0}{2}$	1.9101×10^{-2}	$*$	1.2757×10^{-4}	$*$	1.6356×10^{-5}	$*$
$\dfrac{\Delta t_0}{4}$	4.7841×10^{-3}	1.9973	1.1049×10^{-5}	3.5294	3.4931×10^{-7}	5.5491
$\dfrac{\Delta t_0}{8}$	1.1966×10^{-2}	1.9993	7.5552×10^{-7}	3.8703	5.9125×10^{-9}	5.8846
$\dfrac{\Delta t_0}{16}$	2.9919×10^{-4}	1.9998	4.8296×10^{-8}	3.9675	9.4309×10^{-11}	5.9702

10.8 Conclusions and Discussions

It is known that the KG equation and the Schrödinger equation are two important equations of Quantum Physics. We have derived and analysed a class of time integrators for the semilinear KG equation (10.1)–(10.2) in this chapter. As distinct from traditional approaches, these schemes are based on the operator-variation-of-constants formula (10.10) which is introduced on the Hilbert space $L^2(\Omega)$ using operator spectral theory, and it is in fact an implicit expression of the solution of the semilinear KG equation. Keeping the eventual discretisation in mind, a class of time integration formulae (10.22) has been designed by applying a two-point Hermite interpolation to the nonlinear integrals that appear in the operator-variation-of-constants formula. It has been shown that these formulae can have arbitrary order and are also symmetric. *A significant advantage of this approach is that the requirement of temporal smoothness is reduced compared with the traditional schemes for PDEs in the literature.* In order to approximate the unbounded positive semi-definite spatial differential operator $\mathscr{A}$, we have also discussed the importance of the choice of a positive semi-definite differentiation matrix. Moreover, stability and convergence for the fully discrete scheme have been proved in both linear and nonlinear settings. In particular, the long-time preservation of the discrete energy conservation law has been analysed. Since the fully discrete scheme is implicit, iteration is required, and we have applied the waveform relaxation algorithm (10.80)–(10.81) in practical computations and analysed the convergence of the iteration. Numerical experiments implemented in this chapter demonstrate that the time integrators so constructed have excellent numerical behaviour in comparison with existing standard finite difference and method-of-lines schemes in the science literature.

Note that the methodology presented in this chapter can be extended to a range of other nonlinear wave equations. Some of the more immediate possible extensions are as follows:

1. *High Dimensional Problems.* Although Eq. (10.1) is one-dimensional, the method can be extended to KG equations in a moderate number, d, of space dimensions,

$$u_{tt} - a^2 \Delta u = f(u), \qquad t_0 \leqslant t \leqslant T, \quad x \in [-\pi, \pi]^d, \tag{10.90}$$

   where $u = u(x, t)$ and $\Delta` = \sum_{i=1}^{d} \frac{\partial^2}{\partial x_i^2}$, with periodic boundary conditions. A large dimension d requires combining the time integration formula (10.22) with other spatial approximate techniques, such as *sparse grids* [53] or *discrete FFT* [51, 54].

2. *Neumann and Dirichlet Boundary Problems.* In this chapter we only consider problems (10.1) subject to periodic boundary conditions (10.2). However, the approach presented in this chapter can be extended to problems with Neumann and Dirichlet boundary conditions with domain $\Omega = [0, \pi]^d$. The corresponding spatial discretisation could be the *discrete Fast Sine Transformation* for Dirichlet boundary conditions or *discrete Fast Cosine Transformation* for the Neumann boundary case. Fortunately, much related work on the *discrete Fast Cosine/Sine Transformation* has been widely published in the science literature (see, e.g. [55]). Therefore, we are hopeful of obtaining related new results.

3. Furthermore, the approach presented in this chapter also can be directly applied to the computation of the following problems:

   (a) *The damped semilinear KG equation*

$$\begin{cases} u_{tt} + \alpha(x)u_t - \beta \Delta u + u + f'(u) = 0, & (x, t) \in \Omega \times [t_0, +\infty), \\ \\ u(x, t_0) = \varphi_1(x), \ u_t(x, t_0) = \varphi_2(x), & x \in \bar{\Omega}, \end{cases} \tag{10.91}$$

   where Ω is a C^1 domain in $\mathbb{R}^d$, β represents the amplitude of the diffusion and the damping coefficient $\alpha : \Omega \to [0, \infty)$ is effective uniform in the neighborhood of the spatial infinity,

$$\alpha(x) \geqslant 0, \qquad \alpha \in L^\infty(\Omega), \qquad \liminf_{|x| \to \infty} \alpha(x) > 0.$$

The damper $\alpha(x)$ satisfies appropriate conditions which guarantee that the total energy defined by

$$E(t) = \frac{1}{2} \int_{\Omega} \left[|u_t|^2 + |\nabla u|^2 + |u|^2 + 2f(u) \right] dx$$

decays uniformly.

(b) *The hyperbolic telegraph equation*

$$\begin{cases} u_{tt} + 2\alpha u_t + \beta^2 u = \Delta u + f(x, t), & (x, t) \in \Omega \times [0, +\infty), \\ u(x, t_0) = \varphi_1(x), \ u_t(x, t_0) = \varphi_2(x), & x \in \{\bar{\Omega}, \end{cases}$$

(10.92)

where $\alpha > 0$ and $\beta > 0$ are known constants. This equation has been widely used in many different fields of science and mathematical engineering such as the vibration of structures, the transmission and propagation of electrical signals and random walk theory.

The material in this chapter is based on the work by Liu et al. [56].

References

1. Dodd, R.K., Eilbeck, I.C., Gibbon, J.D., et al.: Solitons and Nonlinear Wave Equations. Academic Press, London (1982)
2. Drazin, P.J., Johnson, R.S.: Solitons: An Introduction. Cambridge University Press, Cambridge (1989)
3. Brenner, P., von Wahl, W.: Global classical solutions of nonlinear wave equations. Math. Z. **176**, 87–121 (1981)
4. Ginibre, J., Velo, G.: The global Cauchy problem for the nonlinear Klein-Gordon equation. Math. Z. **189**, 487–505 (1985)
5. Ibrahim, S., Majdoub, M., Masmoudi, N.: Global solutions for a semilinear, two-dimensional Klein-Gordon equation with exponential-type nonlinearity. Commun. Pure Appl. Math. **59**, 1639–1658 (2006)
6. Kosecki, R.: The unit condition and global existence for a class of nonlinear Klein-Gordon equations. J. Differ. Equations **100**, 257–268 (1992)
7. Strauss, W.A.: Nonlinear Wave Equations. Regional Conference Series in Mathematics. Regional Conference Series in Mathematics, vol. 73 (American Mathematical Society, Providence, 1989)
8. Ablowitz, M.J., Kruskal, M.D., Ladik, J.F.: Solitary wave collisions. SIAM J. Appl. Math. **36**, 428–437 (1979)
9. Bao, W.Z., Dong, X.C.: Analysis and comparison of numerical methods for the Klein-Gordon equation in the nonrelativistic limit regime. Numer. Math. **120**, 189–229 (2012)
10. Duncan, D.B.: Symplectic finite difference approximations of the nonlinear Klein-Gordon equation. SIAM J. Numer. Anal. **34**, 1742–1760 (1997)
11. Li, S., Vu-Quoc, L.: Finite difference calculus invariant structure of a class of algorithms for the nonlinear Klein-Gordon equation. SIAM J. Numer. Anal. **32**, 1839–1875 (1995)
12. Pascual, P.J., Jiménezz, S., Vázquez, L.: Numerical Simulations of a Nonlinear Klein-Gordon Model. Applications. Lecture Notes and Physics **448**, 211–270 (1995)

13. Cao, W., Guo, B.: Fourier collocation method for solving nonlinear Klein-Gordon equation. J. Comput. Phys. **108**, 296–305 (1993)
14. Cohen, D., Hairer, E., Lubich, C.: Conservation of energy, momentum and actions in numerical discretizations of non-linear wave equations. Numer. Math. **110**, 113–143 (2008)
15. Guo, B.Y., Li, X., Vázquez, L.: A Legendre spectral method for solving the nonlinear Klein-Gordon equation. Comput. Appl. Math. **15**, 19–36 (1996)
16. Tourigny, Y.: Product approximation for nonlinear Klein-Gordon equations. IMA J. Numer. Anal. **9**, 449–462 (1990)
17. Cox, S., Matthews, P.: Exponential time differencing for stiff systems. J. Comput. Phys. **176**, 430–455 (2002)
18. Hochbruck, M., Ostermann, A.: Explicit exponential Runge-Kutta methods for semilinear parabolic problems. SIAM J. Numer. Anal. **43**, 1069–1090 (2005)
19. Hochbruck, M., Ostermann, A.: Exponential Runge-Kutta methods for parabolic problems. Appl. Numer. Math. **53**, 323–339 (2005)
20. Hochbruck, M., Ostermann, A.: Exponential integrators. Acta Numer. **19**, 209–286 (2010)
21. Kassam, A.K., Trefethen, L.N.: Fourth-order time stepping for stiff PDEs. SIAM J. Sci. Comput. **26**, 1214–1233 (2005)
22. Bátkai, A., Farkas, B., Csomós, P. et al.: Operator semigroups for numerical analysis. In: 15th Internet Seminar 2011/12 (2011)
23. Liu, C.Y., Wu, X.Y.: Arbitrarily high-order time-stepping schemes based on the operator spectrum theory for high-dimensional nonlinear Klein-Gordon equations. J. Comput. Phys. **340**, 243–275 (2017)
24. Bader, P., Iserles, A., Kropielnicka, K. et al.: Effective approximation for the semiclassical Schrödinger equation. Found. Comput. Math. **14**, 689–720 (2014)
25. Wu, X.Y., Wang, B.: Recent Developments in Structure-Preserving Algorithms for Oscillatory Differential Equations. Springer Nature Singapore Pte Ltd., Singapore (2018)
26. Grundy, R.E.: Hermite interpolation visits ordinary two-point boundary value problems. ANZIAM J. **48**, 533–552 (2007)
27. Phillips, G.M.: Explicit forms for certain Hermite approximations. BIT Numer. Math. **13**, 177–180 (1973)
28. Dyn, N.: On the existence of Hermite-Birkhoff quadrature formulas of Gaussian type. J. Approx. Theor. **31**, 22–32 (1981)
29. Jetter, K.: Uniqueness of Gauss-Birkhoff quadrature formulas. SIAM J. Numer. Anal. **24**, 147–154 (1987)
30. Nikolov, G.: Existence and uniqueness of Hermite-Birkhoff Gaussian quadrature formulas. Calcolo **26**, 41–59 (1989)
31. Hairer, E., Lubich, C., Wanner, G.: Geometric Numerical Integration: Structure-Preserving Algorithms for Ordinary Differential Equations, 2nd edn. Springer, Berlin (2006)
32. Grimm, V.: On error bounds for the Gautschi-type exponential integrator applied to oscillatory second-order differential equations. Numer. Math. **100**, 71–89 (2005)
33. Hairer, E., Lubich, C.: Long-time energy conservation of numerical methods for oscillatory differential equations. SIAM J. Numer. Anal. **38**, 414–441 (2000)
34. Hochbruck, M., Lubich, C.: A Gautschi-type method for oscillatory second-order differential equations. Numer. Math. **83**, 403–426 (1999)
35. Wu, X.Y., You, X., Wang, B.: Structure-Preserving Algorithms for Oscillatory Differential Equations. Springer, Berlin (2013)
36. Iserles, A.: A First Course in the Numerical Analysis of Differential Equations, 2nd edn. Cambridge University Press, Cambridge (2008)
37. Bank, R., Graham, R.L., Stoer, J. et al.: High Order Difference Methods for Time Dependent PDEs. Springer, Berlin (2008)
38. Liu, C.Y., Shi, W., Wu, X.Y.: An efficient high-order explicit scheme for solving Hamiltonian nonlinear wave equations. Appl. Math. Comput. **246**, 696–710 (2014)
39. Sun, Z.Z.: Numerical Methods of Partial Differential Equations, 2nd edn. Science Press, Beijing (2012)

40. Hesthaven, J.S., Gottlieb, S., Gottlieb, D.: Spectral Methods for Time-Dependent Problems. Cambridge University Press, Cambridge (2007)
41. Shen, J., Tang, T., Wang, L.L.: Spectral Methods: Algorithms, Analysis, Applications. Springer, Berlin (2011)
42. Janssen, J., Vandewalle, S.: On SOR waveform relaxation methods. SIAM J. Numer. Anal. **34**, 2456–2481 (1997)
43. Khanamiryan, M.: Quadrature methods for highly oscillatory linear and nonlinear systems of ordinary differential equations, Part I. BIT Numer. Math. **48**, 743–762 (2008)
44. Lubich, C., Ostermann, A.: Multigrid dynamic iteration for parabolic equations. BIT Numer. Math. **27**, 216–234 (1987)
45. Vandewalle, S.: Parallel Multigrid Waveform Relaxation for Parabolic Problems. Teubner Scripts on Numerical Mathematics. Vieweg+Teubner Verlag, Wiesbaden (1993)
46. Wang, B., Liu, K., Wu, X.Y.: A Filon-type asymptotic approach to solving highly oscillatory second-order initial value problems. J. Comput. Phys. **243**, 210–223 (2013)
47. Tang, W.S., Ya, Y.J., Zhang, J.J. High order symplectic integrators based on continuous-stage Runge-Kutta-Nyström methods. Appl. Math. Comput. **361**, 670–679 (2019)
48. Bratsos, A.G.: A modified predictor-corrector scheme for the two-dimensional sine-Gordon equation. Numer. Algorithms **43**, 295–308 (2006)
49. Dehghan, M., Ghesmati, A.: Numerical simulation of two-dimensional sine-Gordon solitons via a local weak meshless technique based on the radial point interpolation method (RPIM). Comput. Phys. Commun. **181**, 772–786 (2010)
50. Sheng, Q., Khaliq, A.Q.M., Voss, D.A.: Numerical simulation of two-dimensional sine-Gordon solitons via a split cosine scheme. Math. Comput. Simul. **68**, 355–373 (2005)
51. Briggs, W.L., Henson, V.E.: The DFT: An Owner's Manual for the Discrete Fourier Transform. SIAM, Philadelphia (2000)
52. Britanak, V., Yip, P.C., Rao, K.R.: Discrete cosine and sine transforms: General properties, fast algorithms and integer approximations. IEEE Trans. Signal Process. **52**, 306–311 (2006)
53. Bungartz, H.J., Griebel, M.: Sparse grids. Acta Numer. **13**, 147–269 (2004)
54. Bueno-Orovio, A., Pérez-García, V.M., Fenton, F.H.: Spectral methods for partial differential equations in irregular domains: The spectral smoothed boundary method. SIAM J. Sci. Comput. **28**, 886–900 (2006)
55. Mulholland, L.S., Huang, W.Z., Sloan, D.M.: Pseudospectral solution of near-singular problems using numerical coordinate transformations based on adaptivity. SIAM J. Sci. Comput. **19**, 1261–1289 (1998)
56. Liu, C.Y., Iserles, A., Wu, X.Y.: Symmetric and arbitrarily high-order Birkhoff-Hermite time integrators and their long-time behaviour for solving nonlinear Klein-Gordon equations. J. Comput. Phys. **356**, 1–30 (2018)

Chapter 11
Symplectic Approximations for Efficiently Solving Semilinear KG Equations

Among typical geometric integrators are multi-symplectic approximations to nonlinear Hamiltonian PDEs. However, it is also an important aspect to analyse the nonlinear stability and convergence when a fully discrete symplectic scheme is designed for nonlinear Hamiltonian PDEs. This chapter presents a symplectic approximation for efficiently solving semilinear Klein–Gordon equations, which can be formulated as an abstract Hamiltonian ordinary differential equation. We first analyse an extended Runge–Kutta–Nyström-type approximation based on the operator-variation-of-constants formula for the abstract Hamiltonian system. We then present the symplectic conditions for the approximation. The most important issue is that we initiate the nonlinear stability and convergence analysis for the symplectic approximation of semilinear Klein–Gordon equations.

11.1 Introduction

On the one hand, symplectic approximation is an important consideration in the design of numerical schemes for solving nonlinear Hamiltonian PDEs. On the other hand, it is also crucial to perform the nonlinear stability and convergence analysis for a fully discrete symplectic scheme when applied to nonlinear Hamiltonian PDEs. *Unfortunately, current analysis of nonlinear stability and convergence is inadequate although multi-symplectic methods for PDEs have been proposed for multi-symplectic Hamiltonian PDEs* (see, e.g. [1–3]). This is the primary concern of this chapter.

The main aim of this chapter is to present an efficient symplectic approximation, accompanying its fundamental theoretical properties for the semilinear Klein–

© The Author(s), under exclusive license to Springer Nature Singapore Pte Ltd. 2021
X. Wu, B. Wang, *Geometric Integrators for Differential Equations with Highly Oscillatory Solutions*, https://doi.org/10.1007/978-981-16-0147-7_11

Gordon (KG) equation in a single space variable:

$$\begin{cases} u_{tt} - a^2 \Delta u = f(u), & t_0 < t \leqslant T, \ x \in \Omega, \\ u(x, t_0) = \varphi_1(x), \ u_t(x, t_0) = \varphi_2(x), & x \in \bar{\Omega}. \end{cases} \tag{11.1}$$

where $u(x, t)$ represents the wave displacement at position x and time t, and $f(u)$ is a nonlinear function of u chosen as the negative derivative of a potential energy $G(u) \geqslant 0$. The KG equation (11.1) is supplemented with the periodic boundary condition on the domain $\Omega = (-\pi, \pi)$:

$$u(x, t) = u(x + 2\pi, t). \tag{11.2}$$

In this chapter, we restrict ourselves to the one dimensional case, since *all issues presented and analysed in this chapter can be easily extended to two-dimensional and high-dimensional KG equations as shown in [4].*

It is well known that a key feature is that the KG equation is a Hamiltonian PDE, which can be formulated as

$$\begin{cases} u_t = v, \\ v_t = a^2 \Delta u + f(u) \end{cases} \tag{11.3}$$

with the Hamiltonian

$$\mathscr{H} = \frac{1}{2} \int \left(u_t^2 + a^2 |\nabla u|^2 + 2G(u) \right) \mathrm{d}x.$$

The semilinear KG equation (11.1), as a relativistic counterpart of the Schröinger equation, is an important model which can be used to simulate a variety of nonlinear phenomena, including the propagation of dislocations in crystals and the behaviour of elementary particles and of Josephson junctions. Its computation, analysis and related topics represent a major challenge. Much effort has been made to derive effective approximations for solving the semilinear KG equation, and we refer the reader to [2, 3, 5–10] and references therein. The finite differences approximation of the KG equation has been researched for a long time. The authors in [11] studied the Perring–Skyrme (PS) approximation of the one-dimensional sine-Gordon equation. As a simple modification of the PS scheme, the Ablowitz–Kruskal–Ladik (AKL) scheme was discussed in [12]. One popular scheme for solving the two-dimensional problem can be found in [13]. Some energy-conserving or symplectic-preserving standard finite difference schemes were analysed in [5, 14]. Other approaches, such as the finite element method and the spectral method, were also studied in [4, 15–19]. Recently, similar physical systems such as the "Good" Boussinesq equations have been developed, and we refer the reader to [20–25] as well as the references contained therein.

There has been much work in recent years on the research of numerical approximations of Hamiltonian ODEs (see, e.g. [26–34]). It is known, that due to the symplectic geometric structure, Hamiltonian systems have important applications in mechanics, celestial and molecular dynamics, and optics. It is of great interest for numerical simulations to preserve the structure and intrinsic properties (see e.g. [35]) of the original continuous system. Hence, in this chapter, we are concerned with the preservation of the symplectic geometric structure of nonlinear Hamiltonian PDEs. In the literature, various symplectic algorithms have been proposed, and we refer the reader to [36–42] and references therein. It is common practice that once suitable space derivative approximations are used, the KG equation is reduced to a Hamiltonian system of ODEs. Here, differently from the multi-symplectic approximation to multi-symplectic Hamiltonian PDEs, this chapter also pays attention to the analysis of nonlinear stability and convergence for the symplectic approximation to the semilinear KG equation (11.1).

It is noted that for Hamiltonian PDEs, Poisson mapping properties generalize the symplectic mapping properties of the exact solution operator and determine the dynamics of the solution (see [38, 43]). Some researchers make good use of these properties and design numerical schemes for Hamiltonian PDEs (see, e.g. [38]). Furthermore, it follows from the work in [14] that numerical methods ultimately reduce to discrete mappings from time level to time level, and symplectic methods reproduce Hamiltonian dynamics. Thus, this chapter only requires knowledge of the symplectic property of discrete maps to study the symplectic approximation to the semilinear KG equation (11.1).

The method of lines is a standard approach to obtaining a symplectic approximation of (11.1). The discretisation process is carried out in two distinct steps. First, approximating the space derivatives in a suitable manner gives a Hamiltonian system of ODEs in time. Second, the Hamiltonian ODEs are solved by an appropriate symplectic method. However, differently from the conventional route, we consider another approach in this chapter.[1] We first formulate the semilinear KG equation (11.1)–(11.2) as an abstract Hamiltonian system of ODEs on an infinite-dimensional Hilbert space $L^2(\Omega)$. We then introduce the operator-variation-of-constants formula (also termed the Duhamel Principle) and symplectic approximation for the abstract Hamiltonian system. The choice of spatial discretisation is flexible at this stage. Moreover, the nonlinear stability and convergence of the symplectic approximation can be analysed in detail after the implementation of a full discretisation, and this represents an important step toward symplectic approximations for solving semilinear KG equations. As is known, the problem of nonlinear stability and convergence is a very essential and crucial issue in numerical solution of PDEs. *Unfortunately, this point has not received enough attention in the study of the geometric numerical integration for PDEs in the literature.*

---

[1] Since this chapter is devoted to the symplectic approximation in time and different choices of spatial discretisation can be used, we use another approach which is more suitable for presenting this chapter succinctly.

11.2 Abstract Hamiltonian System of ODEs

The main purpose of this section is to formulate the semilinear KG equation (11.1)–(11.2) as an abstract Hamiltonian system of ODEs. We then introduce the operator-variation-of-constants formula for the abstract Hamiltonian system.

We first define the linear differential operator $\mathscr{A}$ by (see, e.g. [19])

$$(\mathscr{A}v)(x) = -a^2 v_{xx}(x), \tag{11.4}$$

where $\mathscr{A}$ is a linear, unbounded positive semi-definite operator, whose domain is

$$D(\mathscr{A}) := \left\{ v \in H^1(\Omega) : v(x) = v(x + 2\pi) \right\}.$$

Clearly, the operator $\mathscr{A}$ has a complete system of orthogonal eigenfunctions $\{e^{ikx} : k \in \mathbb{Z}\}$, and the linear span of all these eigenfunctions

$$X := \lin\{e^{ikx} : k \in \mathbb{Z}\} \tag{11.5}$$

is dense in the Hilbert space $L^2(\Omega)$. This means that we obtain an orthonormal basis of eigenvectors of the operator $\mathscr{A}$ with the corresponding eigenvalues $a^2 k^2$ for $k \in \mathbb{Z}$.

In what follows, we introduce the operator-argument functions ϕ_j as follows:

$$\phi_j(\mathscr{A}) := \sum_{k=0}^{\infty} \frac{(-1)^k \mathscr{A}^k}{(2k+j)!} : L^2(\Omega) \to L^2(\Omega), \quad j = 0, 1, 2, \cdots. \tag{11.6}$$

According to the results described in [4, 8], we have the following proposition for these operator-argument functions.

Proposition 11.1 *All the operator-argument functions defined by* (11.6) *are bounded.*

A proof of this proposition can be found in a very recent paper [8].

We next define $q(t)$ as the function that maps x to $u(x, t)$:

$$q(t) = [x \mapsto u(x, t)].$$

In such a way, we formulate the semilinear KG equation (11.1)–(11.2) as the following abstract Hamiltonian system of ODEs on the Hilbert space $L^2(\Omega)$:

$$\begin{cases} q'(t) = p(t), & q(t_0) = \varphi_1(x), \\ p'(t) = -\mathscr{A}q(t) - G'(q(t)), & p(t_0) = \varphi_2(x). \end{cases} \tag{11.7}$$

We are now in a position to present an integral formula for the semilinear KG equation (11.1)–(11.2) on the basis of this background. The solution of the abstract Hamiltonian system (11.7) and its derivative can be represented by the operator-variation-of-constants formula as follows.

Theorem 11.1 *The solution of (11.7) and its derivative satisfy*

$$
\begin{cases}
q(t) = \phi_0\big((t - t_0)^2 \mathscr{A}\big)q(t_0) + (t - t_0)\phi_1\big((t - t_0)^2 \mathscr{A}\big)p(t_0) \\
\qquad - \displaystyle\int_{t_0}^{t} (t - \zeta)\phi_1\big((t - \zeta)^2 \mathscr{A}\big)G'(q(\zeta))\mathrm{d}\zeta, \\
p(t) = -(t - t_0)\mathscr{A}\phi_1\big((t - t_0)^2 \mathscr{A}\big)q(t_0) + \phi_0\big((t - t_0)^2 \mathscr{A}\big)p(t_0) \\
\qquad - \displaystyle\int_{t_0}^{t} \phi_0\big((t - \zeta)^2 \mathscr{A}\big)G'(q(\zeta))\mathrm{d}\zeta,
\end{cases}
\tag{11.8}
$$

for $t \in [t_0, T]$, where both $\phi_0\big((t - t_0)^2 \mathscr{A}\big)$ and $\phi_1\big((t - t_0)^2 \mathscr{A}\big)$ are bounded operators, although $\mathscr{A}$ is an unbounded symmetric positive semi-definite operator (see [8]).

Proof The outline of the proof can be found in [4], and we skip the details for brevity. □

Remark 11.1 It is noted that the operator-variation-of-constants formula (11.8) is an implicit expression of the solution of the semilinear KG equation (11.7), which assists in the analysis of the underlying geometry integration. Even more important is that (11.8) is adapted to different boundary conditions under suitable assumptions (see [44]). Obviously, the formula (11.8) discloses some useful information about the solution which allows us to design and analyse structure-preserving integrators for the abstract Hamiltonian system of ODEs (11.7). This formula also makes it possible to forego the standard steps of first semidiscretising and then deal with the semidiscretisation in a totally different approach. Actually, the semidiscretisation is deferred to the very last moment here. Moreover, this approach provides the possibility to analyse the nonlinear stability and convergence of the symplectic approximation to the semilinear KG equation (11.1)–(11.2).

11.3 Formulation of the Symplectic Approximation

11.3.1 The Time Approximation

It follows from Theorem 11.1 that the solution and its derivative of (11.7) at time $t_{n+1} = t_n + \Delta t$ for $n = 0, 1, 2, \cdots$, are

$$
\begin{cases}
q(t_{n+1}) = \phi_0(\mathscr{V})q(t_n) + \Delta t \phi_1(\mathscr{V})p(t_n) \\
\qquad - \Delta t^2 \int_0^1 (1-z)\phi_1\big((1-z)^2\mathscr{V}\big)G'\big(u(t_n + z\Delta t)\big)dz, \\
p(t_{n+1}) = -\Delta t \mathscr{A}\phi_1(\mathscr{V})q(t_n) + \phi_0(\mathscr{V})p(t_n) \\
\qquad - \Delta t \int_0^1 \phi_0\big((1-z)^2\mathscr{V}\big)G'\big(u(t_n + z\Delta t)\big)dz,
\end{cases}
\tag{11.9}
$$

where Δt is the time stepsize, and $\mathscr{V} = \Delta t^2 \mathscr{A}$.

To design an effective and practical numerical scheme, it is necessary to approximate the integrals appearing in (11.9) with a quadrature formula by choosing suitable nodes c_i for $i = 1, 2, \cdots, s$. This motivates the following definition.

Definition 11.1 An s-stage extended RKN-type time-stepping approximation with time stepsize Δt for solving the nonlinear Hamiltonian system (11.7) is defined by

$$
\begin{cases}
Q_{ni} = \phi_0(c_i^2\mathscr{V})q_n + c_i\Delta t\phi_1(c_i^2\mathscr{V})p_n - \Delta t^2 \sum_{j=1}^{s} \bar{a}_{ij}(\mathscr{V})G'(Q_{nj}), \quad i = 1, 2, \cdots, s, \\
q_{n+1} = \phi_0(\mathscr{V})q_n + \Delta t\phi_1(\mathscr{V})p_n - \Delta t^2 \sum_{i=1}^{s} \bar{b}_i(\mathscr{V})G'(Q_{ni}), \\
p_{n+1} = -\Delta t\mathscr{A}\phi_1(\mathscr{V})q_n + \phi_0(\mathscr{V})p_n - \Delta t \sum_{i=1}^{s} b_i(\mathscr{V})G'(Q_{ni}),
\end{cases}
\tag{11.10}
$$

where $b_i(\mathscr{V})$, $\bar{b}_i(\mathscr{V})$ and $\bar{a}_{ij}(\mathscr{V})$ are operator-argument functions of $\mathscr{V}$.

Remark 11.2 We remark that, altogether differently from the traditional and standard time-stepping integrator, the extended RKN-type time-stepping approximation to the nonlinear Hamiltonian system of ODEs (11.7) is a time-stepping scheme without spatial discretisation. In fact, the traditional and standard approach always requires that the spatial discretisation is implemented before the time discretisation.

Remark 11.3 The above pattern of extended RKN-type approximations for solving the system of second-order ordinary differential equations with highly oscillatory solutions

$$
\begin{cases}
y'' + My = f(y), \quad x \in [x_0, x_{\text{end}}], \\
y(x_0) = y_0, \quad y'(x_0) = y_0'.
\end{cases}
\tag{11.11}
$$

was initially proposed in [45], and further researched in [32, 34, 41, 46, 47]. However, the approximation (11.10) presented in this chapter is based on the operator-variation-of-constants formula (11.9). This approach makes the approximation (11.10) more suitable for the underlying original continuous semilinear KG equations in the spirit of Geometric Integration.

Remark 11.4 Clearly, the semidiscrete time-stepping scheme (11.10) could be employed using symbolic computation such as Mathematica for some PDEs. However, in this chapter we focus on scientific computing with floating point numbers. The operator $\mathscr{A}$ appearing in (11.10) will be approximated in an appropriate way such that the fully discrete scheme is a symplectic algorithm for Hamiltonian PDEs.

11.3.2 Symplectic Conditions for the Fully Discrete Scheme

In practice, the operator $\mathscr{A}$ will be approximated by a symmetric and positive semi-definite differentiation matrix, and this assists in structure preservation for numerical simulations. In what follows, we derive the symplectic conditions for the time-stepping approximation (11.10) after the differential operator $\mathscr{A}$ is replaced by a symmetric and positive semi-definite differentiation matrix. It is noted that a similar result to the extended RKN methods for solving the system of second-order oscillatory ODEs was derived in [34]. We here present a simplified result for the approximation (11.10) to the nonlinear Hamiltonian system (11.7) with a simpler proof.

As mentioned in Introduction, a standard approach to the approximation of (11.1) is the method-of-lines, where the discretisation is carried out in two distinct procedures: the first is to approximate the space derivatives leaving a Hamiltonian system of ODEs in time, and the second is to solve the ODEs by an appropriate numerical method. Of course, there exist many different ways to approximate $\mathscr{A}$. We will consider two types of spatial discretisations.

Theorem 11.2 *According to the method-of-lines, the semilinear KG equation (11.1) can be rewritten as $u_{tt} + Au = f(u)$, where A is a symmetric and positive semi-definite differentiation matrix which approximates the operator $\mathscr{A}$. Accordingly, the approximation to (11.10) reads*

$$
\begin{cases}
Q_{ni} = \phi_0\left(c_i^2 V\right)q_n + c_i \Delta t \phi_1\left(c_i^2 V\right)p_n + \Delta t^2 \sum_{j=1}^{s} \bar{a}_{ij}(V) f(Q_{nj}), \\[2mm]
\quad i = 1, 2, \cdots, s, \\[2mm]
q_{n+1} = \phi_0(V)q_n + \Delta t \phi_1(V)p_n + \Delta t^2 \sum_{i=1}^{s} \bar{b}_i(V) f(Q_{ni}), \\[2mm]
p_{n+1} = -\Delta t A \phi_1(V)q_n + \phi_0(V)p_n + \Delta t \sum_{i=1}^{s} b_i(V) f(Q_{ni}),
\end{cases}
\tag{11.12}
$$

where Δt is the time stepsize and $V = \Delta t^2 A$. The fully discrete scheme (11.12) *is symplectic for the Hamiltonian system* (11.7) *if its coefficients satisfy*

$$
\begin{cases}
\phi_0(V)b_i(V) + V\phi_1(V)\bar{b}_i(V) = d_i\phi_0(c_i^2 V), & d_i \in \mathbb{R}, \quad i = 1, 2, \cdots, s, \\
\phi_1(V)b_i(V) - \phi_0(V)\bar{b}_i(V) = c_i d_i \phi_1(c_i^2 V), & i = 1, 2, \cdots, s, \\
\bar{b}_i(V)b_j(V) + d_i\bar{a}_{ij}(V) = \bar{b}_j(V)b_i(V) + d_j\bar{a}_{ji}(V), & i, \; j = 1, 2, \cdots, s.
\end{cases}
$$

$$(11.13)$$

Here, it is important to remember that V contains the information about the spatial mesh structure with the boundary conditions, and the time step as $V = \Delta t^2 A$, where A denotes the approximation to the operator $\mathscr{A}$ in (11.10)*, and Δt is the time stepsize.*

Proof We begin with the special case where A is a diagonal matrix with nonnegative entries: $A = \text{diag}(m_{11}, m_{22}, \cdots, m_{dd})$. In this case, $\phi_0(V)$, $\phi_1(V)$, $b_i(V)$, $\bar{b}_i(V)$, and $\bar{a}_{ij}(V)$ are all diagonal matrices. We denote $f_i = -G'(Q_{ni})$ and $v_{ii} = \Delta t^2 m_{ii}$, and then the scheme (11.10) is identical to

$$
\begin{cases}
Q_{ni}^J = \phi_0(c_i^2 v_{JJ})q_n^J + c_i\phi_1(c_i^2 v_{JJ})\Delta t p_n^J + \Delta t^2 \sum_{j=1}^{s} \bar{a}_{ij}(v_{JJ})f_j^J, \; i = 1, \cdots, s, \\
q_{n+1}^J = \phi_0(v_{JJ})q_n^J + \phi_1(v_{JJ})\Delta t p_n^J + \Delta t^2 \sum_{i=1}^{s} \bar{b}_i(v_{JJ})f_i^J, \\
p_{n+1}^J = -\Delta t m_{JJ}\phi_1(v_{JJ})q_n^J + \phi_0(v_{JJ})p_n^J + \Delta t \sum_{i=1}^{s} b_i(v_{JJ})f_i^J,
\end{cases}
$$

$$(11.14)$$

where the superscript indices $J = 1, 2, \cdots, d$ denote the J-th entry of a vector.

The symplecticity of the scheme (11.14) is given by

$$
\sum_{J=1}^{d} dq_{n+1}^J \wedge dp_{n+1}^J = \sum_{J=1}^{d} dq_n^J \wedge dp_n^J.
$$

Differentiating q_{n+1}^J and p_{n+1}^J and taking external products, we obtain

$$
dq_{n+1}^J \wedge dp_{n+1}^J = [\phi_0^2(v_{JJ}) + v_{JJ}\phi_1^2(v_{JJ})]dq_n^J \wedge dp_n^J
$$
$$
+ \Delta t \sum_{i=1}^{s} [b_i(v_{JJ})\phi_0(v_{JJ}) + \bar{b}_i(v_{JJ})v_{JJ}\phi_1(v_{JJ})]dq_n^J \wedge df_i^J
$$

$$+ \Delta t^2 \sum_{i=1}^{s} [b_i(v_{JJ})\phi_1(v_{JJ}) - \bar{b}_i(v_{JJ})\phi_0(v_{JJ})] dp_n^J \wedge df_i^J$$

$$+ \Delta t^3 \sum_{i,j=1}^{s} \bar{b}_i(v_{JJ})b_j(v_{JJ})df_i^J \wedge df_j^J.$$

As $\phi_0^2(v_{JJ}) + v_{JJ}\phi_1^2(v_{JJ}) = 1$, this gives

$$dq_{n+1}^J \wedge dp_{n+1}^J = dq_n^J \wedge dp_n^J + \Delta t \sum_{i=1}^{s} \left(d_i\phi_0(c_i^2 v_{JJ}) \right) dq_n^J \wedge df_i^J$$

$$+ \Delta t^2 \sum_{i=1}^{s} \left(c_i d_i \phi_1(c_i^2 v_{JJ}) \right) dp_n^J \wedge df_i^J + \Delta t^3 \sum_{i,j=1}^{s} \bar{b}_i(v_{JJ})b_j(v_{JJ})df_i^J \wedge df_j^J.$$

It then follows from

$$\phi_0(c_i^2 v_{JJ})dq_n^J \wedge df_i^J = dQ_{ni}^J \wedge df_i^J - c_i\phi_1(c_i^2 v_{JJ})\Delta t dp_n^J \wedge df_i^J$$

$$- \Delta t^2 \sum_{j=1}^{s} \bar{a}_{ij}(v_{JJ})df_j^J \wedge df_i^J,$$

that

$$dq_{n+1}^J \wedge dp_{n+1}^J = dq_n^J \wedge dp_n^J + \Delta t \sum_{i=1}^{s} d_i dQ_{ni}^J \wedge df_i^J$$

$$+ \Delta t^2 \sum_{i=1}^{s} \left(d_i \cdot \left(- c_i\phi_1(c_i^2 v_{JJ}) \right) + c_i d_i \phi_1(c_i^2 v_{JJ}) \right) dp_n^J \wedge df_i^J$$

$$+ \Delta t^3 \sum_{i,j=1}^{s} \left(d_i \bar{a}_{ij}(v_{JJ}) + \bar{b}_i(v_{JJ})b_j(v_{JJ}) \right) df_i^J \wedge df_j^J.$$

Summing over all J leads to

$$\sum_{J=1}^{d} dq_{n+1}^J \wedge dp_{n+1}^J = \sum_{J=1}^{d} dq_n^J \wedge dp_n^J + \Delta t \sum_{i=1}^{s} \sum_{J=1}^{d} d_i dQ_{ni}^J \wedge df_i^J$$

$$+ \Delta t^3 \sum_{i,j=1}^{s} \sum_{J=1}^{d} \left(d_i \bar{a}_{ij}(v_{JJ}) + \bar{b}_i(v_{JJ})b_j(v_{JJ}) \right) df_i^J \wedge df_j^J. \tag{11.15}$$

Keeping $f(z) = -G'(z)$ in mind, we obtain

$$\sum_{J=1}^{d} d_i \, \mathrm{d}Q_{ni}^J \wedge \mathrm{d}f_i^J = -d_i \sum_{J,I=1}^{d} \left(\frac{\partial f^J}{\partial q^I}(Q_{ni}) \mathrm{d}Q_{ni}^I \right) \wedge \mathrm{d}Q_{ni}^J$$

$$= -d_i \sum_{J,I=1}^{d} \left(-\frac{\partial^2 G}{\partial q^J \partial q^I} \right) \mathrm{d}Q_{ni}^I \wedge \mathrm{d}Q_{ni}^J = 0.$$

Using the third condition of (11.13), we conclude that the last term of (11.15) vanishes. Thus, we obtain

$$\sum_{J=1}^{d} \mathrm{d}q_{n+1}^J \wedge \mathrm{d}p_{n+1}^J = \sum_{J=1}^{d} \mathrm{d}q_n^J \wedge \mathrm{d}p_n^J.$$

We next consider the general case where A is symmetric and positive semi-definite. This implies that there exist an orthogonal matrix P and a positive semi-definite diagonal matrix Ω so that A can be decomposed into

$$A = P^{\mathsf{T}} \Omega^2 P.$$

Then the semidiscrete system of (11.7) is of the form

$$\begin{cases} q'(t) = p(t), & q(t_0) = \big(\varphi_1(x_1), \varphi_1(x_2), \cdots, \varphi_1(x_M)\big)^{\mathsf{T}}, \\ p'(t) = -P^{\mathsf{T}} \Omega^2 P q(t) - G'(q(t)), & p(t_0) = \big(\varphi_2(x_1), \varphi_2(x_2), \cdots, \varphi_2(x_M)\big)^{\mathsf{T}}, \end{cases}$$
$$(11.16)$$

where $x_1, \cdots, x_M$ are referred to the interior discretised points. With the variable substitution $z(t) = Pq(t)$, the system (11.16) is identical to the following transformed system

$$\begin{cases} z'(t) = Pp(t) := \mathrm{l}(t), & z(t_0) = Pq(t_0), \\ \mathrm{l}'(t) = -\Omega^2 z(t) - PG'(P^{\mathsf{T}} z(t)), & \mathrm{l}(t_0) = Pp(t_0). \end{cases}$$
$$(11.17)$$

It is clear now that the symplectic extended RKN-type approximation for diagonal matrix A with nonnegative entries can be applied to the transformed system. Moreover, the approximation is invariant under linear transformation. This means that the extended RKN-type approximation with symplectic conditions (11.13) can be applied to systems with a symmetric and positive semi-definite matrix A.

To summarise, an extended RKN-type approximation satisfying the conditions (11.13) is a symplectic approximation to the Hamiltonian system (11.7) whose differential operator $\mathscr{A}$ is approximated by a symmetric and positive semi-definite differentiation matrix A. $\qquad \square$

The result of Theorem 11.2 can be further simplified as follows.

Corollary 11.1 *The fully discrete extended RKN-type approximation* (11.12) *is symplectic for* (11.7) *if its coefficients satisfy:*

$$
\begin{cases}
b_i(V) = d_i \phi_0((1 - c_i)^2 V), & d_i \neq 0, \quad i = 1, 2, \cdots, s, \\
\bar{b}_i(V) = d_i(1 - c_i)\phi_1((1 - c_i)^2 V), & i = 1, 2, \cdots, s, \\
\bar{a}_{ij}(V) = \dfrac{1}{d_i}(\bar{b}_j(V)b_i(V) - \bar{b}_i(V)b_j(V)), & i > j, \quad i, \ j = 1, 2, \cdots, s,
\end{cases}
\tag{11.18}
$$

where $V = \Delta t^2 A$ *contains information about the spatial mesh structure* A *and the time stepsize* Δt.

Proof This result follows immediately from solving the symplectic conditions (11.14) of an s-stage scheme. □

Remark 11.5 From (11.18) and Proposition 11.1, it can be verified that the functions $b_i(V)$, $\bar{b}_i(V)$, $\bar{a}_{ij}(V)$, and $\sqrt{V}b_i(V)$ are uniformly bounded and the bounds are independent of $\|V\|$.

11.3.3 Error Analysis of the Extended RKN-Type Approximation

An important issue for numerical approximations is error analysis. We next analyse the local error bounds of the extended RKN-type approximation under the following hypothesis on the nonlinearity f.

Assumption 11.1 Suppose that (11.7) possesses sufficiently smooth solutions, and that $f : D(\mathscr{A}) \to \mathbb{R}$ is sufficiently often Fréchet differentiable in a strip along the exact solution. Moreover, let f be locally Lipschitz-continuous along the exact solution $u(t)$, which implies that there exists a real number L such that

$$
\|f(v(t)) - f(w(t))\| \leqslant L\|v(t) - w(t)\|
$$

for all $t \in [t_0, T]$.

Remark 11.6 Here, the local Lipschitz-continuous condition of the nonlinear function is needed in this chapter. It allows the nonlinear analysis and convergence to go through without any difficulty. For the case where the nonlinear function f does not satisfy the local Lipschitz-continuous condition, the analysis of nonlinear stability and convergence presented in this chapter does not work any more. In this situation, the corresponding results may be dealt with in other suitable ways such as using the linearized stability analysis and the a-priori recovery technique (see [21, 24]).

Theorem 11.3 *It is assumed that* $f_t^{(s)} \in L^\infty(0, T; L^2(\Omega))$. *Under the local assumptions of* $q_n = q(t_n)$, $p_n = p(t_n)$, *if the approximation* (11.10) *to the solution of* (11.7) *satisfies the following conditions:*

$$
\begin{cases}
\displaystyle\sum_{i=1}^{s} b_i(\mathcal{V}) \frac{c_i^j}{j!} = \phi_{j+1}(\mathcal{V}) + \mathcal{O}(\Delta t^{r-j}), \quad j = 0, 1, \cdots, r-1, \\[4mm]
\displaystyle\sum_{i=1}^{s} \bar{b}_i(\mathcal{V}) \frac{c_i^j}{j!} = \phi_{j+2}(\mathcal{V}) + \mathcal{O}(\Delta t^{r-1-j}), \quad j = 0, 1, \cdots, r-2, \\[4mm]
\displaystyle\sum_{k=1}^{s} \bar{a}_{ik}(\mathcal{V}) \frac{c_k^j}{j!} = c_i^{j+2} \phi_{j+2}(c_i^2 \mathcal{V}) + \mathcal{O}(\Delta t^{r-2-j}), \quad j = 0, 1, \cdots, r-3, \\[4mm]
\qquad\qquad\qquad i = 1, 2, \cdots, s,
\end{cases}
\tag{11.19}
$$

then the local error bounds of (11.10) *admit the following inequalities*

$$
\|q(t_{n+1}) - q_{n+1}\| \leqslant \tilde{C}_1 \Delta t^{r+1} \quad and \quad \|p(t_{n+1}) - p_{n+1}\| \leqslant \tilde{C}_1 \Delta t^{r+1}, \tag{11.20}
$$

where r *is a positive integer.*

Proof We will divide the proof into two steps. The first step shows the discrepancies (or residuals) of the approximation (11.10), and the second one presents the local error bounds.

(I) First, inserting the exact solution of (11.7) into the approximation (11.10) yields

$$
\begin{cases}
q(t_n + c_i \Delta t) = \phi_0(c_i^2 \mathcal{V}) q(t_n) + c_i \Delta t \phi_1(c_i^2 \mathcal{V}) p(t_n) \\[2mm]
\qquad\qquad + \Delta t^2 \displaystyle\sum_{j=1}^{s} \bar{a}_{ij}(\mathcal{V}) \hat{f}(t_n + c_j \Delta t) + \Delta_{ni}, \quad i = 1, 2, \cdots, s, \\[4mm]
q(t_{n+1}) = \phi_0(\mathcal{V}) q(t_n) + \Delta t \phi_1(\mathcal{V}) p(t_n) + \Delta t^2 \displaystyle\sum_{i=1}^{s} \bar{b}_i(\mathcal{V}) \hat{f}(t_n + c_i \Delta t) + \delta_{n+1}, \\[4mm]
p(t_{n+1}) = -\Delta t \mathscr{A} \phi_1(\mathcal{V}) q(t_n) + \phi_0(\mathcal{V}) p(t_n) + \Delta t \displaystyle\sum_{i=1}^{s} b_i(\mathcal{V}) \hat{f}(t_n + c_i \Delta t) + \delta'_{n+1},
\end{cases}
\tag{11.21}
$$

where Δ_{ni}, δ_{n+1} and δ'_{n+1} express the discrepancies of the approximation (11.10), and $\hat{f}(t) \equiv f(q(t))$. Then using the operator-variation-of-constants formula (11.8) we have

$$
q(t_n + c_i \Delta t) = \phi_0(c_i^2 \mathcal{V}) q(t_n) + c_i \Delta t \phi_1(c_i^2 \mathcal{V}) p(t_n)
$$
$$
+ \Delta t^2 \int_0^{c_i} (c_i - z) \phi_1((c_i - z)^2 \mathcal{V}) \hat{f}(t_n + \Delta t z) dz. \tag{11.22}
$$

Comparing (11.22) with the first formula in (11.21) gives

$$\Delta_{ni} = \Delta t^2 \int_0^{c_i} (c_i - z)\phi_1((c_i - z)^2 \mathcal{V}) \hat{f}(t_n + \Delta t z) dz - \Delta t^2 \sum_{j=1}^{s} \bar{a}_{ij}(\mathcal{V}) \hat{f}(t_n + c_j \Delta t).$$

We express $\hat{f}$ of the above formula by the Taylor series expansion as follows

$$\begin{aligned}
\Delta_{ni} &= \Delta t^2 \int_0^{c_i} (c_i - z)\phi_1((c_i - z)^2 \mathcal{V}) \sum_{j=0}^{\infty} \frac{\Delta t^j z^j}{j!} \hat{f}^{(j)}(t_n) dz \\
&\quad - \Delta t^2 \sum_{k=1}^{s} \bar{a}_{ik}(\mathcal{V}) \sum_{j=0}^{\infty} \frac{c_k^j \Delta t^j}{j!} \hat{f}^{(j)}(t_n) \\
&= \sum_{j=0}^{\infty} \Delta t^{j+2} c_i^{j+2} \int_0^1 \frac{(1-\xi)\phi_1(c_i^2(1-\xi)^2 \mathcal{V})\xi^j}{j!} d\xi \, \hat{f}^{(j)}(t_n) \\
&\quad - \Delta t^2 \sum_{k=1}^{s} \bar{a}_{ik}(\mathcal{V}) \sum_{j=0}^{\infty} \frac{c_k^j \Delta t^j}{j!} \hat{f}^{(j)}(t_n) \\
&= \sum_{j=0}^{\infty} \Delta t^{j+2} c_i^{j+2} \phi_{j+2}(c_i^2 \mathcal{V}) \hat{f}^{(j)}(t_n) - \Delta t^2 \sum_{k=1}^{s} \bar{a}_{ik}(\mathcal{V}) \sum_{j=0}^{\infty} \frac{c_k^j \Delta t^j}{j!} \hat{f}^{(j)}(t_n) \\
&= \sum_{j=0}^{\infty} \Delta t^{j+2} \left[c_i^{j+2} \phi_{j+2}(c_i^2 \mathcal{V}) - \sum_{k=1}^{s} \bar{a}_{ik}(\mathcal{V}) \frac{c_k^j}{j!} \right] \hat{f}^{(j)}(t_n),
\end{aligned}$$

where $\hat{f}^{(j)}(t)$ denotes the j-th order derivative of $f(q(t))$ with respect to t.

In a similar way, we can obtain

$$\delta_{n+1} = \sum_{j=0}^{\infty} \Delta t^{j+2} \left[\phi_{j+2}(\mathcal{V}) - \sum_{k=1}^{s} \bar{b}_k(\mathcal{V}) \frac{c_k^j}{j!} \right] \hat{f}^{(j)}(t_n),$$

$$\delta'_{n+1} = \sum_{j=0}^{\infty} \Delta t^{j+1} \left[\phi_{j+1}(\mathcal{V}) - \sum_{k=1}^{s} b_k(\mathcal{V}) \frac{c_k^j}{j!} \right] \hat{f}^{(j)}(t_n).$$

It follows from the conditions (11.19) that

$$\begin{aligned}
\|\Delta_{ni}\| &\leqslant C_1 \Delta t^r, \qquad i = 1, 2, \cdots, s, \\
\|\delta_{n+1}\| &\leqslant C_2 \Delta t^{r+1}, \quad \|\delta'_{n+1}\| \leqslant C_3 \Delta t^{r+1}.
\end{aligned}$$

$$(11.23)$$

(II) We denote

$$e_n^q = q(t_n) - q_n, \ e_n^p = p(t_n) - p_n, \ E_{ni} = q(t_n + c_i h) - Q_{ni}.$$

Then subtracting (11.10) from (11.21) yields

$$\begin{cases} E_{ni} = \Delta t^2 \sum_{j=1}^{s} \bar{a}_{ij}(\mathcal{V})\big(\hat{f}(t_n + c_j \Delta t) - f(Q_{nj})\big) + \Delta_{ni}, \quad i = 1, 2, \cdots, s, \\[3mm] e_{n+1}^q = \Delta t^2 \sum_{i=1}^{s} \bar{b}_i(\mathcal{V})\Big(\hat{f}(t_n + c_i \Delta t) - f(Q_{ni})\Big) + \delta_{n+1}, \\[3mm] e_{n+1}^p = \Delta t \sum_{i=1}^{s} b_i(\mathcal{V})\Big(\hat{f}(t_n + c_i \Delta t) - f(Q_{ni})\Big) + \delta_{n+1}'. \end{cases}$$

This results in

$$\|E_{ni}\| \leqslant \Delta t^2 \sum_{j=1}^{s} \|\bar{a}_{ij}(\mathcal{V})\|_{L^2(\Omega) \leftarrow L^2(\Omega)} \|\hat{f}(t_n + c_j \Delta t) - f(Q_{nj})\| + \|\Delta_{ni}\|$$

$$\leqslant \Delta t^2 \alpha L \sum_{j=1}^{s} \|E_{nj}\| + \|\Delta_{ni}\|, \ i = 1, \cdots, s,$$

$$\|e_{n+1}^q\| \leqslant \Delta t^2 \sum_{i=1}^{s} \|\bar{b}_i(\mathcal{V})\|_{L^2(\Omega) \leftarrow L^2(\Omega)} \|\hat{f}(t_n + c_i \Delta t) - f(Q_{ni})\| + \|\delta_{n+1}\|$$

$$\leqslant \Delta t^2 \beta L \sum_{i=1}^{s} \|E_{ni}\| + \|\delta_{n+1}\|,$$

$$\|e_{n+1}^p\| \leqslant \Delta t \sum_{i=1}^{s} \|b_i(\mathcal{V})\|_{L^2(\Omega) \leftarrow L^2(\Omega)} \|\hat{f}(t_n + c_i \Delta t) - f(Q_{ni})\| + \|\delta_{n+1}'\|$$

$$\leqslant \Delta t \gamma L \sum_{i=1}^{s} \|E_{ni}\| + \|\delta_{n+1}'\|,$$

$$(11.24)$$

where α, β, γ are respectively the uniform bounds of $\bar{a}_{ij}(\mathcal{V})$, $\bar{b}_i(\mathcal{V})$, $b_i(\mathcal{V})$ under the norm $\|\cdot\|_{L^2(\Omega) \leftarrow L^2(\Omega)}$. It follows from the first s inequalities of (11.24) that

$$\sum_{i=1}^{s} \|E_{ni}\| \leqslant \Delta t^2 \alpha L s \sum_{j=1}^{s} \|E_{nj}\| + \sum_{i=1}^{s} \|\Delta_{ni}\|, \ i = 1, \cdots, s.$$

Under that assumption that the time stepsize Δt satisfies $\Delta t^2 \alpha L s \leqslant \dfrac{1}{2}$, i.e., $\Delta t \leqslant \sqrt{\dfrac{1}{2\alpha L s}}$, we obtain

$$\sum_{i=1}^{s} \|E_{ni}\| \leqslant 2 \sum_{i=1}^{s} \|\Delta_{ni}\|, \ i = 1, \cdots, s.$$

This leads to

$$\|e_{n+1}^{q}\| \leqslant 2\Delta t^2 \beta L \sum_{i=1}^{s} \|\Delta_{ni}\| + \|\delta_{n+1}\|,$$

$$\|e_{n+1}^{p}\| \leqslant 2\Delta t \gamma L \sum_{i=1}^{s} \|\Delta_{ni}\| + \|\delta_{n+1}'\|.$$

According to the bounds (11.23) of Δ_{ni}, δ_{n+1}, δ_{n+1}', the statement of the theorem is evident. $\qquad\square$

11.4 Analysis of the Nonlinear Stability

In this section, we will present a nonlinear stability analysis for the approximation (11.10). To this end, we consider the following perturbed problem associated with (11.7)

$$\begin{cases} \tilde{q}'(t) = \tilde{p}(t), & \tilde{q}(t_0) = \varphi_1(x) + \tilde{\varphi}_1(x), \\ \tilde{p}'(t) = -\mathscr{A}\tilde{q}(t) + f(\tilde{q}(t)), & \tilde{p}(t_0) = \varphi_2(x) + \tilde{\varphi}_2(x), \end{cases} \tag{11.25}$$

where $\tilde{\varphi}_1(x), \tilde{\varphi}_2(x)$ are perturbation functions. Let

$$\hat{q}(t) = \tilde{q}(t) - q(t), \ \hat{p}(t) = \tilde{p}(t) - p(t).$$

Subtracting (11.7) from (11.25) gives

$$\begin{cases} \hat{q}'(t) = \hat{p}(t), & \hat{q}(t_0) = \tilde{\varphi}_1(x), \\ \hat{p}'(t) = -\mathscr{A}\hat{q}(t) + f(\tilde{q}(t)) - f(q(t)), & \hat{p}(t_0) = \tilde{\varphi}_2(x). \end{cases} \tag{11.26}$$

Applying the approximation (11.10)–(11.7), and (11.25), respectively, and subtracting the first result from the second, we have

$$
\begin{cases}
\tilde{Q}_{ni} - Q_{ni} = \phi_0(c_i^2 \mathcal{V})(\tilde{q}_n - q_n) + c_i \Delta t \phi_1(c_i^2 \mathcal{V})(\tilde{p}_n - p_n) \\
\qquad\qquad + \Delta t^2 \sum_{j=1}^{s} \bar{a}_{ij}(\mathcal{V})\big(f(\tilde{Q}_{nj}) - f(Q_{nj})\big), \quad i = 1, 2, \cdots, s, \\
\tilde{q}_{n+1} - q_{n+1} = \phi_0(\mathcal{V})(\tilde{q}_n - q_n) + \Delta t \phi_1(\mathcal{V})(\tilde{p}_n - p_n) \\
\qquad\qquad + \Delta t^2 \sum_{i=1}^{s} \bar{b}_i(\mathcal{V})\big(f(\tilde{Q}_{ni}) - f(Q_{ni})\big), \\
\tilde{p}_{n+1} - p_{n+1} = -\Delta t \mathscr{A} \phi_1(\mathcal{V})(\tilde{q}_n - q_n) + \phi_0(\mathcal{V})(\tilde{p}_n - p_n) \\
\qquad\qquad + \Delta t \sum_{i=1}^{s} b_i(\mathcal{V})\big(f(\tilde{Q}_{ni}) - f(Q_{ni})\big).
\end{cases}
\tag{11.27}
$$

This provides the following approximation of (11.26)

$$
\begin{cases}
\tilde{Q}_{ni} - Q_{ni} = \phi_0(c_i^2 \mathcal{V})\hat{q}_n + c_i \Delta t \phi_1(c_i^2 \mathcal{V})\hat{p}_n + \Delta t^2 \sum_{j=1}^{s} \bar{a}_{ij}(\mathcal{V})\big(f(\tilde{Q}_{nj}) - f(Q_{nj})\big), \\
\qquad i = 1, 2, \cdots, s, \\
\hat{q}_{n+1} = \phi_0(\mathcal{V})\hat{q}_n + \Delta t \phi_1(\mathcal{V})\hat{p}_n + \Delta t^2 \sum_{i=1}^{s} \bar{b}_i(\mathcal{V})\big(f(\tilde{Q}_{ni}) - f(Q_{ni})\big), \\
\hat{p}_{n+1} = -\Delta t \mathscr{A} \phi_1(\mathcal{V})\hat{q}_n + \phi_0(\mathcal{V})\hat{p}_n + \Delta t \sum_{i=1}^{s} b_i(\mathcal{V})\big(f(\tilde{Q}_{ni}) - f(Q_{ni})\big).
\end{cases}
\tag{11.28}
$$

Because the operator $\mathscr{A}$ is approximated by a symmetric and positive semi-definite differentiation matrix A, there exist an orthogonal matrix P and a positive semi-definite diagonal matrix Ω such that

$$
A = P^{\mathsf{T}} \Omega^2 P = \sqrt{A}^2,
$$

where $\sqrt{A} = P^{\mathsf{T}} \Omega P$. Then similarly to the boundedness of the operator-argument functions, we also have

$$
\|\phi_j(t^2 A)\| = \sqrt{\lambda_{\max}\big(\phi_j^2(t^2 A)\big)} \leqslant \gamma_j, \quad j = 0, 1, 2, \cdots.
\tag{11.29}
$$

We next present the nonlinear stability of our approximation (11.10).

Theorem 11.4 *It is assumed that the nonlinear function f satisfies Assumption 11.1 and the operator $\mathscr{A}$ is approximated by a symmetric and positive semi-definite differentiation matrix A. If the time stepsize Δt satisfies $0 < \Delta t \leqslant \sqrt{\dfrac{1}{2\alpha L s}}$, then we have the following nonlinear stability results*

$$\|\hat{q}_n\| \leqslant \exp\left((1 + 4s\tilde{\gamma}L)T\right)\left(\|\tilde{\varphi}_1\| + \sqrt{\|\sqrt{A}\tilde{\varphi}_1\|^2 + \|\tilde{\varphi}_2\|^2}\right),$$
$$\|\hat{p}_n\| \leqslant \exp\left((1 + 4s\tilde{\gamma}L)T\right)\left(\|\tilde{\varphi}_1\| + \sqrt{\|\sqrt{A}\tilde{\varphi}_1\|^2 + \|\tilde{\varphi}_2\|^2}\right), \tag{11.30}$$

where $\tilde{\gamma} = \max(\tilde{\alpha}, \beta)$ which is independent of $\|V\|$, and $\tilde{\alpha}$ is the uniform bound of $\operatorname{diag}(\Delta t\sqrt{A}\bar{b}_i(V), b_i(V))$.

Proof First, it follows from the penultimate equality of (11.28) that

$$\left\|\hat{q}_{n+1}\right\| \leqslant \left\|\hat{q}_n\right\| + \Delta t\left\|\hat{p}_n\right\| + \Delta t^2\beta\sum_{i=1}^{s}\left\|f(\tilde{Q}_{ni}) - f(Q_{ni})\right\|.$$

We then rewrite the last two equalities of (11.28) in the following compact form:

$$\begin{pmatrix}\sqrt{A}\hat{q}_{n+1} \\ \hat{p}_{n+1}\end{pmatrix} = \begin{pmatrix}\phi_0(V) & \Delta t\sqrt{A}\phi_1(V) \\ -\Delta t\sqrt{A}\phi_1(V) & \phi_0(V)\end{pmatrix}\begin{pmatrix}\sqrt{A}\hat{q}_n \\ \hat{p}_n\end{pmatrix}$$
$$+ \Delta t\sum_{i=1}^{s}\begin{pmatrix}\Delta t\sqrt{A}\bar{b}_i(V) & \\ & b_i(V)\end{pmatrix}\begin{pmatrix}f(\tilde{Q}_{ni}) - f(Q_{ni}) \\ f(\tilde{Q}_{ni}) - f(Q_{ni})\end{pmatrix}.$$

This leads to

$$\sqrt{\|\sqrt{A}\hat{q}_{n+1}\|^2 + \|\hat{p}_{n+1}\|^2} \leqslant \sqrt{\|\sqrt{A}\hat{q}_n\|^2 + \|\hat{p}_n\|^2} + \Delta t\tilde{\alpha}\sum_{i=1}^{s}\left\|f(\tilde{Q}_{ni}) - f(Q_{ni})\right\|,$$

where $\tilde{\alpha}$ is the uniform bound of $\operatorname{diag}(\Delta t\sqrt{A}\bar{b}_i(V), b_i(V))$. Then summing up the above results gives

$$\|\hat{q}_{n+1}\| + \sqrt{\|\sqrt{A}\hat{q}_{n+1}\|^2 + \|\hat{p}_{n+1}\|^2}$$
$$\leqslant \|\hat{q}_n\| + \sqrt{\|\sqrt{A}\hat{q}_n\|^2 + \|\hat{p}_n\|^2} + \Delta t\|\hat{p}_n\| + \Delta t(1 + \Delta t)\tilde{\gamma}\sum_{i=1}^{s}\left\|f(\tilde{Q}_{ni}) - f(Q_{ni})\right\|$$
$$\leqslant \|\hat{q}_n\| + \sqrt{\|\sqrt{A}\hat{q}_n\|^2 + \|\hat{p}_n\|^2} + \Delta t\|\hat{p}_n\| + \Delta t(1 + \Delta t)\tilde{\gamma}L\sum_{i=1}^{s}\left\|\tilde{Q}_{ni} - Q_{ni}\right\|, \tag{11.31}$$

with $\tilde{\gamma} = \max(\tilde{\alpha}, \beta)$. Likewise, it follows from the first s equalities in (11.28) that

$$
\left\| \tilde{Q}_{ni} - Q_{ni} \right\| \leqslant \|\hat{q}_n\| + c_i \Delta t \|\hat{p}_n\| + \Delta t^2 \sum_{j=1}^{s} \|\bar{a}_{ij}(V)\| \left\| f(\tilde{Q}_{nj}) - f(Q_{nj}) \right\|
$$

$$
\leqslant \|\hat{q}_n\| + c_i \Delta t \|\hat{p}_n\| + \Delta t^2 \alpha L \sum_{j=1}^{s} \left\| \tilde{Q}_{nj} - Q_{nj} \right\|, \quad i = 1, \cdots, s.
$$

$$(11.32)$$

Then, summing up the results of (11.32) for i from 1 to s, we have the following result

$$
\sum_{i=1}^{s} \left\| \tilde{Q}_{ni} - Q_{ni} \right\| \leqslant \sum_{i=1}^{s} \left(\|\hat{q}_n\| + c_i \Delta t \|\hat{p}_n\| \right) + \Delta t^2 \alpha L s \sum_{j=1}^{s} \left\| \tilde{Q}_{nj} - Q_{nj} \right\|.
$$

Since the time stepsize Δt satisfies $\Delta t \leqslant \sqrt{\dfrac{1}{2\alpha L s}}$, we obtain

$$
\sum_{i=1}^{s} \left\| \tilde{Q}_{ni} - Q_{ni} \right\| \leqslant 2 \sum_{i=1}^{s} \left(\|\hat{q}_n\| + c_i \Delta t \|\hat{p}_n\| \right).
$$

$$(11.33)$$

Inserting (11.33) into (11.31) leads to

$$
\|\hat{q}_{n+1}\| + \sqrt{\|\sqrt{A}\hat{q}_{n+1}\|^2 + \|\hat{p}_{n+1}\|^2}
$$

$$
\leqslant \|\hat{q}_n\| + \sqrt{\|\sqrt{A}\hat{q}_n\|^2 + \|\hat{p}_n\|^2} + \Delta t \|\hat{p}_n\| + 2\Delta t(1 + \Delta t)\tilde{\gamma} L \sum_{i=1}^{s} \left(\|\hat{q}_n\| + c_i \Delta t \|\hat{p}_n\| \right)
$$

$$
\leqslant \|\hat{q}_n\| + \sqrt{\|\sqrt{A}\hat{q}_n\|^2 + \|\hat{p}_n\|^2} + (\Delta t + 4\Delta t \tilde{\gamma} L s)\left(\|\hat{q}_n\| + \sqrt{\|\sqrt{A}\hat{q}_n\|^2 + \|\hat{p}_n\|^2} \right).
$$

An argument by induction arrives at the following result

$$
\|\hat{q}_{n+1}\| + \sqrt{\|\sqrt{A}\hat{q}_{n+1}\|^2 + \|\hat{p}_{n+1}\|^2}
$$

$$
\leqslant \left(1 + \Delta t(1 + 4s\tilde{\gamma} L) \right)^n \left(\|\hat{q}_0\| + \sqrt{\|\sqrt{A}\hat{q}_0\|^2 + \|\hat{p}_0\|^2} \right)
$$

$$
\leqslant \exp \left(T(1 + 4s\tilde{\gamma} L) \right)\left(\|\tilde{\varphi}_1\| + \sqrt{\|\sqrt{A}\tilde{\varphi}_1\|^2 + \|\tilde{\varphi}_2\|^2} \right),
$$

which shows the conclusions of the theorem. $\square$

Remark 11.7 Here, it is very important to note that, for classical symplectic RKN methods, the bound of the nonlinear analysis will depend on $\|A\|$, whereas the result of the time integrators presented in this chapter is independent of $\|A\|$. The same situation also happens in the analysis of convergence presented in the next section. This point is crucial, in particular, when $\|A\|$ is very large, since as the mesh partition in the space discretisation increases, $\|A\|$ will increase. The reason for this difference is that the time integrators are derived based on the operator-variation-of-constants-formula (11.8). Moreover, it should be pointed out that the formula (11.8) solves the linear system $u_{tt} - a^2 \Delta u = 0$ exactly. The corresponding semidiscrete system inherits an analogous property.

11.5 Convergence

The convergence analysis of fully discrete schemes is a very important issue. This section pays attention to the convergence analysis of the fully discrete symplectic approximation. Under suitable assumptions of smoothness and spatial discretisation strategies, the abstract Hamiltonian system (11.7) can be discretised as follows:

$$\begin{cases} Q'(t) = P(t), & Q(t_0) = \varphi_1(x), \\ P'(t) = -AQ(t) + f(Q(t)) + \hat{\delta}(\Delta x), & P(t_0) = \varphi_2(x), \end{cases} \tag{11.34}$$

where A is a symmetric positive semi-definite differentiation matrix,

$$Q(t) = \big(u(x_1, t), u(x_2, t), \cdots, u(x_M, t)\big)^\mathsf{T}, \quad \varphi_l(x) = \big(\varphi_l(x_1), \varphi_l(x_2), \cdots, \varphi_l(x_M)\big)^\mathsf{T}$$

for $l = 1, 2$, and $\hat{\delta}(\Delta x)$ is the truncation error introduced by approximating the spatial differential operator $\mathscr{A}$ by a symmetric positive semi-definite matrix A.

We insert the exact solution of (11.34) into the numerical approximation (11.10) and obtain

$$\begin{cases} Q(t_n + c_i \Delta t) = \phi_0\big(c_i^2 V\big) Q(t_n) + c_i \Delta t \phi_1\big(c_i^2 V\big) P(t_n) \\ \qquad\qquad + \Delta t^2 \sum_{j=1}^{s} \bar{a}_{ij}(V) f(U(t_n + c_j \Delta t)) + \hat{A}_{ni}, \quad i = 1, 2, \cdots, s, \\[2mm] Q(t_{n+1}) = \phi_0\big(V\big) Q(t_n) + \Delta t \phi_1\big(V\big) P(t_n) + \Delta t^2 \sum_{i=1}^{s} \bar{b}_i(V) f(U(t_n + c_i \Delta t)) + \hat{\delta}_{n+1}, \\[2mm] P(t_{n+1}) = -\Delta t A \phi_1\big(V\big) Q(t_n) + \phi_0\big(V\big) P(t_n) + \Delta t \sum_{i=1}^{s} b_i(V) f(U(t_n + c_i \Delta t)) + \hat{\delta}'_{n+1}, \end{cases}$$
$$\tag{11.35}$$

where $\hat{\Delta}_{ni}$, $\hat{\delta}_{n+1}$ and $\hat{\delta}'_{n+1}$ are the discrepancies. With the conditions of Theorem 11.3, a similar analysis to that described in Sect. 11.3.3 leads to the following bounds for these discrepancies

$$\left\| \hat{\Delta}_{ni} \right\| \leqslant C_1 \Delta t^r + \Delta t^2 \left\| \int_0^1 (1-z)\phi_1\big((1-z)^2 c_i^2 V\big)\hat{\delta}(\Delta x)\mathrm{d}z \right\|,$$

$$\left\| \hat{\delta}_{n+1} \right\| \leqslant C_2 \Delta t^{r+1} + \Delta t^2 \left\| \int_0^1 (1-z)\phi_1\big((1-z)^2 V\big)\hat{\delta}(\Delta x)\mathrm{d}z \right\|,$$

$$\left\| \hat{\delta}'_{n+1} \right\| \leqslant C_3 \Delta t^{r+1} + \Delta t \left\| \int_0^1 \phi_0\big((1-z)^2 V\big)\hat{\delta}(\Delta x)\mathrm{d}z \right\|.$$

This further implies that

$$\begin{cases} \|\hat{\Delta}_{ni}\| \leqslant C_1\Delta t^r + \dfrac{1}{2}\Delta t^2\|\hat{\delta}(\Delta x)\|, \\[2mm] \|\hat{\delta}_{n+1}\| \leqslant C_2\Delta t^{r+1} + \dfrac{1}{2}\Delta t^2\|\hat{\delta}(\Delta x)\|, \\[2mm] \|\hat{\delta}'_{n+1}\| \leqslant C_3\Delta t^{r+1} + \Delta t\|\hat{\delta}(\Delta x)\|. \end{cases} \tag{11.36}$$

We apply the numerical approximation (11.10)–(11.34) and ignore $\hat{\delta}(\Delta x)$, and then obtain

$$\begin{cases} Q_{ni} = \phi_0\big(c_i^2 V\big)Q_n + c_i\Delta t\phi_1\big(c_i^2 V\big)P_n + \Delta t^2 \sum_{j=1}^{s} \bar{a}_{ij}(V)f(Q_{nj}), \ i = 1, 2, \cdots, s, \\[4mm] Q_{n+1} = \phi_0(V)Q_n + \Delta t\phi_1(V)P_n + \Delta t^2 \sum_{i=1}^{s} \bar{b}_i(V)f(Q_{ni}), \\[4mm] P_{n+1} = -\Delta t A\phi_1(V)Q_n + \phi_0(V)P_n + \Delta t \sum_{i=1}^{s} b_i(V)f(Q_{ni}). \end{cases}$$

$$\tag{11.37}$$

We are now in a position to present the convergence result for the fully discrete scheme (11.37).

Theorem 11.5 *Under Assumption 11.1 and the conditions of Theorem 11.3, it is assumed that $u(x,t)$ satisfies some suitable assumptions on smoothness. If the time* *stepsize Δt satisfies $\Delta t \leqslant \sqrt{\dfrac{1}{2\alpha Ls}}$, then there exists a constant C such that*

$$\begin{cases} \|Q(t_n) - Q_n\| \leqslant CT \exp\big((1 + 4s\tilde{\gamma}L)T\big)\big(\Delta t^r + \|\hat{\delta}(\Delta x)\|\big), \\ \|P(t_n) - P_n\| \leqslant CT \exp\big((1 + 4s\tilde{\gamma}L)T\big)\big(\Delta t^r + \|\hat{\delta}(\Delta x)\|\big), \end{cases} \tag{11.38}$$

where $\tilde{\gamma}$ is given in Theorem 11.4 and C is a constant independent of n, Δt and Δx.

Proof Let $e_n^Q = Q(t_n) - Q_n$, $e_n^P = P(t_n) - P_n$ and $E_{ni}^Q = Q(t_n + c_i\Delta t) - Q_{ni}$. Subtracting (11.37) from (11.35) yields the system of error equations

$$\begin{cases} E_{ni}^Q = \phi_0(c_i^2 V)e_n^Q + c_i\Delta t\phi_1(c_i^2 V)e_n^P \\ \qquad + \Delta t^2 \sum_{j=1}^{s} \bar{a}_{ij}(V)\Big(f\big(Q(t_n + c_i\Delta t)\big) - f(Q_{ni})\Big) + \hat{\Delta}_{ni}, \\ e_{n+1}^Q = \phi_0(V)e_n^Q + \Delta t\phi_1(V)e_n^P \\ \qquad + \Delta t^2 \sum_{i=1}^{s} \bar{b}_i(V)\Big(f\big(Q(t_n + c_i\Delta t)\big) - f(Q_{ni})\Big) + \hat{\delta}_{n+1}, \\ e_{n+1}^P = -\Delta t A\phi_1(V)e_n^Q + \phi_0(V)e_n^P \\ \qquad + \Delta t \sum_{i=1}^{s} b_i(V)\Big(f\big(Q(t_n + c_i\Delta t)\big) - f(Q_{ni})\Big) + \hat{\delta}_{n+1}', \end{cases} \tag{11.39}$$

with the initial conditions $e_0^Q = 0$, $e_0^P = 0$. We rewrite the last two equalities of (11.39) as follows

$$\begin{pmatrix} \sqrt{A}e_{n+1}^Q \\ e_{n+1}^P \end{pmatrix} = \begin{pmatrix} \phi_0(V) & \Delta t\sqrt{A}\phi_1(V) \\ -\Delta t\sqrt{A}\phi_1(V) & \phi_0(V) \end{pmatrix} \begin{pmatrix} \sqrt{A}e_n^Q \\ e_n^P \end{pmatrix}$$

$$+ \Delta t \sum_{i=1}^{s} \begin{pmatrix} \Delta t\sqrt{A}\bar{b}_i(V) \\ b_i(V) \end{pmatrix} \begin{pmatrix} f\big(Q(t_n + c_i\Delta t)\big) - f(Q_{ni}) \\ f\big(Q(t_n + c_i\Delta t)\big) - f(Q_{ni}) \end{pmatrix} + \begin{pmatrix} \sqrt{A}\hat{\delta}_{n+1} \\ \hat{\delta}_{n+1}' \end{pmatrix}.$$

This results in

$$\sqrt{\|\sqrt{A}e_{n+1}^Q\|^2 + \|e_{n+1}^P\|^2} \leqslant \sqrt{\|\sqrt{A}e_n^Q\|^2 + \|e_n^P\|^2}$$

$$+ \Delta t\tilde{\alpha} \sum_{i=1}^{s} \|f\big(Q(t_n + c_i\Delta t)\big) - f(Q_{ni})\| + \sqrt{\|\sqrt{A}\hat{\delta}_{n+1}\|^2 + \|\hat{\delta}_{n+1}'\|^2}.$$

It follows from the second equality of (11.39) that

$$\left\| e_{n+1}^Q \right\| \leqslant \left\| e_n^Q \right\| + \Delta t \left\| e_n^P \right\| + \Delta t^2 \beta \sum_{i=1}^s \left\| f\left(Q(t_n + c_i \Delta t) \right) - f(Q_{ni}) \right\| + \left\| \hat{\delta}_{n+1} \right\|.$$

We then have

$$\left\| e_{n+1}^Q \right\| + \sqrt{\left\| \sqrt{A} e_{n+1}^Q \right\|^2 + \left\| e_{n+1}^P \right\|^2} \leqslant \left\| e_n^Q \right\| + \Delta t \left\| e_n^P \right\| + \sqrt{\left\| \sqrt{A} e_n^Q \right\|^2 + \left\| e_n^P \right\|^2}$$

$$+ \Delta t (1 + \Delta t) \tilde{\gamma} L \sum_{i=1}^s \left\| E_{ni}^Q \right\| + \left\| \hat{\delta}_{n+1} \right\| + \sqrt{\left\| \sqrt{A} \hat{\delta}_{n+1} \right\|^2 + \left\| \hat{\delta}_{n+1}' \right\|^2}.$$

$$(11.40)$$

On the other hand, it follows from the first s equalities of (11.39) that

$$\| E_{ni}^Q \| \leqslant \| e_n^Q \| + c_i \Delta t \| e_n^P \| + \Delta t^2 \alpha L \sum_{i=1}^s \| E_{ni}^Q \| + \| \hat{\Delta}_{ni} \|, \quad i = 1, 2, \cdots, s.$$

$$(11.41)$$

Summing up the results of (11.41) for i from 1 to s, we thus obtain

$$\sum_{i=1}^s \| E_{ni}^Q \| \leqslant \sum_{i=1}^s (\| e_n^Q \| + c_i \Delta t \| e_n^P \| + \| \hat{\Delta}_{ni} \|) + \Delta t^2 \alpha L s \sum_{j=1}^s \| E_{nj}^Q \|.$$

As $\Delta t \leqslant \sqrt{\dfrac{1}{2\alpha L s}}$, we have

$$\sum_{i=1}^s \| E_{ni}^Q \| \leqslant 2 \sum_{i=1}^s (\| e_n^Q \| + c_i \Delta t \| e_n^P \|) + 2 \sum_{i=1}^s \| \hat{\Delta}_{ni} \|. \qquad (11.42)$$

Inserting (11.42) into (11.40) yields

$$\left\| e_{n+1}^Q \right\| + \sqrt{\left\| \sqrt{A} e_{n+1}^Q \right\|^2 + \left\| e_{n+1}^P \right\|^2}$$

$$\leqslant \left\| e_n^Q \right\| + \Delta t \left\| e_n^P \right\| + \sqrt{\left\| \sqrt{A} e_n^Q \right\|^2 + \left\| e_n^P \right\|^2} + 2\Delta t (1 + \Delta t) \tilde{\gamma} L \sum_{i=1}^s (\| e_n^Q \| + c_i \Delta t \| e_n^P \|)$$

$$+ \| \hat{\delta}_{n+1} \| + \sqrt{\left\| \sqrt{A} \hat{\delta}_{n+1} \right\|^2 + \left\| \hat{\delta}_{n+1}' \right\|^2} + 2\Delta t (1 + \Delta t) \tilde{\gamma} L \sum_{i=1}^s \| \hat{\Delta}_{ni} \|,$$

and this gives

$$\left\|e_{n+1}^Q\right\| + \sqrt{\|\sqrt{A}e_{n+1}^Q\|^2 + \|e_{n+1}^P\|^2} \leqslant \left(1 + \Delta t(1 + 4s\tilde{\gamma}L)\right)\left(\left\|e_n^Q\right\| + \sqrt{\|\sqrt{A}e_n^Q\|^2 + \|e_n^P\|^2}\right)$$

$$+ \|\hat{\delta}_{n+1}\| + \sqrt{\|\sqrt{A}\hat{\delta}_{n+1}\|^2 + \|\hat{\delta}'_{n+1}\|^2} + 2\Delta t(1 + \Delta t)\tilde{\gamma}L\sum_{i=1}^{s}\|\hat{\Delta}_{ni}\|.$$

$$(11.43)$$

On noting the truncation errors (11.36), there exists a constant C satisfying

$$\|\hat{\delta}_{n+1}\| + \sqrt{\|\sqrt{A}\hat{\delta}_{n+1}\|^2 + \|\hat{\delta}'_{n+1}\|^2} + 2\Delta t(1 + \Delta t)\tilde{\gamma}L\sum_{i=1}^{s}\|\hat{\Delta}_{ni}\| \leqslant C\Delta t\left(\Delta t^r + \|\hat{\delta}(\Delta x)\|\right).$$

We then apply the Gronwall's inequality (see, e.g. [48]) to (11.43) and obtain

$$\left\|e_{n+1}^Q\right\| + \sqrt{\|\sqrt{A}e_{n+1}^Q\|^2 + \|e_{n+1}^P\|^2} \leqslant \exp\left(n\Delta t(1 + 4s\tilde{\gamma}L)\right)\left(\|e_0^Q\| + \sqrt{\|\sqrt{A}e_0^Q\|^2 + \|e_0^P\|^2}\right.$$

$$\left. + Cn\Delta t\left(\Delta t^r + \|\hat{\delta}(\Delta x)\|\right)\right).$$

This confirms (11.38) and the proof of this theorem is complete. □

11.6 Symplectic Extended RKN-Type Approximation Schemes

In what follows, we will construct practical one-stage and two-stage symplectic approximation schemes. The multi-stage symplectic approximation schemes can be obtained in a similar way.

11.6.1 One-Stage Symplectic Approximation Schemes

It follows from Theorem 11.3 that a one-stage symplectic approximation scheme is of order two if the following conditions:

$$\begin{cases} b_1(V) = \phi_1(V) + \mathcal{O}(\Delta t^2), \\ b_1(V)c_1 = \phi_2(V) + \mathcal{O}(\Delta t), \\ \bar{b}_1(V) = \phi_2(V) + \mathcal{O}(\Delta t), \end{cases} \qquad (11.44)$$

are satisfied. These equations determine the parameters c_1, d_1. Substituting the coefficients (11.18) of one-stage approximation into (11.44) yields

$$d_1 = 1, \ c_1 = \frac{1}{2}.$$

This gives a family of one-stage symplectic approximation schemes of order two with arbitrary $\bar{a}_{11}(V)$ and the following additional coefficients

$$c_1 = \frac{1}{2}, \ \ b_1(V) = \phi_0(V/4), \ \ \bar{b}_1(V) = \phi_1(V/4)/2.$$

Case One Let $\bar{a}_{11}(V) = 0$. We then obtain a one-stage explicit symplectic approximation scheme of order two, which is termed ESA1s2.

Case Two We consider a third-order order condition

$$\bar{a}_{11}(V) = c_1^2 \phi_2(c_1^2 V) + \mathcal{O}(\Delta t),$$

and then choose $\bar{a}_{11}(V) = c_1^2 \phi_2(c_1^2 V)$. This results in a one-stage implicit symplectic approximation scheme of order two. This scheme is termed ISA1s2.

11.6.2 Two-Stage Symplectic Approximation Schemes

We next turn to two-stage symplectic approximation schemes.

Case One We first consider two-stage symplectic approximation schemes of order three. It follows from Theorem 11.3 that a two-stage symplectic approximation scheme is of order three if the following conditions

$$\begin{cases} b_1(V) + b_2(V) = \phi_1(V) + \mathcal{O}(\Delta t^3), \\[4pt] b_1(V)c_1 + b_2(V)c_2 = \phi_2(V) + \mathcal{O}(\Delta t^2), \\[4pt] b_1(V)c_1^2 + b_2(V)c_2^2 = 2\phi_3(V) + \mathcal{O}(\Delta t), \\[4pt] \bar{b}_1(V) + \bar{b}_2(V) = \phi_2(V) + \mathcal{O}(\Delta t^2), \\[4pt] \bar{b}_1(V)c_1 + \bar{b}_2(V)c_2 = \phi_3(V) + \mathcal{O}(\Delta t), \\[4pt] \bar{a}_{11}(V) + \bar{a}_{12}(V) = c_1^2 \phi_2(c_1^2 V) + \mathcal{O}(\Delta t), \\[4pt] \bar{a}_{21}(V) + \bar{a}_{22}(V) = c_2^2 \phi_2(c_2^2 V) + \mathcal{O}(\Delta t), \end{cases} \tag{11.45}$$

are satisfied. We substitute the coefficients $b_1(V)$, $b_2(V)$, $\bar{b}_1(V)$, $\bar{b}_2(V)$ of (11.18) into the first four equations of (11.45). This leads to

$$c_2 = \frac{2 - 3c_1}{3 - 6c_1}, \quad d_1 = \frac{1 - 2c_2}{2(c_1 - c_2)}, \quad d_2 = \frac{-1 + 2c_1}{2(c_1 - c_2)}.$$

We here consider diagonally implicit schemes. This gives

$$\bar{a}_{12}(V) = 0.$$

We solve the last equation of (11.18) as well as the last two equations of (11.45) and obtain

$$\bar{a}_{21}(V) = \frac{1}{d_2}\left(b_2(V)\bar{b}_1(V) - \bar{b}_2(V)b_1(V)\right),$$

$$\bar{a}_{11}(V) = c_1^2\phi_2(c_1^2 V),$$

$$\bar{a}_{22}(V) = c_2^2\phi_2(c_2^2 V) - \bar{a}_{21}(V).$$

The choice of $c_1 = \dfrac{3 - \sqrt{3}}{6}$ provides a two-stage diagonally implicit symplectic approximation scheme of order three. This scheme is termed DISA2s3.

Case Two We then consider two-stage symplectic approximation schemes of order four. It follows from Theorem 11.3 that the fourth-order conditions for a two-stage approximation scheme are

$$\begin{cases} b_1(V) + b_2(V) = \phi_1(V) + \mathcal{O}(\Delta t^4), \\[4pt] b_1(V)c_1 + b_2(V)c_2 = \phi_2(V) + \mathcal{O}(\Delta t^3), \\[4pt] b_1(V)c_1^2 + b_2(V)c_2^2 = 2\phi_3(V) + \mathcal{O}(\Delta t^2), \\[4pt] b_1(V)c_1^3 + b_2(V)c_2^3 = 6\phi_4(V) + \mathcal{O}(\Delta t), \\[4pt] \bar{b}_1(V) + \bar{b}_2(V) = \phi_2(V) + \mathcal{O}(\Delta t^3), \\[4pt] \bar{b}_1(V)c_1 + \bar{b}_2(V)c_2 = \phi_3(V) + \mathcal{O}(\Delta t^2), \\[4pt] \bar{b}_1(V)c_1^2 + \bar{b}_2(V)c_2^2 = 2\phi_4(V) + \mathcal{O}(\Delta t), \\[4pt] \bar{a}_{11}(V) + \bar{a}_{12}(V) = c_1^2\phi_2(c_1^2 V) + \mathcal{O}(\Delta t^2), \\[4pt] \bar{a}_{21}(V) + \bar{a}_{22}(V) = c_2^2\phi_2(c_2^2 V) + \mathcal{O}(\Delta t^2), \\[4pt] \bar{a}_{11}(V)c_1 + \bar{a}_{12}(V)c_2 = c_1^3\phi_3(c_1^2 V) + \mathcal{O}(\Delta t), \\[4pt] \bar{a}_{21}(V)c_1 + \bar{a}_{22}(V)c_2 = c_2^3\phi_3(c_2^2 V) + \mathcal{O}(\Delta t). \end{cases} \qquad (11.46)$$

Using the last equation of (11.18) as well as the following equations deduced from (11.46)

$$\bar{a}_{11}(V) + \bar{a}_{12}(V) = c_1^2 \phi_2(c_1^2 V),$$

$$\bar{a}_{21}(V) + \bar{a}_{22}(V) = c_2^2 \phi_2(c_2^2 V),$$

$$\bar{a}_{11}(V)c_1 + \bar{a}_{12}(V)c_2 = c_1^3 \phi_3(c_1^2 V),$$

we have

$$
\begin{cases}
\bar{a}_{11}(V) = \dfrac{c_1^2 \big(c_1 \phi_3(c_1^2 V) - c_2 \phi_2(c_1^2 V)\big)}{c_1 - c_2}, \\[2ex]
\bar{a}_{12}(V) = \dfrac{c_1^3 \big(\phi_2(c_1^2 V) - \phi_3(c_1^2 V)\big)}{c_1 - c_2}, \\[2ex]
\bar{a}_{21}(V) = \dfrac{(c_1 - c_2)(b_2(V)\bar{b}_1(V) - b_1(V)\bar{b}_2(V)) + c_1^3 d_1 \phi_2(c_1^2 V) - c_1^3 d_1 \phi_3(c_1^2 V)}{d_2(c_1 - c_2)}, \\[2ex]
\bar{a}_{22}(V) = c_2^2 \phi_2(c_2^2 V) - \bar{a}_{21}(V).
\end{cases}
$$

$$(11.47)$$

We substitute the coefficients $b_1(V)$, $b_2(V)$, $\bar{b}_1(V)$, $\bar{b}_2(V)$ of (11.18) into the first four equations of (11.45), and obtain

$$c_1 = \frac{3 - \sqrt{3}}{6}, \quad c_2 = \frac{3 + \sqrt{3}}{6}, \quad d_1 = \frac{1 - 2c_2}{2(c_1 - c_2)}, \quad d_2 = \frac{-1 + 2c_1}{2(c_1 - c_2)}.$$

This yields a two-stage symplectic approximation scheme. It can be verified that this scheme satisfies all the fourth-order conditions, and this scheme is termed ISA2s4.

Remark 11.8 Consider explicit two-stage extended RKN-type methods. If we choose

$$c_1 = 0, \ c_2 = 1, \ b_1(V) = \frac{1}{2}\phi_0(V), \ b_2(V) = \frac{1}{2}I,$$

$$\bar{b}_1(V) = \frac{1}{2}\phi_1(V), \ \bar{b}_2(V) = 0, \ \bar{a}_{21}(V) = \frac{1}{2}\phi_1(V),$$

it can be verified that these coefficients satisfy the symplectic conditions (11.18) with $d_1 = d_2 = \dfrac{1}{2}$. We here point out that this explicit symplectic extended RKN-type method reduces to the Deuflhard method which was first given in [49].

11.7 Numerical Experiments

Our sole goal in this section is to demonstrate the efficiency of the symplectic approximation schemes for the semilinear KG equation (11.1)–(11.2). We will use a collection of classical finite difference and the method-of-lines approximations for comparison. These schemes are described as follows.

1. *The Standard Finite Difference Schemes* (see, e.g. [5])

We first consider the following three frequently used *finite difference schemes* to discretise the equation (11.1)–(11.2):

- Explicit finite difference (EFD) scheme

$$\delta_t^2 u_j^n - a^2 \delta_x^2 u_j^n = f(u_j^n);$$

- Semi-implicit finite difference (SIFD) scheme

$$\delta_t^2 u_j^n - \frac{a^2}{2} \left(\delta_x^2 u_j^{n+1} + \delta_x^2 u_j^{n-1} \right) = f(u_j^n);$$

- Compact finite difference (CFD) scheme

$$\left(I + \frac{\Delta x^2}{12} \delta_x^2 \right) \delta_t^2 u_j^n - \frac{a^2}{2} \left(\delta_x^2 u_j^{n+1} + \delta_x^2 u_j^{n-1} \right) = \left(I + \frac{\Delta x^2}{12} \delta_x^2 \right) f(u_j^n).$$

Here u_j^n is the approximation of $u(x_j, t_n)$ for $j = 0, 1, \cdots, M$ and $n = 0, 1, \cdots, N$, and

$$\delta_t^2 u_j^n = \frac{u_j^{n+1} - 2u_j^n + u_j^{n-1}}{\Delta t^2} \quad \text{and} \quad \delta_x^2 u_j^n = \frac{u_{j+1}^n - 2u_j^n + u_{j-1}^n}{\Delta x^2}.$$

2. *The Method-of-lines Schemes*

As is known, a very popular approach to the approximation of (11.1) is the method-of-lines, where the discretisation is carried out in two distinct steps: the first is to approximate the space derivatives leaving a Hamiltonian ODEs in time; the second is to solve the Hamiltonian ODEs by an appropriate numerical method. There exist many different ways to approximate $\mathscr{A}$ in the literature. We here consider the following two types of spatial discretisation.

(I) *Symmetric finite difference (SFD)* (see, e.g. [50])

The operator $\mathscr{A}$ is approximated by the following 9-diagonal differentiation matrix:

$$
A_{sfd} = \frac{-a^2}{\Delta x^2}
\begin{pmatrix}
-\frac{205}{72} & \frac{8}{5} & -\frac{1}{5} & \frac{8}{315} & -\frac{1}{560} & & & -\frac{1}{560} & \frac{8}{315} & -\frac{1}{5} & \frac{8}{5} \\[4pt]
\frac{8}{5} & -\frac{205}{72} & \frac{8}{5} & -\frac{1}{5} & \frac{8}{315} & -\frac{1}{560} & & & -\frac{1}{560} & \frac{8}{315} & -\frac{1}{5} \\[4pt]
-\frac{1}{5} & \frac{8}{5} & -\frac{205}{72} & \frac{8}{5} & -\frac{1}{5} & \frac{8}{315} & -\frac{1}{560} & & & -\frac{1}{560} & \frac{8}{315} \\[4pt]
\frac{8}{315} & -\frac{1}{5} & \frac{8}{5} & -\frac{205}{72} & \frac{8}{5} & -\frac{1}{5} & \frac{8}{315} & -\frac{1}{560} & & & -\frac{1}{560} \\[4pt]
-\frac{1}{560} & \frac{8}{315} & -\frac{1}{5} & \frac{8}{5} & -\frac{205}{72} & \frac{8}{5} & -\frac{1}{5} & \frac{8}{315} & -\frac{1}{560} & & \\[4pt]
 & \ddots & \ddots & \ddots & \ddots & \ddots & \ddots & \ddots & \ddots & \ddots & \\[4pt]
 & & & -\frac{1}{560} & \frac{8}{315} & -\frac{1}{5} & \frac{8}{5} & -\frac{205}{72} & \frac{8}{5} & -\frac{1}{5} & \frac{8}{315} & -\frac{1}{560} \\[4pt]
-\frac{1}{560} & & & & -\frac{1}{560} & \frac{8}{315} & -\frac{1}{5} & \frac{8}{5} & -\frac{205}{72} & \frac{8}{5} & -\frac{1}{5} & \frac{8}{315} \\[4pt]
\frac{8}{315} & -\frac{1}{560} & & & -\frac{1}{560} & \frac{8}{315} & -\frac{1}{5} & \frac{8}{5} & -\frac{205}{72} & \frac{8}{5} & -\frac{1}{5} \\[4pt]
-\frac{1}{5} & \frac{8}{315} & -\frac{1}{560} & & & -\frac{1}{560} & \frac{8}{315} & -\frac{1}{5} & \frac{8}{5} & -\frac{205}{72} & \frac{8}{5} \\[4pt]
\frac{8}{5} & -\frac{1}{5} & \frac{8}{315} & -\frac{1}{560} & & & -\frac{1}{560} & \frac{8}{315} & -\frac{1}{5} & \frac{8}{5} & -\frac{205}{72}
\end{pmatrix}_{M \times M}
$$

The accuracy of this approximation for the space derivative is of order eight with an accuracy of $\mathscr{O}(\Delta x^8)$, and the differentiation matrix A_{sfd} is a positive semi-definite matrix.

(II) *Fourier spectral collocation (FSC)* (see, e.g. [51, 52])

The operator $\mathscr{A}$ is approximated by the second-order Fourier-spectral-collocation differentiation matrix A_{fsc}, and the entries of the $A_{fsc} = (a_{kj})_{M \times M}$ are

$$
a_{kj} = \begin{cases} \dfrac{(-1)^{k+j}}{2} \sin^{-2}\left(\dfrac{(k-j)\pi}{M}\right), & k \neq j, \\[10pt] \dfrac{M^2}{12} + \dfrac{1}{6}, & k = j. \end{cases} \tag{11.48}
$$

According to classical concepts, the spatial discretisation is of infinite order, and A_{fsc} is also a positive semi-definite matrix.

The time solvers chosen for comparisons are listed below:

- SVF: the classical Störmer-Verlet formula;
- GM1s2: the Gautschi's method of order two given in [53];

- ESA1s2: the one-stage explicit symplectic approximation scheme of order two derived in this chapter;
- ISA1s2: the one-stage implicit symplectic approximation scheme of order two proposed in this chapter;
- DISRK2s3: the two-stage diagonally implicit symplectic Runge–Kutta method of order three discussed in [37];
- ISRKN2s4: the two-stage implicit symplectic Runge–Kutta–Nyström method of order four considered in [54];
- DISA2s3: the two-stage diagonally implicit symplectic approximation scheme of order three presented in this chapter;
- ISA2s4: the two-stage implicit symplectic approximation scheme of order four derived in this chapter.

Since some of the above methods are implicit, iterative solutions are needed, and we use a fixed-point iteration in the practical computations. For the implementations of numerical experiments, we set 10^{-15} as the error tolerance and 10 as the maximum number of iterations.

It is noted that throughout the numerical experiments, the efficiency curves are plotted showing the global error versus the computational cost measured by the number of function evaluations required by each scheme, both in logarithmic scale.

Problem 11.1 Consider the semilinear KG equation (see, e.g. [7, 19])

$$
\begin{cases}
\dfrac{\partial^2 u}{\partial t^2} - a^2 \dfrac{\partial^2 u}{\partial x^2} = bu^3 - au, \quad -20 \leqslant x \leqslant 20, \quad 0 \leqslant t \leqslant T, \quad u(-20, t) = u(20, t), \\[2mm]
u(x, 0) = \sqrt{\dfrac{2a}{b}} \operatorname{sech}(\lambda x), \quad u_t(x, 0) = c\lambda \sqrt{\dfrac{2a}{b}} \operatorname{sech}(\lambda x) \tanh(\lambda x)
\end{cases}
$$

with $\lambda = \sqrt{\dfrac{a}{a^2 - c^2}}$ and $a, b, a^2 - c^2 > 0$. The exact solution is given by

$$
u(x, t) = \sqrt{\frac{2a}{b}} \operatorname{sech}(\lambda(x - ct)).
$$

Following [19], we consider the parameters $a = 0.3$, $b = 1$ and $c = 0.25$.

We first solve this equation by using the symplectic approximation schemes ESA1s2, ISA1s2, DISA2s3, ISA2s4 coupled with the eighth-order symmetric finite difference method and the Fourier spectral collocation method. We choose $M = 500$, $T = 10$, $\Delta t = 0.01$ and show the results in Figs. 11.1 and 11.2. It can be observed from the numerical results that the Fourier spectral collocation method to discretise the spatial variable is much better than the eighth-order finite difference method. Hence, we employ the Fourier spectral collocation method for all the spatial approximations.

We then compare our symplectic approximation schemes with the classical finite difference schemes. The problem is integrated with $T = 10$, $\Delta t = 0.1/2^j$ for

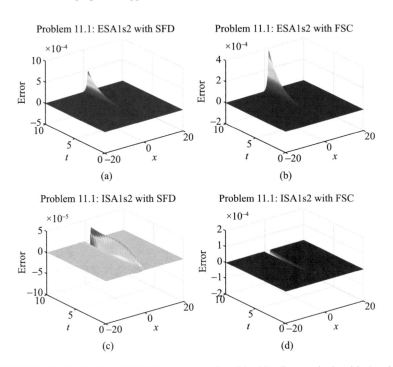

Fig. 11.1 Results for Problem 11.1: The errors produced by blending methods with the eighth-order SFD and FSC spatial discretisations

$j = 0, 1, 2, 3$. We set $M = 1000$ for the finite difference schemes EFD, SIFD, CFD and $M = 500$ for the symplectic approximation schemes ESA1s2, ISA1s2, DISA2s3, ISA2s4 coupled with the Fourier spectral collocation method. The global errors with $N = T/\Delta t$ are presented in Fig. 11.3a.

Finally, in comparison with the method-of-lines schemes, we discretise the spatial derivative by the Fourier spectral collocation method with $M = 500$, and then integrate the semidiscrete system with $\Delta t = 0.2/2^j$ for $j = 0, 1, 2, 3$ by different time-stepping methods. The efficiency curves are shown in Fig. 11.3b. It is remarked that after approximating the operator $\mathscr{A}$ by a positive semi-definite differentiation matrix A, there also exists a corresponding energy conservation law. Hence, we will test the effectiveness of our symplectic approximation schemes to preserve the semidiscrete energy. The energy conservation errors of Problem 11.1 are shown in Fig.11.3c with $M = 200$, $\Delta t = 0.01$ and $T = 20 \times 2^j$ for $j = 0, 1, 2, 3$. These results indicate that our symplectic approximation schemes are most accurate in preserving discrete energy among these underlying numerical methods.

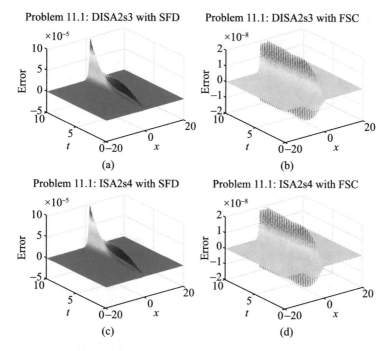

Fig. 11.2 Results for Problem 11.1: the errors produced by blending methods with the eighth-order SFD and FSC spatial discretisations

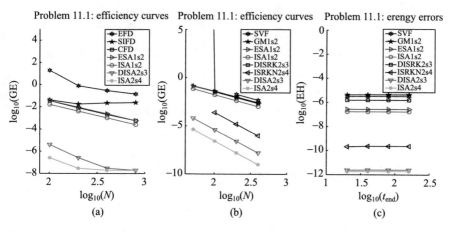

Fig. 11.3 Results for Problem 11.1: efficiency curves and energy errors

Problem 11.2 Consider the following sine-Gordon equation with periodic boundary conditions

$$\begin{cases} \dfrac{\partial^2 u}{\partial t^2} = \dfrac{\partial^2 u}{\partial x^2} - \sin u, & -20 \leqslant x \leqslant 20, \quad 0 \leqslant t \leqslant T, \quad u(-20,t) = u(20,t), \\ u(x,0) = 0, \quad u_t(x,0) = 4\operatorname{sech}\left(x/\sqrt{1+c^2}\right)/\sqrt{1+c^2}, \end{cases}$$

where $\kappa = 1/\sqrt{1+c^2}$. The exact solution of this problem reads

$$u(x,t) = 4\arctan\left(c^{-1}\sin(ct/\sqrt{1+c^2})\operatorname{sech}(x/\sqrt{1+c^2})\right)$$

and we choose $c = 0.5$.

We first solve this equation by using the symplectic approximation schemes coupled with the eighth-order symmetric finite difference method and the Fourier spectral collocation method. We set $M = 200$, $T = 100$, $\Delta t = 0.01$, and Figs. 11.4 and 11.5 show the results for the errors. It follows from the results that the Fourier spectral collocation method to discretise the spatial variable is much better than the

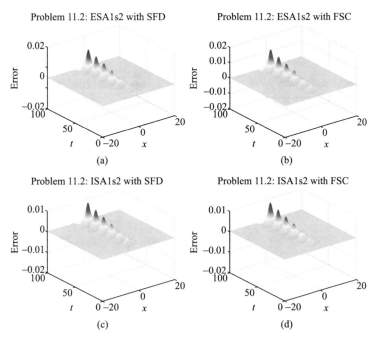

Fig. 11.4 Results for Problem 11.2: The errors produced by blending methods with the eighth-order SFD and FSC spatial discretisations

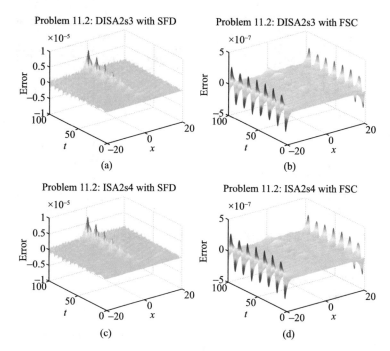

Fig. 11.5 Results for Problem 11.2: The errors produced by blending methods with the eighth-order SFD and FSC spatial discretisations

eighth-order finite difference method. Consequently, we choose the Fourier spectral collocation method for this problem.

We then integrate the problem with $T = 100$, $\Delta t = 0.1/2^j$ for $j = 1, 2, 3, 4$ in order to compare the symplectic approximation schemes with the classical finite difference schemes. We choose $M = 1000$ for the finite difference schemes EFD, SIFD, CFD and $M = 500$ for the symplectic approximation schemes coupled with the Fourier spectral collocation method. The global errors are presented in Fig. 11.6a.

We finally compare the symplectic approximation schemes with the method-of-lines schemes. In order to show the long-time performance of symplectic approximation methods, we discretise the spatial derivative by the Fourier spectral collocation method with $M = 200$, and then integrate the semidiscrete system on $[0, 1000]$ with different $\Delta t = 0.1/2^j$ for $j = 0, 1, 2, 3$. The efficiency curves are shown in Fig. 11.6b. The errors of the semidiscrete energy conservation with $M = 200$, $\Delta t = 0.1$ and $T = 10^j$ for $j = 1, 2, 3, 4$ are presented in Fig. 11.6c. With regard to the long-time analysis of symplectic extended RKN-type methods for PDEs via modulated Fourier expansions, we refer the readers to [55].

Again these results indicate that the symplectic approximation schemes are most efficient among these underlying numerical methods.

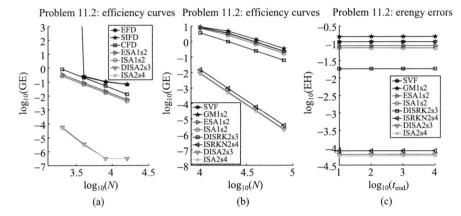

Fig. 11.6 Results for Problem 11.2: Efficiency curves and energy errors

Problem 11.3 Consider the dimensionless relativistic KG equation with a dimensionless parameter $\varepsilon > 0$ (see, e.g. [5, 56])

$$\begin{cases} \varepsilon^2 \dfrac{\partial^2 u}{\partial t^2} - \dfrac{\partial^2 u}{\partial x^2} + \dfrac{1}{\varepsilon^2} u + f(u) = 0, \quad -L \leqslant x \leqslant L, \;\; 0 \leqslant t \leqslant T, \quad u(-L, t) = u(L, t), \\ u(x, 0) = \phi(x), \quad u_t(x, 0) = \dfrac{1}{\varepsilon^2} \gamma(x). \end{cases}$$

Following [5], we here choose

$$f(u) = \lambda u^{p+1}, \quad \phi(x) = \frac{2}{\exp(x^2) + \exp(-x^2)}, \quad \gamma(x) = 0.$$

The solution of Problem 11.3 is highly oscillating in time. The symplectic approximation schemes will be compared with the method-of-lines schemes. We first discretise the spatial derivative by the Fourier spectral collocation method with $M = 200$, and then solve the semidiscrete system with $L = 8$ and $T = 10$. We choose $\Delta t = 0.04/2^j$ for $j = 0, 1, 2, 3$ when the parameter $\varepsilon = 0.2$, and for $j = 3, 4, 5, 6$ when the parameter $\varepsilon = 0.1$. Fig. 11.7a,b presents the coresponding efficiency curves. The errors of the semidiscrete energy conservation with $\Delta t = 0.001$ and $T = 10^j$ for $j = 0, 1, 2, 3$ are displayed in Fig. 11.7c,d. Clearly, it can be observed from Fig. 11.7 that the numerical behaviour of our symplectic approximation schemes is much better than that of the others.

It is noted that the symplectic approximation schemes can be extended to solving high-dimensional KG equations (see, e.g. [4]). As an illustrative example, we next consider a two-dimensional sine-Gordon equation to demonstrate that the symplectic approximation schemes also exhibit very good performance for two-dimensional KG equations.

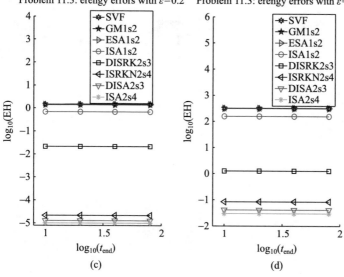

Fig. 11.7 Results for Problem 11.3: Efficiency curves and energy errors

Problem 11.4 Consider the following two-dimensional sine-Gordon equation (see, e.g. [57, 58]):

$$\begin{cases} u_{tt} - (u_{xx} + u_{yy}) = -\sin u, & -10 \leqslant x \leqslant 10, \ -10 \leqslant y \leqslant 10, \ 0 \leqslant t \leqslant T, \\ u(\pm 10, y, t) = u(x, \pm 10, t) = 0, \\ u(x, y, 0) = 4\arctan\left(\exp\left(\dfrac{4 - \sqrt{(x+3)^2 + (y+3)^2}}{0.436}\right)\right), & -10 \leqslant x, y \leqslant 10, \\ u_t(x, y, 0) = \dfrac{4.13}{\cosh\left(\exp\left(\left(4 - \sqrt{(x+3)^2 + (y+3)^2}\right)/0.436\right)\right)}, & -10 \leqslant x, y \leqslant 10. \end{cases}$$

We first integrate this problem over the region $(x, y) \in [-30, 10] \times [-30, 10]$ by the symplectic approximation schemes ESA1s2, ISA2s4 and the RKN method ISRKN2s4. The size of mesh region is 800×800 in space with the time stepsize $\Delta t = 0.1$. The numerical results are shown in Figs. 11.8 and 11.9 in terms of $\sin(u/2)$ at the time points $t = 3, 6, 9$. We then solve the problem over $[0, 2]$ with the stepsizes $\Delta t = 0.2/2^j$ for $j = 0, 1, 2, 3$. The log-log plots of global errors against N and the CPU time are shown in Fig. 11.10.

The numerical results clearly indicate that our symplectic approximation schemes are really very promising as compared with the well-known numerical schemes in the literature.

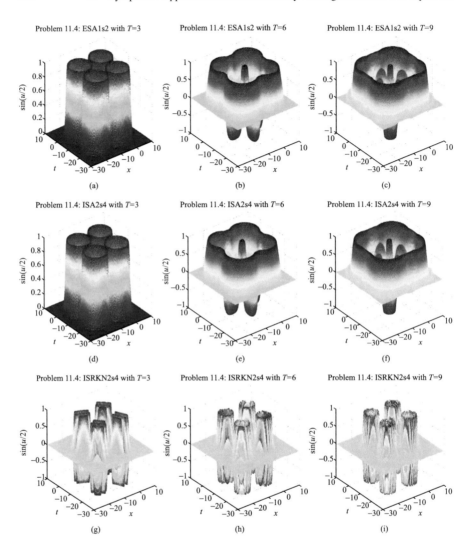

Fig. 11.8 Results for Problem 11.4: Collision of four ring solitons

11.8 Concluding Remarks

It is known that the first significant area where the idea of geometric integration was introduced was in the symplectic integration of Hamiltonian ODEs. However, the study of symplectic methods for Hamiltonian PDEs is far less developed than that for ODEs in the geometric integration literature, although multi-symplectic schemes for a class of PDEs with multi-symplectic structure have been considered (see, e.g. [1–3]). In particular, the important analysis issues, such as nonlinear stability and convergence, are still far from being satisfactory for multi-symplectic

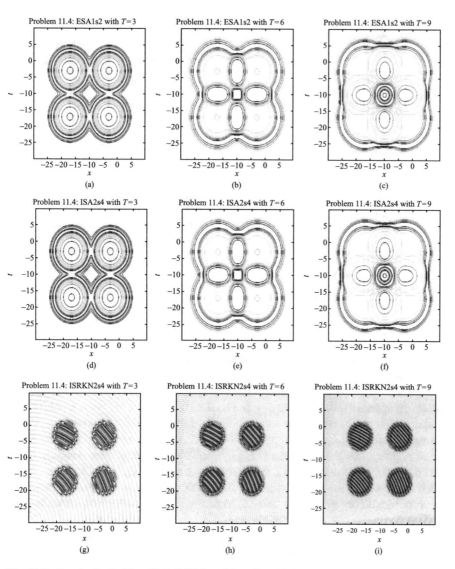

Fig. 11.9 Results for Problem 11.4: Collision of four ring solitons

methods. It is very important to note that the nonlinear stability and convergence for a fully discrete symplectic scheme are essential for the numerical simulation of nonlinear Hamiltonian PDEs. In Chap. 9, we presented the energy-preserving schemes for high-dimensional semilinear KG equations (see also [59]). As is known, the KG equation with nonlinear potential occurs in a wide range of application areas in science and engineering, and its computation and analysis represent a major challenge. In this chapter, we analysed an efficient symplectic approximation for

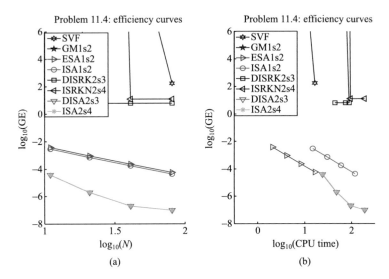

Fig. 11.10 Results for Problem 11.4: The log-log plots of global errors against N and the CPU time

the numerical simulation of the semilinear KG equations. The approximation is based on the operator-variation-of-constants formula of the abstract Hamiltonian (11.7) in the spirit of geometric integration. We showed the symplectic conditions for the fully discrete scheme, and the error bounds of the extended RKN-type time-stepping approximation for solving the nonlinear Hamiltonian system (11.7). The nonlinear stability and the convergence for the fully discrete scheme were analysed in detail. Numerical experiments, including KG equations in the nonrelativistic limit regime, where the solution is highly oscillatory in time, were compared with existing numerical schemes in the literature. Both the analytical and numerical results show that the symplectic approximation respects qualitative features, can capture singularity efficiently and preserve the discrete energy conservation law satisfactorily.

Last but not least, it is important to note that this chapter presented a nonlinear stability and convergence analysis for the symplectic approximation to the semi-linear KG equations, and all essential features of the symplectic approximation were considered and analysed in the one-dimensional case, although the proposed approximation lends itself equally well to high-dimensional semilinear KG equations, as shown by the last numerical experiment in this chapter. It is believed that the methodology presented in this chapter can be extended to a range of other nonlinear Hamiltonian wave equations.

The material in this chapter is based on the recent work by Wang and Wu [60].

References

1. Bridges, T.J.: Multi-symplectic structures and wave propagation. Math. Proc. Cambridge Philos. Soc. **121**, 147–190 (1997)
2. Li, Y.W., Wu, X.: General local energy-preserving integrators for solving multi-symplectic Hamiltonian PDEs. J. Comput. Phys. **301**, 141–166 (2015)
3. Marsden, J.E, Patrick, G.P., Shkoller, S.: Multi-symplectic, variational integrators, and nonlinear PDEs. Commun. Math. Phys. **4**, 351–395 (1999)
4. Liu, C., Wu, X.: Arbitrarily high-order time-stepping schemes based on the operator spectrum theory for high-dimensional nonlinear Klein-Gordon equations. J. Comput. Phys. **340**, 243–275 (2017)
5. Bao, W.Z., Dong, X.C.: Analysis and comparison of numerical methods for the Klein-Gordon equation in the nonrelativistic limit regime. Numer. Math. **120**, 189–229 (2012)
6. Cohen, D., Hairer, E., Lubich, Ch.: Conservation of energy, momentum and actions in numerical discretizations of non-linear wave equations. Numer. Math. **110**, 113–143 (2008)
7. Dehghan, M., Shokri, A.: Numerical solution of the nonlinear Klein-Gordon equation using radial basis functions. J. Comput. Appl. Math. **230**, 400–410 (2009)
8. Liu, C., Wu, X.: The boundness of the operator-valued functions for multidimensional nonlinear wave equations with applications. Appl. Math. Lett. **74**, 60–67 (2017)
9. Wu, X., Liu, C., Mei, L.: A new framework for solving partial differential equations using semi-analytical explicit RK(N)-type integrators. J. Comput. Appl. Math. **301**, 74–90 (2016)
10. Liu, C., Shi, W., Wu, X.: An efficient high-order explicit scheme for solving Hamiltonian nonlinear wave equations. Appl. Math. Comput. **246**, 696–710 (2014)
11. Perring, J.K., Skyrme, T.H.R.: A model unified field equation. Nuclear Phys. **31**, 550–555 (1962)
12. Ablowitz, M.J., Kruskal, M.D., Ladik, J.F.: Solitary wave collisions. SIAM J. Appl. Math. **36**, 428–437 (1979)
13. Dodd, R.K., Eilbeck, I.C., Gibbon, J.D., et al.: Solitons and Nonlinear Wave Equations. Academic Press, London (1982)
14. Duncan, D.B.: Symplectic finite difference approximations of the nonlinear Klein-Gordon equation. SIAM J. Numer. Anal. **34**, 1742–1760 (1997)
15. Dehghan, M., Ghesmati, A.: Application of the dual reciprocity boundary integral equation technique to solve the nonlinear Klein-Gordon equation. Comput. Phys. Commun. **181**, 1410–1418 (2010)
16. Dehghan, M., Mohammadi, V.: Two numerical meshless techniques based on radial basis functions (RBFs) and the method of generalized moving least squares (GMLS) for simulation of coupled Klein-Gordon-Schrodinger (KGS) equations. Comput. Math. Appl. **71**, 892–921 (2016)
17. Guo, B.Y., Li, X., Vázquez, L.: A Legendre spectral method for solving the nonlinear Klein-Gordon equation. Comput. Appl. Math. **15**, 19–36 (1996)
18. Lakestani, M., Dehghan, M.: Collocation and finite difference-collocation methods for the solution of nonlinear Klein-Gordon equation. Comput. Phys. Commun. **181**, 1392–1401 (2010)
19. Liu, C., Iserles, A., Wu, X.: Symmetric and arbitrarily high-order Birkhoff-Hermite time integrators and their long-time behavior for solving nonlinear Klein-Gordon equations. J. Comput. Phys. **356**, 1–30 (2018)
20. Cai, J., Wang, Y.: Local structure-preserving algorithms for the "good" Boussinesq equation. J. Comput. Phys. **239**, 72–89 (2013)
21. Cheng, K., Feng, W., Gottlieb, S., et al.: A Fourier pseudospectral method for the "good" Boussinesq equation with second-order temporal accuracy. Numer. Methods Partial Differ. Equ. **31**, 202–224 (2015)
22. De Frutos, J., Ortega, T.M., Sanz-Serna, J.: Pseudospectiral method for the "good" Boussinesq equation. Math. Comput. **57**, 109–122 (1991)

23. Yan, J., Zhang, Z.: New energy-preserving schemes using Hamiltonian Boundary Value and Fourier pseudospectral methods for the numerical solution of the "good" Boussinesq equation. Comput. Phys. Commun. **201**, 33–42 (2016)
24. Zhang, C., Wang, H., Huang, J., et al.: A second order operator splitting numerical scheme for the "good" Boussinesq equation. Appl. Numer. Math. **119**, 179–193 (2017)
25. Zhang, C., Huang, J., Wang, C., et al.: On the operator splitting and integral equation preconditioned deferred correction methods for the "Good" Boussinesq equation. J. Sci. Comput. **75**, 687–712 (2018)
26. Brugnano, L., Frasca Caccia, G., Iavernaro, F.: Efficient implementation of Gauss collocation and Hamiltonian boundary value methods. Numer. Algor. **65**, 633–650 (2014)
27. Hairer, E., Lubich, C., Wanner, G.: Geometric Numerical Integration: Structure-Preserving Algorithms for Ordinary Differential Equations, 2nd edn. Springer, Berlin (2006)
28. Hochbruck, M., Ostermann, A.: Exponential integrators. Acta Numer. **19**, 209–286 (2010)
29. Mei, L., Wu, X.: Symplectic exponential Runge-Kutta methods for solving nonlinear Hamiltonian systems. J. Comput. Phys. **338**, 567–584 (2017)
30. Wang, B., Wu, X., Meng, F.: Trigonometric collocation methods based on Lagrange basis polynomials for multi-frequency oscillatory second-order differential equations. J. Comput. Appl. Math. **313**, 185–201 (2017)
31. Wang, B., Meng, F., Fang, Y.: Efficient implementation of RKN-type Fourier collocation methods for second-order differential equations. Appl. Numer. Math. **119**, 164–178 (2017)
32. Wu, X., Liu, K., Shi, W.: Structure-Preserving Algorithms for Oscillatory Differential Equations II. Springer, Heidelberg (2015)
33. Wu, X., Wang, B.: Recent Developments in Structure-Preserving Algorithms for Oscillatory Differential Equations. Springer Nature Singapore Pte Ltd., Singapore (2018)
34. Wu, X., You, X., Wang, B.: Structure-Preserving Algorithms for Oscillatory Differential Equations. Springer, Berlin (2013)
35. Olver, P.J.: Applications of Lie Group to Differential Equations. Springer, New York (1986)
36. Channell, P.J., Scovel, C.: Symplectic integration of Hamiltonian systems. Nonlinearity **3**, 231–259 (1990)
37. Feng, K., Qin, M.Z.: The Symplectic Methods for The Computation of Hamiltonian Equations//Numerical Methods for Partial Differential Equations, pp. 1–37. Springer, Berlin (2006)
38. Mclachlan, R.: Symplectic integration of Hamiltonian wave equations. Numer. Math. **66**, 465–492 (1994)
39. Sanz-Serna, J.M.: Symplectic integrators for Hamiltonian problems: An overview. Acta Numer. **1**, 243–286 (1992)
40. Wang, B., Iserles, A., Wu, X.: Arbitrary-order trigonometric Fourier collocation methods for multi-frequency oscillatory systems. Found. Comput. Math. **16**, 151–181 (2016)
41. Wang, B., Yang, H., Meng, F.: Sixth order symplectic and symmetric explicit ERKN schemes for solving multi-frequency oscillatory nonlinear Hamiltonian equations. Calcolo **54**, 117–140 (2017)
42. Wu, X., Wang, B., Xia, J.: Explicit symplectic multidimensional exponential fitting modified Runge-Kutta-Nyström methods. BIT Numer. Math. **52**, 773–795 (2012)
43. Sanz-Serna, J.M., Calvo, M.P.: Numerical Hamiltonian Problems. Chapman and Hall, London (1994)
44. Wu, X., Liu, C.: An integral formula adapted to different boundary conditions for arbitrarily high-dimensional nonlinear Klein-Gordon equations with its applications. J. Math. Phys. **57**, 021504 (2016)
45. Yang, H., Wu, X., You, X., et al.: Extended RKN-type methods for numerical integration of perturbed oscillators. Comput. Phys. Commun. **180**, 1777–1794 (2009)
46. Mei, L., Liu, C., Wu, X.: An essential extension of the finite-energy condition for extended Runge-Kutta-Nyström integrators when applied to nonlinear wave equations. Commun. Comput. Phys. **22**, 742–764 (2017)

47. Wang, B., Wu, X., Xia, J.: Error bounds for explicit ERKN integrators for systems of multifrequency oscillatory second-order differential equations. Appl. Numer. Math. **74**, 17–34 (2013)
48. Hayes, L.J.: Galerkin alternating-direction methods for nonrectangular regions using patch approximations. SIAM J. Numer. Anal. **18**, 627–643 (1987)
49. Deuflhard, P.: A study of extrapolation methods based on multistep schemes without parasitic solutions. Z. Angew. Math. Phys. **30**, 177–189 (1979)
50. Bank, R., Graham, R.L., Stoer, J., et al.: High Order Difference Method for Time Dependent PDE. Springer, Berlin (2008)
51. Hesthaven, J.S., Gottlieb, S., Gottlieb, D.: Spectral Methods for Time-Dependent Problems//Cambridge Monographs on Applied and Computational Mathematics. Cambridge University, Cambridge (2007)
52. Shen, J., Tang, T., Wang, L.: Spectral Methods: Algorithms, Analysis, Applications. Springer, Berlin (2011)
53. Hairer, E., Lubich, C.: Long-time energy conservation of numerical methods for oscillatory differential equations. SIAM J. Numer. Anal. **38**, 414–441 (2000)
54. Tang, W., Sun, Y., Zhang, J.: High order symplectic integrators based on continuous-stage Runge-Kutta Nyström methods. Appl. Math. Comput. **361**, 670–679 (2019)
55. Wang, B., Wu, X.: Long-term analysis of symplectic or symmetric extended RKN methods for nonlinear wave equations (2018). arXiv: 1805. 06679v2
56. Najman, B.: The nonrelativistic limit of the nonlinear Klein-Gordon equation. Nonlinear Anal. **15**(3), 217–228 (1990)
57. Bratsos, A.G.: The solution of the two-dimensional sine-Gordon equation using the method of lines. J. Comput. Appl. Math. **206**, 251–277 (2007)
58. Sheng, Q., Khaliq, A.Q.M., Voss, D.A.: Numerical simulation of two-dimensional sine-Gordon solitons via a split cosine scheme. Math. Comput. Simul. **68**, 355–373 (2005)
59. Wang, B., Wu, X.: The formulation and analysis of energy-preserving schemes for solving high-dimensional nonlinear Klein-Gordon equations. IMA J. Numer. Anal. **39**, 2016–2044 (2019)
60. Wang, B., Wu, X.: A symplectic approximation with nonlinear stability and convergence analysis for efficiently solving semi-linear Klein-Gordon equations. Appl. Numer. Math. **142**, 64–89 (2019)

Chapter 12
Continuous-Stage Leap-Frog Schemes for Semilinear Hamiltonian Wave Equations

The standard leap-frog scheme is a well-known time integration scheme for some nonlinear partial differential equations due to its simplicity and its ease of implementation. The main aim of this chapter is to present continuous-stage modified leap-frog schemes for high-dimensional semilinear Hamiltonian wave equations. We begin with the formulation of the wave equation as an abstract second-order ordinary differential equation (ODE) and its operator-variation-of-constants formula. Then a continuous-stage modified leap-frog scheme is formulated, and its convergence, energy preservation, symplectic conservation, and long-time behaviour are rigorously analysed. The theory is accompanied by numerical results to demonstrate the remarkable advantage and efficiency of the modified leap-frog schemes in comparison with popular numerical schemes in the literature.

12.1 Introduction

Hamiltonian wave equations have many important applications in mathematical physics. Such problems occur frequently in acoustics, solid state physics, fluid dynamics, plasma physics, electromagnetics, nonlinear optics and quantum field theory (see, e.g. [1]). Hamiltonian wave equations have been identified in a variety of nonlinear partial differential equations (PDEs) such as the sine-Gordon (SG) equation, the Klein–Gordon (KG) equation, and the Korteweg de Vries equation. The efficient and accurate numerical solution of nonlinear wave equations is of fundamental importance and has received much attention in recent decades. Finite element methods (see, e.g. [2–7]), trigonometric methods (see, e.g. [8, 9]), energy-preserving methods (see, e.g. [10–14]), waveform relaxations (see, e.g. [15]), symplectic methods (see, e.g. [16]), spectrally accurate space-time solutions (see, e.g. [17]) and other methods (see, e.g. [18–22]) have been proposed and analysed.

© The Author(s), under exclusive license to Springer Nature Singapore Pte Ltd. 2021   393
X. Wu, B. Wang, *Geometric Integrators for Differential Equations with Highly
Oscillatory Solutions*, https://doi.org/10.1007/978-981-16-0147-7_12

Among typical time discretisation techniques for PDEs is the standard leap-frog (LF) scheme, and it has been investigated in many publications (see, e.g. [23–27]). As is known, LF discretisations of PDEs may result in unbounded solutions for any choice of mesh-sizes even for choices satisfying conditions for linear stability [26]. This algorithm, however, does not behave well for wave equations with some space discretisations with a high degree of precision. In this chapter, we present an improved version, which is termed the *modified* LF scheme. The modified LF scheme makes it possible to preserve different structures of the underlying system. As it is known that the energy conservation and symplecticity are two key features of Hamiltonian systems, a numerical algorithm should respect them as much as possible in the sprit of Geometric Numerical Integration. In this chapter, after showing the convergence of the modified LF scheme, different modified LF versions will be derived to preserve different geometric or physical properties such as energy preservation, symplecticity and long-time numerical energy conservation of explicit methods.

We next will present and analyse a modified LF scheme for the high-dimensional semilinear Hamiltonian wave equation of the form

$$\begin{cases} u_{tt} - a^2 \Delta u = f(u), & 0 < t \leqslant T, \ x \in \Omega, \\ u(x, 0) = \varphi_1(x), \ u_t(x, 0) = \varphi_2(x), \ x \in \bar{\Omega}, \end{cases} \tag{12.1}$$

where $u(x, t)$ denotes the wave displacement at time t and position $x \in \Omega$ with $\Omega := (0, X_1) \times \cdots \times (0, X_d) \subset \mathbb{R}^d$, a is a real parameter,

$$\Delta = \sum_{j=1}^{d} \frac{\partial^2}{\partial x_j^2},$$

and $f(u)$ is the negative derivative of a smooth potential energy $G(u) \geqslant 0$ which can be expressed as

$$f(u) = -G'(u).$$

It is assumed that Eq. (12.1) is supplemented with the following periodic boundary conditions

$$u(x, t)|_{\partial\Omega \cap \{x_j = 0\}} = u(x, t)|_{\partial\Omega \cap \{x_j = X_j\}}, \quad j = 1, 2, \cdots, d. \tag{12.2}$$

Clearly, problem (12.1) is a Hamiltonian PDE of the form

$$\begin{cases} u_t = v, \\ v_t = a^2 \Delta u + f(u) \end{cases} \tag{12.3}$$

with the Hamiltonian

$$\mathscr{H}[u, v] = \frac{1}{2} \int_\Omega \left(v^2 + a^2 |\nabla u|^2 + 2G(u) \right) dx, \qquad (12.4)$$

where $dx = dx_1 dx_2 \cdots dx_d$.

In order to improve the standard LF scheme, numerical integrators having continuous stages will be considered in this chapter. In the literature, continuous-stage numerical methods have been proposed and studied for a long time, such as continuous Runge–Kutta (RK) methods (see, e.g. [28–33]), continuous Runge–Kutta–Nystörm (RKN) methods (see, e.g. [34–36]) and continuous-stage energy-preserving methods (see, e.g. [37–45]). The idea of continuous-stage numerical methods suggests an improved framework of classical Runge–Kutta-type methods. In this chapter, we will make full use of the idea of continuous-stage methods and exponential integrators (see, e.g. [46–48]) for the modification. We here remark that compared with classical discontinuous-stage methods, the main advantage of continuous-stage methods is that they can generate numerical methods having different kinds of structure-preserving properties. For instance, it will be shown that the continuous-stage method presented in this chapter can generate not only energy-preserving methods but also symplectic methods. This is a significant advantage of continuous-stage methods over discontinuous stage methods.

12.2   A Continuous-Stage Modified Leap-Frog Scheme

In this section, we commence by expressing the semilinear wave equation (12.1) as the following abstract Hamiltonian system of ODEs on the Hilbert space $L^2(\Omega)$:

$$\begin{cases} q'(t) = p(t), & q(t_0) = \varphi_1(x), \\ p'(t) = -\mathscr{A}q(t) + f(q(t)), & p(t_0) = \varphi_2(x), \end{cases} \qquad (12.5)$$

where $q(t)$ maps x to $u(x, t)$: $q(t) = [x \mapsto u(x, t)]$, $\mathscr{A}$ is a linear, unbounded positive semi-definite operator defined by (see, e.g. [9])

$$(\mathscr{A}v)(x) = -a^2 \Delta v(x), \qquad (12.6)$$

with domain

$$D(\mathscr{A}) = \{ u \in H^1(\Omega) : u(x, t)|_{\partial\Omega \cap \{x_j = 0\}} = u(x, t)|_{\partial\Omega \cap \{x_j = X_j\}},$$
$$\nabla u(x, t)|_{\partial\Omega \cap \{x_j = 0\}} = \nabla u(x, t)|_{\partial\Omega \cap \{x_j = X_j\}}, \ j = 1, 2, \cdots, d \}. \qquad (12.7)$$

It is obvious that the operator $\mathscr{A}$ defined by (12.6) is a positive semi-definite operator, i.e.,

$$\left(\mathscr{A}u(x,t), u(x,t)\right) = \int_{\Omega} \mathscr{A}u(x,t) \cdot u(x,t)\mathrm{d}x = a^2 \int_{\Omega} |\nabla u(x,t)|^2 \mathrm{d}x \geqslant 0,$$

(12.8)

for all $u(x,t) \in D(\mathscr{A})$. Hence, the energy (12.4) can be expressed in the following form:

$$\mathscr{H}[u,v](t) \equiv \frac{1}{2}\left(v(x,t), v(x,t)\right) + \frac{1}{2}\left(\mathscr{A}u(x,t), u(x,t)\right) + \int_{\Omega} G\big(u(x,t)\big)\mathrm{d}x$$

$$= \mathscr{H}[u,v](0).$$

(12.9)

It then follows from the operator-variation-of-constants formula (also termed the Duhamel Principle) of (12.5) that

$$\begin{cases} q(t_n + \tau h) = \phi_0\big(\tau^2 \mathscr{V}\big)q(t_n) + \tau h\phi_1\big(\tau^2 \mathscr{V}\big)p(t_n) \\ \qquad\qquad + \tau^2 h^2 \int_0^1 (1-z)\phi_1\big((1-z)^2\tau^2\mathscr{V}\big) f\big(q(t_n + \tau hz)\big)\mathrm{d}z, \\ p(t_n + \tau h) = -\tau h\mathscr{A}\phi_1\big(\tau^2\mathscr{V}\big)q(t_n) + \phi_0\big(\tau^2\mathscr{V}\big)p(t_n) \\ \qquad\qquad + \tau h \int_0^1 \phi_0\big((1-z)^2\tau^2\mathscr{V}\big) f\big(q(t_n + \tau hz)\big)\mathrm{d}z, \end{cases}$$

(12.10)

for $\tau \in [0,1]$, where h is a time stepsize, $\mathscr{V} = h^2\mathscr{A}$ and the operator-argument functions ϕ_j are defined by

$$\phi_j(\mathscr{A}) := \sum_{k=0}^{\infty} \frac{(-1)^k \mathscr{A}^k}{(2k+j)!} : L^2(\Omega) \to L^2(\Omega), \quad j = 0, 1, 2, \cdots.$$

(12.11)

Using this definition, it can be easily verified that $\phi_0(x) = \cos(\sqrt{x})$ and $\phi_1(x) = x^{-1/2}\sin(\sqrt{x})$. We remark that these operator-argument functions have been researched for different boundary conditions in [9] and the following propositions are needed in this chapter.

Proposition 12.1 (See [9]) As far as the inner product of the space $L^2(\Omega)$: $(p,q) = \int_{\Omega} p(x)\overline{q(x)}\mathrm{d}x$, is concerned, all the operator-argument functions ϕ_j for $j \in \mathbb{N}$ are symmetric operators and the norm of the function in $L^2(\Omega)$ can be characterized by

$$\|q\|^2 = (q,q) = \int_{\Omega} |q(x)|^2\mathrm{d}x.$$

(12.12)

In particular, all the functions ϕ_j are bounded by $\|\phi_j(t\mathscr{A})\|_* \leqslant \gamma_j$, $j \in \mathbb{N}$, $t \geqslant 0$, where $\| \cdot \|_*$ is the Sobolev norm and γ_j for $j = 0, 1, \cdots$ are the bounds of the functions $\phi_j(x)$ with $x \geqslant 0$. Clearly, we have

$$\|\phi_j(t\mathscr{A})\|_* \leqslant 1, \quad j = 0, 1, \qquad \|\phi_2(t\mathscr{A})\|_* \leqslant \frac{1}{2}.$$

Here, it is noted that the norm $\| \cdot \|$ used in this chapter is referred to the norm which is defined by (12.12), and the Sobolev norm is given by

$$\|\phi_j(t\mathscr{A})\|_* = \sup_{\|u\| \neq 0} \frac{\|\phi_j(t\mathscr{A})u\|}{\|u\|}.$$

As is known, the standard LF scheme for (12.5) can be written as the following two-step formulation with time stepsize h

$$\frac{q_{n+1} - 2q_n + q_{n-1}}{h^2} = g(q_n), \qquad (12.13)$$

where the function $g(q) = -\mathscr{A}q + f(q)$ and the approximation p_n is given by

$$p_n = \frac{q_{n+1} - q_{n-1}}{2h}.$$

In what follows, we present a modification for the standard LF scheme, which is termed *a continuous-stage modified LF scheme*. It follows from the operator-variation-of-constants formula (12.10) that the term $q_{n+1} - 2q_n + q_{n-1}$ can be modified to $q_{n+1} - 2\phi_0(\mathscr{V})q_n + q_{n-1}$ and $p_n = \dfrac{q_{n+1} - q_{n-1}}{2h}$ can be changed to $\phi_1(\mathscr{V})p_n = \dfrac{q_{n+1} - q_{n-1}}{2h}$ as well as one additional term. Moreover, we work with $f(q_n)$ by the idea of continuous-stage methods. In such a way, we obtain the modified version of the standard LF scheme as follows.

Algorithm 12.1 (A Continuous-Stage Modified LF Scheme) A continuous-stage modified LF (CSMLF) scheme is defined by

$$\frac{q_{n+1} - 2\phi_0(\mathscr{V})q_n + q_{n-1}}{h^2} = \int_0^1 \bar{b}_\tau(\mathscr{V})(f(Q_{n,\tau}^+) + f(Q_{n,\tau}^-))d\tau, \qquad (12.14)$$

where $\bar{b}_\tau(\mathscr{V})$ is the bounded operator-argument function of τ and $\mathscr{V}$, and

$$Q_{n,\tau}^\pm = C_\tau(\mathscr{V})q_n \pm hD_\tau(\mathscr{V})p_n + h^2 \int_0^1 \bar{A}_{\tau,\sigma}(\mathscr{V})f(Q_{n,\sigma}^\pm)d\sigma, \quad 0 \leqslant \tau \leqslant 1,$$

$$(12.15)$$

with a bounded operator-argument function $\bar{A}_{\tau,\sigma}(\mathcal{V})$ depending on τ, σ and $\mathcal{V}$. The bounded operator-argument functions $C_\tau(\mathcal{V})$ and $D_\tau(\mathcal{V})$ are required to satisfy

$$C_{c_i}(\mathcal{V}) = \phi_0(c_i^2 \mathcal{V}), \quad D_{c_i}(\mathcal{V}) = c_i\phi_1(c_i^2 \mathcal{V}) \quad \text{for } i = 0, \cdots, s, \qquad (12.16)$$

where c_i for $i = 0, \cdots, s$ are the fitting nodes, and it is assumed that $0 = c_0 \leqslant c_1 < \cdots \leqslant c_s = 1$. Accordingly, the approximation p_n now becomes

$$\phi_1(\mathcal{V})p_n = \frac{q_{n+1} - q_{n-1}}{2h} - \frac{h}{2}\int_0^1 \bar{b}_\tau(\mathcal{V})(f(Q_{n,\tau}^+) - f(Q_{n,\tau}^-))\mathrm{d}\tau. \qquad (12.17)$$

Starting Values In the light of the operator-variation-of-constants formula (12.10), the starting values q_1 and p_1 are chosen as

$$
\begin{aligned}
q_1 &= \phi_0(\mathcal{V})q_0 + h\phi_1(\mathcal{V})p_0 + h^2\int_0^1 \bar{b}_\tau(\mathcal{V})f(Q_{0,\tau}^+)\mathrm{d}\tau, \\
p_1 &= -h\mathscr{A}\phi_1(\mathcal{V})q_0 + \phi_0(\mathcal{V})p_0 + h\int_0^1 b_\tau(\mathcal{V})f(Q_{0,\tau}^+)\mathrm{d}\tau,
\end{aligned}
\qquad (12.18)
$$

where $b_\tau(\mathcal{V})$ is a bounded operator-argument function of τ and $\mathcal{V}$ which will be determined by symmetry conditions of a one-step map.

One-step Map $(p_n, q_n) \mapsto (p_{n+1}, q_{n+1})$ The continuous-stage modified LF scheme with the starting values (12.18) can be written as symmetric one-step map of a form that is motivated by the operator-variation-of-constants formula (12.10)

$$
\left\{
\begin{aligned}
Q_{n,\tau}^+ &= C_\tau(\mathcal{V})q_n + hD_\tau(\mathcal{V})p_n + h^2\int_0^1 \bar{A}_{\tau,\sigma}(\mathcal{V})f(Q_{n,\sigma}^+)\mathrm{d}\sigma, \quad 0 \leqslant \tau \leqslant 1, \\
q_{n+1} &= \phi_0(\mathcal{V})q_n + h\phi_1(\mathcal{V})p_n + h^2\int_0^1 \bar{b}_\tau(\mathcal{V})f(Q_{n,\tau}^+)\mathrm{d}\tau, \\
p_{n+1} &= -h\mathscr{A}\phi_1(\mathcal{V})q_n + \phi_0(\mathcal{V})p_n + h\int_0^1 b_\tau(\mathcal{V})f(Q_{n,\tau}^+)\mathrm{d}\tau.
\end{aligned}
\right.
\qquad (12.19)
$$

Moreover, this one-step map is symmetric if and only if the following conditions are satisfied

$$
\left\{
\begin{aligned}
&\phi_1(\mathcal{V})b_\tau(\mathcal{V}) - \phi_0(\mathcal{V})\bar{b}_\tau(\mathcal{V}) = \bar{b}_{1-\tau}(\mathcal{V}), \\
&\phi_0(\mathcal{V})b_\tau(\mathcal{V}) + \mathcal{V}\phi_1(\mathcal{V})\bar{b}_\tau(\mathcal{V}) = b_{1-\tau}(\mathcal{V}), \\
&C_\tau(\mathcal{V})\phi_0(\mathcal{V}) + \mathcal{V}D_\tau(\mathcal{V})\phi_1(\mathcal{V}) = C_{1-\tau}(\mathcal{V}), \\
&C_\tau(\mathcal{V})\phi_1(\mathcal{V}) - D_\tau(\mathcal{V})\phi_0(\mathcal{V}) = D_{1-\tau}(\mathcal{V}), \\
&C_\tau(\mathcal{V})\bar{b}_{1-\sigma}(\mathcal{V}) - D_\tau(\mathcal{V})b_{1-\sigma}(\mathcal{V}) + \bar{A}_{\tau,\sigma}(\mathcal{V}) = \bar{A}_{1-\tau,1-\sigma}(\mathcal{V}),
\end{aligned}
\right.
\qquad (12.20)
$$

which are straightforwardly obtained by exchanging $n + 1 \leftrightarrow n$ and replacing h by $-h$ in (12.19). It is important to note that under these conditions, we obtain $Q_{n+1,\tau}^- = Q_{n,1-\tau}^-$, which is used in the equivalence of two-step form and one-step form.

Remark 12.2.1 The functions $\bar{A}_{\tau,\sigma}(\mathcal{V})$, $\bar{b}_\tau(\mathcal{V})$, $b_\tau(\mathcal{V})$ are assumed to be uniformly bounded. In this chapter, we use α, β, γ respectively to express the uniformly bounds of $\bar{A}_{\tau,\sigma}(\mathcal{V})$, $\bar{b}_\tau(\mathcal{V})$, $b_\tau(\mathcal{V})$ under the norm $\|\cdot\|_{L^2(\Omega)\leftarrow L^2(\Omega)}$.

Remark 12.2.2 It is important to note that the above continuous-stage modified LF scheme solves the homogeneous linear wave equation $u_{tt} - a^2 \Delta u = 0$ exactly. Moreover, if $a = 0$, this integrator reduces to the continuous-stage Runge–Kutta–Nyström method which has been researched for ODEs in [35]. Very recently, continuous-stage trigonometric integrators have also been studied in [49] for solving second-order ODEs. However, that paper only discussed continuous-stage extended Runge–Kutta–Nyström methods for ODEs and no explicit scheme was derived. Moreover, convergence analysis, symplecticity-preservation and long term energy conservation of explicit methods were not shown there. In this chapter, we not only present the convergence, but also derive different kinds of modified LF schemes to achieve different structure-preserving properties.

Generally, the operator $\mathscr{A}$ will be approximated by a differentiation matrix A on an M-dimensional space. This process converts the underlying PDE into a set of coupled ODEs in time, which may then be integrated.

Fully Discrete Scheme Using suitable spatial discretisation strategies such as the Fourier spectral collocation, and under some suitable assumptions of smoothness, the original continuous system (12.5) can be discretised as

$$
\begin{cases}
Q'(t) = P(t), & Q(t_0) = \varphi_1(x), \\
P'(t) = -AQ(t) + f(Q(t)) + \Theta(\Delta x), & P(t_0) = \varphi_2(x),
\end{cases} \tag{12.21}
$$

where A is a differentiation matrix, $\Theta(\Delta x)$ denotes the spatial discretisation error ($\Theta(\Delta x)$ satisfies $\Theta(\Delta x_j) \to 0$ as $\Delta x \to 0$) introduced by approximating the spatial differential operator $\mathscr{A}$ through the differentiation matrix A, and

$$
Q(t) = \left(u(\tilde{x}_1, t)^\mathsf{T}, u(\tilde{x}_2, t)^\mathsf{T}, \cdots, u(\tilde{x}_d, t)^\mathsf{T} \right)^\mathsf{T},
$$

$$
\varphi_l(x) = \left(\varphi_l(\tilde{x}_1)^\mathsf{T}, \varphi_l(\tilde{x}_2)^\mathsf{T}, \cdots, \varphi_l(\tilde{x}_d)^\mathsf{T} \right)^\mathsf{T} \text{ for } l = 1, 2,
$$

in which

$$
u(\tilde{x}_j, t) = \left(u(\tilde{x}_{j,0}, t), u(\tilde{x}_{j,1}, t), \cdots, u(\tilde{x}_{j,M-1}, t) \right)^\mathsf{T} \text{ for } j = 1, 2, \cdots, d,
$$

$$
\varphi_l(\tilde{x}_j) = \left(\varphi_l(\tilde{x}_{j,0}), \varphi_l(\tilde{x}_{j,1}), \cdots, \varphi_l(\tilde{x}_{j,M-1}) \right)^\mathsf{T} \text{ for } l = 1, 2, \quad j = 1, 2, \cdots, d,
$$

and M is a positive integer which represents the number of interior discretised points for each spatial variable. In one-dimensional case, i.e., $d = 1$, A is a positive semi-definite differentiation matrix, and

$$Q(t) = \left(u(\tilde{x}_0, t), u(\tilde{x}_1, t), \cdots, u(\tilde{x}_{M-1}, t)\right)^{\mathsf{T}},$$

$$\varphi_l(x) = \left(\varphi_l(\tilde{x}_0), \varphi_l(\tilde{x}_1), \cdots, \varphi_l(\tilde{x}_{M-1})\right)^{\mathsf{T}} \text{ for } l = 1, 2.$$

This approach leads to the following fully discrete continuous-stage modified LF scheme corresponding to (12.19)

$$
\begin{cases}
Q_{n,\tau}^{+} = C_\tau(V)Q_n + hD_\tau(V)P_n + h^2 \displaystyle\int_0^1 \bar{A}_{\tau,\sigma}(V)f(Q_{n,\sigma}^{+})\mathrm{d}\sigma, \\[2ex]
Q_{n+1} = \phi_0(V)Q_n + h\phi_1(V)P_n + h^2 \displaystyle\int_0^1 \bar{b}_\tau(V)f(Q_{n,\tau}^{+})\mathrm{d}\tau, \\[2ex]
P_{n+1} = -hA\phi_1(V)Q_n + \phi_0(V)P_n + h \displaystyle\int_0^1 b_\tau(V)f(Q_{n,\tau}^{+})\mathrm{d}\tau,
\end{cases}
\tag{12.22}
$$

where $V = h^2 A$.

In what follows, the fully discrete continuous-stage modified LF scheme (12.22) is termed fully discrete CSMLF scheme.

Remark 12.2.3 For the actual computation in applications, the operator $\mathscr{A}$ is usually approximated by a symmetric positive semi-definite differentiation matrix A because this is essential for structure preservation, and then the continuous-stage modified LF scheme needs the computation of some matrix-valued functions. At first sight it seems that this brings additional cost in comparison with the standard LF scheme. However, it is important to note that all the matrix-valued functions only need to be computed once, and when the matrix A is symmetric positive semi-definite, they can be implemented by the functions sine and cosine acting on diagonal matrices. Moreover, the nonlinearity for the modified LF scheme becomes $f(q)$ but for the standard LF scheme is $g(q) = -\mathscr{A}q + f(q)$. This difference can reduce some of the cost of the modified LF scheme in the computation of the nonlinearity. It can be observed from the numerical results shown in Sect. 12.7 that compared with the standard LF scheme, the modified LF scheme can be implemented inexpensively and has competitive advantages such as accuracy and long-time energy-preserving property.

12.3   Convergence

This section concerns the convergence of the fully discrete CSMLF scheme (12.22) for the wave equation (12.5). We begin with the following hypothesis on the nonlinearity f, which has been considered in many publications (see, e.g. [13, 46]).

Assumption 12.1 It is assumed that the function $f(u) = -G'(u) : D(\mathscr{A}) \to \mathbb{R}$ is sufficiently often Fréchet differentiable in a strip along the exact solution and is sufficiently smooth with respect to time. Moreover, let $f(u)$ be locally Lipschitz-continuous in a strip along the exact solution for the L^2-norm, and we denote the Lipschitz constant by K.

With the Assumption 12.1, we have following convergence theorem for the fully discrete CSMLF integrator (12.22).

Theorem 12.1 (Convergence) *Under the conditions of Assumption 12.1, if the stepsize h satisfies $h \leqslant \sqrt{\dfrac{1}{2\alpha K}}$ and the following order conditions are satisfied:*

$$\begin{cases} \displaystyle\int_0^1 b_\tau(V)\frac{\tau^j}{j!}d\tau - \phi_{j+1}(V) = \mathscr{O}(h^{r-j}), & j = 0, 1, \cdots, r-1, \\[2mm] \displaystyle\int_0^1 \bar{b}_\tau(V)\frac{\tau^j}{j!}d\tau - \phi_{j+2}(V) = \mathscr{O}(h^{r-1-j}), & j = 0, 1, \cdots, r-2, \\[2mm] \displaystyle\int_0^1\int_0^1 \bar{a}_{\tau,\sigma}(V)\frac{\sigma^j}{j!}d\tau d\sigma - \int_0^1 \tau^{j+2}\phi_{j+2}(\tau^2 V)d\tau = \mathscr{O}(h^{r-2-j}), & j = 0, 1, \cdots, r-3, \end{cases}$$

$$(12.23)$$

where $V = h^2 A$, then we have the convergence of the fully discrete CSMLF scheme (12.22):

$$\begin{cases} \|Q(t_n) - Q_n\| \leqslant CT \exp\left((1 + 4\tilde{\gamma}K)T\right)\left(h^m + \|\Theta(\Delta x)\|\right), \\[2mm] \|P(t_n) - P_n\| \leqslant CT \exp\left((1 + 4\tilde{\gamma}K)T\right)\left(h^m + \|\Theta(\Delta x)\|\right), \end{cases}$$

$$(12.24)$$

where $\| \cdot \|$ denotes the L^2-norm, $\Theta(\Delta x)$ is the spatial discretisation error defined in (12.21), C is a constant independent of n, h and Δx, $\tilde{\gamma} = \max(\tilde{\alpha}, \beta)$ which is independent of $\|V\|$, and $\tilde{\alpha}$ is the uniform bound of diag $(h\sqrt{A}\bar{b}_\tau(V), b_\tau(V))$. *Here $m = \min(r, s+1)$ with the positive integer s given in (12.16) and the positive integer r determined by (12.23). In particular, if $C_\tau(V)$ and $D_\tau(V)$ are chosen as*

$$C_\tau(V) = \phi_0(\tau^2 V), \quad D_\tau(V) = \tau\phi_1(\tau^2 V),$$

then $m = r$ in the result.

Proof The proof is divided into two parts. The first part is concerned with local errors and the second part presents global error bounds.

(I) Bounds of Local Errors

Inserting the exact solution of (12.21) into the scheme (12.22), we obtain

$$
\left\{
\begin{aligned}
Q(t_n + \tau h) &= C_\tau(V)Q(t_n) + hD_\tau(V)P(t_n) \\
&\quad + h^2 \int_0^1 \bar{A}_{\tau,\sigma}(\mathscr{V})f(Q(t_n + \sigma h))\mathrm{d}\sigma + \hat{\Delta}_\tau, \\
Q(t_{n+1}) &= \phi_0(V)Q(t_n) + h\phi_1(V)P(t_n) \\
&\quad + h^2 \int_0^1 \bar{b}_\tau(\mathscr{V})f(Q(t_n + \tau h))\mathrm{d}\tau + \hat{\delta}_{n+1}, \\
P(t_{n+1}) &= -hA\phi_1(V)Q(t_n) + \phi_0(V)P(t_n) \\
&\quad + h \int_0^1 b_\tau(\mathscr{V})f(Q(t_n + \tau h))\mathrm{d}\tau + \hat{\delta}'_{n+1},
\end{aligned}
\right.
\tag{12.25}
$$

with the local errors $\hat{\Delta}_\tau$, $\hat{\delta}_{n+1}$ and $\hat{\delta}'_{n+1}$. These errors are bounded by

$$
\begin{aligned}
\left\| \hat{\Delta}_\tau \right\| &\leqslant C_1 h^m + \frac{1}{2}h^2 \|\Theta(\Delta x)\|, \\
\left\| \hat{\delta}_{n+1} \right\| &\leqslant C_2 h^{r+1} + \frac{1}{2}h^2 \|\Theta(\Delta x)\|, \\
\left\| \hat{\delta}'_{n+1} \right\| &\leqslant C_3 h^{r+1} + h \|\Theta(\Delta x)\|,
\end{aligned}
\tag{12.26}
$$

where C_1, C_2, C_3 are constants depending on the constants symbolised by $\mathscr{O}$ in (12.23) and the bound of the remainder of Lagrange interpolation.

It follows from the matrix-variation-of-constants-formula of (12.21) that

$$
\left\{
\begin{aligned}
Q(t_n + \tau h) =&\, \phi_0(\tau^2 V)Q(t_n) + \tau h \phi_1(\tau^2 V)P(t_n) \\
&+ \tau^2 h^2 \int_0^1 (1-z)\phi_1((1-z)^2\tau^2 V)f(Q(t_n + \tau hz))\mathrm{d}z \\
&+ \tau^2 h^2 \int_0^1 (1-z)\phi_1((1-z)^2\tau^2 V)\Theta(\Delta x)\mathrm{d}z, \\
P(t_n + \tau h) =&- \tau h A\phi_1(\tau^2 V)Q(t_n) + \phi_0(\tau^2 V)P(t_n) \\
&+ \tau h \int_0^1 \phi_0((1-z)^2\tau^2 V)f(Q(t_n + \tau hz))\mathrm{d}z \\
&+ \tau h \int_0^1 \phi_0((1-z)^2\tau^2 V)\Theta(\Delta x)\mathrm{d}z.
\end{aligned}
\right.
\tag{12.27}
$$

Comparing the first equality of (12.27) with the first one of (12.25) gives

$$\hat{\Delta}_\tau = \Big(\phi_0(\tau^2 V) - C_\tau(V)\Big) Q(t_n) + \Big(h\tau\phi_1(\tau^2 V) - h D_\tau(V)\Big) P(t_n)$$

$$+ \tau^2 h^2 \int_0^1 (1-z)\phi_1\big((1-z)^2\tau^2 V\big)\Theta(\Delta x)dz$$

$$+ \tau^2 h^2 \int_0^1 (1-\sigma)\phi_1\big((1-\sigma)^2\tau^2 V\big)\hat{f}(t_n + \sigma\tau h)d\sigma$$

$$- h^2 \int_0^1 \bar{A}_{\tau,\sigma}(V)\hat{f}(t_n + \sigma h)d\sigma,$$

where $\hat{f}(t) = f(Q(t))$. It follows from the condition (12.16) and the results of Lagrange interpolation that

$$\phi_0(\tau^2 V) - C_\tau(V) = \mathcal{O}(h^{s+1}), \quad h\tau\phi_1(\tau^2 V) - h D_\tau(V) = \mathcal{O}(h^{s+1}). \quad (12.28)$$

Using the Taylor series expansion of $\hat{f}$ at t_n and the above results, we rewrite Δ_τ as

$$\hat{\Delta}_\tau = \mathcal{O}(h^{s+1}) + \tau^2 h^2 \int_0^1 (1-z)\phi_1\big((1-z)^2\tau^2 V\big)\Theta(\Delta x)dz$$

$$+ \tau^2 h^2 \int_0^1 (1-\sigma)\phi_1\big((1-\sigma)^2\tau^2 V\big)\sum_{j=0}^r \frac{(\sigma\tau h)^j}{j!}\hat{f}^{(j)}(t_n)d\sigma$$

$$- h^2 \int_0^1 \bar{A}_{\tau,\sigma}(V)\sum_{j=0}^r \frac{\sigma^j h^j}{j!}\hat{f}^{(j)}(t_n)d\sigma + \mathcal{O}(h^{r+2})$$

$$= \mathcal{O}(h^{s+1}) + \tau^2 h^2 \int_0^1 (1-z)\phi_1\big((1-z)^2\tau^2 V\big)\Theta(\Delta x)dz$$

$$+ \sum_{j=0}^r h^{j+2}\Big[\tau^{j+2}\phi_{j+2}(\tau^2 V) - \int_0^1 \bar{A}_{\tau,\sigma}(V)\frac{\sigma^j}{j!}d\sigma\Big]\hat{f}^{(j)}(t_n)d\sigma + \mathcal{O}(h^{r+2}),$$

where $\hat{f}^{(j)}(t)$ denotes the j-th order derivative of $f(Q(t))$ with respect to t, and the following result has been used here

$$\int_0^1 (1-\sigma)\phi_1\big((1-\sigma)^2\tau^2 V\big)\frac{\sigma^j}{j!}d\sigma = \phi_{j+2}(\tau^2 V)d\sigma.$$

In a similar way, we can obtain the results for δ_{n+1} and δ'_{n+1}. Under the conditions (12.23), it is clear that the bounds given in (12.26) are true.

(II) Global Error Bounds

We define that

$$e_n^Q = Q(t_n) - Q_n, \ e_n^P = P(t_n) - P_n, \ E_\tau^Q = Q(t_n + \tau h) - Q_{n,\tau}^+.$$

Subtracting (12.22) from (12.25) leads to the error equations

$$
\begin{cases}
E_\tau^Q = \phi_0(\tau^2 V)e_n^Q + \tau h \phi_1(\tau^2 V)e_n^P \\
\qquad + h^2 \displaystyle\int_0^1 \bar{A}_{\tau,\sigma}(V)\Big(f\big(Q(t_n + \sigma h)\big) - f(Q_{n,\sigma}^+)\Big)\mathrm{d}\sigma + \hat{\Delta}_\tau + \mathcal{O}(h^{s+1}), \\
e_{n+1}^Q = \phi_0(V)e_n^Q + h\phi_1(V)e_n^P \\
\qquad + h^2 \displaystyle\int_0^1 \bar{b}_\tau(V)\Big(f\big(Q(t_n + \tau h)\big) - f(Q_{n,\tau}^+)\Big)\mathrm{d}\tau + \hat{\delta}_{n+1}, \\
e_{n+1}^P = -hA\phi_1(V)e_n^Q + \phi_0(V)e_n^P \\
\qquad + h \displaystyle\int_0^1 b_\tau(V)\Big(f\big(Q(t_n + \tau h)\big) - f(Q_{n,\tau}^+)\Big)\mathrm{d}\tau + \hat{\delta}_{n+1}',
\end{cases}
$$

$$(12.29)$$

where the initial conditions are $e_0^Q = 0$, $e_0^P = 0$. We here replace $C_\tau(V)$ and $D_\tau(V)$ by $\phi_0(\tau^2 V)$ and $\tau\phi_1(\tau^2 V)$, respectively, and this generates the $\mathcal{O}(h^{s+1})$ term[1] in (12.29).

Because A is a symmetric positive semi-definite differentiation matrix, we reformulate the last two equations of (12.29) as

$$
\begin{pmatrix} \sqrt{A}e_{n+1}^Q \\ e_{n+1}^P \end{pmatrix} = \begin{pmatrix} \phi_0(V) & h\sqrt{A}\phi_1(V) \\ -h\sqrt{A}\phi_1(V) & \phi_0(V) \end{pmatrix} \begin{pmatrix} \sqrt{A}e_n^Q \\ e_n^P \end{pmatrix}
$$
$$
+ h \int_0^1 \begin{pmatrix} h\sqrt{A}\bar{b}_\tau(V) \\ & b_\tau(V) \end{pmatrix} \begin{pmatrix} f\big(Q(t_n + \tau h)\big) - f(Q_{n,\tau}^+) \\ f\big(Q(t_n + \tau h)\big) - f(Q_{n,\tau}^+) \end{pmatrix} \mathrm{d}\tau
$$
$$
+ \begin{pmatrix} \sqrt{A}\hat{\delta}_{n+1} \\ \hat{\delta}_{n+1}' \end{pmatrix}.
$$

Using the result in [50], we have

$$\left\| \begin{pmatrix} \phi_0(V) & h\sqrt{A}\phi_1(V) \\ -h\sqrt{A}\phi_1(V) & \phi_0(V) \end{pmatrix} \right\| = 1.$$

---

[1]This result is clear from (12.28).

Furthermore, it follows from Assumption 12.1 that $\left\| f\big(Q(t_n + \tau h)\big) - f(Q_{n,\tau}^+) \right\| \leqslant K \left\| E^Q \right\|$. We then obtain

$$\sqrt{\|\sqrt{A}e_{n+1}^Q\|^2 + \|e_{n+1}^P\|^2} \leqslant \sqrt{\|\sqrt{A}e_n^Q\|^2 + \|e_n^P\|^2} + h\tilde{\alpha}K\|E^Q\|_c$$
$$+ \sqrt{\|\sqrt{A}\hat{\delta}_{n+1}\|^2 + \|\hat{\delta}_{n+1}'\|^2}, \quad (12.30)$$

where $\|\cdot\|_c$ denotes the maximum norm for continuous functions which is defined as

$$\|E\|_c = \max_{\tau \in [0,1]} \|E_\tau\| \quad (12.31)$$

for a continuous $\mathbb{R}^M$-valued function E_τ on $[0, 1]$. Using the second formula of (12.29), we obtain

$$\left\| e_{n+1}^Q \right\| \leqslant \left\| e_n^Q \right\| + h \left\| e_n^P \right\| + h^2 \beta K \|E^Q\|_c + \left\| \hat{\delta}_{n+1} \right\|. \quad (12.32)$$

It then follows from (12.30) and (12.32) that

$$\left\| e_{n+1}^Q \right\| + \sqrt{\|\sqrt{A}e_{n+1}^Q\|^2 + \|e_{n+1}^P\|^2}$$
$$\leqslant \left\| e_n^Q \right\| + h \left\| e_n^P \right\| + \sqrt{\|\sqrt{A}e_n^Q\|^2 + \|e_n^P\|^2} + h(1+h)\tilde{\gamma}K\|E^Q\|_c$$
$$+ \|\hat{\delta}_{n+1}\| + \sqrt{\|\sqrt{A}\hat{\delta}_{n+1}\|^2 + \|\hat{\delta}_{n+1}'\|^2}. \quad (12.33)$$

Taking advantage of the first equation of (12.29) and (12.31), we have

$$\|E_\tau^Q\| \leqslant \|e_n^Q\| + \tau h\|e_n^P\| + h^2\alpha K\|E^Q\|_c + \|\hat{\Delta}_\tau\| + C_4 h^{s+1}. \quad (12.34)$$

This results in

$$\|E^Q\|_c \leqslant \|e_n^Q\| + h\|e_n^P\| + h^2\alpha K\|E^Q\|_c + \|\hat{\Delta}_\tau\|_c + C_4 h^{s+1},$$

where the constant C_4 depends on the constant symbolised by $\mathscr{O}$ in the first formula of (12.29). Under the condition that $h \leqslant \sqrt{\dfrac{1}{2\alpha K}}$, we deduce that

$$\|E^Q\|_c \leqslant 2(\|e_n^Q\| + h\|e_n^P\|) + 2\|\hat{\Delta}_\tau\|_c + 2C_4 h^{s+1}. \quad (12.35)$$

Then inserting (12.35) into (12.33) yields

$$
\left\| e_{n+1}^Q \right\| + \sqrt{\|\sqrt{A} e_{n+1}^Q\|^2 + \|e_{n+1}^P\|^2}
$$

$$
\leqslant \left\| e_n^Q \right\| + h \left\| e_n^P \right\| + \sqrt{\|\sqrt{A} e_n^Q\|^2 + \|e_n^P\|^2} + 2h(1+h)\tilde{\gamma}K(\|e_n^Q\| + h\|e_n^P\|)
$$

$$
+ \|\hat{\delta}_{n+1}\| + \sqrt{\|\sqrt{A}\hat{\delta}_{n+1}\|^2 + \|\hat{\delta}_{n+1}'\|^2} + 2h(1+h)\tilde{\gamma}K\|\hat{\Delta}_\tau\|_c + 2C_4 h^{s+1}
$$

$$
\leqslant \left(1 + h(1+4\tilde{\gamma}K)\right)\left(\left\| e_n^Q \right\| + \sqrt{\|\sqrt{A} e_n^Q\|^2 + \|e_n^P\|^2}\right)
$$

$$
+ \|\hat{\delta}_{n+1}\| + \sqrt{\|\sqrt{A}\hat{\delta}_{n+1}\|^2 + \|\hat{\delta}_{n+1}'\|^2} + 2h(1+h\tilde{\gamma}K)\|\hat{\Delta}_\tau\|_c + 2C_4 h^{s+1}.
$$

Finally, using the Gronwall's inequality, we obtain

$$
\left\| e_{n+1}^Q \right\| + \sqrt{\|\sqrt{A} e_{n+1}^Q\|^2 + \|e_{n+1}^P\|^2}
$$

$$
\leqslant \exp\left(nh(1+4\tilde{\gamma}K)\right)\left(\|e_0^Q\| + \sqrt{\|\sqrt{A} e_0^Q\|^2 + \|e_0^P\|^2} + Cnh\left(h^m + \|\Theta(\Delta x)\|\right)\right),
$$

which confirms (12.24).

The conclusion of this theorem follows.   □

12.4 Energy-Preserving Continuous-Stage Modified LF Schemes

This section concerns energy preservation of continuous-stage modified LF schemes. We first present the energy-preserving conditions for these schemes.

Theorem 12.2 (Energy-Preserving Conditions) *Let*

$$
\bar{A}_{0,\sigma}(\mathscr{V}) = 0, \quad \bar{A}_{1,\sigma}(\mathscr{V}) = \bar{b}_\sigma(\mathscr{V}). \tag{12.36}
$$

If the following conditions

$$
\begin{cases}
\mathscr{V}\phi_0(\mathscr{V})\bar{b}_\tau(\mathscr{V}) - \mathscr{V}\phi_1(\mathscr{V})b_\tau(\mathscr{V}) = C_\tau'(\mathscr{V}), \\
\phi_0(\mathscr{V})b_\tau(\mathscr{V}) + \mathscr{V}\phi_1(\mathscr{V})\bar{b}_\tau(\mathscr{V}) = D_\tau'(\mathscr{V}), \\
b_\tau(\mathscr{V})b_\sigma(\mathscr{V}) + \mathscr{V}\bar{b}_\tau(\mathscr{V})\bar{b}_\sigma(\mathscr{V}) = \bar{A}_{\tau,\sigma}'(\mathscr{V}) + \bar{A}_{\sigma,\tau}'(\mathscr{V}),
\end{cases} \tag{12.37}
$$

are satisfied, where $C'_\tau(\mathscr{V}) = \dfrac{\mathrm{d}}{\mathrm{d}\tau}C_\tau(\mathscr{V})$, $D'_\tau(\mathscr{V}) = \dfrac{\mathrm{d}}{\mathrm{d}\tau}D_\tau(\mathscr{V})$ *and* $\bar{A}'_{\tau,\sigma}(\mathscr{V}) = \dfrac{\partial}{\partial\tau}\bar{A}_{\tau,\sigma}(\mathscr{V})$, *then the CSMLF scheme* (12.19) *preserves the energy* (12.9) *exactly, i.e., it is true that*

$$\mathscr{H}[p_{n+1}, q_{n+1}] = \mathscr{H}[p_n, q_n], \quad n = 0, 1, \cdots.$$

Moreover, under these conditions with V, *instead of* $\mathscr{V}$, *the fully discrete CSMLF scheme* (12.22) *exactly preserves the semidiscrete energy*

$$H(P, Q) = \frac{1}{2}P^\mathsf{T}P + \frac{1}{2}Q^\mathsf{T}AQ + G(Q).$$

Proof We let

$$I_1 = \int_0^1 \bar{b}_\tau(\mathscr{V})f(Q^+_{n,\tau})\mathrm{d}\tau, \quad I_2 = \int_0^1 b_\tau(\mathscr{V})f(Q^+_{n,\tau})\mathrm{d}\tau.$$

Inserting the continuous-stage modified LF scheme (12.19) into $\mathscr{H}[p_{n+1}, q_{n+1}]$ determined in (12.9) gives

$$\mathscr{H}[p_{n+1}, q_{n+1}]$$

$$= \frac{1}{2}\Big(\big(\phi_0^2(\mathscr{V}) + \mathscr{V}\phi_1^2(\mathscr{V})\big)p_n, p_n\Big) + \frac{1}{2}\Big(\mathscr{A}\big(\phi_0^2(\mathscr{V}) + \mathscr{V}\phi_1^2(\mathscr{V})\big)q_n, q_n\Big)$$

$$+ \big(\mathscr{V}\phi_0(\mathscr{V})q_n, I_1\big) - \big(\mathscr{V}\phi_1(\mathscr{V})q_n, I_2\big) + h\big(\phi_0(\mathscr{V})p_n, I_2\big) + h\big(\mathscr{V}\phi_1(\mathscr{V})p_n, I_1\big)$$

$$+ \frac{1}{2}h^2\big(I_2, I_2\big) + \frac{1}{2}h^2\big(\mathscr{V}I_1, I_1\big) + \int_\Omega G\big(q_{n+1}\big)\mathrm{d}x$$

$$= \frac{1}{2}\big(p_n, p_n\big) + \frac{1}{2}\big(\mathscr{A}q_n, q_n\big) + \int_\Omega G\big(q_{n+1}\big)\mathrm{d}x + \big(q_n, \mathscr{V}\phi_0(\mathscr{V})I_1 - \mathscr{V}\phi_1(\mathscr{V})I_2\big)$$

$$+ h\big(p_n, \phi_0(\mathscr{V})I_2 + \mathscr{V}\phi_1(\mathscr{V})I_1\big) + \frac{1}{2}h^2\big(I_2, I_2\big) + \frac{1}{2}h^2\big(\mathscr{V}I_1, I_1\big).$$

$$(12.38)$$

On noticing the requirements (12.16) and (12.36) and the conditions $c_0 = 0$ and $c_s = 1$, we have

$$Q^+_{n,0} = q_n, \quad Q^+_{n,1} = q_{n+1}. \tag{12.39}$$

We then obtain

$$\mathscr{H}[p_{n+1}, q_{n+1}] - \mathscr{H}[p_n, q_n]$$

$$= \int_\Omega \big(G(q_{n+1}) - G(q_n)\big)\mathrm{d}x + \big(q_n, \mathscr{V}\phi_0(\mathscr{V})I_1 - \mathscr{V}\phi_1(\mathscr{V})I_2\big)$$

$$+ h\big(p_n, \phi_0(\mathcal{V})I_2 + \mathcal{V}\phi_1(\mathcal{V})I_1\big) + \frac{1}{2}h^2\big(I_2, I_2\big) + \frac{1}{2}h^2\big(\mathcal{V}I_1, I_1\big)$$

$$= \big(q_n, \mathcal{V}\phi_0(\mathcal{V})I_1 - \mathcal{V}\phi_1(\mathcal{V})I_2 - \int_0^1 C_\tau'(\mathcal{V})f(Q_{n,\tau}^+)\mathrm{d}\tau\big)$$

$$+ h\big(p_n, \phi_0(\mathcal{V})I_2 + \mathcal{V}\phi_1(\mathcal{V})I_1 - \int_0^1 D_\tau'(\mathcal{V})f(Q_{n,\tau}^+)\mathrm{d}\tau\big) + \frac{1}{2}h^2\big(I_2, I_2\big)$$

$$+ \frac{1}{2}h^2\big(\mathcal{V}I_1, I_1\big) - \int_\Omega \left(\int_0^1 \int_0^1 f^{\mathsf{T}}(Q_{n,\tau}^+)\bar{A}_{\tau\sigma}'(\mathcal{V})f(Q_{n,\sigma}^+)\mathrm{d}\sigma\,\mathrm{d}\tau\right)\mathrm{d}x.$$

$$\tag{12.40}$$

Using the first two equations of (12.37) and exchanging $\tau \leftrightarrow \sigma$, we rewrite (12.40) as

$$2\mathcal{H}[p_{n+1}, q_{n+1}] - 2\mathcal{H}[p_n, q_n] = h^2 \int_\Omega \int_0^1 \int_0^1 f^{\mathsf{T}}(Q_{n,\tau}^+)\big(b_\tau(\mathcal{V})b_\sigma(\mathcal{V})$$

$$+ \mathcal{V}\bar{b}_\tau(\mathcal{V})\bar{b}_\sigma(\mathcal{V}) - \bar{A}_{\tau\sigma}'(\mathcal{V}) - \bar{A}_{\sigma\tau}'(\mathcal{V})\big)f(Q_{n,\sigma}^+)\mathrm{d}\sigma\,\mathrm{d}\tau\mathrm{d}x.$$

It then follows from the third equation of (12.37), that $\mathcal{H}[p_{n+1}, q_{n+1}] - \mathcal{H}[p_n, q_n] = 0$. The conclusion of this theorem is confirmed. □

Using the energy-preserving conditions (12.37) and the order conditions (12.23), we next derive a practical energy-preserving CSMLF scheme (12.19) for solving Hamiltonian wave equations. We here only consider a scheme of order two for brevity. Higher-order energy-preserving CSMLF schemes can be constructed in a similar way.

Algorithm 12.2 (Energy-Preserving Scheme) Consider the special case where $s = 1$. We define a practical continuous-stage modified LF scheme (12.19) with the coefficients

$$C_\tau(\mathcal{V}) = (1 - \tau)I + \tau\phi_0(\mathcal{V}), \qquad D_\tau(\mathcal{V}) = \tau\phi_1(\mathcal{V}),$$

$$\bar{b}_\tau(\mathcal{V}) = \phi_2(\mathcal{V}), \quad b_\tau(\mathcal{V}) = \phi_1(\mathcal{V}), \quad \bar{A}_{\tau,\sigma}(\mathcal{V}) = \tau\phi_2(\mathcal{V}).$$

$$\tag{12.41}$$

It can be shown that this integrator is energy-preserving, symmetric and of order two. The one-step pattern of this scheme is given by

$$\begin{cases} Q_{n,\tau}^+ = ((1 - \tau)I + \tau\phi_0(\mathcal{V}))q_n + h\tau\phi_1(\mathcal{V})p_n + h^2 \int_0^1 \tau\phi_2(\mathcal{V})f(Q_{n,\sigma}^+)\mathrm{d}\sigma, \quad 0 \leqslant \tau \leqslant 1, \\[2ex] q_{n+1} = \phi_0(\mathcal{V})q_n + h\phi_1(\mathcal{V})p_n + h^2 \int_0^1 \phi_2(\mathcal{V})f(Q_{n,\tau}^+)\mathrm{d}\tau, \\[2ex] p_{n+1} = -h\mathcal{A}\phi_1(\mathcal{V})q_n + \phi_0(\mathcal{V})p_n + h \int_0^1 \phi_1(\mathcal{V})f(Q_{n,\tau}^+)\mathrm{d}\tau. \end{cases}$$

Remark 12.4.1 We remark that this method was firstly proposed in [51] for solving second-order ODEs and was further studied in [49, 52, 53]. However, no convergence analysis of this scheme was made in these publications.

Using the above energy-preserving conditions with a careful calculation, another energy-preserving continuous-stage modified LF scheme can be obtained as follows:

$$c_0 = 0, \quad c_1 = 1/2, \quad c_2 = 1,$$
$$\bar{b}_\tau(\mathscr{V}) = \bar{b}_1(\mathscr{V}) + \bar{b}_2(\mathscr{V})\tau, \quad b_\tau(\mathscr{V}) = b_1(\mathscr{V}) + b_2(\mathscr{V})\tau, \tag{12.42}$$
$$\bar{A}_{\tau,\sigma}(\mathscr{V}) = a_{11}(\mathscr{V})\tau + a_{12}(\mathscr{V})\tau\sigma + a_{21}(\mathscr{V})\tau^2 + a_{22}(\mathscr{V})\tau^2\sigma,$$

where

$$a_{11}(\mathscr{V}) = 4\phi_2(\mathscr{V}/4) - 3\phi_2(\mathscr{V}), \quad a_{12}(\mathscr{V}) = -\phi_1^2(\mathscr{V}/16)(I + \mathscr{V}\phi_2(\mathscr{V}/4)/4),$$
$$a_{21}(\mathscr{V}) = 1/2\phi_1^2(\mathscr{V}/16)(I - 3\mathscr{V}\phi_2(\mathscr{V}/4)/4), \quad a_{22}(\mathscr{V}) = \mathscr{V}\phi_1^4(\mathscr{V}/16)/4,$$
$$\bar{b}_1(\mathscr{V}) = 3\phi_2(\mathscr{V}) - \phi_2(\mathscr{V}/4), \quad \bar{b}_2(\mathscr{V}) = 2\phi_2(\mathscr{V}/4) - 4\phi_2(\mathscr{V}),$$
$$b_1(\mathscr{V}) = -2\phi_1(\mathscr{V}/4) + 3\phi_1(\mathscr{V}), \quad b_2(\mathscr{V}) = 4\phi_1(\mathscr{V}/4) - 4\phi_1(\mathscr{V}).$$

12.5 Symplectic Continuous-Stage Modified LF Scheme

In this section, we design a symplectic fully discrete CSMLF scheme, which preserves the symplecticity of Hamiltonian systems. To this end, we first study the symplectic conditions of fully discrete CSMLF schemes and then derive some practical symplectic methods.

Theorem 12.3 (Symplecticity) *Let*

$$C_\tau(V) = \phi_0(\tau^2 V), \quad D_\tau(V) = \tau\phi_1(\tau^2 V), \tag{12.43}$$

where $V = h^2 A$ and A a symmetric and positive semi-definite matrix. If the following conditions

$$\begin{cases} \phi_0(V)b_\tau(V) + V\phi_1(V)\bar{b}_\tau(V) = d_\tau\phi_0(\tau^2 V), \\ \phi_1(V)b_\tau(V) - \phi_0(V)\bar{b}_\tau(V) = \tau d_\tau\phi_1(\tau^2 V), \\ \bar{b}_\tau(V)b_\sigma(V) + d_\tau\bar{a}_{\tau,\sigma}(V) = \bar{b}_\sigma(V)b_\tau(V) + d_\sigma\bar{a}_{\sigma,\tau}(V), \end{cases} \tag{12.44}$$

are fulfilled, where $d_\tau \in \mathbb{R}$ is a function of τ, then the fully discrete CSMLF scheme (12.22) is symplectic, i.e., it is true that (see, e.g. [54])

$$dQ_{n+1} \wedge dP_{n+1} = dQ_n \wedge dP_n.$$

Proof We divide the proof into two parts.

(I) The result is firstly proved for the special case where $A = \text{diag}(a_{11}, a_{22}, \cdots, a_{MM})$. Let $f_\tau = -G'(Q_{n,\tau}^+)$ and $v_{ii} = h^2 a_{ii}$ for $i = 1, \cdots, M$. Denoting the J-th component of a vector by the superscript $(\cdot)^J$, we obtain

$$dQ_{n+1}^J \wedge dP_{n+1}^J = dQ_n^J \wedge dP_n^J$$

$$+ h \int_0^1 \Big(\big(\phi_0(v_{JJ}) \bar{b}_\tau(v_{JJ}) + v_{JJ}\phi_1(v_{JJ}) \bar{b}_\tau(v_{JJ}) \big) \phi_0^{-1}(\tau^2 v_{JJ}) \Big) \big(d(Q_{n,\tau}^+)^J \wedge df_\tau^J \big) d\tau$$

$$+ h^2 \int_0^1 \Big(\phi_1(v_{JJ}) b_\tau(v_{JJ}) - \phi_0(v_{JJ}) \bar{b}_\tau(v_{JJ}) - \big(\phi_0(v_{JJ}) b_\tau(v_{JJ}) + v_{JJ}\phi_1(v_{JJ}) \bar{b}_\tau(v_{JJ}) \big)$$

$$\cdot \phi_0^{-1}(\tau^2 v_{JJ}) \tau \phi_1(\tau^2 v_{JJ}) \Big) \big(dp_n^J \wedge df_\tau^J \big) d\tau$$

$$+ h^3 \int_0^1 \int_0^1 \Big(\big(\phi_0(v_{JJ}) b_\tau(v_{JJ}) + v_{JJ}\phi_1(v_{JJ}) \bar{b}_\tau(v_{JJ}) \big) \phi_0^{-1}(\tau^2 v_{JJ}) \bar{A}_{\tau,\sigma}(v_{JJ})$$

$$+ \bar{b}_\tau(v_{JJ}) b_\sigma(v_{JJ}) \Big) \big(df_\tau^J \wedge df_\sigma^J \big) d\tau d\sigma.$$

According to the first two conditions of (12.44), the above result can be reduced to

$$dQ_{n+1}^J \wedge dP_{n+1}^J = dQ_n^J \wedge dP_n^J + h \int_0^1 d_\tau \big(d(Q_{n,\tau}^+)^J \wedge df_\tau^J \big) d\tau$$

$$+ h^3 \int_0^1 \int_0^1 \Big(d_\tau \bar{A}_{\tau,\sigma}(v_{JJ}) + \bar{b}_\tau(v_{JJ}) b_\sigma(v_{JJ}) \Big) \big(df_\tau^J \wedge df_\sigma^J \big) d\tau d\sigma.$$

Then the fact that $f(z) = -G'(z)$ vanishes the term $\sum_{J=1}^M \big(d(Q_{n,\tau}^+)^J \wedge df_\tau^J \big)$ and the third condition of (12.44) implies that

$$\sum_{J=1}^M \Big(d_\tau \bar{A}_{\tau,\sigma}(v_{JJ}) + \bar{b}_\tau(v_{JJ}) b_\sigma(v_{JJ}) \Big) \big(df_\tau^J \wedge df_\sigma^J \big) = 0.$$

Consequently, we have

$$\sum_{J=1}^M dQ_{n+1}^J \wedge dP_{n+1}^J = \sum_{J=1}^M dQ_n^J \wedge dP_n^J, \tag{12.45}$$

which proves the result.

(II) If A is symmetric and positive semi-definite, the results can be proved by considering the fact that the approximation is invariant under linear transformations.

The proof is complete.   □

Remark 12.5.1 We here remark that d_τ is an artificial function created to facilitate the proof. The only requirement of this function is that $d_\tau \in \mathbb{R}$. It can be determined by the construction of a symplectic algorithm.

Algorithm 12.3 (Symplectic Scheme) Define a class of CSMLF schemes by (12.22) with the coefficients (12.43) and

$$\bar{b}_\tau(V) = (1-\tau)\phi_1((1-\tau)^2 V), \quad b_\tau(V) = \phi_0((1-\tau)^2 V),$$

$$\bar{A}_{\sigma,\tau}(V) = \left(k(0,0) + \sum_{i=1}^{N} \left(k(i,0)P_i(\tau) + k(0,i)P_i(\sigma) + k(i,i)P_i(\tau)P_i(\sigma)\right)\right)$$

$$\cdot \phi_1((\tau-\sigma)^2 V), \tag{12.46}$$

where $N \geqslant 1$ and P_i is the normalized shifted Legendre polynomial of degree i. This class of methods is symmetric and symplectic if $k(0,0)$ is arbitrary,

$$k(1,0) = -k(0,1) = \frac{\sqrt{3}}{12}, \tag{12.47}$$

and other parameters are symmetric, i.e. $k(i,j) = k(j,i)$ for $i+j > 1$. In this case, the corresponding schemes are of order two at least.

12.6 Explicit Continuous-Stage Modified LF Scheme

All the modified LF schemes derived in the previous two sections are implicit and iteration is needed in a practical implementation. It is known that explicit schemes can avoid the complicated iterative process. Therefore, in this section, we present an explicit fully discrete CSMLF scheme.

Algorithm 12.4 (Explicit Scheme) Let

$$C_\tau(V) = \phi_0(\tau^2 V), \quad D_\tau(V) = \tau\phi_1(\tau^2 V), \quad \bar{A}_{\sigma,\tau}(V) = 0,$$

$$\bar{b}_\tau(V) = (1-\tau)\phi_1((1-\tau)^2 V), \quad b_\tau(V) = \phi_0((1-\tau)^2 V), \tag{12.48}$$

and then we obtain a second-order explicit fully discrete CSMLF scheme (12.22). Before presenting the time integration scheme, we remark that some quadrature rule is needed in actual computations. Here, we consider the midpoint rule and this leads to

$$
\begin{cases}
Q^{+}_{n,\frac{1}{2}} = C_{\frac{1}{2}}(V)q_n + hD_{\frac{1}{2}}(V)p_n, \\[2mm]
Q_{n+1} = \phi_0(V)Q_n + h\phi_1(V)P_n + \dfrac{1}{2}h^2\phi_1\left(\dfrac{1}{4}V\right) f\left(Q^{+}_{n,\frac{1}{2}}\right), \\[2mm]
P_{n+1} = -hA\phi_1(V)Q_n + \phi_0(V)P_n + h\phi_0\left(\dfrac{1}{4}V\right) f\left(Q^{+}_{n,\frac{1}{2}}\right).
\end{cases}
\tag{12.49}
$$

We next analyse the numerical energy conservation of this method when the corresponding fully discrete scheme (12.21), ignoring Θ, is used, where the matrix $A = \Omega^2$ is assumed to be diagonal[2] with $\Omega = \mathrm{diag}(\omega_j)$ for $|j| \leqslant M$ and $d = 1$. Then, the system (12.21) is a finite-dimensional complex Hamiltonian system with the energy

$$
H_M(P, Q) = \frac{1}{2} \sum_{|j|\leqslant M} \left(|P_j|^2 + \omega_j^2|Q_j|^2\right) + G(q).
$$

In this section, we use the notation

$$
|k| = (|k_l|)_{l=0}^{\infty}, \quad \|k\| = \sum_{l=0}^{\infty} |k_l|, \quad k \cdot \lambda = \sum_{l=0}^{\infty} k_l\lambda_l, \quad \lambda^{\sigma|k|} = \Pi_{l=0}^{\infty}\lambda_l^{\sigma|k_l|}
$$

$$
\tag{12.50}
$$

for real σ, $k = (k_l)_{l=0}^{\infty}$ and $\lambda = (\lambda_l)_{l=0}^{\infty}$. Denote by $\langle j \rangle = (0, \cdots, 0, 1, 0, \cdots, 0)^{\mathsf{T}}$ the vector, where the only entry 1 at the $|j|$-th position for $j \in \mathbb{Z}$ and all other entries are zero. For $s \in \mathbb{R}^{+}$, H^s is referred to the Sobolev space of $2M$-periodic sequences $Q = (Q_j)$ endowed with the weighted norm $\|Q\|_s = \left(\sum_{|j|\leqslant M} \omega_j^{2s}|Q_j|^2\right)^{1/2}$.

The numerical energy conservation of the explicit modified LF scheme (12.49) is given by the following theorem.

Theorem 12.4 (Numerical Energy Conservation)   *Let the initial values of* (12.21) *satisfy*

$$
\left(\|Q(t_0)\|_{s+1}^2 + \|P(t_0)\|_s^2\right)^{1/2} \leqslant \varepsilon
\tag{12.51}
$$

with a small parameter ε. Assume that all the assumptions given in [8] are true. These conditions essentially imply that the continuous problem (12.21) has a "small" nonlinear term with "small" initial conditions. Then the explicit fully

---

[2]Since A is a symmetric positive semi-definite differentiation matrix, it can be diagonalized and the approximation is invariant under linear transformation.

discrete CSMLF scheme (12.49) *has the near-conservation estimate of the energy*

$$\frac{|H_M(P_n, Q_n) - H_M(P_0, Q_0)|}{\varepsilon^2} \leqslant C\varepsilon, \tag{12.52}$$

where $0 \leqslant t = nh \leqslant \varepsilon^{-N+1}$ *and the constant C is independent of* ε, M, h *and the time* $t = nh$. *Here N is a positive integer appearing in the modulated Fourier expansion given in the proof.*

Proof

(I) Modulated Fourier Expansion

Using the technique of modulated Fourier expansions (see [54–58]), we can show that the numerical solution (P_n, Q_n) admits the following multi-frequency modulated Fourier expansion

$$\tilde{Q}(t) = \sum_{\|k\| \leqslant 2N} e^{i(k \cdot \omega)t} \zeta^k(\varepsilon t), \quad \tilde{P}(t) = \sum_{\|k\| \leqslant 2N} e^{i(k \cdot \omega)t} \eta^k(\varepsilon t)$$

such that

$$\left\| Q_n - \tilde{Q}(t) \right\|_{s+1} + \left\| P_n - \tilde{P}(t) \right\|_s \leqslant C\varepsilon^N \quad \text{for} \quad 0 \leqslant t = nh \leqslant \varepsilon^{-1}, \tag{12.53}$$

where we have used the notation introduced in (12.50) with $k_l = 0$ for $l > M$. The expansion is bounded by

$$\left\| \tilde{Q}(t) \right\|_{s+1} + \left\| \tilde{P}(t) \right\|_s \leqslant C\varepsilon \quad \text{for} \quad 0 \leqslant t \leqslant \varepsilon^{-1}. \tag{12.54}$$

For $|j| \leqslant M$, we have

$$\tilde{Q}_j(t) = \zeta_j^{\langle j \rangle}(\varepsilon t)e^{i\omega_j t} + \zeta_j^{-\langle j \rangle}(\varepsilon t)e^{-i\omega_j t} + r_j \quad \text{with} \quad \|r\|_{s+1} \leqslant C\varepsilon^2. \tag{12.55}$$

The bound of the modulation functions ζ^k is

$$\sum_{\|k\| \leqslant 2N} \left(\frac{\omega^{|k|}}{\varepsilon^{\|k\|}} \left\| \zeta^k(\varepsilon t) \right\|_s \right)^2 \leqslant C. \tag{12.56}$$

For any fixed number of derivatives of ζ^k with respect to the slow time $\tau = \varepsilon t$, they have the same bounds. Moreover, it is deduced that $\zeta_{-j}^{-k} = \overline{\zeta_j^k}$ and the constant C is independent of ε, M, h and $t \leqslant \varepsilon^{-1}$.

These results follow from the techniques and tools developed in [55, 56, 59] and the recent analysis given in [14]. We skip the proof for brevity.

(II) An Almost-Invariant

It follows from the construction of the modulated Fourier expansion that

$$\frac{1}{h^2}\tilde{L}^k\zeta_j^k + \nabla_{-j}^{-k}\mathscr{U}(\zeta) = d_j^k,$$

where

$$d_j^k = \frac{1}{h^2}\tilde{L}^k\zeta_j^k + \sum_{m=2}^{N}\frac{f^{(m)}(0)}{m!}\sum_{k^1+\cdots+k^m=k}\sum_{j_1+\cdots+j_m\equiv j \bmod 2M}\left(\zeta_{j_1}^{k^1}\cdots\cdots\zeta_{j_m}^{k^m}\right),$$

$\nabla_{-j}^{-k}\mathscr{U}(y)$ is the partial derivative with respect to y_{-j}^{-k} of the extended potential [8, 59],

$$\mathscr{U}(\zeta) = \sum_{l=-N}^{N}\mathscr{U}_l(\zeta),$$

$$\mathscr{U}_l(\zeta) = \sum_{m=2}^{N}\frac{G^{(m+1)}(0)}{(m+1)!}\sum_{k^1+\cdots+k^{m+1}=0}\sum_{j_1+\cdots+j_{m+1}=2Ml}\left(\zeta_{j_1}^{k^1}\cdots\cdots\zeta_{j_{m+1}}^{k^{m+1}}\right),$$

and $\tilde{L}^k$ denotes the truncation of the operator L^k after the ε^N term.

An almost-invariant is obtained on noticing that

$$\varepsilon\frac{\mathrm{d}}{\mathrm{d}\tau}\mathscr{U}(\zeta) = \sum_{\|k\|\leqslant K}\sum_{|j|\leqslant M}\dot{\zeta}_{-j}^{-k}\nabla_{-j}^{-k}\mathscr{U}(\zeta)$$

$$= \sum_{\|k\|\leqslant K}\sum_{|j|\leqslant M}\left(\mathrm{i}(k\cdot\omega)\zeta_{-j}^{-k} + \varepsilon\dot{\zeta}_{-j}^{-k}\right)\left(-\frac{1}{h^2}\tilde{L}^k\zeta_j^k + d_j^k\right).$$

Similarly to the analysis in Sect. 7.3 of [8], it can be deduced that there is a function $\varepsilon\mathscr{H}[\zeta](\tau)$ such that

$$-\varepsilon\frac{\mathrm{d}}{\mathrm{d}\tau}\mathscr{H}[\zeta](\tau) = \sum_{\|k\|\leqslant K}\sum_{|j|\leqslant M}\left(\mathrm{i}(k\cdot\omega)\zeta_{-j}^{-k} + \varepsilon\dot{\zeta}_{-j}^{-k}\right)d_j^k. \tag{12.57}$$

It follows from the smallness of the right-hand side in (12.57) that $\left|\frac{\mathrm{d}}{\mathrm{d}\tau}\mathscr{H}[\zeta](\tau)\right| \leqslant C\varepsilon^{N+1}$ for $\tau \leqslant 1$.

(III) Relationship with the Energy

Under the conditions of Theorem 12.4, along the numerical solution and the associated modulation sequence $\zeta(\varepsilon t)$, we obtain

$$\mathcal{H}[\zeta](\varepsilon t_n) = H_M(P_n, Q_n) + \mathcal{O}(\varepsilon^3),$$

where the constant is independent of ε, M, h, and n.

On the basis of the results stated above, the statement of Theorem 12.4 is proved by patching together many intervals of length ε^{-1} (see [55, 56, 59] for details). □

12.7 Numerical Experiments

It is known that a study of numerical schemes remains incomplete without computational experiments. Our prime purpose in this section is to show the efficiency of continuous-stage modified LF schemes by implementing them in comparison with some efficient methods found in the literature. The following methods are chosen for comparison:

- LFS: the well-known standard leap-frog scheme;
- GAM: the Gautschi method given in [58];
- MLFS1: the energy-preserving CSMLF scheme presented in (12.41);
- MLFS2: the symplectic CSMLF scheme presented in (12.46) with $N = 1$, $k(0, 0) = \dfrac{1}{2}$ and $k(1, 1) = -\dfrac{1}{15}$;[3]
- MLFS3: the explicit CSMLF scheme presented in (12.48).

In the development and design of numerical schemes, established methods are constantly being improved, and the numerical comparison of schemes is a never-ending activity. In what follows, we describe the comparisons in detail. Since MLFS1 and MLFS2 are implicit schemes, iterative solutions are considered for them. In order to show that these methods can perform well even for the simplest iteration method, we choose standard fixed-point iteration in this section and set 10^{-16} as the error tolerance of each iteration. The maximum number of iterations is 10. In the actual implementation of these schemes, Gaussian quadrature formulas are applied to the definite integrals appearing in the continuous-stage modified LF schemes.

---

[3] These choices are made such that the method can satisfy more order conditions (12.23).

Problem 12.1 We first consider the semilinear Klein–Gordon equation

$$\begin{cases} \dfrac{\partial^2 u}{\partial t^2} - a^2 \dfrac{\partial^2 u}{\partial x^2} = bu^3 - au, & -L \leqslant x \leqslant L, \ \ 0 \leqslant t \leqslant T, \ \ u(-L,t) = u(L,t), \\[2mm] u(x,0) = \sqrt{\dfrac{2a}{b}}\,\mathrm{sech}(\lambda x), \ \ u_t(x,0) = c\lambda\sqrt{\dfrac{2a}{b}}\,\mathrm{sech}(\lambda x)\tanh(\lambda x), \end{cases}$$

where $\lambda = \sqrt{\dfrac{a}{a^2 - c^2}}$ and $a, b, a^2 - c^2 > 0$, and the exact solution is

$$u(x,t) = \sqrt{\frac{2a}{b}}\,\mathrm{sech}(\lambda(x - ct)).$$

For solving this problem, we approximate the operator $\mathscr{A}$ by *Fourier spectral collocation (FSC)* (see, e.g. [60, 61])

$$A_{fsc} = (a_{kj})_{M \times M} \quad \text{with} \quad a_{kj} = \begin{cases} \dfrac{(-1)^{k+j}}{2}\,\sin^{-2}\left(\dfrac{(k-j)\pi}{M}\right), & k \neq j, \\[3mm] \dfrac{M^2}{12} + \dfrac{1}{6}, & k = j. \end{cases}$$

$$\tag{12.58}$$

In this experiment, we choose $a = 0.3$, $b = 1$, $c = 0.25$, and $M = 200$. This system is integrated on $[0, 10]$ with $h = 0.2/2^j$ for $j = 0, 1, \cdots, 4$. The global errors plotted against the logarithm of CPU time and h are shown in Fig. 12.1. We here

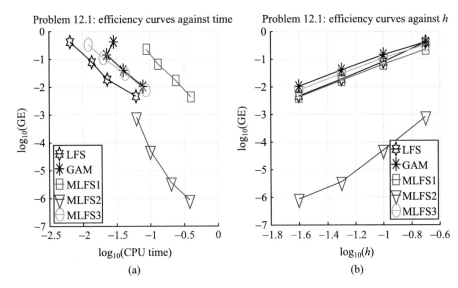

Fig. 12.1 The logarithm of the global errors against the logarithm of CPU time and h

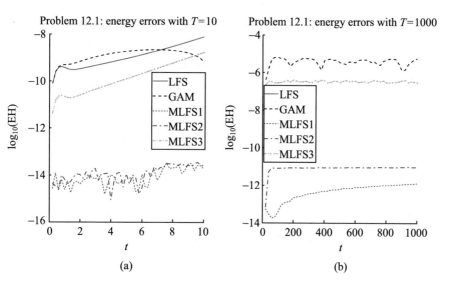

Fig. 12.2   The logarithm of the energy errors against the time t

remark that the global errors are measured by the l^{∞}-norm of the error between the numerical solution $u_N(x_j)$ and the exact (or reference) solution $u(x_j, T)$: GE$(T) = \max_j \left| u_N(x_j) - u(x_j, T) \right|$, where $N = T/h$.

We then solve the problem with $h = 0.01$ and $T = 10, 1000$, and the errors of the semidiscrete energy conservation are presented in Fig. 12.2. It can be observed from these results that MLFS3 behaves similarly to LFS and GAM, MLFS1 has similar accuracy to them but has better energy-preserving property, MLFS2 performs better than LFS and GAM not only in accuracy but also in energy conservation. Similar numerical behaviour of these continuous-stage modified LF schemes can be observed in the following three numerical experiments.

Problem 12.2   Consider the well-known sine-Gordon equation (see, e.g. [62])

$$\frac{\partial^2 u}{\partial t^2} = \frac{\partial^2 u}{\partial x^2} - \sin u, \qquad x \in [-20, 20], \qquad t \in [0, T]$$

with the initial conditions:

$$u(x, 0) \equiv 0, \quad u_t(x, 0) = 4/\gamma \operatorname{sech}\left(x/\gamma\right), \quad \gamma \geqslant 0.$$

The exact solution is given by

$$u(x, t) = 4 \arctan\left(\psi(t, \gamma) \operatorname{sech}(x/\gamma)\right),$$

where

$$\psi(t, \gamma) = \begin{cases} \sinh\left(\dfrac{\sqrt{1 - \gamma^2}t}{\gamma}\right) \bigg/ \sqrt{1 - \gamma^2}, & 0 < \gamma < 1, \\ t, & \gamma = 1, \\ \sin\left(\dfrac{\sqrt{\gamma^2 - 1}t}{\gamma}\right) \bigg/ \sqrt{\gamma^2 - 1}, & \gamma > 1. \end{cases}$$

In this experiment, we discretise the spatial derivative by the Fourier spectral collocation method (12.58) with $M = 200$. The problem is solved on $[0, 50]$ with $h = 0.2/2^j$ for $j = 0, 1, 2, 3$. The efficiency curves are shown in Fig. 12.3 for different γ. We then integrate the equation with $h = 0.01$. The detailed results for the errors of the semidiscrete energy conservation are presented in Fig. 12.4.

Problem 12.3 We consider the two-dimensional semilinear wave equation (see, e.g. [63])

$$\frac{\partial^2 u(x, y, t)}{\partial t^2} = \Delta u(x, y, t) - u^3(x, y, t), \quad (x, y) \in [-1, 1] \times [-1, 1], \ t > 0$$

with periodic boundary conditions. The initial conditions are given by

$$u(x, y, 0) = \text{sech}(10x)\,\text{sech}(10y), \ u_t(x, y, 0) = 0.$$

Similarly to [63], in this experiment, we use the spectral element method to semi-discretise the wave equation, where we discretise the space with a tensor product Lagrange quadrature formula based on $p + 1$ Gauss–Lobatto–Legendre (GLL) quadrature nodes in each space direction. This problem is integrated on the interval $[0, 10]$ with $p = 5, h = \dfrac{0.1}{2^i}$ for $i = 0, \cdots, 3$. See Fig. 12.5 for the global errors. We then solve the system with $h = 0.1$ and $T = 10, 1000$, and the results of energy conservation are indicated in Fig. 12.6.

Problem 12.4 Finally, we consider the following two-dimensional sine-Gordon equation:

$$\begin{cases} u_{tt} - (u_{xx} + u_{yy}) = -\sin u, \ (x, y) \in [-1, 1] \times [-1, 1], \ t > 0, \\ u(x, y, 0) = 4\arctan\left(\exp\left(\dfrac{4 - \sqrt{(x+3)^2 + (y+3)^2}}{0.436}\right)\right), \\ u_t(x, y, 0) = \dfrac{4.13}{\cosh\left(\exp\left((4 - \sqrt{(x+3)^2 + (y+3)^2})/0.436\right)\right)}, \end{cases}$$

with periodic boundary conditions. We use the same spectral element method to semidiscretise the wave equation as in Problem 12.3. The problem is solved on the

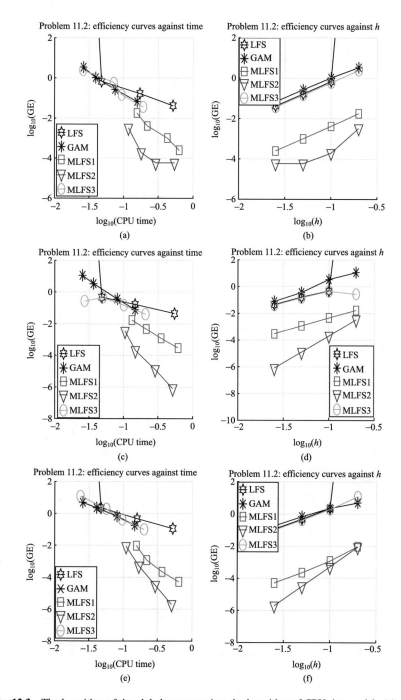

Fig. 12.3 The logarithm of the global errors against the logarithm of CPU time and h. (**a**), (**b**): $\gamma = 0.99$. (**c**), (**d**): $\gamma = 1.01$. (**e**), (**f**): $\gamma = 1$

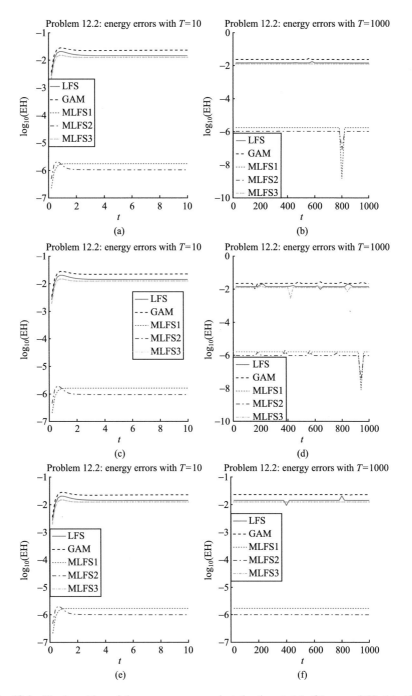

Fig. 12.4   The logarithm of the energy errors against the time t. (**a**), (**b**): $\gamma = 0.99$. (**c**), (**d**): $\gamma = 1.01$. (**e**), (**f**): $\gamma = 1$

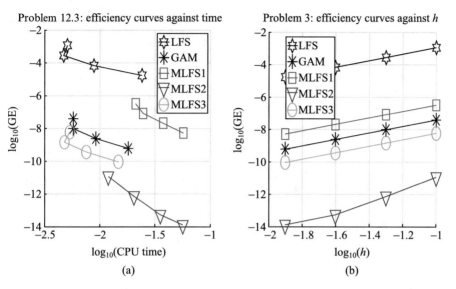

Fig. 12.5 The logarithm of the global errors against the logarithm of CPU time and h

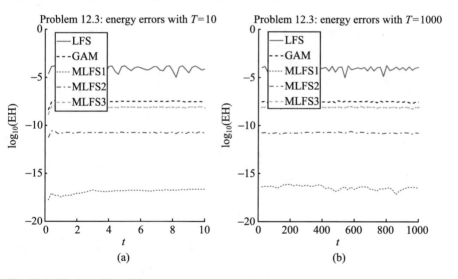

Fig. 12.6 The logarithm of the energy errors against the time t

interval $[0, 10]$ with $h = \dfrac{0.1}{2^i}$ for $i = 0, 1, 2, 3$. Figure 12.7 presents the global errors. We then integrate the system with $h = 0.01$ and $T = 10, 1000$. The results of energy conservation are presented in Fig. 12.8.

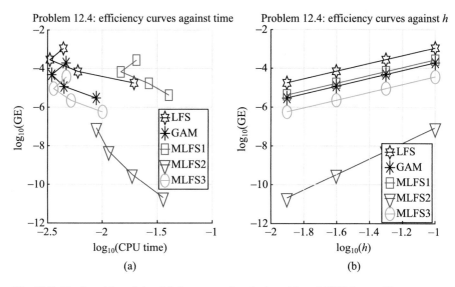

Fig. 12.7 The logarithm of the global errors against the logarithm of CPU time and h

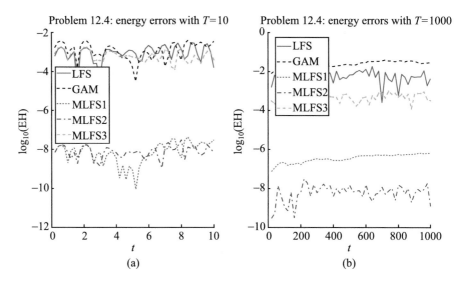

Fig. 12.8 The logarithm of the energy errors against the time t

12.8 Conclusions and Discussions

As is known, a most welcome feature of the standard LF scheme is that it is easy to use for the time integration of PDEs. However, the standard LF scheme needs to be improved in the sense of structure-preserving algorithms. Therefore, in this chapter, we formulated and analysed improved LF schemes for efficiently solving high-dimensional semilinear Hamiltonian wave equations. These improved schemes take full advantage of the operator-variation-of-constants formula (also termed the Duhamel Principle) of the underlying problem and their formulations incorporate the idea of continuous-stage methods and exponential integrators. The properties of the improved scheme were analysed including its convergence, energy preservation, symplecticity conservation, and the long-time behaviour of the explicit method. The results of numerical experiments are very promising and demonstrate that these continuous-stage modified LF schemes are significantly more efficient compared with the standard leap-frog scheme and a Gautschi-type method in the literature. It is believed that this approach to the improvement of the standard LF scheme could be extended to other classes of PDEs as well.

The material in this chapter is based on the work by Wang et al. [64].

References

1. Dodd, R.K., Eilbeck, I.C., Gibbon, J.D., et al.: Solitons and Nonlinear Wave Equations. Academic Press, London (1982)
2. Cao, W., Li, D., Zhang, Z.: Optimal super convergence of energy conserving local discontinuous Galerkin methods for wave equations. Commun. Comput. Phys. **21**, 211–236 (2017)
3. Cheng, Y., Chou, C.S., Li, F., et al.: L2-stable discontinuous Galerkin methods for one dimensional two-way wave equations. Math. Comput. **86**, 121–155 (2017)
4. Cohen, G., Joly, P., Roberts, J.E., et al.: Higher order triangular finite elements with mass lumping for the wave equation. SIAM J. Numer. Anal. **38**, 2047–2078 (2001)
5. Grote, M.J., Schneebeli, A., Schötzau, D.: Discontinuous Galerkin finite element method for the wave equation. SIAM J. Numer. Anal. **44**, 2408–2431 (2006)
6. Wang, S., Nissen, A., Kreiss, G.: Convergence of finite difference methods for the wave equation in two space dimensions. Math. Comput. **87**, 2737–2763 (2017)
7. Yi, N., Liu, H.: An energy conserving local discontinuous Galerkin method for a nonlinear variational wave equation. Commun. Comput. Phys. **23**, 747–772 (2018)
8. Cohen, D., Hairer, E., Lubich, C.: Conservation of energy, momentum and actions in numerical discretizations of non-linear wave equations. Numer. Math. **110**, 113–143 (2008)
9. Liu, C., Wu, X.: The boundness of the operator-valued functions for multidimensional nonlinear wave equations with applications. Appl. Math. Lett. **74**, 60–67 (2017)
10. Chou, C.S., Shu, C.W., Xing, Y.: Optimal energy conserving local discontinuous Galerkin methods for second-order wave equation in heterogeneous media. J. Comput. Phys. **272**, 88–107 (2014)
11. Diaz, J., Grote, M.J.: Energy conserving explicit local time stepping for second-order wave equations. SIAM J. Sci. Comput. **31**, 1985–2014 (2009)
12. Grote, M.J., Mehlin, M., Sauter, S.: Convergence analysis of energy conserving explicit local time-stepping methods for the wave equation. SIAM J. Numer. Anal. **56**, 994–1021 (2017)

13. Wang, B., Wu, X.: The formulation and analysis of energy-preserving schemes for solving high-dimensional nonlinear Klein-Gordon equations. IMA J. Numer. Anal. **39**, 2016–2044 (2019)
14. Wang, B., Wu, X.: Long-time momentum and actions behaviour of energy-preserving methods for semilinear wave equations via spatial spectral semi-discretizations. Adv. Comput. Math. **45**, 2921–2952 (2019)
15. Gander, M.J., Halpern, L., Nataf, F.: Optimal Schwarz waveform relaxation for the one dimensional wave equation. SIAM J. Numer. Anal. **41**, 1643–1681 (2003)
16. Wang, B., Wu, X.: A symplectic approximation with nonlinear stability and convergence analysis for efficiently solving semi-linear Klein-Gordon equations. Appl. Numer. Math. **142**, 64–89 (2019)
17. Brugnano, L., Iavernaro, F., Montijano, J. I., et al.: Spectrally accurate space-time solution of Hamiltonian PDEs. Numer. Algor. **81**, 1183–1202 (2019)
18. Bao, W.Z., Dong, X.C.: Analysis and comparison of numerical methods for the Klein-Gordon equation in the nonrelativistic limit regime. Numer. Math. **120**, 189–229 (2012)
19. Bao, G., Lai, J., Qian, J.: Fast multiscale Gaussian beam methods for wave equations in bounded convex domains. J. Comput. Phys. **261**, 36–64 (2014)
20. Collino, F., Fouquet, T., Joly, P.: A conservative space-time mesh refinement method for the 1-D wave equation. I. Construction. Numer. Math. **95**, 197–221 (2003)
21. Joly, P., Rodríguez, J.: An error analysis of conservative space-time mesh refinement methods for the one-dimensional wave equation. SIAM J. Numer. Anal. **43**, 825–859 (2005)
22. Wu, F., Cheng, X., Li, D., et al.: A two-level linearized compact ADI scheme for two dimensional nonlinear reaction-diffusion equations. Comput. Math. Appl. **75**, 2835–2850 (2018)
23. Huang, Y., Chen, M., Li, J., et al.: Numerical analysis of a leapfrog ADI-FDTD method for Maxwell's equations in lossy media. Comput. Math. Appl. **76**, 938–956 (2018)
24. Hurl, N., Layton, W., Li, Y., et al.: Stability analysis of the Crank-Nicolson-Leapfrog method with the Robert-Asselin-Williams time filter. BIT Numer. Math. **54**, 1009–1021 (2014)
25. Jiang, N., Kubacki, M., Layton, W., et al.: A Crank-Nicolson-Leapfrog stabilization: Unconditional stability and two applications. J. Comput. Appl. Math. **281**, 263–276 (2015)
26. Sanz-Serna, J.: Studies in numerical nonlinear instability. I. Why do leapfrog schemes go unstable? SIAM J. Sci. Statist. Comput. **6**, 923–938 (1985)
27. Shi, W., Wu, X., Xia, J.: Explicit multi-symplectic extended leap-frog methods for Hamiltonian wave equations. J. Comput. Phys. **231**, 7671–7694 (2012)
28. Amodio, P., Brugnano, L., Iavernaro, F.: A note on the continuous-stage Runge-Kutta(-Nyström) formulation of Hamiltonian Boundary Value Methods (HBVMs) (2019). arXiv: 1906. 04071
29. Baker, T.S., Dormand, J.R., Gilmore, J.P., et al.: Continuous approximation with embedded Runge-Kutta methods. Appl. Numer. Math. **22**, 51–62 (1996)
30. Owren, B., Zennaro, M.: Derivation of efficient, continuous, explicit Runge-Kutta methods. SIAM J. Sci. Stat. Comput. **13**, 1488–1501 (1992)
31. Owren, B., Zennaro, M.: Order barriers for continuous explicit Runge-Kutta methods. Math. Comput. **56**, 645–661 (1991)
32. Papakostas, S.N., Tsitouras, C.: Highly continuous interpolants for one-step ODE solvers and their application to Runge-Kutta methods. SIAM J. Numer. Anal. **34**, 22–47 (1997)
33. Verner, J.H., Zennaro, M.: The orders of embedded continuous explicit Runge-Kutta methods. BIT Numer. Math. **35**, 406–416 (1995)
34. Marthinsen, A.: Continuous extensions to Nyström methods for second order initial value problems. BIT Numer. Math. **36**, 309–332 (1996)
35. Tang, W., Zhang, J.: Symplecticity-preserving continuous-stage Runge-Kutta-Nyström methods. Appl. Math. Comput. **323**, 204–219 (2018)
36. Tang, W., Sun, Y., Zhang, J.: High order symplectic integrators based on continuous-stage Runge-Kutta-Nyström methods. Appl. Math. Comput. **361**, 670–679 (2019)

37. Brugnano, L., Calvo, M., Montijano, J.I., et al.: Energy preserving methods for Poisson systems. J. Comput. Appl. Math. 236, 3890–3904 (2012)
38. Brugnano, L., Iavernaro, F., Trigiante, D.: Hamiltonian boundary value methods (Energy preserving discrete line integral methods). J. Numer. Anal. Ind. Appl. Math. 5, 7–17 (2010)
39. Brugnano, L., Iavernaro, F., Trigiante, D.: A simple framework for the derivation and analysis of effective one-step methods for ODEs. Appl. Math. Comput. 218, 8475–8485 (2012)
40. Cohen, D., Hairer, E.: Linear energy-preserving integrators for Poisson systems. BIT Numer. Math. 51, 91–101 (2011)
41. Hairer, E.: Energy-preserving variant of collocation methods. J. Numer. Anal. Ind. Appl. Math. 5, 73–84 (2010)
42. Li, Y.W., Wu, X.: Functionally fitted energy-preserving methods for solving oscillatory nonlinear Hamiltonian systems. SIAM J. Numer. Anal. 54, 2036–2059 (2016)
43. Miyatake, Y.: An energy-preserving exponentially-fitted continuous stage Runge-Kutta methods for Hamiltonian systems. BIT Numer. Math. 54, 777–799 (2014)
44. Miyatake, Y.: A derivation of energy-preserving exponentially-fitted integrators for Poisson systems. Comput. Phys. Commun. 187, 156–161 (2015)
45. Wang, B., Wu, X.: Functionally-fitted energy-preserving integrators for Poisson systems. J. Comput. Phys. 364, 137–152 (2018)
46. Hochbruck, M., Ostermann, A.: Exponential integrators. Acta Numer. 19, 209–286 (2010)
47. Wang, B., Iserles, A., Wu, X.: Arbitrary-order trigonometric Fourier collocation methods for multi-frequency oscillatory systems. Found. Comput. Math. 16, 151–181 (2016)
48. Wu, X., Wang, B.: Recent Developments in Structure-Preserving Algorithms for Oscillatory Differential Equations. Springer Nature Singapore Pte Ltd., Singapore (2018)
49. Li, J., Wu, X.: Energy-preserving continuous stage extended Runge-Kutta-Nyström method for oscillatory Hamiltonian systems. Appl. Numer. Math. 145, 469–487 (2019)
50. Liu, C., Wu, X.: Arbitrarily high-order time-stepping schemes based on the operator spectrum theory for high-dimensional nonlinear Klein-Gordon equations. J. Comput. Phys. 340, 243–275 (2017)
51. Wang, B., Wu, X.: A new high precision energy preserving integrator for system of oscillatory second-order differential equations. Phys. Lett. A 376, 1185–1190 (2012)
52. Liu, C., Wu, X.: An energy-preserving and symmetric scheme for nonlinear Hamiltonian wave equations. J. Math. Anal. Appl. 440, 167–182 (2016)
53. Wu, X., Wang, B., Shi, W.: Efficient energy preserving integrators for oscillatory Hamiltonian systems. J. Comput. Phys. 235, 587–605 (2013)
54. Hairer, E., Lubich, C., Wanner, G.: Geometric Numerical Integration: Structure-Preserving Algorithms for Ordinary Differential Equations, 2nd edn. Springer, Berlin (2006)
55. Cohen, D., Hairer, E., Lubich, C.: Numerical energy conservation for multi-frequency oscillatory differential equations. BIT Numer. Math. 45, 287–305 (2005)
56. Cohen, D., Hairer, E., Lubich, C.: Long-time analysis of nonlinearly perturbed wave equations via modulated Fourier expansions. Arch. Ration. Mech. Anal. 187, 341–368 (2008)
57. Gauckler, L., Hairer, E., Lubich, C.: Long-term analysis of semilinear wave equations with slowly varying wave speed. Commun. Partial Differential Eq. 41, 1934–1959 (2016)
58. Hairer, E., Lubich, C.: Long-time energy conservation of numerical methods for oscillatory differential equations. SIAM J. Numer. Anal. 38, 414–441 (2000)
59. Hairer, E., Lubich, C.: Spectral semi-discretisations of weakly nonlinear wave equations over long times. Found. Comput. Math. 8, 319–334 (2008)
60. Hesthaven, J.S., Gottlieb, S., Gottlieb, D.: Spectral Methods for Time-Dependent Problems, Cambridge Monographs on Applied and Computational Mathematics. Cambridge University, Cambridge (2007)
61. Shen, J., Tang, T., Wang, L.: Spectral Methods: Algorithms, Analysis, Applications. Springer, Berlin (2011)
62. Brugnano, L., Frasca Caccia, G., Iavernaro, F.: Energy conservation issues in the numerical solution of the semilinear wave equation. Appl. Math. Comput. 270, 842–870 (2015)

63. Celledoni, E., Grimm, V., McLachlan, R.I., et al.: Preserving energy resp. dissipation in numerical PDEs using the "Average Vector Field" method. J. Comput. Phys. **231**, 6770–6789 (2012)
64. Wang, B., Wu, X., Fang, Y.: A continuous-stage modified leap-frog schemes for high dimensional semi-linear Hamiltonian wave equations. Numer. Math. Theor. Meth. Appl. **13**, 814–844 (2020)

Chapter 13
Semi-Analytical ERKN Integrators for Solving High-Dimensional Nonlinear Wave Equations

Incorporating the operator-variation-of-constants formula for high-dimensional nonlinear wave equations with Fast Fourier Transform techniques in this chapter, we present a class of semi-analytical ERKN integrators, which can nearly preserve the spatial continuity as well as the oscillations of the underlying nonlinear waves equations. Standard ERKN methods require, in every time step, the computation of the matrix-vector product whose computational complexity, in terms of basic multiplication is $\mathcal{O}(N^2)$, once a direct calculation procedure is implemented, where N is the dimension of the underlying differentiation matrix. We design and analyse efficient algorithms which are incorporated with the Fast Fourier Transform in the implementation of ERKN integrators, so that these algorithms reduce the computational cost from $\mathcal{O}(N^2)$ to $\mathcal{O}(N \log N)$ in terms of basic multiplication.

13.1 Introduction

This chapter concerns the numerical simulation of high-dimensional nonlinear wave equations. Although all of the ideas, algorithms and analysis in this chapter can be straightforwardly extended to the solution of nonlinear wave equations in a moderate number of space dimensions, we begin with the nonlinear one-dimensional Hamiltonian wave equation

$$u_{tt} - a^2 u_{xx} = f(u), \qquad (13.1)$$

with 2π-periodic boundary condition ($x \in \Omega = \mathbb{R}/(2\pi\mathbb{Z})$) and initial values

$$u(x, t_0) = \varphi(x) \in H^{s+1}(\Omega), \quad u_t(x, t_0) = \psi(x) \in H^s(\Omega), \qquad (13.2)$$

© The Author(s), under exclusive license to Springer Nature Singapore Pte Ltd. 2021
X. Wu, B. Wang, *Geometric Integrators for Differential Equations with Highly Oscillatory Solutions*, https://doi.org/10.1007/978-981-16-0147-7_13

where $s \geqslant 0$, and $H^s(\Omega)$ is the Sobolev space on Ω. We consider the domain $\Omega = [0, 2\pi]$ for simplicity. (13.1) is a conservative system due to the conservation of the Hamiltonian energy

$$H = H(t) = \frac{1}{2} \int_{\Omega} \left((u_t)^2 + a^2(u_x)^2 + 2V(u(x, t)) \right) dx, \tag{13.3}$$

where $f(u) = -\dfrac{dV(u)}{du}$.

We first define the formal series

$$\phi_j(x) := \sum_{k=0}^{\infty} \frac{(-1)^k x^k}{(2k + j)!}, \quad j = 0, 1, \cdots, \tag{13.4}$$

for any $x \geqslant 0$, and the differential operator $\mathscr{A}$

$$(\mathscr{A}v)(x) = -a^2 v_{xx}(x).$$

This leads to the following operator-variation-of-constants formula for the initial-boundary-value problem of (13.1) (see, e.g. [1])

$$\begin{cases} u(x, t) = \phi_0\big((t - t_0)^2 \mathscr{A}\big)\varphi(x) + (t - t_0)\phi_1\big((t - t_0)^2 \mathscr{A}\big)\psi(x) \\ \qquad + \int_{t_0}^{t} (t - \zeta)\phi_1\big((t - \zeta)^2 \mathscr{A}\big) f(u(x, \zeta)) d\zeta, \\ u_t(x, t) = -(t - t_0)\mathscr{A}\phi_1\big((t - t_0)^2 \mathscr{A}\big)\varphi(x) + \phi_0\big((t - t_0)^2 \mathscr{A}\big)\psi(x) \\ \qquad + \int_{t_0}^{t} \phi_0\big((t - \zeta)^2 \mathscr{A}\big) f(u(x, \zeta)) d\zeta, \end{cases} \tag{13.5}$$

where both $\phi_0\big((t - t_0)^2 \mathscr{A}\big)$ and $\phi_1\big((t - t_0)^2 \mathscr{A}\big)$ are bounded operators as stated in Chap. 1 (see also [2]), although $\mathscr{A}$ is a linear, unbounded positive semi-definite operator.

We remark that the formula (13.5) exactly provides an implicit expression for the solution to (13.1). In particular, for the special case where $f(u) = 0$, (13.5) yields the closed-form solution to (13.1). Moreover, with the help of (13.5), we are hopeful of obtaining semi-analytical integrators for (13.1), which preserve the continuity of the spatial variable x, and only discretise the time variable t. An interesting example is the energy-preserving and symmetric scheme presented in Chap. 9 (see also [3]), which can exactly preserve the true continuous energy (13.3), not a discrete energy after spatial discretisations as is typically the case for other methods. It is noted that the extended Runge–Kutta–Nyström (ERKN) methods have been well developed for highly oscillatory systems of ordinary differential equations (see, e.g. [4–8])

$$\begin{cases} y'' + My = f(y), & t \in [t_0, T], \\ y(t_0) = y_0, \ y'(t_0) = y'_0, \end{cases} \tag{13.6}$$

where M is a (symmetric) positive semi-definite matrix and $\|M\| \gg \max\left\{1, \left\|\dfrac{\partial f}{\partial y}\right\|\right\}$. This line of research for (13.6) will assist in the design and development of numerical schemes for (13.1).

For the formulation of semi-analytical ERKN integrators for (13.1), we first rewrite (13.5) as

$$\begin{cases} u(x, t_n + \tau h) = \phi_0(\tau^2 V)u(x, t_n) + \tau h \phi_1(\tau^2 V)u_t(x, t_n) \\ \qquad\qquad + h^2 \displaystyle\int_0^\tau (\tau - \zeta)\phi_1((\tau - \zeta)^2 V) f(u(x, t_n + \zeta h)) \mathrm{d}\zeta, \\ u_t(x, t_n + \tau h) = -\tau h M \phi_1(\tau^2 V)u(x, t_n) + \phi_0(\tau^2 V)u_t(x, t_n) \\ \qquad\qquad + h \displaystyle\int_0^\tau \phi_0((\tau - \zeta)^2 V) f(u(x, t_n + \zeta h)) \mathrm{d}\zeta, \end{cases} \tag{13.7}$$

where $V = h^2 \mathscr{A}$ and $h > 0$ is the time stepsize. We assume that $U_i \approx u(x, t_n + C_i h)$, $u_{n+1} \approx u(x, t_n + h)$ and $u'_{n+1} \approx u_t(x, t_n + h)$. We set $\tau = C_i$ satisfying $0 < C_i < 1$ for $i = 1, \cdots, s$, and approximate the first integral appearing in (13.7) by a suitable quadrature formula with the weights $A_{ij}(V)$. This leads to the internal stages of ERKN integrators for (13.1). Likewise, by setting $\tau = 1$, the updates of ERKN integrators for (13.1) follow from the approximations to the two integrals appearing in (13.7) by suitable quadrature formulae with the weights $\overline{B}_i(V)$ and $B_i(V)$, respectively. Then we are in a position to define a semi-analytical ERKN integrator for (13.1).

An s-stage semi-analytical ERKN integrator for (13.1) reads

$$\begin{cases} U_i = \phi_0(C_i^2 V)u_n + C_i h \phi_1(C_i^2 V)u'_n + h^2 \displaystyle\sum_{j=1}^s A_{ij}(V) f(U_j), \quad i = 1, \cdots, s, \\[2mm] u_{n+1} = \phi_0(V)u_n + h \phi_1(V)u'_n + h^2 \displaystyle\sum_{i=1}^s \overline{B}_i(V) f(U_i), \\[2mm] u'_{n+1} = -h \mathscr{A} \phi_1(V)u_n + \phi_0(V)u'_n + h \displaystyle\sum_{i=1}^s B_i(V) f(U_i), \end{cases} \tag{13.8}$$

where the constants $C_1, \cdots, C_s$, and the operator-argument coefficients $A_{ij}(V)$, $\overline{B}_i(V)$ and $B_i(V)$ for $i, j = 1, \cdots, s$ are determined to ensure that the numerical scheme (13.8) is convergent and stable. It can be observed from (13.8) that the ERKN integrator defines a time-stepping procedure, and the initial conditions are exactly (13.2), i.e., $u_0 = u(x, t_0)$ and $u'_0 = u_t(x, t_0)$. This class of integrators possesses the superior property of preserving spatial continuity. Therefore, we call them semi-analytical integrators (see, e.g. [9]). It should be noted that the ERKN method for (13.6) can also be expressed in the form of (13.8) by replacing the operator $\mathscr{A}$ in V with a suitable differentiation matrix M, and remember that the ERKN method for (13.6) is oscillation preserving as stated in Chap. 1. For convenience, we denote this ERKN integrator corresponding to (13.8) by a partitioned Butcher tableau

$$
\begin{array}{c|ccc}
C_1 & A_{11}(V) & \cdots & A_{1s}(V) \\
\vdots & \vdots & & \vdots \\
\frac{C}{\bar{B}(V)^\mathsf{T}} \;\; \frac{A(V)}{} = & C_s \;\; A_{s1}(V) & \cdots & A_{ss}(V) \\
\hline
 & \bar{B}_1(V) & \cdots & \bar{B}_s(V) \\
\hline
B(V)^\mathsf{T} & B_1(V) & \cdots & B_s(V)
\end{array}
\cdot
\tag{13.9}
$$

Despite the superior properties we have mentioned, the semi-analytical ERKN integrators (13.8) as well as the energy-preserving and symmetric scheme in [3] could not be easily applied to (13.1) for general nonlinear cases, since it is difficult to calculate and implement the operator-argument functions involved in these integrators. This fact greatly confines the potential and further application of these semi-analytical integrators, although they have been successfully applied to some linear or homogeneous wave equations (see, e.g. [10]). A feasible approach to implementing (13.8) in practice is approximating the operator $\mathscr{A}$ by a suitable differentiation matrix M, once the spatial discretisation is carried out. When sufficient spatial mesh grids are appropriately chosen, the spatial discretisation error will be smaller than the roundoff error in theory (see, e.g. [11, 12]). Hence, in the sense of numerical computation, we can expect and consider that the spatial precision and continuity are nearly preserved by the ERKN integrators (13.8), because it turns out that the global error of ERKN integrators is independent of the spatial refinement (see, e.g. [13]). Moreover, it has been emphasised that the global error bounds of the ERKN integrators are completely independent of the differentiation matrix M in Chap. 3.

However, for the sake of the near preservation of spatial continuity, the dimension of the differentiation matrix M will be selected so large that the error of spatial discretisations can be almost ignored. This results in the following three difficulties in the practical implementation of (13.8). First, the computation of matrix-valued functions $A_{ij}(V)$, $B_i(V)$, and $\bar{B}_i(V)$ will be of high complexity for such a high-dimensional matrix M, since they are in fact expressed in the series of M. Second, the multiplication at each time step between these matrices and vectors is also highly costly. For instance, if we denote N as the dimension of the matrix M, the multiplication between $A_i(V)$ and $f(Y_i)$ contains N^2 basic scalar multiplication and

$N(N-1)$ basic scalar addition, which are an order of magnitude $\mathcal{O}(N^2)$. Hence, the computational cost of basic operation in each time step will be rapidly increased in the magnitude of $\mathcal{O}(N^2)$ as N increases. Third, the necessary computer memory to store $A_{ij}(V)$, $B_i(V)$ and $\bar{B}_i(V)$ will also sharply increase, which may result in the computer running out of memory, in particular for high-dimensional problems. In order to obtain the near preservation of the spatial continuity in the application of semi-analytical ERKN integrators (13.8), and overcome the above mentioned obstacles in the practical implementation after a possibly highly refined spatial discretisation, it is wise to avoid the calculation and storage of such matrix-valued functions, as well as the direct multiplication between matrix-valued functions and the corresponding vectors. Consequently, in this chapter, we consider solving the nonlinear wave equations in Fourier space. Then the system of ordinary differential equations with respect to the Fourier coefficients will have a natural harmony with the ERKN method, i.e., the matrix-vector multiplication will disappear when the ERKN method is used to solve this system. Since the Fourier coefficients can be obtained by the Fast Fourier Transform (FFT) with $\mathcal{O}(N \log N)$ operations (see, e.g. [14]), we are hopeful of obtaining a fast implementation approach to ERKN integrators when applied to the nonlinear wave equations, even if N is very large. This motivates the presentation of Algorithm 1 in this chapter. Furthermore, making use of the equivalence between the splitting method and an important class of symplectic ERKN methods, we present Algorithm 2, which is shown to be more efficient than Algorithm 1.

The finite difference method could also yield the near preservation of spatial continuity, in theory, for the nonlinear wave equation (13.1), once the spatial stepsize $\Delta x \rightarrow 0$ and the convergence of the numerical solution to the exact solution is satisfied. Unfortunately, however, it follows from the Courant–Friedrichs–Lewy (CFL) condition in the literature (see, e.g. [15–17]) that the mesh ratio should satisfy $h/\Delta x \leqslant \gamma$ for the sake of numerical stability, where γ is a positive constant depending only on the selected difference scheme. This implies that the time stepsize h would be restricted to a very tiny magnitude, once the Δx is selected as so small that it can ensure near preservation of the spatial continuity. This fact greatly confines the application of the finite difference method. Fortunately, the stability analysis of the ERKN integrator (13.8) in [18] shows that the time stepsize is independent of the spatial stepsize, but dependent only on the coefficients of ERKN integrator (13.8) and the Lipschitz constant of the function $f(u)$. This advantage admits the use of a large time stepsize even though Δx is very small after the requirement of spatial-mesh refinement, once the semi-analytical ERKN integrator (13.8) is applied to solve the nonlinear wave equations.

Another noteworthy aspect of scientific research related to the theme of this chapter is the application of the Fourier spectral method and the FFT techniques. In the literature (see, e.g. [19–22]), the authors respectively discussed the applications of Fourier spectral discretisation for different types of partial differential equations. In particular, in [19, 20, 22] and Chap. 2 in [23], the authors also incorporated the FFT into the implementation to try to achieve smaller computational cost and lower memory storage. However, all such methods used a difference scheme or a Runge–Kutta-type discretisation for the time derivative. It is very clear that these procedures are not suitable for oscillatory nonlinear wave equations due to the high oscillation of the semidiscrete system. Moreover, the CFL condition is also required and crucial for these methods (see, e.g. [22]), which results in the same fatal defect of a tiny time stepsize for the finite difference method. On noticing that ERKN methods can efficiently solve a highly oscillatory system of ordinary differential equations, the two proposed algorithms combined with the FFT technique are really useful and promising in the implementation of semi-analytical ERKN integrator (13.8) for efficiently solving a nonlinear wave equations. Apart from the exponential integrators studied in this chapter we also note that there exist alternative approaches for effectively providing numerical solutions for (13.1) in the literature (see, e.g. [24–29]).

13.2   Preliminaries

We begin by considering initial value problems of second-order differential equations

$$
\begin{cases}
y'' = f(y), & t \in [t_0, T], \\
y(t_0) = y_0, \ y'(t_0) = y_0'.
\end{cases}
\tag{13.10}
$$

The standard Runge–Kutta–Nyström (RKN) method (see, e.g. [30, 31]) for (13.10) is given by

$$
\begin{cases}
Y_i = y_n + c_i h y_n' + h^2 \displaystyle\sum_{j=1}^{s} a_{ij} f(Y_j), & i = 1, \cdots, s, \\[2mm]
y_{n+1} = y_n + h y_n' + h^2 \displaystyle\sum_{i=1}^{s} \bar{b}_i f(Y_i), \\[2mm]
y_{n+1}' = y_n' + h \displaystyle\sum_{i=1}^{s} b_i f(Y_i),
\end{cases}
\tag{13.11}
$$

where $a_{ij}, \bar{b}_i, b_i, c_i$ for $i, j = 1, \cdots, s$ are real constants. An intrinsic relation between ERKN methods and RKN methods has been explored in [4], in which the authors revealed the underlying extension from the RKN method to the ERKN method. We summarise the following three theorems, which are useful for our subsequent analysis and the details can be found in [4].

Theorem 13.1 (See [4]) *Let a RKN method be of order r for (13.10) with coefficients c_i, b_i, $\bar{b}_i$ and a_{ij} for $i, j = 1, \cdots, s$. Then the ERKN method determined by* *the mapping:*

$$\begin{cases} C_i = c_i, \\ A_{ij}(V) = a_{ij}\phi_1((c_i - c_j)^2 V), \\ \bar{B}_i(V) = \bar{b}_i\phi_1((1 - c_i)^2 V), \\ B_i(V) = b_i\phi_0((1 - c_i)^2 V), \end{cases} \tag{13.12}$$

is also of order r for (13.6).

Theorem 13.2 (See [4]) *Let a RKN method be symplectic for* (13.10) *with coefficients* c_i, b_i, $\bar{b}_i$ *and* a_{ij} *for* $i, j = 1, \cdots, s$. *Then the ERKN method determined by* (13.12) *is also symplectic for* (13.6).

Theorem 13.3 (See [4]) *Let a RKN method be symmetric for* (13.10), *whose coefficients* c_i, b_i, $\bar{b}_i$ *and* a_{ij} *satisfy the simplifying assumption* $\bar{b}_i = b_i(1 - c_i)$ *for* $i, j = 1, \cdots, s$. *Then the ERKN method determined by* (13.12) *is also symmetric for* (13.6).

During the implementation of ERKN integrators, the matrix-valued coefficients $A_{ij}(V)$, $\bar{B}_i(V)$ and $B_i(V)$ should be calculated in advance. Due to their complicated computation, we note that the coefficients of the ERKN integrators in this chapter share the form in (13.12), since the special cases where $\phi_0(x) = \cos\sqrt{x}$ and $\phi_1(x) = \sin\sqrt{x}/\sqrt{x}$ can highly simplify the calculation of $A_{ij}(V)$, $\bar{B}_i(V)$ and $B_i(V)$. Another advantage of the formula (13.12) is that the ERKN integrator obtained from (13.12) and the corresponding RKN method nearly has the best structure-preserving properties among its congruence class, which will reduce to the same RKN method (see [4]).

A preliminary step to simplifying the calculation of the computational cost of ERKN integrators will be made, provided we carry out the following transformation. If we set $F_i = f(U_i)$, then the first formula of (13.8) can be rewritten as

$$U_i = \phi_0(C_i^2 V)u_n + C_i h\phi_1(C_i^2 V)u_n' + h^2 \sum_{j=1}^{s} A_{ij}(V)F_j, \quad i = 1, \cdots, s,$$

by replacing $f(U_i)$ with F_i. Using $F_i = f(U_i)$ once again with the above equation, we obtain

$$F_i = f\left(\phi_0(C_i^2 V)u_n + C_i h\phi_1(C_i^2 V)u_n' + h^2 \sum_{j=1}^{s} A_{ij}(V)F_j\right), \quad i = 1, \cdots, s. \tag{13.13}$$

Finally, replacing $f(U_i)$ with F_i in the last two equations of (13.8) and combining with (13.13) we can deduce the equivalent formulation

$$
\begin{cases}
F_i = f\!\left(\phi_0(C_i^2 V)u_n + C_i h\phi_1(C_i^2 V)u_n' + h^2 \sum_{j=1}^{s} A_{ij}(V)F_j\right), \quad i = 1, \cdots, s, \\[2ex]
u_{n+1} = \phi_0(V)u_n + h\phi_1(V)u_n' + h^2 \sum_{i=1}^{s} \bar{B}_i(V)F_i, \\[2ex]
u_{n+1}' = -h\mathscr{A}\phi_1(V)u_n + \phi_0(V)u_n' + h \sum_{i=1}^{s} B_i(V)F_i.
\end{cases}
$$

$$\tag{13.14}$$

It is clear that, in comparison with (13.8), the ERKN integrator rewritten in the form of (13.14) reduces the number of function evaluations of $f(u)$ from $s^2 + 2s$ to s for each time step, while it maintains the same number $(s + 2)^2$, of matrix-vector multiplications. Therefore, the practical formulation of ERKN integrators in applications should be (13.14), rather than (13.8) for the nonlinear wave equation (13.1).

Since the error analysis of ERKN integrators as well as of Gauschi-type methods for nonlinear wave equations has been made in [13, 18, 32, 33], we will not consider this issue further, but pay attention to the practical implementation of these semi-analytical ERKN integrators for nonlinear wave equations.

13.3   Fast Implementation of ERKN Integrators

For $s \geqslant 0$, we have $H^{s+1}(\Omega) \subset L^2(\Omega)$, where $L^2(\Omega)$ is the complex Hilbert space equipped with the inner product and the norm

$$
(u, v) = \frac{1}{2\pi} \int_{\Omega} u(x)\bar{v}(x)\mathrm{d}x, \quad \|u\| = (u, u). \tag{13.15}
$$

Consider the Fourier series of $u(x, t)$ as follows

$$
u_*(x, t) = \sum_{k=-\infty}^{+\infty} \hat{u}_k(t)\mathrm{e}^{\mathrm{i}kx}, \tag{13.16}
$$

where the Fourier coefficients $\hat{u}_k(t)$ are determined by

$$
\hat{u}_k(t) = (u, \mathrm{e}^{\mathrm{i}kx}) = \frac{1}{2\pi} \int_{\Omega} u(x, t)\mathrm{e}^{-\mathrm{i}kx}\mathrm{d}x, \quad k \in \mathbb{Z}. \tag{13.17}
$$

Taking account of the completeness of the Fourier/trigonometric system $\{e^{ikx} : k \in \mathbb{Z}\}$ in $L^2(\Omega)$ (see, e.g. [34]), we obtain that $||u(x, t) - u_*(x, t)||_{L^2(\Omega)} = 0$. In the sense of $L^2(\Omega)$-norm, it is sufficient to find $u_*(x, t)$ instead of the exact $u(x, t)$. Note that the complete orthonormal set $\{e^{ikx} : k \in \mathbb{Z}\}$ constitutes the set of orthogonal eigenfunctions of the operator $\mathscr{A}$ in the Hilbert space $L^2(\Omega)$, i.e.,

$$\mathscr{A}(e^{ikx}) = a^2 k^2 e^{ikx}, \quad k \in \mathbb{Z}. \tag{13.18}$$

With the formulae (13.4) and (13.18), we consequently have that

$$\phi_j(h^2 \mathscr{A})e^{ikx} = \phi_j(a^2 k^2 h^2)e^{ikx}. \tag{13.19}$$

Two special cases of (13.19) are $j = 0$ and $j = 1$:

$$\phi_0(h^2 \mathscr{A})e^{ikx} = \cos(akh)e^{ikx}, \quad \phi_1(h^2 \mathscr{A})e^{ikx} = \frac{\sin(akh)}{akh}e^{ikx}. \tag{13.20}$$

Let N denote the number of spatial mesh grids after spatial discretisation, and we only consider the even integer case for N. Since u is a real function, the Fourier coefficients $\hat{u}_k$ satisfy $\hat{u}_{-k} = \overline{\hat{u}_k}$. Let $X_N = \text{span}\{e^{ikx} : -N/2 \leq k \leq N/2\}$, and $P_N : L^2(\Omega) \to X_N$ be the L^2-orthogonal projection. It is obvious that $P_N u$ will be the truncated Fourier series

$$(P_N u)(x) = \sum_{k=-N/2}^{N/2} \hat{u}_k e^{ikx}, \tag{13.21}$$

which is also the best approximation to $u(x)$ in L^2-norm. The truncated Fourier series (13.21) provides us an efficient way to approximate $u(x)$. However, it is clear that the Fourier coefficients $\hat{f}_k(u(x))$ of $f(u)$ are hard to obtain, on noticing the integral in (13.17), since the expression of $f(u(x))$ with respect to x is always unknown. Therefore, we will not use the truncated Fourier series (13.21) in practice, but consider the following Fourier/trigonometric interpolation instead.

Let

$$x_j = jh = j\frac{2\pi}{N}, \quad 1 \leq j \leq N, \tag{13.22}$$

be N equispaced points in $[0, 2\pi]$. Since N is even, we set

$$Y_N = \left\{ u(x) = \sum_{k=-N/2}^{N/2} \tilde{u}_k e^{ikx} : \tilde{u}_{-N/2} = \tilde{u}_{N/2} \right\}. \tag{13.23}$$

Then the Fourier/trigonometric interpolation polynomial on the equispaced points x_j ($1 \leqslant j \leqslant N$) can be obtained by the interpolation operator $I_N : L^2(\Omega) \to Y_N$ as follows:

$$(I_N u)(x) = \sum_{k=-N/2}^{N/2} \tilde{u}_k e^{ikx}. \tag{13.24}$$

This satisfies

$$(I_N u)(x_j) = u(x_j), \quad 1 \leqslant j \leqslant N. \tag{13.25}$$

Differently from the Fourier coefficients $\hat{u}_k$ in (13.16), the interpolation coefficients $\tilde{u}_k$ can be effectively obtained from $u_j = u(x_j)$ ($1 \leqslant j \leqslant N$) by the Discrete Fourier Transform (DFT)

$$\tilde{u}_k = \frac{1}{\omega_k N} \sum_{j=1}^{N} u_j e^{-ikx_j}, \quad k = -N/2, \cdots, N/2, \tag{13.26}$$

where $\omega_k = 1$ for $|k| < N/2$, and $\omega_k = 2$ for $k = \pm N/2$. Meanwhile, it follows from (13.24) and (13.25) that u_j can also be obtained by the inverse DFT

$$u_j = \sum_{k=-N/2}^{N/2} \tilde{u}_k e^{ikx_j}, \quad j = 1, \cdots, N. \tag{13.27}$$

The DFT (13.26) and inverse DFT (13.27) can be carried out by the Fast Fourier Transform (FFT) and inverse Fast Fourier Transform (IFFT) with only $\mathcal{O}(N \log N)$ operations (see, e.g. [14]), rather than by the direct matrix-vector multiplication with $\mathcal{O}(N^2)$ operations. This computational process for FFT and IFFT can be easily accomplished by using MATLAB (see, e.g. [12]).

Although the Fourier interpolation approximation $I_N u$ in (13.24) to $u(x)$ is usually not better than the truncation $P_N u$ in (13.21), the rigorous error analysis in [11] shows that *'the penalty for using interpolation instead of truncation is at worst a factor of two'*. Hence, we can still have the spectral accuracy of exponential convergence of the Fourier/trigonometric interpolation approximation (see, e.g. [12]). With regard to more details on the truncation error of the Fourier truncation and the interpolation error of the Fourier/trigonometric interpolation, readers are referred to [23, 35, 36].

Now replacing $u(x, t)$ with the trigonometric interpolation polynomial $I_N u$ in (13.1) leads to the nonlinear system

$$\tilde{u}_k'' + a^2 k^2 \tilde{u}_k = \tilde{f}_k(t), \quad k = -N/2, \cdots, N/2. \tag{13.28}$$

Algorithm 1 Fast implementation of explicit ERKN integrator (13.14) with large N

1: set $U_0^j = \varphi(x_j)$, $V_0^j = \psi(x_j)$
2: **for** $n = 1$ to N_T **do**
3:    FFT: $U_{n-1} \longrightarrow \widehat{U}_{n-1}$, $V_{n-1} \longrightarrow \widehat{V}_{n-1}$
4:    set $\widehat{U} = \cos(c_1 hK)\widehat{U}_{n-1} + K^{-1}\sin(c_1 hK)\widehat{V}_{n-1}$
5:    inverse FFT: $\widehat{U} \longrightarrow U$
6:    set $F_1 = f(U)$
7:    FFT: $F_1 \longrightarrow \widehat{F}_1$
8:    set $\widehat{U}_n = \cos(hK)\widehat{U}_{n-1} + K^{-1}\sin(hK)\widehat{V}_{n-1} + h^2 \bar{b}_1 \cdot ((1-c_1)hK)^{-1}\sin((1-c_1)hK)\cdot \widehat{F}_1$
      $\widehat{V}_n = -K\sin(hK)\widehat{U}_{n-1} + \cos(hK)\widehat{V}_{n-1} + hb_1 \cos((1-c_1)hK)\cdot \widehat{F}_1$
9:    **for** $i = 2$ to s **do**
10:        set $\widehat{U} = \cos(c_i hK)\widehat{U}_{n-1} + K^{-1}\sin(c_i hK)\widehat{V}_{n-1}$
11:        **for** $k = 1$ to $i - 1$ **do**
12:            $\widehat{U} = \widehat{U} + h^2 a_{ik} \cdot ((c_i - c_k)hK)^{-1}\sin((c_i - c_k)hK)\cdot \widehat{F}_k$
13:        **end for**
14:        inverse FFT: $\widehat{U} \longrightarrow U$
15:        set $F_i = f(U)$
16:        FFT: $F_i \longrightarrow \widehat{F}_i$
17:        $\widehat{U}_n = \widehat{U}_n + h^2 \bar{b}_i \cdot ((1-c_i)hK)^{-1}\sin((1-c_i)hK)\cdot \widehat{F}_i$
          $\widehat{V}_n = \widehat{V}_n + hb_i \cos((1-c_i)hK)\cdot \widehat{F}_i$
18:    **end for**
19:    inverse FFT: $\widehat{U}_n \longrightarrow U_n$, $\widehat{V}_n \longrightarrow V_n$
20: **end for**

---

where the $\tilde{f}_k(t)$ for $k = -N/2, \cdots, N/2$ are the trigonometric interpolation coefficients of $f(u(x))$ at time t. Since $\tilde{u}_k$ and $\tilde{f}_k$ are easily obtained by the FFT, by means of the variation-of-constants formula for (13.28) and the FFT, we propose an algorithm to implement the explicit ERKN integrator determined by (13.12). With the notation $x_j = j\Delta x$, $t_n = t_0 + nh$, $U_n^j \approx u(x_j, t_n)$, $V_n^j \approx u_t(x_j, t_n)$, $N_T = (T - t_0)/h$ and $K = |a \cdot (-N/2, \cdots, N/2)|^{\mathsf{T}}$, this algorithm is stated in Algorithm 1.

Note that in Algorithm 1, we do not directly apply the ERKN formula to thenonlinear system (13.28). That is, the wave equation (13.1) cannot be solved merely in Fourier space, since $\tilde{f}_k(t)$ cannot be directly expressed by $\tilde{u}_k$ for general nonlinear functions $f(u)$. This differs from [32], where the author considered the particular nonlinear case of $f(u) = u^p$ for $p \geqslant 2$. In that case, the wave equation (13.1) can be converted into a nonlinear system for $\tilde{u}_k$ with respect to t, where all the $\tilde{f}_k$ are obtained by the discrete convolution

$$\tilde{f}(u) = \underbrace{u * u * \cdots * u}_{p \text{ times}}, \quad (y * z)_j = \sum_{k+l \equiv j \bmod 2N} y_k z_l, \quad j = -N/2, \cdots, N/2.$$

$$(13.29)$$

That is why we carry out the FFT for each internal stage F_i in Algorithm 1. Here, it is important to note that we consider the general nonlinear function $f(u)$, which is the negative derivative of a potential energy $V(u)$. It is also worth mentioning that for the special case of $c_1 = 0$, we can simplify Step 4, 5, 6 in Algorithm 1 as $F_1 = f(U_{n-1})$, since $U = U_{n-1}$ under this situation.

13.4   The Case of Symplectic ERKN Integrators

As is known, the symplectic structure has many important physical and mathematical consequences, and it is therefore usually important to preserve it if possible.

We have presented Algorithm 1 in the previous section, which is used for the fast implementation of explicit ERKN integrators (13.8) whose coefficients are determined by (13.12). However, taking into account an important class of explicit symplectic ERKN integrators, we can design another fast implementation algorithm apart from Algorithm 1 on the basis of the equivalence between this important class of explicit symplectic ERKN methods and the corresponding splitting methods. The equivalence is stated below, and a similar result can be found in [37], where the author conducted the presentation and the proof in a different manner.

Theorem 13.4   *Let Ψ be an explicit symplectic ERKN method whose coefficients C_i, B_i, $\bar{B}_i$ and A_{ij} are determined by (13.12), and c_i, b_i, $\bar{b}_i$ and a_{ij} satisfy*

$$
\begin{cases}
\bar{b}_i = b_i(1 - c_i), \\[4pt]
a_{ij} = b_j(c_i - c_j), \\[4pt]
c_i = \displaystyle\sum_{k=1}^{i} b_k - \frac{1}{2}b_i,
\end{cases}
\tag{13.30}
$$

for all $i, j = 1, \cdots, s$. Then the ERKN method is equivalent to a splitting method (see, e.g. [30])

$$
\Phi_h = \varphi^{[1]}_{\alpha_{s+1}h} \circ \varphi^{[2]}_{\beta_s h} \circ \varphi^{[1]}_{\alpha_s h} \circ \cdots \circ \varphi^{[2]}_{\beta_2 h} \circ \varphi^{[1]}_{\alpha_2 h} \circ \varphi^{[2]}_{\beta_1 h} \circ \varphi^{[1]}_{\alpha_1 h},
\tag{13.31}
$$

where

$$
\begin{cases}
\beta_i = b_i, \quad i = 1, \cdots, s, \\[4pt]
\alpha_1 = \dfrac{1}{2}b_1, \quad \alpha_{s+1} = \dfrac{1}{2}b_s, \quad \alpha_j = \dfrac{1}{2}(b_j + b_{j-1}), \quad j = 2, \cdots, s,
\end{cases}
\tag{13.32}
$$

$\varphi^{[1]}_t$ and $\varphi^{[2]}_t$ respectively denote the exact phase flows of the following first-order systems

$$
\begin{cases}
q' = p, \\[4pt]
p' = -Mq,
\end{cases}
\tag{13.33}
$$

and

$$\begin{cases} q' = 0, \\ p' = f(q), \end{cases} \tag{13.34}$$

by denoting $q = y$ and $p = y'$.

Proof We will complete the proof by showing that such a splitting method of the type (13.31) can be equivalently expressed by an ERKN method as stated in the theorem. Since $\varphi_t^{[1]}$ denotes the exact phase flow of (13.33), we then derive from the group property of the exact phase flow that

$$\varphi_{\alpha_1 h}^{[1]} = \varphi_{\beta_1 h/2}^{[1]}, \quad \varphi_{\alpha_{s+1} h}^{[1]} = \varphi_{\beta_s h/2}^{[1]}, \quad \varphi_{\alpha_i h}^{[1]} = \varphi_{\beta_{i+1} h/2}^{[1]} \circ \varphi_{\beta_i h/2}^{[1]}, \quad i = 2, \cdots, s, \tag{13.35}$$

due to the equalities in (13.32). Using the associativity of the combination operation $\circ$, we can write the splitting method Φ_h (13.31) as

$$\Phi_h = \Psi_{\beta_s h} \circ \Psi_{\beta_{s-1} h} \circ \cdots \circ \Psi_{\beta_2 h} \circ \Psi_{\beta_1 h}, \tag{13.36}$$

where $\Psi_{\beta_i h} = \varphi_{b_i h/2}^{[1]} \circ \varphi_{b_i h}^{[2]} \circ \varphi_{b_i h/2}^{[1]}$ for all $i = 1, \cdots, s$.

We now show that $\Psi_{\beta_i h}$ is equivalent to an ERKN method with the stepsize $b_i h$. Let (p_0, q_0) and (p_1, q_1) be initial values and the corresponding numerical solutions after applying $\Psi_{\beta_i h}$ to the initial values, respectively. We can derive the scheme from the formulation of $\Psi_{\beta_i h}$ as follows:

$$\begin{cases} Q_1 = \phi_0\left(\frac{1}{4}b_i^2 V\right) q_0 + \frac{1}{2}b_i h \phi_1\left(\frac{1}{4}b_i^2 V\right) p_0, \\[2mm] P_1 = -\frac{1}{2}b_i h M \phi_1\left(\frac{1}{4}b_i^2 V\right) q_0 + \phi_0\left(\frac{1}{4}b_i^2 V\right) p_0, \\[2mm] Q_2 = Q_1, \\[2mm] P_2 = P_1 + b_i h f(Q_2), \\[2mm] q_1 = \phi_0\left(\frac{1}{4}b_i^2 V\right) Q_2 + \frac{1}{2}b_i h \phi_1\left(\frac{1}{4}b_i^2 V\right) P_2, \\[2mm] p_1 = -\frac{1}{2}b_i h M \phi_1\left(\frac{1}{4}b_i^2 V\right) Q_2 + \phi_0\left(\frac{1}{4}b_i^2 V\right) P_2, \end{cases} \tag{13.37}$$

where $V \equiv h^2 M$. It follows from the identities

$$\begin{cases} \lambda \phi_0(\kappa^2 V)\phi_1(\lambda^2 V) + \kappa \phi_0(\lambda^2 V)\phi_1(\kappa^2 V) = (\lambda + \kappa)\phi_1((\lambda + \kappa)^2 V), \\ \phi_0(\lambda^2 V)\phi_0(\kappa^2 V) + \lambda \kappa V \phi_1(\kappa^2 V)\phi_1(\lambda^2 V) = \phi_0((\lambda - \kappa)^2 V), \end{cases} \tag{13.38}$$

that the direct calculation by eliminating P_1, P_2, and Q_1 from (13.37) leads to

$$
\begin{cases}
Q_1 = \phi_0\left(\frac{1}{4}b_i^2 V\right) q_0 + \frac{1}{2}b_i h \phi_1\left(\frac{1}{4}b_i^2 V\right) p_0, \\[2mm]
q_1 = \phi_0(b_i^2 V)q_0 + b_i h \phi_1(b_i^2 V)p_0 + \frac{1}{2}(b_i h)^2 \phi_1\left(\frac{1}{4}b_i^2 V\right) f(Q_1), \qquad (13.39) \\[2mm]
p_1 = -b_i h M \phi_1(b_i^2 V)q_0 + \phi_0(b_i^2 V)p_0 + b_i h \phi_0\left(\frac{1}{4}b_i^2 V\right) f(Q_1).
\end{cases}
$$

It can be easily verified that (13.39) is just a particular ERKN method with the stepsize $b_i h$, whose Butcher tableau reads

$$
\begin{array}{c|c}
1/2 & 0 \\
\hline
 & \phi_1(V/4)/2 \\
\hline
 & \phi_0(V/4)
\end{array}
\qquad (13.40)
$$

Let $(p_{n+1}^{(1)}, q_{n+1}^{(1)}) = \Psi_{\beta_1 h}(p_n, q_n)$, $(p_{n+1}^{(i+1)}, q_{n+1}^{(i+1)}) = \Psi_{\beta_{i+1} h}(p_{n+1}^{(i)}, q_{n+1}^{(i)})$ for $i = 1, \cdots, s - 1$, and $(p_{n+1}, q_{n+1}) = (p_{n+1}^{(s)}, q_{n+1}^{(s)})$. In consequence, we have $(p_{n+1}, q_{n+1}) = \Phi_h(p_n, q_n)$. In what follows, we aim at showing that the transformation $\Phi_h : (p_n, q_n) \mapsto (p_{n+1}, q_{n+1})$ can be explicitly expressed by an ERKN method, which has exactly the property required in the theorem. To complete the proof, we just need to prove the following proposition by induction on the superscript i.

The mapping $\Psi_{\beta_i h} \circ \Psi_{\beta_{i-1} h} \circ \cdots \circ \Psi_{\beta_2 h} \circ \Psi_{\beta_1 h} : (p_n, q_n) \mapsto (p_{n+1}^{(i)}, q_{n+1}^{(i)})$ *can be expressed by an ERKN scheme as follows*

$$
\begin{cases}
Q_k = \phi_0(c_k^2 V)q_n + c_k h \phi_1(c_k^2 V)p_n + h^2 \sum_{j=1}^{k-1} A_{kj}^{(i)} f(Q_j), \quad k = 1, \cdots, i, \\[2mm]
q_{n+1}^{(i)} = \phi_0(\mu_i^2 V)q_n + \mu_i h \phi_1(\mu_i^2 V)p_n + h^2 \sum_{k=1}^{i} \bar{B}_k^{(i)} f(Q_k), \\[2mm]
p_{n+1}^{(i)} = -\mu_i h M \phi_1(\mu_i^2 V)q_n + \phi_0(\mu_i^2 V)p_n + h \sum_{k=1}^{i} B_k^{(i)} f(Q_k),
\end{cases}
$$

$$(13.41)$$

where $\mu_i = \sum_{j=1}^{i} b_i$, $c_i = \mu_i - \dfrac{b_i}{2}$, $A_{kj}^{(i)} = b_j(c_k - c_j)\phi_1((c_k - c_j)^2 V)$, $\bar{B}_k^{(i)} = b_k(\mu_i - c_k)\phi_1((\mu_i - c_k)^2 V)$ *and* $B_k^{(i)} = b_k \phi_0((\mu_i - c_k)^2 V)$.

For $i = 1$, (13.41) naturally holds on account of (13.39). Suppose that (13.41) holds for any $i < s$. Then we turn to showing that it also holds for the case of $i + 1$.

On noticing the fact that $(p_{n+1}^{(i+1)}, q_{n+1}^{(i+1)}) = \Psi_{\beta_{i+1}h}(p_{n+1}^{(i)}, q_{n+1}^{(i)})$ and $\Psi_{\beta_{i+1}h}$ is also an ERKN method, it follows from the composition law for ERKN method in [4] that the mapping $\Psi_{\beta_{i+1}h} \circ \Psi_{\beta_i h} \circ \cdots \circ \Psi_{\beta_2 h} \circ \Psi_{\beta_1 h} : (p_n, q_n) \mapsto (p_{n+1}^{(i+1)}, q_{n+1}^{(i+1)})$ really can be expressed in an ERKN method, whose coefficients read

$$
\begin{cases}
A_{kj}^{(i+1)} = A_{kj}^{(i)}, \quad k = 1, \cdots, i, \ j = 1, \cdots, k-1, \\[2mm]
A_{i+1,j}^{(i+1)} = \phi_0\left(\dfrac{b_{i+1}^2}{4}V\right)\bar{B}_j^{(i)} + \dfrac{b_{i+1}}{2}\phi_1\left(\dfrac{b_{i+1}^2}{4}V\right)B_j^{(i)}, \quad j = 1, \cdots, i, \\[3mm]
\bar{B}_j^{(i+1)} = \phi_0(b_{i+1}^2 V)\bar{B}_j^{(i)} + b_{i+1}\phi_1(b_{i+1}^2 V)B_j^{(i)}, \quad j = 1, \cdots, i, \\[3mm]
\bar{B}_{i+1}^{(i+1)} = \dfrac{b_{i+1}^2}{2}\phi_1\left(\dfrac{b_{i+1}^2}{4}V\right), \\[3mm]
B_j^{(i+1)} = \phi_0(b_{i+1}^2 V)B_j^{(i)} - b_{i+1}V\phi_1(b_{i+1}^2 V)\bar{B}_j^{(i)}, \quad j = 1, \cdots, i, \\[3mm]
B_{i+1}^{(i+1)} = b_{i+1}\phi_1\left(\dfrac{b_{i+1}^2}{4}V\right).
\end{cases}
\tag{13.42}
$$

Then with the help of the induction hypothesis and the identities in (13.38), it follows from (13.42) that

$$
\begin{cases}
A_{kj}^{(i+1)} = b_j(c_k - c_j)\phi_1((c_k - c_j)^2 V), \quad k = 1, \cdots, i+1, \ j = 1, \cdots, k-1, \\[2mm]
\bar{B}_j^{(i+1)} = b_j(\mu_{i+1} - c_j)\phi_1((\mu_{i+1} - c_j)^2 V), \quad j = 1, \cdots, i+1, \\[2mm]
B_j^{(i+1)} = b_j\phi_0((\mu_{i+1} - c_j)^2 V), \quad j = 1, \cdots, i+1,
\end{cases}
\tag{13.43}
$$

which confirms that the result also holds for the case of $i + 1$. Moreover, the consistency of the splitting method means that $\mu_s = \sum_{j=1}^{s} b_j = 1$. By setting $i = s$ in (13.41) we finally conclude that the splitting method in (13.31) can be written as an ERKN method, whose coefficients satisfy (13.12) and (13.30). This completes the proof. $\qquad\square$

We set $q = y$, $p = y'$, and then the phase flow $\varphi_h^{[1]}$ and $\varphi_h^{[2]}$ can be respectively expressed as

$$
\varphi_h^{[1]} : (y_0, y_0') \mapsto (\phi_0(V)y_0 + h\phi_1(V)y_0', \ -hM\phi_1(V)y_0 + \phi_0(V)y_0'), \tag{13.44}
$$

and

$$
\varphi_h^{[2]} :. (y_0, y_0') \mapsto (y_0, \ y_0' + hf(y_0)). \tag{13.45}
$$

Algorithm 2 Fast implementation of special symplectic ERKN integrator

---

1: set $U_0^j = \varphi(x_j)$, $V_0^j = \psi(x_j)$
2: **for** $n = 1$ to N_T **do**
3:     set $U = U_{n-1}$, $V = V_{n-1}$
4:     **for** $k = 1$ to s **do**
5:         FFT: $U \longrightarrow \widehat{U}$, $V \longrightarrow \widehat{V}$
6:         $\widehat{U}_{(0)} = \cos(\alpha_k h K)\widehat{U} + K^{-1}\sin(\alpha_k h K)\widehat{V}$,
           $\widehat{V}_{(0)} = -K\sin(\alpha_k h K)\widehat{U} + \cos(\alpha_k h K)\widehat{V}$
7:         inverse FFT: $\widehat{U}_{(0)} \longrightarrow U$, $\widehat{V}_{(0)} \longrightarrow V$
8:         $V = V + \beta_k h f(U)$
9:     **end for**
10:    FFT: $U \longrightarrow \widehat{U}$, $U \longrightarrow \widehat{V}$
11:    $\widehat{U}_{(0)} = \cos(\alpha_{s+1} h K)\widehat{U} + K^{-1}\sin(\alpha_{s+1} h K)\widehat{V}$,
       $\widehat{V}_{(0)} = -K\sin(\alpha_{s+1} h K)\widehat{U} + \cos(\alpha_{s+1} h K)\widehat{V}$
12:    inverse FFT: $\widehat{U}_{(0)} \longrightarrow U$, $\widehat{V}_{(0)} \longrightarrow V$
13:    set $U_n = U$, $V_n = V$
14: **end for**

With the help of Theorem 13.4, if we solve (13.44) and (13.45) in the Fourier space with the FFT and IFFT, we can obtain Algorithm 2, which is specially designed for the implementation of the symplectic ERKN integrators stated in Theorem 13.4. Note that the multiplication of the two vectors occurring in both Algorithms 1 and 2 is in the componentwise sense.

13.5 Analysis of Computational Cost and Memory Usage

13.5.1 Computational Cost at Each Time Step

In this section, we focus on the analysis of computational cost at each time step for the three implementation approaches, i.e., the direct calculation approach of (13.14) with matrix-vector multiplication, Algorithm 1 for general ERKN integrators determined by (13.12) and Algorithm 2 for symplectic ERKN integrators that are equivalent to splitting methods. We estimate the computational cost for an explicit ERKN integrator denoted by N_d and N_f, which respectively denote the basic scalar operations (multiplication or addition) and function evaluations of $f(u)$. Note that the multiplication between an N-dimensional matrix-valued function and a corresponding vector contains N^2 basic scalar multiplications and $N(N-1)$ basic scalar additions. Since the FFT (or IFFT) can be accomplished with $\mathscr{O}(N \log N)$ operations for an N-dimensional vector, we assume that $\mathscr{O}(N \log N) = \tilde{C} \cdot N \log N$, where $\tilde{C}$ is a positive constant independent of N.

We assume that the underlying ERKN integrator is of s-stages. The computational cost for each approach is shown in Table 13.1. This shows that the number of function evaluations in one time step is the same for the three different approaches, i.e., s. However, they differ greatly in the number of basic scalar operations. The

Table 13.1 The number of floating point of calculation for the three implementation approaches

Approach	Number N_d	N_f
Direct calculation	$(s^2 + 7s + 6)N^2 - (s+2)N$	s
Algorithm 1	$(2s + 3)\tilde{C} \cdot N \log N + \left(\dfrac{3s^2}{2} + \dfrac{15s}{2} + 4 \right) N$	s
Algorithm 2	$4(s + 1)\tilde{C} \cdot N \log N + (9s + 8)N$	s

primary difference is that the cost of the direct calculation approach is $\mathcal{O}(N^2)$, whereas the other two algorithms proposed in this chapter are $\mathcal{O}(N \log N)$, which will sharply decrease the calculation cost once the spatial grid number N isso large that the spatial continuity can be nearly preserved by ERKN integrators, in the sense of numerical computation. A careful observation shows that this advantage essentially derives from the faster calculation of spectral derivatives by the FFT.

We now turn to the comparison between Algorithms 1 and 2. A rough estimate may give that Algorithm 2 takes more basic operations than Algorithm 1, since the coefficient of the dominant part $N \log N$ of the former is $4(s + 1)$, which is more than that of the latter, i.e., $2s + 3$. However, our numerical simulations in Sect. 13.6 show that for a symplectic ERKN integrator of the type stated in Theorem 13.4, Algorithm 2 consumes less CPU time than Algorithm 1. In order to explain this phenomenon, we make a detailed comparison between the two algorithms as follows.

For a fixed integer N, let $\Theta = \tilde{C} \log N > 0$. Then a comparison between the computational cost of the two algorithms reduces to the comparison between

$$(2s + 3)\Theta + \left(\frac{3s^2}{2} + \frac{15s}{2} + 4 \right) = \frac{3s^2}{2} + \left(\frac{15}{2} + 2\Theta \right) s + (4 + 3\Theta) \text{ and}$$

$$4(s + 1)\Theta + (9s + 8) = (9 + 4\Theta)s + (8 + 4\Theta). \text{ Since } s \text{ is positive, the only}$$

zero point of $\left(\frac{3s^2}{2} + \left(\frac{15}{2} + 2\Theta \right) s + (4 + 3\Theta) \right) - \left((9 + 4\Theta)s + (8 + 4\Theta) \right) =$

$$\frac{3s^2}{2} - \left(\frac{3}{2} + 2\Theta \right) s - (4 + \Theta) \text{ is } s_0 = \left(\frac{1}{2} + \frac{2}{3}\Theta \right) + \frac{1}{3}\sqrt{\left(\frac{3}{2} + 2\Theta \right)^2 + (24 + 6\Theta)}.$$

Therefore, the computational cost for Algorithm 2 will be less than that for Algorithm 1 provided $s \geqslant s_0$. On the contrary, Algorithm 2 costs more once $s < s_0$. Here, we list some possible values of s_0 for different Θ in Table 13.2. On noticing that s denotes the stage of ERKN integrator and larger s always implies higher order, we can roughly conclude that Algorithm 2 will be more efficient than Algorithm 1 for high-order symplectic ERKN integrators. Though the constant $\tilde{C}$ of $\mathcal{O}(N \log N)$ cannot be precisely determined, the numerical experiments in the following section confirm that Algorithm 2 consumes less CPU time than Algorithm 1, even for the symplectic ERKN integrator of 3 stages, which clearly supports the higher efficiency of Algorithm 2.

Table 13.2 Values of s_0 for different Θ

Θ	1	2	3	4
s_0	3.3333	4.5465	5.8040	7.0860

Here it is remarked that we are concerned only with explicit ERKN integrators in Algorithm 1. For implicit integrators, a similar analysis can be made, in which iterative solutions are needed in the implementation. Likewise, the computational cost for implicit ERKN integrators will have an order of magnitude $\mathcal{O}(N \log N)$, which is also much smaller than that of direct calculation, i.e., an order of magnitude $\mathcal{O}(N^2)$.

13.5.2 Occupied Memory and Maximum Number of Spatial Mesh Grids

For the numerical experiments in Sect. 13.6, the program runs in MATLAB 2012a on a computer Lenovo Yangtian A6860f (CPU: Intel (R) Core (TM) i5-6500 CPU @ 3.20 GHz (4CPUs), Memory: 8 GB, Os: Microsoft Windows 7 with 64bit). Hence, the maximum possible occupied memory is set as 8 GB to avoid memory overflow. Besides, each real number is stored in the double-precision floating-point format, which occupies 8 Bytes of memory.

Concerning the direct calculation procedure, all the coefficients $A_{ij}(V)$, $B_i(V)$, $\bar{B}_i(V)$, $\phi_0(C_i^2 V)$, $\phi_1(C_i^2 V)$, $\phi_0(V)$ and $\phi_1(V)$ should be calculated and stored in advance. Since N denotes the number of spatial mesh grids, the numerical solutions u_{n+1} and u'_{n+1} are all N-dimensional vectors. This implies that each coefficient of the ERKN integrators will be an $N \times N$ matrix, which needs N times more memory storage than that of u_{n+1} or u'_{n+1}. Thus, we count up the occupied memory by mainly considering the coefficients of the ERKN integrator due to the large magnitude of N. We now estimate the maximum value of N under the environment of 8 GB memory storage for an s-stage explicit ERKN integrator determined by (13.12). It should be noted that both $\phi_0(C_1^2 V)$ and $\phi_1(C_1^2 V)$ do not need be calculated once $C_1 = 0$ for some explicit ERKN integrators.

For the one-dimensional case of the nonlinear wave equation (13.1), the least required Bytes of memory storage is $8 \left(\dfrac{s(s-1)}{2} + 4s \right) N^2 = 4s(s+7)N^2$. This implies that N should satisfy

$$4s(s+7)N^2 \leqslant 8 \times 1024^3, \tag{13.46}$$

which yields

$$N \leqslant N_{\max} = 32768 \sqrt{\frac{2}{s(s+7)}}. \tag{13.47}$$

For the two-dimensional case, we admit the assumption that the number N_x of grids in the x-direction equals to the number N_y of grids in the y-direction, i.e., $N_x = N_y$. In this case, u_{n+1} and u'_{n+1} will be $N_x N_y$-dimensional vectors, and V is an $N_x N_y \times N_x N_y$ matrix. Then we can obtain that $N_x (= N_y) \leqslant \sqrt{N_{\max}}$. Likewise, the result $N_x (= N_y = N_z) \leqslant \sqrt[3]{N_{\max}}$ can be obtained in a similar manner under the assumption $N_x = N_y = N_z$.

However, it follows from the description of Algorithms 1 and 2 that the only stored values are u_{n+1}, u'_{n+1} and some other intermediate variables. Thus, we can obtain the estimations as follows:

$$N \leqslant \widetilde{N}_{\max} = \frac{1024^3}{2 + \Lambda},$$

$$N_x (= N_y) \leqslant \sqrt{\widetilde{N}_{\max}}, \tag{13.48}$$

$$N_x (= N_y = N_z) \leqslant \sqrt[3]{\widetilde{N}_{\max}},$$

respectively for the one-dimensional, two-dimensional and three-dimensional cases. Here the positive integer Λ denotes the number of intermediate variables during the implementation of the two algorithms, and $\Lambda = s + 2$ for Algorithm 1 while $\Lambda = 4$ for Algorithm 2.

In Table 13.3, we list some values of $N_{\max}$ (or $\widetilde{N}_{\max}$) for different approaches in all the three-dimensional cases with $s = 4, 10$, and 16. Some points can be concluded from this table. First, the value of $\widetilde{N}_{\max}$ is much larger than that of $N_{\max}$, which indicates that the two algorithms presented in this chapter can admit more dense spatial grids than the direct calculation procedure in order to nearly preserve the spatial continuity. Second, for the one-dimensional case, all the values of $N_{\max}$ and $\widetilde{N}_{\max}$ are larger than 1024, which means that all the three approaches can nearly preserve the spatial continuity with sufficient spatial mesh grids. Third, for the two-dimensional and three-dimensional cases, the direct calculation procedure onlyallows a mesh grid number of which is no more than 100. In particular, for the three-dimensional case, the admissible value of $N_{\max}$ is less than 20. This implies that ERKN integrators can hardly preserve the spatial continuity once the direct calculation procedure is applied. However, for the algorithms presented in this chapter, $\widetilde{N}_{\max}$ is at least of magnitude of 512, which is nearly sufficient for the most nonlinear wave equations to nearly preserve the spatial continuity. Finally, a comparison between Algorithms 1 and 2 shows that the former will occupy a little more memory than the latter.

13.6 Numerical Experiments

In this section, we conduct the numerical experiments with different ERKN integrators in order to show the remarkable efficiency of the algorithms presented

Table 13.3 The value of N_{max} (or $\widetilde{N}_{max}$) for different approaches in all the three-dimensional cases with $s = 4, 10, 16$

Approach		Dimension		
		One	Two	Three
$s = 4$	Direct calculation	6986	83	19
	Algorithm 1	128×1024^2	11.31×1024	512
	Algorithm 2	170.67×1024^2	13.06×1024	1.10×512
$s = 10$	Direct calculation	3554	59	15
	Algorithm 1	73.14×1024^2	8.55×1024	0.83×512
	Algorithm 2	170.67×1024^2	13.06×1024	1.10×512
$s = 16$	Direct calculation	2415	49	13
	Algorithm 1	51.20×1024^2	7.16×1024	0.74×512
	Algorithm 2	170.67×1024^2	13.06×1024	1.10×512

in this chapter when applied to nonlinear wave equations. The selected ERKN integrators are as follows:

- ERKN3s4: the 3-stage fourth-order symmetric and symplectic ERKN integrator [4, 38] that can be written as a splitting method ;
- ERKN3s4b: the 3-stage fourth-order ERKN integrator obtained by the RKN method in [31];
- ERKN4s5: the 4-stage fifth-order ERKN integrator obtained by the RKN method in [31];
- ERKN7s6: the 7-stage sixth-order symmetric and symplectic ERKN integrator [38] that can be written as a splitting method;
- ERKN7s6b: the 7-stage sixth-order symplectic ERKN integrator proposed in [4];
- ERKN17s8: the 17-stage eighth-order symmetric and symplectic ERKN integrator derived in [4, 38] that can be written as a splitting method.

We remark that, except for ERKN3s4, ERKN7s6 and ERKN17s8, the other three methods cannot be written as splitting methods. For the three implementation approaches, we respectively use the symbols **D**, **A**, and **F** to denote the direct calculation procedure, Algorithms 1 and 2. For instance, the three implementations of ERKN3s4 are respectively denoted by **D3s4**, **A3s4** and **F3s4**. During the numerical experiments, the numerical solution computed by ERKN16s10 (16-stage ERKN method of order 10 derived in [4]) with sufficiently small stepsize is thought of as the reference solution, when the analytical solution is not available. Note that the Fourier spectral discretisation is used for all the problems considered in this section. Hence, the discrete Hamiltonian energy corresponding to (13.3) has the following form

$$H_n = \frac{1}{2}u'^{\mathsf{T}}_n u'_n + \frac{1}{2}u^{\mathsf{T}}_n M u_n + V(u_n),$$

where M is the spectral differentiation matrix. In this sense, the global Hamiltonian energy error is measured by $\text{GHE}_n = |H_n - H_0|$.

Problem 13.1 (Breather Soliton)   We first consider the well-known sine-Gordon equation (see, e.g. [13, 18])

$$\frac{\partial^2 u}{\partial t^2} = \frac{\partial^2 u}{\partial x^2} - \sin u,$$

on the region $(x, t) \in [-L, L] \times [0, T]$, with the initial conditions

$$u(x, 0) = 0, \quad u_t(x, 0) = 4\kappa \operatorname{sech}(\kappa x),$$

and the boundary conditions

$$u(-L, t) = u(L, t) = 4 \arctan \left(c^{-1} \operatorname{sech}(\kappa L) \sin(c\kappa t) \right),$$

where $\kappa = 1/\sqrt{1 + c^2}$. The exact solution is given by

$$u(x, t) = 4 \arctan \left(c^{-1} \operatorname{sech}(\kappa x) \sin(c\kappa t) \right), \tag{13.49}$$

which is known as the breather solution of the sine-Gordon equation.

We set $L = 40$, $T = 40$ and $c = 0.5$ for this problem in the numerical experiment. We use this problem having the exact solution (13.49) to verify and show that the algorithms presented in this chapter work very well. The global error (GE) results for each ERKN integrator with $N = 512$ and the time stepsize $h = 0.2$ are shown in Fig. 13.1, which are implemented by Algorithm 1. It confirms that these algorithms perform perfectly for this problem. Meanwhile, the CPU times taken by **A3s4**, **A3s4b**, **A4s5**, **A7s6**, **A7s6b**, and **A17s8** are respectively 0.112, 0.114, 0.126, 0.178, 0.184, 0.429 s. These small values really support the theoretical prediction of the high efficiency of Algorithm 1. Furthermore, for the three ERKN integrators that are equivalent to splitting methods, we also implement them by Algorithm 2. The difference between a numerical solution obtained by Algorithm 2 and its counterpart Algorithm 1 is in the magnitude of $\mathcal{O}(10^{-11})$ for all the three ERKN integrators. This fact confirms the consistency between the two algorithms. The CPU times of **F3s4**, **F7s6**, and **F17s8** are respectively 0.105, 0.147, and 0.242 s, which are less than that of Algorithm 1. This fact strongly supports the earlier claim of the higher efficiency of Algorithm 2 than Algorithm 1. In addition, the global Hamiltonian energy errors (GHE) corresponding to Fig. 13.1 are shown in Fig. 13.2. It can be observed from Fig. 13.2 that these ERKN integrators preserve the energy very well.

Note that since the mesh ratio $h/\Delta x = 1.28 > 1$, the finite difference scheme may be numerical unstable for such a large mesh ratio. To illustrate this point, we conduct further numerical experiments with the compact fourth-order centraldifference scheme [39] for the spatial derivative. Meanwhile, we use the 3-stage fourth-order RKN method in [31] to discretise the temporal derivative.

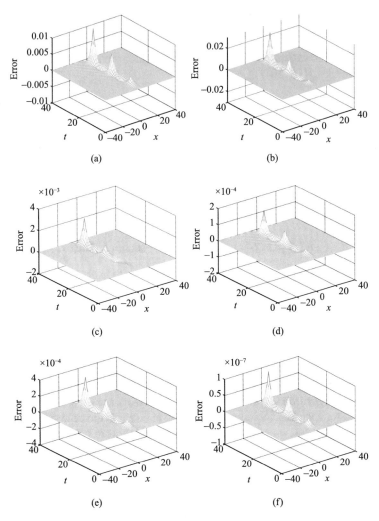

Fig. 13.1 Global errors (GE) of different methods on the region $(x, t) \in [-40, 40] \times [0, 40]$. (**a**) GE of ERKN3s4. (**b**) GE of ERKN3s4b. (**c**) GE of ERKN4s5. (**d**) GE of ERKN7s6. (**e**) GE of ERKN7s6b. (**f**) GE of ERKN17s8

Exactly as we predicted, this method is unstable for these fixed h and Δx due to the numerical overflow of solutions. This fact clearly shows the broader applicability of the algorithms presented in this chapter than the finite difference method in solving nonlinear wave equations, since the former admit large time stepsizes. To numerically check the convergence of the integrators presented in this chapter, we list their global errors at the final time $T = 40$ with different spatial stepsize Δx and time stepsize h in Table 13.4, from which it can be observed that more dense mesh grids indicate smaller global errors. This clearly supports the numerical convergence of the semi-analytical integrators.

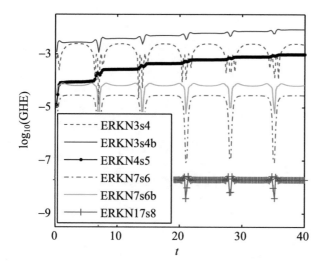

Fig. 13.2   Global Hamiltonian energy errors (GHE) of different methods

Table 13.4   Global errors at the final time $T = 40$ with different mesh grids

$(\Delta x, h)$	(5/4,1/5)	(5/8,1/10)	(5/16,1/20)	(5/32,1/40)
ERKN3s4	4.2905	5.0385×10^{-3}	1.0681×10^{-4}	1.4253×10^{-4}
ERKN3s4b	4.2633	5.0056×10^{-3}	2.1863×10^{-4}	1.6716×10^{-4}
ERKN4s5	4.3140	6.0142×10^{-3}	9.4580×10^{-6}	4.1583×10^{-7}
ERKN7s6	4.3101	5.8759×10^{-3}	8.7842×10^{-7}	2.8212×10^{-9}
ERKN7s6b	4.3104	5.8812×10^{-3}	9.5823×10^{-7}	6.6574×10^{-9}
ERKN17s8	4.3098	5.8720×10^{-3}	8.6836×10^{-7}	1.2319×10^{-11}

Problem 13.2   Consider the nonlinear Klein–Gordon equation (see, e.g. [40, 41])

$$\begin{cases} u_{tt} - u_{xx} + u + u^3 = 0, & 0 < x < L, \quad t \in (0, T), \\ u(0, t) = u(L, t), \end{cases}$$

with the periodic boundary condition. The initial conditions are given by

$$u(x, 0) = A\left[1 + \cos\left(\frac{2\pi}{L}x\right)\right], \quad u_t(x, 0) = 0,$$

where $L = 1.28$ and A is the amplitude.

We set $A = 20$ for this problem in the numerical experiment. Such a large amplitude makes this problem challenging for its numerical solution (see, e.g. [40, 41]), since the solution will have an abrupt change in both time and space directions. This phenomenon can be observed from Fig. 13.3, where we plot the reference numerical solution in the region $(x, t) \in [0, 1.28] \times [0, 10]$. To show the

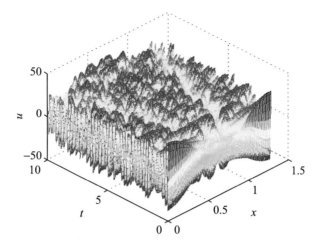

Fig. 13.3 Reference solution with ERKN10

effectiveness of the ERKN integrators, we plot the global errors of these methods in Fig. 13.4 with the time stepsize $h = 0.001$. It can be observed from Fig. 13.4 that ERKN integrators really can solve this problem with some accuracy. It is noted that, the difference between numerical solutions obtained from the three implementation approaches for each ERKN integrator are of magnitude $\mathcal{O}(10^{-12})$, which confirms that the two algorithms presented in this chapter are promising. The global Hamiltonian energy errors corresponding to Fig. 13.4 are presented in Fig. 13.5, which show the good preservation of the ERKN integrator presented in this chapter.

To compare the efficiency of the three implementation approaches, we carry out these methods with different numbers N of spatial grid points and the fixed time stepsize $h = 0.001$. The numerical results for the consumed CPU time are shown in Tables 13.5, and 13.6 indicates the detailed ratio of CPU time of the direct calculation procedure to that of Algorithm 1 (or Algorithm 2). It is clear from the two tables that the two algorithms cost much less time than the direct calculation procedure for all the six ERKN integrators. This fact supports the theoretical analysis that the algorithms described in this chapter really can nearly preserve the spatial continuity with a large number N of spatial grid points and reasonable computational cost. In Table 13.6, for the underlying ERKN methods larger N always implies a larger ratio, which confirms that the superiority of these algorithms over the direct calculation procedure will be more marked for a larger grid number N. Moreover, the comparison between Algorithms 1 and 2 in the two tables shows that Algorithm 2 always consumes less CPU time than Algorithm 1. In particular, for a fixed N, Algorithm 2 becomes more efficient (larger ratio in Table 13.6) than Algorithm 1 for ERKN methods of higher order (hence with more stages, i.e., a bigger s).

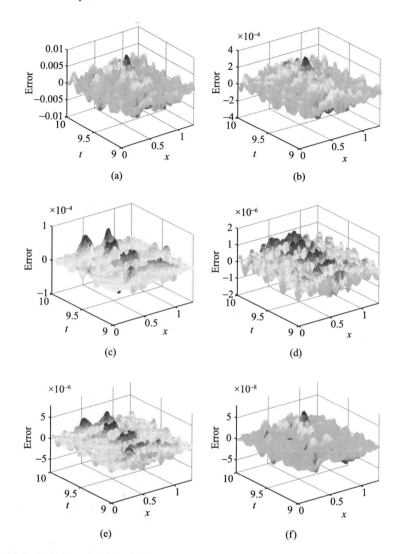

Fig. 13.4 Global errors (GE) of different methods on the region $(x, t) \in [0, 1.28] \times [9, 10]$. **(a)** GE of ERKN3s4 with $N = 2^7$. **(b)** GE of ERKN3s4b with $N = 2^8$. **(c)** GE of ERKN3s5 with $N = 2^8$. **(d)** GE of ERKN7s6 with $N = 2^8$. **(e)** GE of ERKN7s6b with $N = 2^8$. **(f)** GE of ERKN17s8 with $N = 2^9$

Problem 13.3 We then consider the two-dimensional sine-Gordon equation

$$u_{tt} - (u_{xx} + u_{yy}) = -\sin(u), \quad t > 0$$

with the homogeneous Neumann boundary condition

$$u_x(\pm 14, y, t) = 0, \quad u_y(x, \pm 14, t) = 0$$

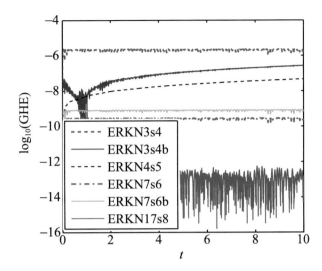

Fig. 13.5 Global energy errors (GHE) of different methods corresponding to Fig. 13.4

Table 13.5 CPU time
(seconds) consumed by each
method for different N with
the time stepsize $h = 0.001$

	$N = 64$	$N = 128$	$N = 256$	$N = 512$
D3s4	3.180	8.807	39.573	241.768
A3s4	1.266	2.144	3.626	6.286
F3s4	0.935	1.706	3.106	5.569
D3s4b	2.534	10.096	38.341	242.901
A3s4b	1.241	2.147	3.569	6.337
D4s5	2.796	12.767	46.291	310.244
A4s5	1.487	2.792	4.344	7.432
D7s6	7.228	25.079	83.279	568.355
A7s6	2.374	4.347	6.665	11.645
F7s6	1.714	3.222	5.571	10.422
D7s6b	7.236	24.720	70.033	574.093
A7s6b	2.366	4.319	6.685	11.857
D17s8	30.047	102.216	319.589	2093.492
A17s8	6.663	10.357	16.204	29.152
F17s8	3.616	6.665	11.654	21.877

in the region $(x, y) \in [-14, 14] \times [-14, 14]$. The initial conditions are given by

$$u(x, y, 0) = 4 \arctan \left(\exp \left(3 - \sqrt{x^2 + y^2} \right) \right), \quad u_t(x, y, 0) = 0,$$
$$(x, y) \in [-14, 14] \times [-14, 14].$$

The solutions of this problem are circular ring solitons (see, e.g. [18, 42, 43]).

For solving this problem, the eighth-order integrator ERKN17s8 implemented by **F17s8** is used to make a comparison with the method GLC4 in [18]. Note that

Table 13.6 Ratio of CPU time of direct calculation approach to that of the two algorithms

	$N = 64$	$N = 128$	$N = 256$	$N = 512$
D3s4/A3s4	2.5118	4.1077	10.9137	38.4613
D3s4/F3s4	3.4011	5.1624	12.7408	43.4132
D3s4b/A3s4b	2.0419	4.7024	10.7428	38.3306
D4s5/A4s5	1.8803	4.5727	10.6563	41.7443
D7s6/A7s6	3.0447	5.7693	12.4950	48.8068
D7s6/F7s6	4.2170	7.7837	14.9487	54.5342
D7s6b/F7s6b	3.0583	5.7235	10.761	48.4181
D17s8/A17s8	4.5085	9.8693	19.7228	71.8130
D17s8/F17s8	8.3095	15.3362	27.4231	95.6937

GLC4 is also of order eight. Here, we select the same time stepsize $h = 0.1$ and mesh region size 400×400 as those in [18]. Numerical results of $\sin(u/2)$ and the corresponding contour plots at the time points $t = 0, 4, 8, 11.5, 13$, and 15 are shown in Figs. 13.6 and 13.7, which are nearly the same as the corresponding figures in [18]. This shows the effectiveness of **F17s8** in solving this problem. In particular, for $t = 15$ the CPU time of **F17s8** is only 38.67 s, which is much less than that of GLC4, i.e. 668.05 s (see [18]). This fact again gives a further support for the high efficiency of the algorithms presented in this chapter. Finally, we display the good energy conservation of ERKN17s8 in Fig. 13.8.

With regard to the formulation and analysis of energy-preserving schemes for Klein–Gordon equations, see Chap. 9 (see also [44]). The framework of semi-analytical integrators for solving partial differential equations was initially proposed in [10], and this chapter focuses on the efficient implementation issue of the semi-analytical integrators.

13.7 Conclusions and Discussions

Although there has been far less numerical treatment of PDEs in the structure-preserving literature than that of ODEs, the recent growth of geometric integration for nonlinear Hamiltonian PDEs has led to the development of numerical schemes which systematically incorporate qualitative features of the underlying problem into their structure. In general, the qualitative characteristics of structure-preserving integrators are mainly concerned with the symmetry, the symplecticity, the multi-symplecticity, the conservation of energy or first integrals, the high oscillation or stiffness, and so on (see, e.g. [30, 45–47]). In this chapter, by presenting a class of semi-analytical ERKN integrators and their implementation approaches for solving nonlinear wave equations, we incorporated the spatial continuity into the structure-preserving property as well. These ERKN integrators can nearly preserve both the spatial continuity and the high oscillation of the original problem, in theory. In order to effectively realize the ERKN integrators on a computer, we presented

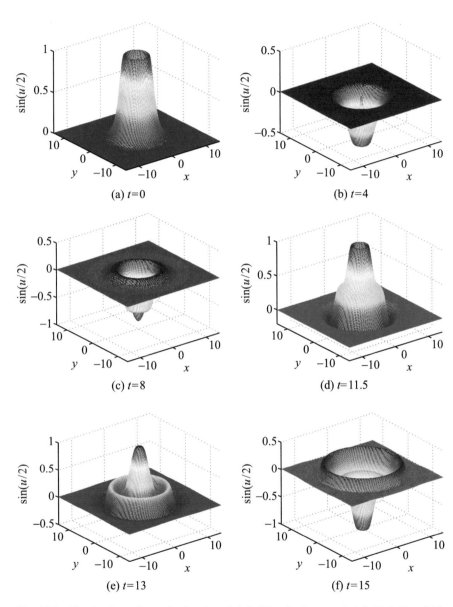

Fig. 13.6 Circular ring solitons: the function of $\sin(u/2)$ at the time $t = 0, 4, 8, 11.5, 13$ and 15

two algorithms accompanied with FFT technique, besides the direct calculation procedure. It follows from the detailed analysis of the computational cost and the memory storage that the algorithms presented in this chapter possess the superiority of smaller computational cost and lower memory storage over the direct calculation procedure. In particular, for nonlinear wave equations of high dimension, the direct

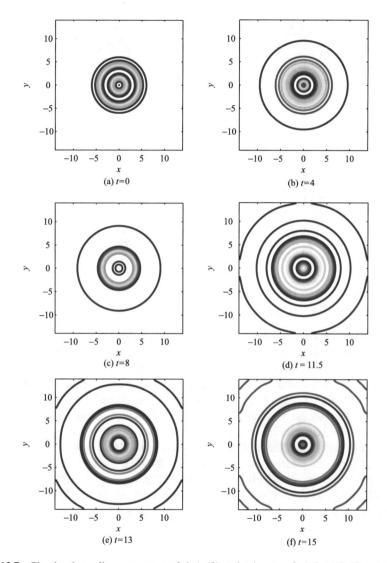

Fig. 13.7 Circular ring solitons: contours of $\sin(u/2)$ at the time $t = 0, 4, 8, 11.5, 13$ and 15

calculation procedure could hardly be used to preserve the spatial continuity, due to the crucial restriction on the dimension of the spatial grid points, while the two algorithms, Algorithms 1 and 2, do not suffer from this trouble, due to the larger maximum number of admissible spatial grid point. Moreover, Algorithm 2, which is suitable for important symplectic ERKN integrators for Hamiltonian systems, is a bit more efficient than Algorithm 1. Finally, we conducted numerical experiments including one-dimensional and two-dimensional wave equations in Sect. 13.6, and the numerical results show strong support for the theoretical analysis

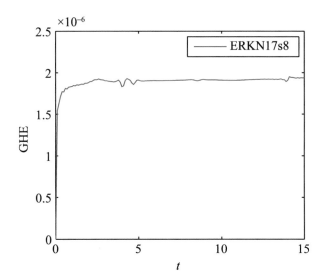

Fig. 13.8 Global energy errors (GHE) of ERKN17s8

presented in this chapter. The numerical results also demonstrate an important fact that, in comparison with the finite difference method, the algorithms presented in this chapter, which can nearly preserve both the spatial continuity and the high oscillation, are robust and efficient when applied to nonlinear wave equations.

It is noted that for nonlinear wave equations equipped with other boundary conditions, such as the homogeneous Dirichlet or Neumann boundary conditions, we can also design such kind of efficient algorithms by just replacing the FFT with the Discrete Fast Cosine/Sine Transform, once the Fourier spectral discretisation is replaced by the cosine/sine scheme [18, 42, 43]. Furthermore, for general boundary conditions, the Chebyshev spectral discretisation accompanied with the FFT technique is also possible, and a further research in this area is needed.

The material in this chapter is based on the work by Mei et al. [48].

References

1. Wu, X., Mei, L., Liu, C.: An analytical expression of solutions to nonlinear wave equations in higher dimensions with Robin boundary conditions. J. Math. Anal. Appl. **426**, 1164–1173 (2015)
2. Liu, C., Wu, X.: The boundness of the operator-valued functions for multidimensional nonlinear wave equations with applications. Appl. Math. Lett. **74**, 60–67 (2017)
3. Liu, C., Wu, X.: An energy-preserving and symmetric scheme for nonlinear Hamiltonian wave equations. J. Math. Anal. Appl. **440**, 167–182 (2016)
4. Mei, L., Wu, X.: The construction of arbitrary order ERKN methods based on group theory for solving oscillatory Hamiltonian systems with applications. J. Comput. Phys. **323**, 171–190 (2016)

5. Wang, B., Iserles, A., Wu, X.: Arbitrary-order trigonometric Fourier collocation methods for multi-frequency oscillatory systems. Found. Comput. Math. **16**, 151–181 (2016)
6. Wu, X., Liu, K., Shi, W.: Structure-Preserving Algorithms for Oscillatory Differential Equations II. Springer, Heidelberg (2015)
7. Wu, X., Wang, B.: Recent Developments in Structure-Preserving Algorithms for Oscillatory Differential Equations. Springer Nature Singapore Pte Ltd., Singapore (2018)
8. Wu, X., You, X., Wang, B.: Structure-Preserving Algorithms for Oscillatory Differential Equations. Springer, Berlin (2013)
9. Jung, C., Nguyen, T.B.: Semi-analytical time differencing methods for stiff problems. J. Sci. Comput. **63**, 355–373 (2015)
10. Wu, X., Liu, C., Mei, L.: A new framework for solving partial differential equations using semi-analytical explicit RK(N)-type integrators. J. Comput. Appl. Math. **301**, 74–90 (2016)
11. Boyd, J.P.: Chebyshev and Fourier Spectral Methods. 2nd edn. Dover Publications, New York (2013)
12. Trefethen, L.N.: Spectral Methods in MATLAB. SIAM, Philadelphia (2000)
13. Mei, L., Liu, C., Wu, X.: An essential extension of the finite-energy condition for extended Runge-Kutta-Nyström integrators when applied to nonlinear wave equations. Commun. Comput. Phys. **22**, 742–764 (2017)
14. Cooley, J., Tukey, J.: An algorithm for the machine calculation of complex Fourier series. Math. Comput. **19**, 297–301 (1965)
15. Bao, W., Dong, X.: Analysis and comparison of numerical methods for the Klein-Gordon equation in the nonrelativistic limit regime. Numer. Math. **120**, 189–229 (2012)
16. Lax Peter, D.: Hyperbolic Partial Differential Equations. American Mathematical Society/Courant Institute of Mathematical Sciences, New York (2006)
17. Morton, K.W., Mayers, D.F.: Numerical Solution of Partial Differential Equations, 2nd edn. Cambridge University, Cambridge (2005)
18. Liu, C., Wu, X.: Arbitrarily high-order time-stepping schemes based on the operator spectrum theory for high-dimensional nonlinear Klein-Gordon equations. J. Comput. Phys. **340**, 243–275 (2017)
19. Bao, W., Cai, Y., Zhao, X.: A uniformly accurate multiscale time integrator pseudospectral method for the Klein-Gordon equation in the nonrelativistic limit regime. SIAM J. Numer. Anal. **52**, 2488–2511 ((2014))
20. Caliari, M., Zuccher, S.: Reliability of the time splitting Fourier method for singular solutions in quantum fluids. Comput. Phys. Commun. **222**, 46–58 (2018)
21. Cao, W., Guo, B.: Fourier collocation method for solving nonlinear Klein-Gordon equation. J. Comput. Phys. **108**, 296–305 (1993)
22. Cheng, K., Feng, W., Gottlieb, S., et al.: A Fourier pseudospectral method for the "good" boussinesq equation with second-order temporal accuracy. Numer. Methods Partial Differ. Equ. **31**, 202–224 (2015)
23. Shen, J., Tang, T., Wang, L.: Spectral Methods: Algorithms, Analysis and Applications. Springer, Berlin (2011)
24. Dehghan, M., Mohebbi, A., Asgari, Z.: Fourth-order compact solution of the nonlinear Klein-Gordon equation. Numer. Algor. **52**, 523–540 (2009)
25. Dehghan, M., Shokri, A.: A numerical method for solution of the two-dimensional sine-Gordon equation using the radial basis functions. Math. Comput. Simul. **79**, 700–715 (2008)
26. Dehghan, M., Shokri, A.: Numerical solution of the nonlinear Klein-Gordon equation using radial basis functions. J. Comput. Appl. Math. **230**, 400–410 (2009)
27. Mirzaei, D., Dehghan, M.: Boundary element solution of the two-dimensional sine-Gordon equation using continuous linear elements. Eng. Anal. Bound. Elem. **33**, 12–24 (2009)
28. Moghaderi, H., Dehghan, M.: A multigrid compact finite difference method for solving the one-dimensional nonlinear sine-Gordon equation. Math. Method. Appl. Sci. **38**, 3901–3922 (2015)
29. Taleei, A., Dehghan, M.: A pseudo-spectral method that uses an overlapping multidomain technique for the numerical solution of sine-Gordon equation in one and two spatial dimensions. Math. Method. Appl. Sci. **37**, 1909–1923 (2014)

30. Hairer, E., Lubich, C., Wanner, G.: Geometric Numerical Integration: Structure-Preserving Algorithms for Ordinary Differential Equations, 2nd edn. Springer, Berlin (2006)
31. Hairer, E., Nørsett, S.P., Wanner, G.: Solving Ordinary Differential Equations I: Nonstiff Problems, 2nd edn. Springer, Berlin (1993)
32. Gauckler, L.: Error analysis of trigonometric integrators for semilinear wave equations. SIAM J. Numer. Anal. **53**, 1082–1106 (2015)
33. Liu, C., Iserles, A., Wu, X.: Symmetric and arbitrarily high-order Birkhoff-Hermite time integrators and their long-time behavior for solving nonlinear Klein-Gordon equations. J. Comput. Phys. 356, 1–30 (2018)
34. Hochbruck, M., Ostermann, A.: Exponential integrators. Acta Numer. **19**, 209–286 (2010)
35. Stein, E., Shakarchi, R.: Fourier Analysis: An Introduction, Princeton Lectures in Analysis, vol 1. Princeton University, Princeton, NJ (2003)
36. Zygmund, A.: Trigonometric Series, 3rd edn. Cambridge University, Cambridge (2002)
37. Blanes, S.: Explicit symplectic RKN methods for perturbed non-autonomous oscillators: splitting, extended and exponentially fitting methods. Comput. Phys. Commun. **21**, 10–18 (2015)
38. Liu, K., Wu, X.: High-order symplectic and symmetric composition methods for multifrequency and multi-dimensional oscillatory Hamiltonian systems. J. Comput. Math. **33**, 356–378 (2015)
39. Lele, S.K.: Compact finite difference schemes with spectral-like resolution. J. Comput. Phys. **103**, 16–42 (1992)
40. Jiménez, S., Vázquez, L.: Analysis of four numerical schemes for a nonlinear Klein-Gordon equation. Appl. Math. Comput. **35**, 61–94 (1990)
41. Wang, Y., Wang, B.: High-order multi-symplectic schemes for the nonlinear Klein-Gordon equation. Appl. Math. Comput. **166**, 608–632 (2005)
42. Bratsos, A.G.: The solution of the two-dimensional sine-Gordon equation using the method of lines. J. Comput. Appl. Math. 206, 251–277 (2007)
43. Sheng, Q., Khaliq, A.Q.M., Voss, D.A.: Numerical simulation of two-dimensional sine-Gordon solitons via a split cosine scheme. Math. Comput. Simul. **68**, 355–373 (2005)
44. Wang, B., Wu, X.: The formulation and analysis of energy-preserving schemes for solving high-dimensional nonlinear Klein-Gordon equations. IMA J. Numer. Anal. **39**, 2016–2044 (2019)
45. Mei, L., Wu, X.: Symplectic exponential Runge-Kutta methods for solving nonlinear Hamiltonian systems. J. Comput. Phys. **338**, 567–584 (2017)
46. Okunbor, D.I., Skeel, R.D.: Canonical Runge-Kutta-Nyström methods of orders five and six. J. Comp. Appl. Math. **51**, 375–382 (1994)
47. Sanz-Serna, J.M., Calvo, M.P.: Numerical Hamiltonian Problems. Chapman and Hall, London (1994)
48. Mei, L., Li, H., Wu, X., et al.: Semi-analytical exponential RKN integrators for efficiently solving high-dimensional nonlinear wave equations based on FFT techniques. Comput. Phys. Commun. **243**, 68–80 (2019)

Chapter 14
Long-Time Momentum and Actions Behaviour of Energy-Preserving Methods for Wave Equations

Wave equations have physically very important properties which should be respected by numerical schemes in order to predict correctly the solution over a long-time period. In this chapter, the long-time behaviour of momentum and actions for energy-preserving methods are analysed in detail for semilinear wave equations.

14.1 Introduction

The main theme of this chapter is the long-time behaviour of energy-preserving (EP) methods when applied to the following one-dimensional semilinear wave equation (see [1–3])

$$\partial_t^2 u - \partial_x^2 u + \rho u + g(u) = 0, \quad -\pi \leqslant x \leqslant \pi, \ t > 0, \tag{14.1}$$

where g is a nonlinear and smooth real function with $g(0) = g'(0) = 0$ and ρ is a positive number. Following the Refs. [1–3], we assume that the initial values $u(\cdot, 0)$ and $\partial_t u(\cdot, 0)$ for this equation are bounded by a small parameter ε, which provides small initial data in appropriate Sobolev norms. Here, we consider 2π-periodic boundary condition $u(x, t) = u(x + 2\pi, t)$ for (14.1).

As is known, several important quantities are conserved by the solution of (14.1). Firstly, the total energy

$$H(u, v) = \frac{1}{2\pi} \int_{-\pi}^{\pi} \left(\frac{1}{2}(v^2 + (\partial_x u)^2 + \rho u^2) + U(u) \right) dx$$

© The Author(s), under exclusive license to Springer Nature Singapore Pte Ltd. 2021
X. Wu, B. Wang, *Geometric Integrators for Differential Equations with Highly Oscillatory Solutions*, https://doi.org/10.1007/978-981-16-0147-7_14

is exactly preserved along the solution, where $v = \partial_t u$ and $U(u)$ is the potential such that $U'(u) = g(u)$. Secondly, the solution of (14.1) also conserves the momentum

$$K(u, v) = \frac{1}{2\pi} \int_{-\pi}^{\pi} \partial_x u(x) v(x) \mathrm{d}x.$$

Thirdly, the harmonic actions

$$I_j(u, v) = \frac{\omega_j}{2} |u_j|^2 + \frac{1}{2\omega_j} |v_j|^2, \; j \in \mathbb{Z}$$

are conserved for the linear wave equation, i.e., $g(u) \equiv 0$, where $\omega_j = \sqrt{\rho + j^2}$ for $j \in \mathbb{Z}$. In the nonlinear case, it has been proved in [2, 4] that, for smooth and small initial data and for almost all values of $\rho > 0$, the actions $I_j(u, v)$ remain constant up to small deviations over a long-time period.

In the past decades it has become increasingly important to design numerical integrators for wave equations aiming at respecting qualitative properties of the solution (see, e.g. [5–14]). Among others, long-time conservation properties of numerical methods when applied to wave equations have been well studied [1–3, 15, 16]. All these analyses are achieved by the technique of modulated Fourier expansions, which was developed by Hairer and Lubich in [17] and has been frequently used in the long-term analysis (see, e.g. [18–22]). On the other hand, as an important kind of method, energy-preserving (EP) methods have also been the subject of many investigations for wave equations. EP methods can exactly preserve the energy of the system under consideration. Concerning some examples of this topic, we refer the readers to [23–31]. Unfortunately, it seems that the study of the long-time behaviour of EP methods in other structure-preserving aspects is quite inadequate for wave equations in the literature, e.g. the numerical conservation of momentum and actions. This chapter focuses on this point.

14.2   Full Discretisation

This section presents a full discretisation for solving the semilinear wave equation (14.1). We begin with a spectral semidiscretisation in space introduced in [1, 3], and then use EP methods in time.

14.2.1   Spectral Semidiscretisation in Space

We here choose equidistant collocation points $x_k = k\pi/M$, $k = -M, -M + 1, \cdots, M - 1$, for the pseudospectral semidiscretisation in space and consider a pair of real-valued trigonometric polynomials as an approximation for the solution of (14.1)

$$u^M(x,t) = \sideset{}{'}\sum_{|j|\leqslant M} q_j(t)e^{ijx}, \quad v^M(x,t) = \sideset{}{'}\sum_{|j|\leqslant M} p_j(t)e^{ijx}, \quad i = \sqrt{-1},$$

$$(14.2)$$

where $p_j(t) = \dfrac{d}{dt}q_j(t)$ and the prime indicates that the first and last terms in the summation are taken with the factor $1/2$. We collect all the q_j in a $2M$-periodic coefficient vector $q(t) = (q_j(t))$, which is a solution of the $2M$-dimensional system of oscillatory ODEs

$$\frac{d^2q}{dt^2} + \Omega^2 q = f(q), \tag{14.3}$$

where $f(q) = -\mathscr{F}_{2M} g(\mathscr{F}_{2M}^{-1} q)$, Ω is diagonal with entries ω_j, and $\mathscr{F}_{2M}$ denotes the discrete Fourier transform $(\mathscr{F}_{2M} w)_j = \dfrac{1}{2M} \sum_{k=-M}^{M-1} w_k e^{-ijx_k}$ for $|j| \leqslant M$. It is noted that the system (14.3) is a finite-dimensional complex Hamiltonian system with the energy

$$H_M(p,q) = \frac{1}{2} \sideset{}{'}\sum_{|j|\leqslant M} \left(|p_j|^2 + \omega_j^2 |q_j|^2\right) + V(q), \tag{14.4}$$

where $V(q) = \dfrac{1}{2M} \sum_{k=-M}^{M-1} U((\mathscr{F}_{2M}^{-1} q)_k)$. Accordingly, the actions (for $|j| \leqslant M$) and the momentum of (14.3) are respectively given by

$$I_j(p,q) = \frac{\omega_j}{2}|q_j|^2 + \frac{1}{2\omega_j}|p_j|^2, \quad K(p,q) = -\sideset{}{''}\sum_{|j|\leqslant M} ij q_{-j} p_j, \quad i = \sqrt{-1},$$

where the double prime indicates that the first and last terms in the summation are taken with the factor $1/4$. We are interested only in real approximation (14.2) throughout this chapter, and hence it holds that $q_{-j} = \bar{q}_j$, $p_{-j} = \bar{p}_j$ and $I_{-j} = I_j$.

It is important to note that the energy (14.4) is exactly preserved along the solution of (14.3). For the momentum and actions in the semidiscretisation, the following results have been proved in [3].

Theorem 14.1 (See [3]) *Under the non-resonance condition* (14.10) *and the Assumption* (14.7) *which are stated in Sect. 14.3.1, it holds that*

$$\sum_{l=0}^{M} \omega_l^{2s+1} \frac{|I_l(p(t), q(t)) - I_l(p(0), q(0))|}{\varepsilon^2} \leqslant C\varepsilon,$$

$$\frac{|K(p(t), q(t)) - K(p(0), q(0))|}{\varepsilon^2} \leqslant Ct\varepsilon M^{-s+1},$$

where $0 \leqslant t \leqslant \varepsilon^{-N+1}$ *and the constant* C *is independent of* ε, M, h *and* t.

14.2.2   EP Methods in Time

It is known that among typical EP integrators is the average vector field (AVF) method (see [32]). Unfortunately, however, it has been pointed out in Chap. 1 that the AVF method cannot efficiently solve the highly oscillatory system (14.3) (see also [33, 34]) since the AVF method is not oscillation preserving. Moreover, the integral appearing in the AVF formula is dependent on the frequency matrix Ω. This fact leads to the following definition.

Definition 14.1 (See [33, 34]) For efficiently solving the oscillatory system (14.3), the *adapted average vector field* (AAVF) method has the form

$$\begin{cases} q_{n+1} = \phi_0(V)q_n + h\phi_1(V)p_n + h^2\phi_2(V)\displaystyle\int_0^1 f((1-\sigma)q_n + \sigma q_{n+1})\mathrm{d}\sigma, \\[2mm] p_{n+1} = -h\Omega^2\phi_1(V)q_n + \phi_0(V)p_n + h\phi_1(V)\displaystyle\int_0^1 f((1-\sigma)q_n + \sigma q_{n+1})\mathrm{d}\sigma, \end{cases}$$

$$(14.5)$$

where h is the stepsize, and

$$\phi_l(V) := \sum_{k=0}^{\infty} \frac{(-1)^k V^k}{(2k+l)!}, \quad l = 0, 1, 2 \tag{14.6}$$

are matrix-valued functions of $V = h^2\Omega^2$.

According to (14.6), it is clear that

$$\phi_0(V) = \cos(h\Omega), \quad \phi_1(V) = \sin(h\Omega)(h\Omega)^{-1}, \quad \phi_2(V) = (I - \cos(h\Omega))(h\Omega)^{-2}.$$

It is interesting to note that as $V \to 0$ the method (14.5) reduces to the well-known AVF method. The following properties of the AAVF method have been shown in [33, 34].

Theorem 14.2 (See [33, 34]) *The AAVF method is symmetric and exactly preserves the energy* (14.4), *which means that*

$$H_M(p_{n+1}, q_{n+1}) = H_M(p_n, q_n) \quad for \quad n = 0, 1, \cdots .$$

Theorem 14.2 ensures that the energy-preserving AAVF method does not exclude symmetry structure, and, as is known, preserving the energy and symmetry of the system simultaneously at the discrete level is important for geometric integrators.

14.3 Main Result and Numerical Experiment

In what follows, we shall use the following notations (see [1]). We denote

$$|k| = (|k_l|)_{l=0}^M, \quad \|k\| = \sum_{l=0}^M |k_l|, \quad k \cdot \omega = \sum_{l=0}^M k_l \omega_l, \quad \omega^{\sigma|k|} = \prod_{l=0}^M \omega_l^{\sigma|k_l|}.$$

for sequences of integers $k = (k_l)_{l=0}^M$, $\omega = (\omega_l)_{l=0}^M$ and a real number σ. We also denote by $\langle j \rangle$ the unit coordinate vector $(0, \cdots, 0, 1, 0, \cdots, 0)^\mathsf{T}$ with 1 in the j-th entry 1 and 0 elsewhere. For $s \in \mathbb{R}^+$, the space of $2M$-periodic sequences $q = (q_j)$ endowed with the weighted norm $\|q\|_s = \left(\sum_{|j| \leqslant M}'' \omega_j^{2s} |q_j|^2 \right)^{1/2}$ is denoted by H^s.

Furthermore, we set

$$[[k]] = \begin{cases} (\|k\| + 1)/2, & k \neq \mathbf{0}, \\ 3/2, & k = \mathbf{0}. \end{cases}$$

14.3.1 Main Result

In this subsection we first present the main result of this chapter, which will be illustrated by numerical experiments. The following assumptions (see [1]) are needed for the main result.

Assumption 14.1 It is assumed that the initial values of (14.3) are bounded by

$$\left(\|q(0)\|_{s+1}^2 + \|p(0)\|_s^2 \right)^{1/2} \leqslant \varepsilon \tag{14.7}$$

with a small parameter $\varepsilon > 0$.

Assumption 14.2 The following non-resonance condition holds for a given step-size h:

$$\left| \sin\left(\frac{h}{2}(\omega_j - k \cdot \boldsymbol{\omega})\right) \cdot \sin\left(\frac{h}{2}(\omega_j + k \cdot \boldsymbol{\omega})\right) \right| \geqslant \varepsilon^{1/2} h^2 (\omega_j + |k \cdot \boldsymbol{\omega}|). \tag{14.8}$$

If this condition is not true, we define a set of near-resonant indices

$$\mathscr{R}_{\varepsilon,h} = \{(j,k) : |j| \leqslant M, \ \|k\| \leqslant 2N, \ k \neq \pm\langle j \rangle, \ \text{not satisfying (14.8)}\}, \tag{14.9}$$

where $N \geqslant 1$ is the truncation number of the expansion (14.15) which will be presented in the next section. Moreover, we assume that there exist $\sigma > 0$ and a constant C_0 such that

$$\sup_{(j,k)\in\mathscr{R}_{\varepsilon,h}} \frac{\omega_j^\sigma}{\boldsymbol{\omega}^{\sigma|k|}} \varepsilon^{\|k\|/2} \leqslant C_0 \varepsilon^N, \tag{14.10}$$

for the set $\mathscr{R}_{\varepsilon,h}$.

Assumption 14.3 Assume that the following numerical non-resonance condition

$$|\sin(h\omega_j)| \geqslant h\varepsilon^{1/2} \quad \text{for } |j| \leqslant M, \tag{14.11}$$

is satisfied.

Assumption 14.4 Suppose that, for a positive constant $c > 0$, another non-resonance condition

$$\left| \sin\left(\frac{h}{2}(\omega_j - k \cdot \boldsymbol{\omega})\right) \cdot \sin\left(\frac{h}{2}(\omega_j + k \cdot \boldsymbol{\omega})\right) \right| \geqslant ch^2 |2\phi_2(h^2\omega_j^2)| \tag{14.12}$$

for (j,k) of the form $j = j_1 + j_2$ and $k = \pm\langle j_1 \rangle \pm \langle j_2 \rangle$,

is also fulfilled, which leads to improved conservation estimates.

The following theorem represents the main result of this chapter.

Theorem 14.3 *We define the following modified momentum and actions, respectively*

$$\hat{I}_j(p,q) = \frac{\cos\left(\frac{1}{2}h\omega_j\right)}{\operatorname{sinc}\left(\frac{1}{2}h\omega_j\right)} I_j(p,q), \quad \hat{K}(p,q) = -\sum_{|j|\leqslant M}^{\prime\prime} ij \frac{\cos\left(\frac{1}{2}h\omega_j\right)}{\operatorname{sinc}\left(\frac{1}{2}h\omega_j\right)} q_{-j} p_j,$$

and choose the stepsize h such that

$$\left| \frac{\cos\left(\frac{1}{2}h\omega_j\right)}{\mathrm{sinc}\left(\frac{1}{2}h\omega_j\right)} \right| \leqslant C_1 \qquad for \quad |j| \leqslant M. \qquad (14.13)$$

Then under the conditions of Assumptions 14.1–14.4 with $s \geqslant \sigma + 1$, for the AAVF method (14.5) and $0 \leqslant t = nh \leqslant \varepsilon^{-N+1}$, the following near-conservation estimates of the modified momentum and actions

$$\sum_{l=0}^{M} \omega_l^{2s+1} \frac{|\hat{I}_l(p_n, q_n) - \hat{I}_l(p_0, q_0)|}{\varepsilon^2} \leqslant C\varepsilon,$$

$$\frac{|\hat{K}(p_n, q_n) - \hat{K}(p_0, q_0)|}{\varepsilon^2} \leqslant C(\varepsilon + M^{-s} + \varepsilon t M^{-s+1})$$

hold with a constant C, depending on s, N, C_0 and C_1, but not on ε, M, h and time t. If (14.12) is not satisfied, then the bound $C\varepsilon$ is weakened to $C\varepsilon^{1/2}$.

The proof of this theorem will be shown in detail in Sect. 14.4 based on the technique of multi-frequency modulated Fourier expansions. It is remarked that the above result for the AAVF method with the integral is also true for the AAVF method with a suitable quadrature rule instead of the integral, and this point will be explained briefly in Sect. 14.5.

An interesting study of the long-time behaviour of a symmetric and symplectic trigonometric integrator for solving wave equations was made by Cohen et al. in [1], and it was shown that this integrator has a near-conservation of energy, momentum and actions in numerical discretisations. However, it is noted that the method studied in [1] cannot preserve the energy (14.4) exactly. Fortunately, it follows from Theorems 14.2 and 14.3 that the AAVF method not only preserves the energy (14.4) exactly but also has a near-conservation of modified momentum and actions over long terms.

Remark 14.1   Theorem 14.3 claims that the AAVF method has a near-conservation of a modified momentum and modified actions over long terms. We here remark that we have tried to prove long-time conservation for natural discretisations. However, after the whole procedure of the proof using modulated Fourier expansion, it turns out that artificial coefficients $\cos(h\omega_j)/\sin(h\omega_j/2)$ form part of each term of the summation of the natural discretisation. Therefore, we only obtain the conservation of the modified momentum and modified actions. Similar results have also been shown in some other publications. For example, the authors in [19] proved long-time conservation of modified energy and modified action for the Störmer-Verlet method and in [35], conservation of the modified energy and modified magnetic moment were shown for a variational integrator. In both publications, long-time conservation of natural invariants was not given. We also note that although the result cannot be obtained for the momentum K and actions I_j, K and I_j are no longer exactly conserved quantities in the semidiscretisation, which can be seen from Theorem 14.1. Moreover, it will be shown in the next subsection that, in comparison with the near-conservation of K and I_j, the modified momentum and modified actions are preserved rather well by the AAVF method. This supports the result of Theorem 14.3.

14.3.2   Numerical Experiments

In what follows, we implement two numerical experiments to show the behaviour of the AAVF method. Since the AAVF method is implicit, iteration solutions are needed. Here, we use fixed-point iteration in practical computation. We set 10^{-16} as the error tolerance and 100 as the maximum number of iterates.

Problem 14.1   Consider the semilinear wave equation (14.1), where $\rho = 0.5$ and $g(u) = -u^2$. The initial conditions are given by (see [1])

$$u(x, 0) = 0.1 \left(\frac{x}{\pi} - 1 \right)^3 \left(\frac{x}{\pi} + 1 \right)^2, \quad \partial_t u(x, 0) = 0.01 \frac{x}{\pi} \left(\frac{x}{\pi} - 1 \right) \left(\frac{x}{\pi} + 1 \right)^2,$$

for $-\pi \leqslant x \leqslant \pi$. We carry out the spatial discretisation[1] with the dimension $2M = 2^7$ and apply the midpoint rule to the integral[2] appearing in the AAVF formula (14.5), which yields

$$\begin{cases} q_{n+1} = \phi_0(V)q_n + h\phi_1(V)p_n + h^2\phi_2(V)f((q_n + q_{n+1})/2), \\ p_{n+1} = -h\Omega^2\phi_1(V)q_n + \phi_0(V)p_n + h\phi_1(V)f((q_n + q_{n+1})/2). \end{cases} \quad (14.14)$$

It is easily verified that the assumption (14.7) holds for $s = 2$ with $\varepsilon \approx 0.1$. We solve this problem with the stepsize $h = 0.05$ on [0, 10000], and the relative errors of momentum/modified momentum and actions/modified actions against t are shown

---

[1]It is noted that for wave equations, the spatial discretisation with the dimension $2M = 2^7$ has been considered in [1, 8, 36] and it worked well in those publications. That is the reason why we use the spatial discretisation with $2M = 2^7$ here.

[2]From the analysis of Sect. 14.5, it follows that the main result is still true for the AAVF method with some quadrature rule.

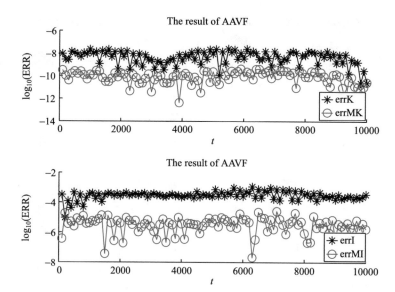

Fig. 14.1 The logarithm of the errors against t

in Fig. 14.1. We here adopt the following notations:

$$\text{errK} = \frac{|K(p_n, q_n) - K(p_0, q_0)|}{|K(p_0, q_0)|}, \qquad \text{errMK} = \frac{|\hat{K}(p_n, q_n) - \hat{K}(p_0, q_0)|}{|\hat{K}(p_0, q_0)|},$$

$$\text{errI} = \frac{\sum\limits_{l=0}^{M} \omega_l^5 |I_l(p_n, q_n) - I_l(p_0, q_0)|}{\sum\limits_{l=0}^{M} \omega_l^5 |I_l(p_0, q_0)|}, \qquad \text{errMI} = \frac{\sum\limits_{l=0}^{M} \omega_l^5 |\hat{I}_l(p_n, q_n) - \hat{I}_l(p_0, q_0)|}{\sum\limits_{l=0}^{M} \omega_l^5 |\hat{I}_l(p_0, q_0)|}.$$

It follows from Fig. 14.1 that the modified momentum and modified actions are better conserved than the momentum and actions, which supports the results given in Theorem 14.3.

We next show the efficiency of the AAVF method in comparison with some other methods. To this end, we consider the classical Störmer-Verlet formula (denoted by SV), Gautschi's method of order two (denoted by GM1s2) given in [17] and the two-stage diagonally implicit symplectic Runge–Kutta method of order three (denoted by RK2s3) presented in [37]. With regard to Gautschi's method, its coefficient functions are chosen as $\phi(\xi) = 1$ and $\psi(\xi) = (\sin(\xi)/\xi)^2$. The long-time behaviour of this method has been shown in [17], and the non-resonance conditions given in [1] are satisfied for this method. We first solve the system on [0, 10] with $h = 0.2/2^j$ for $j = 2, 3, 4, 5$, and the errors GE $= \left(\|q_n - q\|_3^2 + \|p_n - p\|_2^2 \right)^{1/2}$ measured at the final time against the CPU time are presented in Fig. 14.2a. We then integrate the

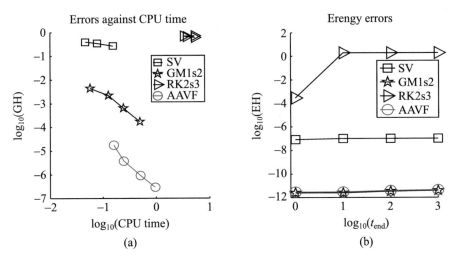

Fig. 14.2 (**a**) The logarithm of the errors against the logarithm of CPU time. (**b**) The logarithm of the energy errors against the logarithm of time

problem on $[0, t_{\text{end}}]$ with $h = 0.01$ and $t_{\text{end}} = 10^j$ for $j = 0, 1, 2, 3$. The errors of the semidiscrete energy conservation are presented in Fig. 14.2b. It can be observed from Fig. 14.2 that the AAVF method shows good overall efficiency.

Problem 14.2 Consider the semilinear Klein–Gordon equation

$$\begin{cases} \partial_t^2 u - a^2 \partial_x^2 u = bu^3 - au, & -\pi \leqslant x \leqslant \pi, \ u(-\pi, t) = u(\pi, t), \ 0 \leqslant t \leqslant T, \\ u(x, 0) = \sqrt{\dfrac{2a}{b}} \text{sech}(\lambda x), \ u_t(x, 0) = c\lambda\sqrt{\dfrac{2a}{b}} \text{sech}(\lambda x) \tanh(\lambda x) \end{cases}$$

where $\lambda = \sqrt{\dfrac{a}{a^2 - c^2}}$ and $a, b, a^2 - c^2 > 0$. The exact solution is

$$u(x, t) = \sqrt{\dfrac{2a}{b}} \text{sech}(\lambda(x - ct)).$$

The choice of parameters $a = 1, b = 0.01, c = 0.25$ makes this problem fit into the form (14.1).

Likewise, the spatial variable is discretised with the dimension $2M = 2^7$, and it can be verified that the assumption (14.7) is true for $s = 1$ with $\varepsilon \approx 0.015$. This problem is solved on $[0, 10000]$ with $h = 0.05$, and the relative errors of momentum/modified momentum and actions/modified actions against t are shown in Fig. 14.3.

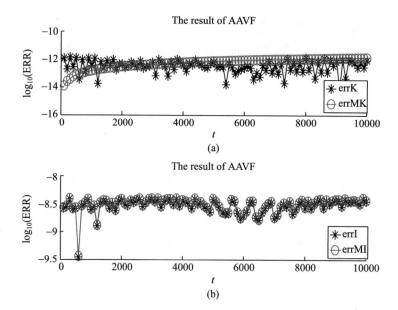

Fig. 14.3 The logarithm of the errors against t

We then apply the AAVF method as well as the methods SV, GM1s2, and RK2s3 to the problem on $[0, 100]$ with $h = 0.2/2^j$ for $j = 0, 1, 2, 3$. The errors measured at the final time against the CPU time are given in Fig. 14.4a. Finally we solve the problem on $[0, t_{end}]$ with $h = 0.01$ and $t_{end} = 10^j$ for $j = 0, 1, 2, 3$, and present the errors of the semidiscrete energy conservation in Fig. 14.4b. Here, it is remarked that for this problem, the conservation of modified momentum and modified actions seems to be similar to those for the natural discretisations of momentum and actions. The reason is that, for some problems, it can be checked that the modified momentum and modified actions are very close to the natural ones of the considered system. Apart from this, according to Fig. 14.4, it is clear that Gautschi's method behaves at least as well as AAVF since both methods behave similarly with respect to the conservation of invariants, but Gautschi's method is explicit while AAVF is implicit although both methods are of order two.

14.4 The Proof of the Main Result

This section concerns the proof of Theorem 14.3. We first present the outline of the proof and then show the key points one by one since the proof is a bit long.

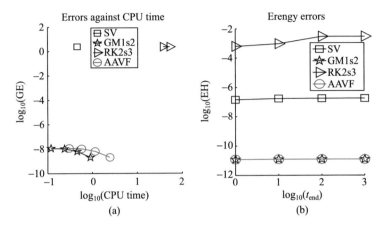

Fig. 14.4 (**a**) The logarithm of the errors against the logarithm of CPU time. (**b**) The logarithm of the energy errors against the logarithm of time

14.4.1   The Outline of the Proof

The proof relies on a careful study of a modulated Fourier expansion of the AAVF method (14.5). It is assumed that the conditions of Theorem 14.3 are true. For the numerical solution (p_n, q_n), determined by (14.5), we will consider the following truncated multi-frequency modulated Fourier expansion (with N from (14.9))

$$\tilde{q}(t) = \sum_{\|k\|\leqslant 2N} e^{i(k\cdot\omega)t} \zeta^k(\varepsilon t), \quad \tilde{p}(t) = \sum_{\|k\|\leqslant 2N} e^{i(k\cdot\omega)t} \eta^k(\varepsilon t), \tag{14.15}$$

where $t = nh$ and $\zeta_{-j}^{-k} = \overline{\zeta_j^k}$, $\eta_{-j}^{-k} = \overline{\eta_j^k}$. For this modulated Fourier expansion, the following key points will be addressed one by one in the rest of this section.

- Formal modulation equations for the modulation functions are derived in Sect. 14.4.2.
- We consider an iterative construction of the functions using reverse Picard iteration in Sect. 14.4.3.
- We then work with a more convenient rescaling and study the estimation of nonlinear terms in Sect. 14.4.4.
- Abstract reformulation of the iteration is presented in Sect. 14.4.5.
- We control the size of the numerical solution by studying the bounds of modulation functions in Sect. 14.4.6.
- The bound of the defect is estimated in Sect. 14.4.7.
- We study the difference between the numerical solution and its modulated Fourier expansion in Sect. 14.4.8.
- We show two invariants of the modulation system and establish their relationship with the modified momentum and modified actions in Sect. 14.4.9.

- Finally, the previous results that are valid only on a short time interval are extended to a long-time interval in Sect. 14.4.10.

It is noted that the above procedure is a standard approach to the study of the long-time behaviour of numerical methods of Hamiltonian partial differential equations by using modulated Fourier expansions (see, e.g. [1–3, 15, 16]). Although the proof presented here closely follows these previous publications, there are novel modifications adapted to the AAVF method in each part. The differences in the analysis arise due to the implicitness of the AAVF method and the integral appearing in the method.

Throughout the proof, denote by C a generic constant which is independent of ε, M, h and $t = nh$. The following lemma, presented in [2], will be needed in the analysis of this chapter.

Lemma 14.1 (See [2]) *For $s > 1/2$, one has* $\sum\limits_{\|k\| \leqslant K} \omega^{-2s|k|} \leqslant C_{K,s} \leqslant \infty$. *For $s > 1/2$ and $m \geqslant 2$, it is true that*

$$\sup_{\|k\| \leqslant K} \sum_{k^1 + \cdots + k^m = k} \frac{\omega^{-2s(|k^1| + \cdots + |k^m|)}}{\omega^{-2s|k|}} \leqslant C_{m,K,s} < \infty,$$

where the sum is taken over $(k^1, \cdots, k^m)$ satisfying $\|k^i\| \leqslant K$. For $s \geqslant 1$, it is further true that $\sup_{\|k\| \leqslant K} \dfrac{\sum\limits_{l \geqslant 0} |k_l| \, \omega_l^{2s+1}}{\omega^{2s|k|}(1 + |k \cdot \omega|)} \leqslant C_{K,s} < \infty$.

14.4.2 Modulation Equations

We commence from the formulation of the modulation equations for the modulated functions. To this end, we first define five operators by

$$L_1^k := e^{i(k \cdot \omega)h} e^{\varepsilon hD} - 2\cos(h\Omega) + e^{-i(k \cdot \omega)h} e^{-\varepsilon hD},$$

$$L_2^k := e^{\frac{1}{2}i(k \cdot \omega)h} e^{\frac{1}{2}\varepsilon hD} + e^{-\frac{1}{2}i(k \cdot \omega)h} e^{-\frac{1}{2}\varepsilon hD},$$

$$L_3^k := (e^{i(k \cdot \omega)h} e^{\varepsilon hD} - 1)(e^{i(k \cdot \omega)h} e^{\varepsilon hD} + 1)^{-1},$$

$$L_4^k(\sigma) := (1 - \sigma)e^{-\frac{1}{2}i(k \cdot \omega)h} e^{-\frac{h}{2}\varepsilon D} + \sigma e^{\frac{1}{2}i(k \cdot \omega)h} e^{\frac{h}{2}\varepsilon D},$$

$$L^k := (L_2^k)^{-1} L_1^k,$$

where D is the differential operator (see [20]). Then the following results for these operators are essential in the analysis.

Proposition 14.1 *The operator L^k can be expressed in Taylor expansions as follows:*

$$L^{\pm\langle j\rangle}(hD)\alpha_j^{\pm\langle j\rangle}(\varepsilon t) = \pm 2i\varepsilon h s_{\langle j\rangle}\dot{\alpha}_j^{\pm\langle j\rangle}(\varepsilon t) + \frac{1}{2}\varepsilon^2 h^2 \sec\left(\frac{1}{2}h\omega_j\right)\ddot{\alpha}_j^{\pm\langle j\rangle}(\varepsilon t) + \cdots,$$

$$L^k(hD)\alpha_j^k(\varepsilon t) = 2\frac{s_{\langle j\rangle}+ks_{\langle j\rangle-k}}{c_k}\alpha_j^k(\varepsilon t) + i\varepsilon h\frac{s_k(1+c_{\langle j\rangle}+kc_{\langle j\rangle-k})}{c_k^2}\dot{\alpha}_j^k(\varepsilon t) + \cdots,$$

$$\tag{14.16}$$

for $|j| > 0$ and $k \neq \pm\langle j\rangle$, where $s_k = \sin\left(\frac{h}{2}(k\cdot\boldsymbol{\omega})\right)$ and $c_k = \cos\left(\frac{h}{2}(k\cdot\boldsymbol{\omega})\right)$. The Taylor expansions of L_3^k are of the forms

$$L_3^k\alpha_j^k(\varepsilon t) = i\tan\left(\frac{1}{2}h(k\cdot\boldsymbol{\omega})\right)\alpha_j^k(\varepsilon t) + \frac{h\varepsilon}{1+c_{2k}}\dot{\alpha}_j^k(\varepsilon t) + \cdots,$$

for $|j| > 0$ and $\|k\| \leqslant 2N$. Moreover, for the operator $L_4^k(\sigma)$ with $\|k\| \leqslant 2N$, we have

$$L_4^k\left(\frac{1}{2}\right) = \cos\left(\frac{h(k\cdot\boldsymbol{\omega})}{2}\right) + \frac{1}{2}\sin\left(\frac{h(k\cdot\boldsymbol{\omega})}{2}\right)(ih\varepsilon D) + \cdots.$$

Theorem 14.4 (Modulation Equations) *The formal modulation equations of the modulated functions ζ^k are given by*

$$L^{\pm\langle j\rangle}\zeta_j^{\pm\langle j\rangle} = -h^2\phi_2(h^2\omega_j^2)\sum_{m\geqslant 2}\frac{g^{(m)}(0)}{m!}\sum_{k^1+\cdots+k^m=\pm\langle j\rangle}\sideset{}{'}\sum_{j_1+\cdots+j_m\equiv j \bmod 2M}$$

$$\cdot\int_0^1\left[(\xi_{j_1}^{k^1}\cdot\cdots\cdot\xi_{j_m}^{k^m})(t\varepsilon,\sigma)\right]d\sigma,$$

$$L^k\zeta_j^k = -h^2\phi_2(h^2\omega_j^2)\sum_{m\geqslant 2}\frac{g^{(m)}(0)}{m!}\sum_{k^1+\cdots+k^m=k}\sideset{}{'}\sum_{j_1+\cdots+j_m\equiv j \bmod 2M}$$

$$\cdot\int_0^1\left[(\xi_{j_1}^{k^1}\cdot\cdots\cdot\xi_{j_m}^{k^m})(t\varepsilon,\sigma)\right]d\sigma, \qquad \text{for } k \neq \pm\langle j\rangle,$$

$$\tag{14.17}$$

where L^k is defined by (14.16) and

$$\xi^k(\varepsilon t,\sigma) = L_4^k(\sigma)\zeta^k(\varepsilon t).$$

The modulation equations for η^k are determined by

$$\eta_j^{\pm\langle j\rangle} = \pm i\omega_j \zeta_j^{\pm\langle j\rangle} + \mathcal{O}(h\varepsilon), \quad \eta_j^k = \frac{\tan\left(\dfrac{1}{2}h(k\cdot\boldsymbol{\omega})\right)}{\tan\left(\dfrac{1}{2}h\omega_j\right)} i\omega_j \zeta_j^k + \mathcal{O}(h\varepsilon)$$

(14.18)

for $k \neq \pm\langle j\rangle$.

Proof The proof will be divided into two parts.

The first part is the proof of (14.17).

Using the symmetry of the AAVF method and the following property

$$\int_0^1 f((1-\sigma)q_n + \sigma q_{n-1})d\sigma = \int_0^1 f((1-\sigma)q_{n-1} + \sigma q_n)d\sigma,$$

leads to

$$q_{n+1} - 2\cos(h\Omega)q_n + q_{n-1}$$
$$= h^2\phi_2(V)\left[\int_0^1 f((1-\sigma)q_n + \sigma q_{n+1})d\sigma + \int_0^1 f((1-\sigma)q_{n-1} + \sigma q_n)d\sigma\right].$$

(14.19)

We then seek for a modulated Fourier expansion of the form

$$\tilde{q}_h\left(t + \frac{h}{2}, \sigma\right) = \sum_{\|k\|\leqslant 2N} e^{i(k\cdot\boldsymbol{\omega})\left(t+\frac{h}{2}\right)} \xi^k\left(\varepsilon\left(t + \frac{h}{2}\right), \sigma\right)$$

for the term $(1-\sigma)q_n + \sigma q_{n+1}$ appearing in (14.19). This implies that

$$\xi^k\left(\varepsilon\left(t + \frac{h}{2}\right), \sigma\right) = \left((1-\sigma)e^{-\frac{1}{2}i(k\cdot\boldsymbol{\omega})h}e^{-\frac{h}{2}\varepsilon D} + \sigma e^{\frac{1}{2}i(k\cdot\boldsymbol{\omega})h}e^{\frac{h}{2}\varepsilon D}\right)\xi^k\left(\varepsilon\left(t + \frac{h}{2}\right)\right)$$

$$= L_4^k(\sigma)\xi^k\left(\varepsilon\left(t + \frac{h}{2}\right)\right).$$

(14.20)

Likewise, for $(1-\sigma)q_{n-1} + \sigma q_n$, we can obtain the following modulated Fourier expansion

$$\tilde{q}_h\left(t - \frac{h}{2}, \sigma\right) = \sum_{\|k\|\leqslant 2N} e^{i(k\cdot\boldsymbol{\omega})\left(t-\frac{h}{2}\right)} \xi^k\left(\varepsilon\left(t - \frac{h}{2}\right), \sigma\right)$$

where

$$\xi^k \left(\varepsilon \left(t - \frac{h}{2} \right), \sigma \right) = L_4^k(\sigma) \zeta^k \left(\varepsilon \left(t - \frac{h}{2} \right) \right). \tag{14.21}$$

Inserting the modulated Fourier expansions (14.15), (14.20), and (14.21) into (14.19) yields

$$\tilde{q}(t+h) - 2\cos(h\Omega)\tilde{q}(t) + \tilde{q}(t-h)$$

$$= h^2 \phi_2(V) \left[\int_0^1 f\left(\tilde{q}_h \left(t + \frac{h}{2}, \sigma \right) \right) d\sigma + \int_0^1 f\left(\tilde{q}_h \left(t - \frac{h}{2}, \sigma \right) \right) d\sigma \right],$$

which can be formulated as

$$(e^{\frac{1}{2}hD} + e^{-\frac{1}{2}hD})^{-1}(e^{hD} - 2\cos(h\Omega) + e^{-hD})\tilde{q}(t) = h^2 \phi_2(V) \int_0^1 f(\tilde{q}_h(t, \sigma)) d\sigma. \tag{14.22}$$

We next rewrite this equation by using the approach introduced in [3]. We begin with the following notation. For a 2π-periodic function $w(x)$, denote by $(\mathcal{Q}w)(x)$ the trigonometric interpolation polynomial to $w(x)$ at the points x_k. If $w(x)$ is of the form $w(x) = \sum\limits_{j=-\infty}^{\infty} w_j e^{ijx}$, then we have that

$$(\mathcal{Q}w)(x) = \sum_{|j| \leqslant M}'' \left(\sum_{l=-\infty}^{\infty} w_{j+2Ml} \right) e^{ijx},$$

where $x_k = \dfrac{k\pi}{M}$. For a $2M$-periodic coefficient sequence $q = (q_j)$, $(\mathcal{P}q)(x)$ is referred to the trigonometric polynomial with coefficients q_j, i.e.,

$$(\mathcal{P}q)(x) = \sum_{|j| \leqslant M}' q_j e^{ijx}.$$

With these new denotations, (14.22) is identical to

$$(e^{\frac{1}{2}hD} + e^{-\frac{1}{2}hD})^{-1}(e^{hD} - 2\cos(h\Omega) + e^{-hD})\mathcal{P}\tilde{q}(t) = h^2 \phi_2(V) \int_0^1 \mathcal{Q}g(\mathcal{P}\tilde{q}_h(t, \sigma)) d\sigma. \tag{14.23}$$

The Taylor expansion of the non-linearity $\mathscr{Q}g$ at 0 is given by[3]

$$\mathscr{Q}g(\mathscr{P}\tilde{q}_h(t,\sigma)) = \sum_{m\geqslant 2} \frac{g^{(m)}(0)}{m!}\mathscr{Q}(\mathscr{P}\tilde{q}_h(t,\sigma))^m$$

$$= \sum_{m\geqslant 2} \frac{g^{(m)}(0)}{m!}\left(\sideset{}{''}\sum_{|j_1|\leqslant M}\sum_{l=-\infty}^{\infty}\sideset{}{'}\sum_{\|k^1\|\leqslant 2N} e^{i(k^1\cdot\omega)t}\xi^{k^1}_{j_1+2Ml}(\tau,\sigma)e^{ij_1x}\right)$$

$$\cdots\left(\sideset{}{''}\sum_{|j_m|\leqslant M}\sum_{l=-\infty}^{\infty}\sideset{}{'}\sum_{\|k^m\|\leqslant 2N} e^{i(k^m\cdot\omega)t}\xi^{k^m}_{j_m+2Ml}(\tau,\sigma)e^{ij_mx}\right)$$

$$= \sum_{m\geqslant 2} \frac{g^{(m)}(0)}{m!}\sideset{}{''}\sum_{|j|\leqslant M}\sum_{j_1+\cdots+j_m\equiv j \bmod 2M}\sum_{\|k^1\|\leqslant 2N,\cdots,\|k^m\|\leqslant 2N}(\xi^{k^1}_{j_1}\cdots\xi^{k^m}_{j_m})(\tau,\sigma)$$

$$e^{i((k^1+\cdots+k^m)\cdot\omega)t}e^{ijx},$$

where $\tau = h\varepsilon$ and the prime on the sum indicates that a factor $1/2$ is included in the appearance of $\xi^{k^i}_{j_i}$ with $j_i = \pm M$. Inserting this into (14.23), considering the j-th Fourier coefficient and comparing the coefficients of $e^{i(k\cdot\omega)t}$, we obtain (14.17).

On the other hand, we need to derive the initial values for $\zeta^{\pm\langle j\rangle}_j$ appearing in (14.17). On noticing the fact that $\tilde{q}(0) = q(0)$, we obtain

$$\zeta^{\langle j\rangle}_j(0) + \zeta^{-\langle j\rangle}_j(0) = q_j(0) - \sum_{k\neq\pm\langle j\rangle}\zeta^k_j(0). \tag{14.24}$$

Furthermore, it follows from $\tilde{p}(0) = p(0)$ that

$$\eta^{\langle j\rangle}_j(0) + \eta^{-\langle j\rangle}_j(0) = p_j(0) - \sum_{k\neq\pm\langle j\rangle}\eta^k_j(0),$$

which results in

$$i\omega_j(\zeta^{\langle j\rangle}_j(0) - \zeta^{-\langle j\rangle}_j(0)) = p_j(0) - \sum_{k\neq\pm\langle j\rangle}\eta^k_j(0)$$

$$= p_j(0) - \sum_{k\neq\pm\langle j\rangle}\frac{\tan\left(\frac{1}{2}h(k\cdot\omega)\right)}{\tan\left(\frac{1}{2}h\omega_j\right)}i\omega_j\zeta^k_j(0) + \mathcal{O}(h\varepsilon). \tag{14.25}$$

The formulae (14.24) and (14.25) determine the initial values for $\zeta^{\pm\langle j\rangle}_j$.

---

[3]It is noted that $g(0) = 0$ and $g'(0) = 0$ are used here.

We now turn to the second part, the proof of (14.18).

For the modulation equations of η^k, it follows from (14.5) that

$$q_{n+1} - q_n = \Omega^{-1} \tan\left(\frac{1}{2}h\Omega\right)(p_{n+1} + p_n). \tag{14.26}$$

According to the definition of L_3, this relation can be expressed as

$$L_3^k \zeta^k = \Omega^{-1} \tan\left(\frac{1}{2}h\Omega\right)\eta^k.$$

It then follows from the Taylor series of L_3^k that the relationship between η^k and ζ^k can be established by (14.18). The proof then is complete.   □

14.4.3 Reverse Picard Iteration

In what follows, we consider the reverse Picard iteration (see [1, 3]) of the functions ζ^k such that after $4N$ iteration steps, the defects in (14.17), (14.24), and (14.25) are of magnitude $\mathcal{O}(\varepsilon^{N+1})$ in the H^s norm.

We here denote by $[\cdot]^{(n)}$ the nth iterate. For $k = \pm\langle j\rangle$ and under the condition (14.17), we design the iteration procedure as follows:

$$\pm 2\mathrm{i}s_{\langle j\rangle}h\varepsilon\left[\dot{\zeta}_j^{\pm\langle j\rangle}\right]^{(n+1)} = \left[-h^2\phi_2(h^2\omega_j^2)\sum_{m\geqslant 2}\frac{g^{(m)}(0)}{m!}\sum_{k^1+\cdots+k^m=k}{\sum_{j_1+\cdots+j_m\equiv j\ \mathrm{mod}\ 2M}}' \right.$$
$$\left. \cdot\int_0^1\left[\left(\xi_{j_1}^{k^1}\cdots\xi_{j_m}^{k^m}\right)(t\varepsilon,\sigma)\right]\mathrm{d}\sigma - \left(\frac{1}{2}\varepsilon^2h^2\sec(\frac{1}{2}h\omega_j)\ddot{\zeta}_j^{\pm\langle j\rangle} + \cdots\right)\right]^{(n)}. \tag{14.27}$$

For $k \neq \pm\langle j\rangle$ and j subject to the non-resonant condition (14.8), the iteration procedure is of the form

$$2\frac{s_{\langle j\rangle}+ks_{\langle j\rangle}-k}{c_k}\left[\zeta_j^k\right]^{(n+1)} = \left[-h^2\phi_2(h^2\omega_j^2)\sum_{m\geqslant 2}\frac{g^{(m)}(0)}{m!}\sum_{k^1+\cdots+k^m=k}{\sum_{j_1+\cdots+j_m\equiv j\ \mathrm{mod}\ 2M}}' \right.$$
$$\left. \cdot\int_0^1\left[\left(\xi_{j_1}^{k^1}\cdots\xi_{j_m}^{k^m}\right)(t\varepsilon,\sigma)\right]\mathrm{d}\sigma - \left(\mathrm{i}\varepsilon h\frac{s_k(1+c_{\langle j\rangle}+kc_{\langle j\rangle}-k)}{c_k^2}\dot{\zeta}_j^k + \cdots\right)\right]^{(n)}, \tag{14.28}$$

where $\zeta_j^k = 0$ for $k \neq \pm\langle j \rangle$ in the near-resonant set $\mathcal{R}_{\varepsilon,h}$. For the initial values (14.24) and (14.25), the iteration procedure is given by

$$
\left[\zeta_j^{\langle j \rangle}(0) + \zeta_j^{-\langle j \rangle}(0)\right]^{(n+1)} = \left[q_j(0) - \sum_{k \neq \pm\langle j \rangle} \zeta_j^k(0)\right]^{(n)},
$$

$$
i\omega_j \left[\zeta_j^{\langle j \rangle}(0) - \zeta_j^{-\langle j \rangle}(0)\right]^{(n+1)} = \left[p_j(0) - \sum_{k \neq \pm\langle j \rangle} \frac{\tan\left(\frac{1}{2}h(k \cdot \omega)\right)}{\tan\left(\frac{1}{2}h\omega_j\right)} i\omega_j \zeta_j^k(0) + \mathcal{O}(h\varepsilon)\right]^{(n)}.
$$

$$(14.29)$$

It is assumed that $\|k\| \leqslant K := 2N$ and $\|k^i\| \leqslant K$ for $i = 1, \cdots, m$, in these iterations. We here remark that the procedure includes an initial value problem of first-order ODEs for $\zeta_j^{\pm\langle j \rangle}$ (for $|j| \leqslant M$) and algebraic equations for ζ_j^k with $k \neq \pm\langle j \rangle$ at each iteration step. The starting iterates ($n = 0$) are chosen as $\zeta_j^k(\tau) = 0$ for $k \neq \pm\langle j \rangle$, and $\zeta_j^{\pm\langle j \rangle}(\tau) = \zeta_j^{\pm\langle j \rangle}(0)$, where $\zeta_j^{\pm\langle j \rangle}(0)$ are determined by (14.29). Obviously, the iteration procedure is well defined.

14.4.4 Rescaling and Estimation of the Nonlinear Terms

In a similar way to Sect. 3.5 of [2] and Sect. 6.3 of [1], we next consider a more convenient rescaling

$$
c\zeta_j^k = \frac{\omega^{|k|}}{\varepsilon^{[[k]]}} \zeta_j^k, \quad c\zeta^k = \left(c\zeta_j^k\right)_{|j| \leqslant M} = \frac{\omega^{|k|}}{\varepsilon^{[[k]]}} \zeta^k
$$

in the space $H^s = (H^s)^{\mathcal{K}} = \{c\zeta = (c\zeta^k)_{k \in \mathcal{K}} : c\zeta^k \in H^s\}$. The norm of this space is defined by $\||c\zeta\||_s^2 = \sum_{k \in \mathcal{K}} \|c\zeta^k\|_s^2$, where the set $\mathcal{K}$ is given by $\mathcal{K} = \{k = (k_l)_{l=0}^M$ with integers $k_l : \|k\| \leqslant K\}$ with $K = 2N$. Likewise, we use the notation $c\xi^k \in H^s$ having the same meaning.

With regard to the expression of the non-linearity for (14.17) in these rescaled variables, we define the nonlinear function $f = (f_j^k)$ by

$$
f_j^k\left(c\xi(\tau)\right) = \frac{\omega^{|k|}}{\varepsilon^{[[k]]}} \sum_{m=2}^N \frac{g^{(m)}(0)}{m!} \sum_{k^1 + \cdots + k^m = k} \frac{\varepsilon^{[[k^1]] + \cdots + [[k^m]]}}{\omega^{|k^1| + \cdots + |k^m|}}
$$

$$
\cdot \sideset{}{'}\sum_{j_1 + \cdots + j_m \equiv j \bmod 2M} \int_0^1 \left(c\xi_{j_1}^{k^1} \cdots \cdot c\xi_{j_m}^{k^m}\right)(\tau, \sigma) \, d\sigma.
$$

Concerning this nonlinear function, we have the following bounds, which can be proved by using the similar arguments presented in [1, 2].

Proposition 14.2 (Estimation of the Nonlinear Terms) *It is true that*

$$
\sum_{k \in \mathcal{K}} \left\| f^k(c\xi) \right\|_s^2 \leqslant C\varepsilon P(\||c\tilde{\xi}\||_s^2), \quad \sum_{|j| \leqslant M} \left\| f^{\pm\langle j\rangle}(c\xi) \right\|_s^2 \leqslant C\varepsilon^3 P_1(\||c\tilde{\xi}\||_s^2),
$$

$$(14.30)$$

where $c\tilde{\xi}(\tau) := \sup_{0 \leqslant \sigma \leqslant 1}\{c\xi(\tau, \sigma)\}$ *and* P *and* P_1 *are polynomials with coefficients bounded independently of* $\varepsilon, h,$ *and* M.

Similarly, we can consider different rescaling

$$
\hat{c}\zeta_j^k = \frac{\omega^{s|k|}}{\varepsilon^{[[k]]}}\zeta_j^k, \quad \hat{c}\zeta^k = \left(\hat{c}\zeta_j^k\right)_{|j| \leqslant M} = \frac{\omega^{s|k|}}{\varepsilon^{[[k]]}}\zeta^k \tag{14.31}
$$

in $H^1 = (H^1)^{\mathcal{K}}$ with norm $\||\hat{c}\zeta\||_1^2 = \sum_{\|k\| \leqslant K} \left\|\hat{c}\zeta^k\right\|_1^2$, where $\hat{f}_j^k$ is exactly the same as f_j^k, but with $\omega^{|k|}$ replaced by $\omega^{s|k|}$. We use similar notations $\hat{c}\xi^k \in H^1$ and also obtain similar bounds

$$
\sum_{k \in \mathcal{K}} \left\| \hat{f}^k(\hat{c}\xi) \right\|_1^2 \leqslant C\varepsilon \hat{P}(\||\hat{c}\tilde{\xi}\||_1^2), \quad \sum_{|j| \leqslant M} \left\| \hat{f}^{\pm\langle j\rangle}(\hat{c}\xi) \right\|_1^2 \leqslant C\varepsilon^3 \hat{P}_1(\||\hat{c}\tilde{\xi}\||_1^2),
$$

with other functions $\hat{P}$ and $\hat{P}_1$.

14.4.5   Reformulation of the Reverse Picard Iteration

This subsection concerns the reverse Picard iteration. On the basis of the two cases: $k = \pm\langle j\rangle$ and $k \neq \pm\langle j\rangle$, we split $c\zeta$ into two parts as follows:

$$
\begin{cases} a\zeta_j^k = c\zeta_j^k & \text{if } k = \pm\langle j\rangle, \quad \text{and } 0 \text{ else,} \\ b\zeta_j^k = c\zeta_j^k & \text{if (14.8) is satisfied,} \quad \text{and } 0 \text{ else.} \end{cases} \tag{14.32}
$$

It is noted that for $a\zeta = (a\zeta_j^k) \in H^s$ and $b\zeta = (b\zeta_j^k) \in H^s$, we have $a\zeta + b\zeta = c\zeta$ and $\||a\zeta\||_s^2 + \||b\zeta\||_s^2 = \||c\zeta\||_s^2$. Here, the same notation and property are used for $c\xi$.

We now rewrite the iterations (14.27) and (14.28) in an abstract form

$$\begin{cases} a\dot\zeta^{(n+1)} = \Omega^{-1}F(a\zeta^{(n)}, b\zeta^{(n)}) - Aa\zeta^{(n)}, \\ b\dot\zeta^{(n+1)} = \Omega^{-1}\Psi G(a\zeta^{(n)}, b\zeta^{(n)}) - Bb\zeta^{(n)}, \end{cases} \tag{14.33}$$

where

$$(\Omega x)_j^k = (\omega_j + |k\cdot\omega|)x_j^k, \quad (\Psi x)_j^k = 2\phi_2(h^2\omega_j^2)\cos\left(\frac{1}{2}h(k\cdot\omega)\right)x_j^k,$$

and the operators A, B are respectively given by

$$(Aa\zeta)_j^{\pm\langle j\rangle}(\tau) = \frac{1}{\pm 2is_{\langle j\rangle}h\varepsilon}\left(\frac{1}{2}\varepsilon^2 h^2\sec\left(\frac{1}{2}h\omega_j\right)a\ddot\zeta_j^{\pm\langle j\rangle} + \cdots\right),$$

$$(Bb\zeta)_j^k(\tau) = \frac{c_k}{2s_{\langle j\rangle+k}s_{\langle j\rangle-k}}\left(i\varepsilon h\frac{s_k(1 + c_{\langle j\rangle+k}c_{\langle j\rangle-k})}{c_k^2}b\dot\zeta_j^k + \cdots\right)$$

for (j, k) subject to (14.8).

The functions $F = (F_j^k)$ and $G = (G_j^k)$ are defined respectively by

$$F_j^{\pm\langle j\rangle}(a\zeta, b\zeta) = \frac{1}{\mp i\varepsilon}\frac{2\phi_2(h^2\omega_j^2)}{\text{sinc}\left(\frac{1}{2}h\omega_j\right)}f_j^{\pm\langle j\rangle}(c\xi), \quad G_j^k(a\zeta, b\zeta) = -\frac{h^2(\omega_j + |k\cdot\omega|)}{4s_{\langle j\rangle+k}s_{\langle j\rangle-k}}f_j^k(c\xi)$$

for (j, k) subject to (14.8).

Theorem 14.5 *The operators A and B are bounded by*

$$|||(Aa\zeta)(\tau)|||_s \leqslant C\sum_{l=2}^{N}h^{l-2}\varepsilon^{l-3/2}\left|\left|\left|\frac{d^l}{d\tau^l}(a\zeta)(\tau)\right|\right|\right|_s,$$

$$|||(Bb\zeta)(\tau)|||_s \leqslant C\varepsilon^{1/2}|||(b\dot\zeta)(\tau)|||_s + C\sum_{l=2}^{N}h^{l-2}\varepsilon^{l-1/2}\left|\left|\left|\frac{d^l}{d\tau^l}(b\zeta)(\tau)\right|\right|\right|_s.$$

Moreover, we have

$$|||F|||_s \leqslant C\varepsilon^{1/2}, \quad |||G|||_s \leqslant C, \quad |||\Psi^{-1}\Omega^{-1}F|||_s \leqslant C.$$

Proof The bound of A follows from

$$\left| \frac{1}{\pm 2is_{\langle j \rangle} h\varepsilon} \frac{1}{2} \varepsilon^2 h^2 \sec\left(\frac{1}{2} h\omega_j\right) \right| = \left| \frac{\frac{1}{2} h\varepsilon}{\sin(h\omega_j)} \right| \leqslant \frac{1}{2} \varepsilon^{1/2}.$$

We compute

$$\left| \frac{c_k}{2s_{\langle j \rangle + k} s_{\langle j \rangle - k}} i\varepsilon h \frac{s_k(1 + c_{\langle j \rangle} + k c_{\langle j \rangle - k})}{c_k^2} \right| \leqslant \left| \frac{\varepsilon h}{\varepsilon^{1/2} h^2(\omega_j + |k \cdot \boldsymbol{\omega}|)} \frac{s_k(1 + c_{\langle j \rangle} + k c_{\langle j \rangle - k})}{c_k} \right|$$

$$\leqslant \frac{\varepsilon^{1/2}}{h} \frac{\frac{h}{2}|k \cdot \boldsymbol{\omega}|}{\omega_j + |k \cdot \boldsymbol{\omega}|} \left| \frac{1 + c_{\langle j \rangle} + k c_{\langle j \rangle - k}}{c_k} \right| \leqslant C\varepsilon^{1/2},$$

where $|s_k| \leqslant \dfrac{h}{2}|k \cdot \boldsymbol{\omega}|$ is used. Hence, we obtain the bound of B.

It follows from

$$\left| \frac{2\phi_2(h^2 \omega_j^2)}{\mathrm{sinc}\left(\frac{1}{2} h\omega_j\right)} \right| = \left| \mathrm{sinc}\left(\frac{1}{2} h\omega_j\right) \right| \leqslant 1$$

and (14.30) that $|||F|||_s \leqslant C\varepsilon^{1/2}$. Then using (14.8) and (14.30) yields $|||G|||_s \leqslant C$. Furthermore, according to (14.11), we obtain

$$|||\Psi^{-1}\Omega^{-1}F|||_s^2 = \sum_{k \in \mathcal{K}} \sideset{}{''}\sum_{|j| \leqslant M} \omega_j^{2s} \left| (\Psi^{-1}\Omega^{-1}F)_j^k \right|^2 = \sum_{k \in \mathcal{K}} \sideset{}{''}\sum_{|j| \leqslant M} \omega_j^{2s} \left| \frac{h/2}{\varepsilon \sin(h\varepsilon)} \right|^2 \left| f_j^{\pm\langle j \rangle} \right|^2$$

$$\leqslant C \sum_{k \in \mathcal{K}} \sideset{}{''}\sum_{|j| \leqslant M} \omega_j^{2s} \left| \frac{1}{\varepsilon^{3/2}} \right|^2 \left| f_j^{\pm\langle j \rangle} \right|^2 = C\frac{1}{\varepsilon^3} |||f^{\pm\langle j \rangle}|||_s^2 \leqslant C.$$

This shows $|||\Psi^{-1}\Omega^{-1}F|||_s \leqslant C$. The proof is complete. $\square$

With regard to the initial value condition (14.29), it can be rewritten as

$$a\zeta^{(n+1)}(0) = v + Pb\zeta^{(n)}(0) + Qb\zeta^{(n)}(0), \tag{14.34}$$

where $v_j^{\pm\langle j \rangle} = \dfrac{\omega_j}{\varepsilon}\left(\dfrac{1}{2}q_j(0) \mp \dfrac{i}{2\omega_j}p_j(0)\right)$ and the operators P and Q are given by

$$(Pb\zeta)_j^{\pm\langle j \rangle}(0) = -\frac{1}{2}\frac{\omega_j}{\varepsilon} \sum_{k \neq \pm\langle j \rangle} \frac{\varepsilon^{[[k]]}}{\boldsymbol{\omega}^{|k|}} b\zeta_j^k(0),$$

$$(Qb\zeta)_j^{\pm\langle j \rangle}(0) = \mp\frac{1}{2\omega_j}\frac{\omega_j}{\varepsilon} \sum_{k \neq \pm\langle j \rangle} \frac{\varepsilon^{[[k]]}}{\boldsymbol{\omega}^{|k|}} b\eta_j^k(0).$$

It can be verified from (14.7) that v is bounded in H^s. For the bounds of the operators P and Q, we have

$$|||Pb\zeta(0)|||_s^2 = \sum_{k\in\mathscr{K}} \sideset{}{''}\sum_{|j|\leqslant M} \omega_j^{2s} \left| \frac{1}{2}\frac{\omega_j}{\varepsilon} \sum_{k\neq\pm\langle j\rangle} \frac{\varepsilon^{[[k]]}}{\omega^{|k|}} b\zeta_j^k(0) \right|^2$$

$$\leqslant \frac{1}{4\varepsilon^2} \sum_{k\in\mathscr{K}} \sideset{}{''}\sum_{|j|\leqslant M} \omega_j^{2s+2} \left(\sum_{k\neq\pm\langle j\rangle} \frac{\varepsilon^{2[[k]]}}{\omega^{2|k|}} \right) \left(\sum_{k\neq\pm\langle j\rangle} b\zeta_j^k(0)^2 \right)$$

$$\leqslant \frac{1}{4} \sum_{k\in\mathscr{K}} \sideset{}{''}\sum_{|j|\leqslant M} \omega_j^{2s+2} \left(\sum_{k\neq\pm\langle j\rangle} \omega^{-2|k|} \right) \left(\sum_{k\neq\pm\langle j\rangle} b\zeta_j^k(0)^2 \right)$$

$$\leqslant C|||\Omega b\zeta(0)|||_s^2 \leqslant C|||b\zeta(0)|||_{s+1}^2.$$

Likewise, we can obtain

$$|||(Qb\zeta)(0)|||_s^2 \leqslant C|||b\eta(0)|||_s^2.$$

Therefore, the bounds $|||(Pb\zeta)(0)|||_s \leqslant C$ and $|||(Qb\zeta)(0)|||_s \leqslant C$ are confirmed. Finally, we remark that the starting iterates of (14.34) are chosen as $a\zeta^{(0)}(\tau) = v$ and $b\zeta^{(0)}(\tau) = 0$, respectively.

14.4.6   Bounds of the Coefficient Functions

Theorem 14.6 (Bounds of the Modulation Functions) *The modulation functions ζ^k of (14.15) are bounded by*

$$\sum_{\|k\|\leqslant 2N} \left(\frac{\omega^{|k|}}{\varepsilon^{[[k]]}} \left\| \zeta^k(\varepsilon t) \right\|_s \right)^2 \leqslant C \tag{14.35}$$

and the same bound holds for any fixed number of derivatives of ζ^k with respect to the slow time $\tau = \varepsilon t$.

Proof According to the analysis stated above and by induction, we can prove that the iterates $a\zeta^{(n)}$, $b\zeta^{(n)}$ and their derivatives with respect to τ are bounded in H^s for $0 \leqslant \tau \leqslant 1$ and $n \leqslant 4N$. These bounds show that $c\zeta^{(n)} = a\zeta^{(n)} + b\zeta^{(n)}$ is bounded in H^s, and then the bound (14.35) follows. □

Theorem 14.7 (Bounds of the Expansion)   *The expansion (14.15) is bounded by*

$$\|\tilde{q}(t)\|_{s+1} + \|\tilde{p}(t)\|_s \leqslant C\varepsilon \quad for \quad 0 \leqslant t \leqslant \varepsilon^{-1}. \tag{14.36}$$

For $|j| \leqslant M$, it further holds that

$$\tilde{q}_j(t) = \zeta_j^{\langle j \rangle}(\varepsilon t)e^{i\omega_j t} + \zeta_j^{-\langle j \rangle}(\varepsilon t)e^{-i\omega_j t} + r_j, \quad where \quad \|r\|_{s+1} \leqslant C\varepsilon^2. \tag{14.37}$$

If the condition (14.12) is not satisfied, then the bound becomes $\|r\|_{s+1} \leqslant C\varepsilon^{3/2}$.

Proof The following bounds for the $(4N)$-th iterates can be obtained

$$|||a\zeta(0)|||_s \leqslant C, \quad |||\Omega a\dot{\zeta}(\tau)|||_s \leqslant C\varepsilon^{1/2},$$

$$|||\Psi^{-1}a\dot{\zeta}(\tau)|||_s \leqslant C, \quad |||\Psi^{-1}\Omega b\zeta(\tau)|||_s \leqslant C, \tag{14.38}$$

where C depends on N, but not on ε, h, M. It then follows from (14.38) that

$$|||a\dot{\zeta}|||_{s+1} = |||\Omega a\dot{\zeta}|||_s \leqslant C\varepsilon^{1/2},$$

$$|||b\zeta|||_{s+1}^2 = \sum_{k\in\mathcal{K}}^{''} \sum_{|j|\leqslant M} \omega_j^{2s+2} |b\zeta_j|^2 = \sum_{k\in\mathcal{K}}^{''} \sum_{|j|\leqslant M} \omega_j^{2s} \frac{\omega_j^2}{(\omega_j + |k\cdot\boldsymbol{\omega}|)^2} |(\omega_j + |k\cdot\boldsymbol{\omega}|)b\zeta_j|^2$$

$$\leqslant |||\Omega b\zeta(\tau)|||_s^2 \leqslant C.$$

We thus obtain

$$|||c\zeta(\tau) - a\zeta(0)|||_{s+1} = |||a\zeta(\tau) + b\zeta(\tau) - a\zeta(0)|||_{s+1} \leqslant |||a\dot{\zeta}|||_{s+1} + |||b\zeta|||_{s+1} \leqslant C.$$

On noticing the fact that $\zeta_j^k = \dfrac{\varepsilon^{[[k]]}}{\boldsymbol{\omega}^{|k|}}(c\zeta_j^k - a\zeta_j^k(0) + a\zeta_j^k(0))$, we have

$$|||\tilde{q}|||_{s+1}^2 = \sum_{k\in\mathcal{K}}^{''} \sum_{|j|\leqslant M} \omega_j^{2s+2} \left| \sum_{\|k\|\leqslant 2N} e^{i(k\cdot\omega)t} \zeta_j^k \right|^2$$

$$\leqslant \sum_{k\in\mathcal{K}}^{''} \sum_{|j|\leqslant M} \omega_j^{2s+2} \left[\frac{\varepsilon}{\omega_j}\left(\left|a\zeta_j^{\langle j \rangle}(0)\right| + \left|a\zeta_j^{-\langle j \rangle}(0)\right|\right) + \sum_{\|k\|\leqslant 2N} \frac{\varepsilon^{[[k]]}}{\boldsymbol{\omega}^{|k|}} \left|c\zeta_j^k - a\zeta_j^k(0)\right| \right]^2$$

$$\leqslant 2\varepsilon^2 \sum_{k\in\mathcal{K}}^{''} \sum_{|j|\leqslant M} \omega_j^{2s} \left(\left|a\zeta_j^{\langle j \rangle}(0)\right| + \left|a\zeta_j^{-\langle j \rangle}(0)\right|\right)^2$$

$$+ 2\sum_{k\in\mathcal{K}}^{''} \sum_{|j|\leqslant M} \omega_j^{2s+2} \left(\sum_{\|k\|\leqslant 2N} \frac{\varepsilon^{[[k]]}}{\boldsymbol{\omega}^{|k|}} \left|c\zeta_j^k - a\zeta_j^k(0)\right|\right)^2$$

$$\leqslant 4\varepsilon^2 |||a\zeta(0)|||_s^2 + 2\sum_{k\in\mathcal{K}}^{''} \sum_{|j|\leqslant M} \omega_j^{2s+2} \left(\sum_{\|k\|\leqslant 2N} \frac{\varepsilon^{2[[k]]}}{\boldsymbol{\omega}^{2|k|}}\right)\left(\sum_{\|k\|\leqslant 2N} \left|c\zeta_j^k - a\zeta_j^k(0)\right|^2\right)$$

$$\leqslant 4\varepsilon^2 |||a\zeta(0)|||_s^2 + 2C_{K,1}\varepsilon^2 |||c\zeta - a\zeta(0)|||_{s+1}^2 \leqslant C\varepsilon^2.$$

According to (14.26), with a similar analysis, it can be proved that $|||\tilde{p}|||_s \leqslant C\varepsilon$. Hence, the bound (14.36) holds.

It then follows from (14.30) and (14.33) that $\left(\sum_{\|k\|=1} \left\| (\Psi^{-1}\Omega b\zeta)^k \right\|_s^2 \right)^{1/2} \leqslant C\varepsilon$

for $b\zeta = (b\zeta)^{(4N)}$. Furthermore, using (14.12), we obtain that

$$\sum_{|j|\leqslant M} \sum_{j_1+j_2=j} \sum_{k=\pm\langle j_1\rangle\pm\langle j_2\rangle} \omega_j^{2(s+1)} |b\zeta_j^k|^2 \leqslant C\varepsilon.$$

These bounds as well as (14.38) lead to (14.37). The proof is complete.   □

Concerning the alternative scaling (14.31), we can obtain the same bounds

$$|||\hat{a}\zeta(0)|||_1 \leqslant C, \quad |||\Omega\hat{a}\dot\zeta(\tau)|||_1 \leqslant C\varepsilon^{1/2}, \quad |||\Psi^{-1}\Omega\hat{b}\zeta(\tau)|||_1 \leqslant C.$$
$$\tag{14.39}$$

Moreover, the following bound is also true for this scaling:

$$\left(\sum_{\|k\|=1} \left\| (\Psi^{-1}\Omega\hat{b}\zeta)^k \right\|_1^2 \right)^{1/2} \leqslant C\varepsilon. \tag{14.40}$$

14.4.7   Defects

In this subsection, we pay attention to the so-called defect. It follows from (14.5) that the defect can be put in another form

$$\delta_j(t) = \frac{\tilde{q}_j(t+h) - 2\cos(h\omega_j)\tilde{q}_j(t) + \tilde{q}_j(t-h)}{h^2\phi_2(h^2\omega_j^2)}$$
$$- \left[\int_0^1 f_j((1-\sigma)\tilde{q}_h(t) + \sigma\tilde{q}_h(t+h))\mathrm{d}\sigma + \int_0^1 f_j((1-\sigma)\tilde{q}_h(t-h) + \sigma\tilde{q}_h(t))\mathrm{d}\sigma \right],$$
$$\tag{14.41}$$

where $\tilde{q}_j$ is determined in (14.15) with $\zeta_j^k = (\zeta_j^k)^{(4N)}$ obtained after $4N$ iterations of the procedure in Sect. 14.4.3. Here, $\delta_j(t)$ can also be rewritten as

$$\delta_j(t) = \sum_{\|k\|\leqslant NK} \mathrm{d}^k(\varepsilon t)\mathrm{e}^{\mathrm{i}(k\cdot\omega)t} + R(t),$$

where

$$
d_j^k = \frac{1}{h^2 \phi_2(h^2 \omega_j^2)} \tilde{L}_j^k \zeta_j^k + \sum_{m=2}^{N} \frac{g^{(m)}(0)}{m!} \sum_{k^1 + \cdots + k^m = k} \sum_{j_1 + \cdots + j_m \equiv j \bmod 2M}'
$$

$$
\cdot \int_0^1 \left[(\xi_{j_1}^{k^1} \cdots \xi_{j_m}^{k^m})(t\varepsilon, \sigma) \right] d\sigma. \tag{14.42}
$$

It is remarked that we consider $\|k\| \leqslant NK$ for d_j^k, and assume that $\zeta_j^k = \eta_j^k = 0$ for $\|k\| > K := 2N$. We denote by $\tilde{L}_j^k$ the truncation of the operator L_j^k after the ε^N term. The remainder terms of the Taylor expansion of f after N terms are absorbed in $R(t)$. Then it can be confirmed by the bound (14.36) and the estimates (14.38) that

$$
\|R(t)\|_{s+1} \leqslant C\varepsilon^{N+1}.
$$

Furthermore, using the Cauchy–Schwarz inequality and Lemma 14.1 results in

$$
\left\| \sum_{\|k\| \leqslant NK} d^k(\varepsilon t) e^{i(k \cdot \omega)t} \right\|_s^2 = \sum_{|j| \leqslant M}'' \omega_j^{2s} \left| \sum_{\|k\| \leqslant NK} d_j^k e^{i(k \cdot \omega)t} \right|^2
$$

$$
= \sum_{|j| \leqslant M}'' \omega_j^{2s} \left| \sum_{\|k\| \leqslant NK} \omega^{-|k|} (\omega^{|k|} d_j^k e^{i(k \cdot \omega)t}) \right|^2
$$

$$
\leqslant \sum_{|j| \leqslant M}'' \omega_j^{2s} \left(\sum_{\|k\| \leqslant NK} \omega^{-2|k|} \right) \left(\sum_{\|k\| \leqslant NK} (\omega^{|k|} d_j^k)^2 \right)
$$

$$
\leqslant C_{NK,1} \sum_{\|k\| \leqslant NK} \left\| \omega^{|k|} d^k(\varepsilon t) \right\|_s^2.
$$

This result leads to bounds on the defects. In fact, the right-hand side of this inequality can be estimated as follows.

Theorem 14.8 (Bounds of the Defects) *It can be deduced that* $\sum_{\|k\| \leqslant NK} \left\| \omega^{|k|} d^k(\varepsilon t) \right\|_s^2 \leqslant C\varepsilon^{2(N+1)}$, *and then the defect (14.41) implies the bound* $\|\delta(t)\|_s \leqslant C\varepsilon^{N+1}$.

Proof To prove this result we will consider three different cases: truncated, near-resonant and non-resonant modes.

- **Truncated and near-resonant modes.** The result for these two cases can be obtained by using the similar analysis given in Sect. 6.8 of [1].

- **Non-resonant mode.** For the non-resonant mode ($\|k\| > K$ and (j, k) satisfies (14.8)), we first reformulate the defect in the scaled variables of Sect. 14.4.4 as

$$\omega^{|k|} d_j^k = \varepsilon^{[[k]]} \left(\frac{1}{h^2 \phi_2(h^2 \omega_j^2)} \tilde{L}_j^k c \zeta_j^k + f_j^k(c\xi) \right).$$

Then splitting them into $k = \pm\langle j\rangle$ and $k \neq \pm\langle j\rangle$ yields

$$\omega_j d_j^{\pm\langle j\rangle} = \varepsilon \left(\pm i \varepsilon \omega_j \frac{\mathrm{sinc}(h\omega_j/2)}{\phi_2(h^2\omega_j^2)} \left(a \dot{\zeta}_j^{\pm\langle j\rangle} + (Aa\zeta)_j^{\pm\langle j\rangle} \right) + f_j^{\pm\langle j\rangle}(c\xi) \right),$$

$$\omega^{|k|} d_j^k = \varepsilon^{[[k]]} \left(\frac{2s_{\langle j\rangle} + ks_{\langle j\rangle} - k}{h^2 c_k \phi_2(h^2 \omega_j^2)} \left(b\zeta_j^k + (Bb\zeta)_j^k \right) + f_j^k(c\xi) \right).$$

We remark that the functions here are actually the $4N$-th iterates of the iteration in Sect. 14.4.3. Expressing $f_j^{\pm\langle j\rangle}$ and f_j^k in terms of F, G and inserting them from (14.33) into this defect, we obtain

$$\omega_j d_j^{\pm\langle j\rangle} = 2\omega_j \alpha_j^{\pm\langle j\rangle} \left([a\dot{\zeta}_j^{\pm\langle j\rangle}]^{(4N)} - [a\dot{\zeta}_j^{\pm\langle j\rangle}]^{(4N+1)} \right), \quad \alpha_j^{\pm\langle j\rangle} = \pm i\varepsilon^2 \frac{\mathrm{sinc}(h\omega_j/2)}{2\phi_2(h^2\omega_j^2)},$$

$$\omega^{|k|} d_j^k = \beta_j^k \left([b\zeta_j^k]^{(4N)} - [b\zeta_j^k]^{(4N+1)} \right), \qquad \beta_j^k = \varepsilon^{[[k]]} \frac{2s_{\langle j\rangle} + ks_{\langle j\rangle} - k}{h^2 c_k \phi_2(h^2\omega_j^2)}.$$

Looking closer at these expressions, we introduce new variables as follows:

$$\tilde{a}\zeta_j^{\pm\langle j\rangle} = \alpha_j^{\pm\langle j\rangle} a\zeta_j^{\pm\langle j\rangle}, \quad \tilde{b}\zeta_j^k = \beta_j^k b\zeta_j^k$$

and then rewrite the iteration (14.33) in these variables as

$$\tilde{a}\dot{\zeta}^{(n+1)} = \Omega^{-1} \tilde{F}(\tilde{a}\zeta^{(n)}, \tilde{b}\zeta^{(n)}) - A\tilde{a}\zeta^{(n)},$$

$$\tilde{b}\zeta^{(n+1)} = \tilde{G}(\tilde{a}\zeta^{(n)}, \tilde{b}\zeta^{(n)}) - B\tilde{b}\zeta^{(n)}.$$

In such a way, the transformed functions are determined by

$$\tilde{F}_j^{\pm\langle j\rangle}(\tilde{a}\zeta, \tilde{b}\zeta) = \alpha_j^{\pm\langle j\rangle} F_j^{\pm\langle j\rangle}(\alpha^{-1}\tilde{a}\zeta, \beta^{-1}\tilde{b}\zeta) = -\varepsilon f_j^{\pm\langle j\rangle}(\alpha^{-1}\tilde{a}\zeta + \beta^{-1}\tilde{b}\zeta),$$

$$\tilde{G}_j^k(\tilde{a}\zeta, \tilde{b}\zeta) = \beta_j^k(\Psi\Omega^{-1}G)_j^k(\alpha^{-1}\tilde{a}\zeta, \beta^{-1}\tilde{b}\zeta) = -\varepsilon^{[[k]]} f_j^k(\alpha^{-1}\tilde{a}\zeta + \beta^{-1}\tilde{b}\zeta).$$

As for the initial values of the iteration, we have

$$\tilde{a}\zeta^{(n+1)}(0) = \alpha v + \tilde{P}\tilde{b}\zeta^{(n)}(0) + \tilde{Q}\tilde{b}\zeta^{(n)}(0),$$

where $\tilde{P} = \alpha P \beta^{-1}$, $\tilde{Q} = \alpha Q \beta^{-1}$. For the bound of $\tilde{P}$, we obtain

$$|||\tilde{P}\tilde{b}\zeta(0)|||_s^2$$

$$= \sum_{k\in\mathcal{K}}\sum_{|j|\leqslant M}{}'' \omega_j^{2s}\left| i\varepsilon^2 \frac{\mathrm{sinc}(h\omega_j/2)}{2\phi_2(h^2\omega_j^2)}\frac{1}{2}\frac{\omega_j}{\varepsilon}\sum_{k\neq\pm\langle j\rangle}\frac{h^2 c_k \phi_2(h^2\omega_j^2)}{\varepsilon^{[[k]]}2s\langle j\rangle+ks\langle j\rangle-k}\frac{\varepsilon^{[[k]]}}{\omega^{|k|}}\tilde{b}\zeta_j^k(0)\right|^2$$

$$\leqslant \frac{\varepsilon^2 h^4}{64}\sum_{k\in\mathcal{K}}\sum_{|j|\leqslant M}{}'' \omega_j^{2s}\left(\sum_{k\neq\pm\langle j\rangle}\frac{\omega_j}{\left|s\langle j\rangle+ks\langle j\rangle-k\right|}\omega^{-|k|}\tilde{b}\zeta_j^k(0)\right)^2$$

$$\leqslant \frac{\varepsilon^2 h^4}{64}\sum_{k\in\mathcal{K}}\sum_{|j|\leqslant M}{}'' \omega_j^{2s}\left(\sum_{k\neq\pm\langle j\rangle}\frac{1}{\varepsilon^{1/2}h^2}\omega^{-|k|}\tilde{b}\zeta_j^k(0)\right)^2$$

$$\leqslant \frac{\varepsilon}{64}\sum_{k\in\mathcal{K}}\sum_{|j|\leqslant M}{}'' \omega_j^{2s}\left(\sum_{k\neq\pm\langle j\rangle}\omega^{-2|k|}\sum_{k\neq\pm\langle j\rangle}(\tilde{b}\zeta_j^k(0))^2\right)\leqslant C\varepsilon|||\tilde{b}\zeta(0)|||_s^2.$$

In a similar way, the following result can be achieved:

$$|||\tilde{Q}\tilde{b}\zeta(0)|||_s^2 \leqslant C\varepsilon|||\tilde{b}\zeta(0)|||_s^2.$$

Clearly, it can be verified that in an H^s-neighbourhood of 0 where the bounds (14.38) hold, the partial derivatives of $\tilde{F}$ with respect to $\tilde{a}\zeta$ and $\tilde{b}\zeta$ are bounded by $\mathcal{O}(\varepsilon^{1/2})$. Moreover, the partial derivative of $\tilde{G}$ with respect to $\tilde{b}\zeta$ is bounded by $\mathcal{O}(\varepsilon^{1/2})$ but that of $\tilde{G}$ with respect to $\tilde{a}\zeta$ is only $\mathcal{O}(1)$. In fact, these results are the same as those described in Sect. 6.9 of [1]. Similarly, we can obtain

$$|||\Omega(\tilde{a}\dot{\zeta}^{(4N+1)} - \tilde{a}\dot{\zeta}^{(4N)})|||_s \leqslant C\varepsilon^{N+2},$$

$$|||\tilde{b}\zeta^{(4N+1)} - \tilde{b}\zeta^{(4N)})|||_s \leqslant C\varepsilon^{N+2},$$

$$|||\tilde{a}\zeta(0)^{(4N+1)} - \tilde{a}\zeta(0)^{(4N)})|||_s \leqslant C\varepsilon^{N+2}.$$

Hence, for $\tau \leqslant 1$ and $(j,k) \in \mathcal{R}_{\varepsilon,h}$, these results yield the bound

$$\left(\sum_{\|k\|\leqslant K}\left\|\omega^{|k|}d^k(\tau)\right\|_s^2\right)^{1/2} \leqslant C\varepsilon^{N+1}. \tag{14.43}$$

It then follows from (14.43) that the defect (14.41) has the bound $\|\delta(t)\|_s \leqslant C\varepsilon^{N+1}$ for $t \leqslant \varepsilon^{-1}$. Concerning the defect in the initial conditions (14.24) and (14.25), it is true that

$$\|q(0) - \tilde{q}(0)\|_{s+1} + \|p(0) - \tilde{p}(0)\|_s \leqslant C\varepsilon^{N+1}.$$

Finally, we turn to the alternative scaling (14.31). For this case, we can obtain

$$\left(\sum_{\|k\|\leqslant K}\left\|\omega^{s|k|}d^k(\tau)\right\|_1^2\right)^{1/2}\leqslant C\varepsilon^{N+1}.\tag{14.44}$$

The proof is complete.   □

14.4.8   Remainders

In this subsection, we are concerned with the difference between the numerical solution and its modulated Fourier expansion.

Theorem 14.9 (Remainders) *The bound on the difference between the numerical solution and its modulated Fourier expansion satisfies*

$$\|q_n - \tilde{q}(t)\|_{s+1} + \|p_n - \tilde{p}(t)\|_s \leqslant C\varepsilon^N \quad for \quad 0 \leqslant t = nh \leqslant \varepsilon^{-1}.\tag{14.45}$$

Proof Let $\Delta q_n = \tilde{q}(t_n) - q_n$, $\Delta p_n = \tilde{p}(t_n) - p_n$. We have

$$\begin{pmatrix}\Delta q_{n+1}\\\Omega^{-1}\Delta p_{n+1}\end{pmatrix}=\begin{pmatrix}\cos(h\Omega) & \sin(h\Omega)\\-\sin(h\Omega) & \cos(h\Omega)\end{pmatrix}\begin{pmatrix}\Delta q_n\\\Omega^{-1}\Delta p_n\end{pmatrix}+h\begin{pmatrix}h\Omega\phi_2(V)\Omega^{-1}(\Delta f+\delta)\\\phi_1(V)\Omega^{-1}(\Delta f+\delta)\end{pmatrix},$$

where

$$\Delta f = \int_0^1 \left(f((1-\sigma)q_n + \sigma q_{n+1}) - f((1-\sigma)\tilde{q}(t_n) + \sigma\tilde{q}(t_n+h))\right)d\sigma.$$

According to the Lipschitz bound given in Sect. 4.2 of [3] and Sect. 6.10 of [1], it is clear that

$$\left\|\Omega^{-1}\Delta f\right\|_{s+1} = \|\Delta f\|_s \leqslant \varepsilon(\|\Delta q_n\|_s + \|\Delta q_{n+1}\|_s).$$

Moreover, we have $\left\|\Omega^{-1}\delta(t)\right\|_{s+1} = \|\delta(t)\|_s \leqslant C\varepsilon^{N+1}$. We then obtain

$$\left\|\begin{pmatrix}\Delta q_{n+1}\\\Omega^{-1}\Delta p_{n+1}\end{pmatrix}\right\|_{s+1} \leqslant \left\|\begin{pmatrix}\Delta q_n\\\Omega^{-1}\Delta p_n\end{pmatrix}\right\|_{s+1} + h\left(C\varepsilon\|\Delta q_n\|_s + C\varepsilon\|\Delta q_{n+1}\|_s + C\varepsilon^{N+1}\right).$$

This leads to $\|\Delta q_n\|_{s+1} + \left\|\Omega^{-1}\Delta p_n\right\|_{s+1} \leqslant C(1 + t_n)\varepsilon^{N+1}$ for $t_n \leqslant \varepsilon^{-1}$. This proves (14.45).   □

14.4.9   Almost Invariants

This subsection concerns almost-invariants of the modulated Fourier expansions.

According to the analysis presented in Sect. 14.4.7, we can rewrite the defect formula (14.42) as

$$\frac{1}{h^2 \phi_2(h^2 \omega_j^2)} \tilde{L}_j^k \zeta_j^k + \nabla_{-j}^{-k} \mathscr{U}(\xi(t)) = d_j^k, \tag{14.46}$$

where $\nabla_{-j}^{-k} \mathscr{U}(y)$ is the partial derivative with respect to y_{-j}^{-k} of the extended potential (see, e.g. [1, 3])

$$\mathscr{U}(\xi(t, \sigma)) = \sum_{l=-N}^{N} \mathscr{U}_l(\xi(t, \sigma)),$$

$$\mathscr{U}_l(\xi(t, \sigma)) = \sum_{m=2}^{N} \frac{U^{(m+1)}(0)}{(m+1)!} \sum_{k^1 + \cdots + k^{m+1} = 0} \sum_{j_1 + \cdots + j_{m+1} = 2Ml}' \int_0^1 \left(\xi_{j_1}^{k^1} \cdots \cdots \xi_{j_{m+1}}^{k^{m+1}} \right)(t, \sigma) d\sigma.$$

We define (see [1])

$$S_{\boldsymbol{\mu}}(\theta) y = \left(e^{i(k \cdot \boldsymbol{\mu})\theta} y_j^k \right)_{|j| \leqslant M, \|k\| \leqslant K}$$

and

$$T(\theta) y = \left(e^{ij\theta} y_j^k \right)_{|j| \leqslant M, \|k\| \leqslant K},$$

where $\boldsymbol{\mu} = (\mu_l)_{l \geqslant 0}$ is an arbitrary real sequence for $\theta \in R$. Using the results given in [1], we obtain $\mathscr{U}(S_{\boldsymbol{\mu}}(\theta) y) = \mathscr{U}(y)$ and $\mathscr{U}_0(T(\theta) y) = \mathscr{U}_0(y)$ for $\theta \in \mathbb{R}$. Hence,

$$0 = \frac{d}{d\theta} \Big|_{\theta=0} \mathscr{U}(S_{\boldsymbol{\mu}}(\theta) \xi(t, \sigma)), \quad 0 = \frac{d}{d\theta} \Big|_{\theta=0} \mathscr{U}_0(T(\theta) \xi(t, \sigma)). \tag{14.47}$$

Theorem 14.10 (Two Almost-Invariants) *There exist two functions $\mathscr{J}_l[\boldsymbol{\zeta}, \boldsymbol{\eta}](\tau)$ and $\mathscr{K}[\boldsymbol{\zeta}, \boldsymbol{\eta}](\tau)$ such that*

$$\sum_{l=1}^{M} \omega_l^{2s+1} \left| \frac{d}{d\tau} \mathscr{J}_l[\boldsymbol{\zeta}, \boldsymbol{\eta}](\tau) \right| \leqslant C\varepsilon^{N+1},$$

$$\left| \frac{d}{d\tau} \mathscr{K}[\boldsymbol{\zeta}, \boldsymbol{\eta}](\tau) \right| \leqslant C(\varepsilon^{N+1} + \varepsilon^2 M^{-s+1}) \tag{14.48}$$

for τ ⩽ 1. Moreover, it is true that

$$\mathscr{J}_l[\zeta,\eta](\varepsilon t_n) = \hat{J}_l(p_n, q_n) + \gamma_l(t_n)\varepsilon^3,$$

$$\mathscr{K}[\zeta,\eta](\varepsilon t_n) = \hat{K}(p_n, q_n) + \mathcal{O}(\varepsilon^3) + \mathcal{O}(\varepsilon^2 M^{-s}), \tag{14.49}$$

where

$$\hat{J}_l = \hat{I}_l + \hat{I}_{-l} = 2\hat{I}_l \quad \text{for} \quad 0 < l < M, \quad \hat{J}_0 = \hat{I}_0, \quad \hat{J}_M = \hat{I}_M.$$

*Here, all the constants in (14.48) and (14.49) are independent of ε, M, h, and n,
and $\sum\limits_{l=0}^{M} \omega_l^{2s+1} \gamma_l(t_n) \leqslant C$ for $t_n \leqslant \varepsilon^{-1}$.*

Proof

- **Proof of** (14.48).

  It follows from the first equality of (14.47) that

$$0 = \frac{d}{d\theta}\Big|_{\theta=0} \mathscr{U}(S_\mu(\theta)\xi(t,\sigma)) = \sum_{\|k\|\leqslant K} \sum_{|j|\leqslant M}{}' i(k\cdot\mu)\xi_{-j}^{-k}(t,\sigma)\nabla_{-j}^{-k}\mathscr{U}(\xi(t,\sigma))$$

$$= \sum_{\|k\|\leqslant K} \sum_{|j|\leqslant M}{}' i(k\cdot\mu)L_4^{-k}(\sigma)\zeta_{-j}^{-k}$$

$$\times \left(\frac{1}{h^2\phi_2(h^2\omega_j^2)}\tilde{L}_j^k\zeta_j^k - d_j^k\right). \tag{14.50}$$

It is noted that the right-hand side is independent of σ. We thus choose $\sigma = 1/2$ in the following analysis. In this case, (14.50) gives

$$\sum_{\|k\|\leqslant K} \sum_{|j|\leqslant M}{}' i(k\cdot\mu)L_4^{-k}\left(\frac{1}{2}\right)\zeta_{-j}^{-k}\frac{1}{h^2\phi_2(h^2\omega_j^2)}\tilde{L}_j^k\zeta_j^k$$

$$= \sum_{\|k\|\leqslant K} \sum_{|j|\leqslant M}{}' i(k\cdot\mu)L_4^{-k}\left(\frac{1}{2}\right)\zeta_{-j}^{-k}d_j^k. \tag{14.51}$$

It then follows from the expansions of $L_4^{-k}\left(\frac{1}{2}\right)$ and $\tilde{L}_j^k$ and the "magic formulas" on p. 508 of [20] that the left-hand side of (14.51) is a total derivative of function $\varepsilon \mathscr{J}_\mu[\zeta,\eta](\tau)$ which depends on $\zeta(\tau), \eta(\tau)$ and their up to $(N-1)$th order derivatives. This implies that (14.51) is identical to the following equation

$$-\varepsilon\frac{d}{d\tau}\mathscr{J}_\mu[\zeta,\eta](\tau) = \sum_{\|k\|\leqslant K}\sum_{|j|\leqslant M}{}' i(k\cdot\mu)L_4^{-k}\left(\frac{1}{2}\right)\zeta_{-j}^{-k}d_j^k.$$

In what follows, we consider the special case where $\mu = \langle l\rangle$. Let $z_j^k = L_4^k(1/2)\zeta_j^k$. It follows from the property of $L_4^k(1/2)$ that the bounds on z_j^k and ζ_j^k are of the same magnitude. Splitting $d = ad + bd$ into two parts: the diagonal ($k = \pm\langle j\rangle$) and nondiagonal ($k \neq \pm\langle j\rangle$), gives

$$|||ad|||_s^2 + \sum_{\|k\|\leqslant K}|||\omega^{s|k|}bd|||_0^2 = \sum_{\|k\|\leqslant K}\left\|\omega^{s|k|}d^k\right\|_0^2 \leqslant C\varepsilon^{2N+2},$$

where (14.44) is used. Using Lemma 3 of [2] and the facts that

- $z_j^k = \frac{\varepsilon}{\omega_j^s}\hat{a}z_j^k + \frac{\varepsilon^{[[k]]}}{\omega^{s|k|}}\hat{a}z_j^k,$
- $|||\hat{a}z_\wedge^k|||_1 \leqslant C,$
- $|||\Omega\hat{b}z^k|||_1 \leqslant$ from (14.39),

we obtain

$$\sum_{l=1}^M\omega_l^{2s+1}\left|\frac{d}{d\tau}\mathscr{J}_l[\zeta,\eta](\tau)\right| = \frac{1}{\varepsilon}\sum_{l=1}^M\omega_l^{2s+1}\left|\sum_{\|k\|\leqslant K}k_l\sum_{j=-\infty}^\infty\zeta_j^kd_j^k\right|$$

$$\leqslant\frac{1}{\varepsilon}\left[|||\frac{\varepsilon}{\omega_j^s}\hat{a}\zeta_j^k|||_{s+1}|||ad|||_s + \left(\sum_{\|k\|\leqslant K}\left\|\omega^{s|k|}(1+|k\cdot\omega|)\frac{\varepsilon^{[[k]]}}{\omega^{s|k|}}\hat{a}\zeta_j^k\right\|_0^2\right)^{1/2}\right.$$

$$\left.\left(\sum_{\|k\|\leqslant K}\left\|\omega^{s|k|}bd^k\right\|_0^2\right)^{1/2}\right]$$

$$\leqslant C\varepsilon^{N+1}.$$

The first statement of (14.48) is proved.

In a similar way, using the second equality of (14.47), we obtain

$$\sum_{\|k\|\leqslant K}\sum_{|j|\leqslant M}{}' ijL_4^{-k}\left(\frac{1}{2}\right)\zeta_{-j}^{-k}\frac{1}{h^2\phi_2(h^2\omega_j^2)}\tilde{L}_j^k\zeta_j^k$$

$$= \sum_{\|k\|\leqslant K}\sum_{|j|\leqslant M}{}' ijL_4^{-k}\left(\frac{1}{2}\right)\zeta_{-j}^{-k}\left(d_j^k - \sum_{l\neq 0}\nabla_{-j}^{-k}(\mathscr{U}_l(\xi(t,\sigma)))\right). \tag{14.52}$$

A careful analysis shows that the left-hand side of (14.52) can be written as a total derivative of function $\varepsilon\mathscr{K}[\zeta,\eta](\tau)$, which yields

$$-\varepsilon\frac{\mathrm{d}}{\mathrm{d}\tau}\mathscr{K}[\zeta,\eta](\tau)=\sum_{\|k\|\leqslant K}\sideset{}{'}\sum_{|j|\leqslant M}ijL_4^{-k}\Big(\frac{1}{2}\Big)\varsigma_{-j}^{-k}\Big(d_j^k-\sum_{l\neq0}\nabla_{-j}^{-k}\big(\mathscr{U}_l(\xi(t,\sigma))\big)\Big).$$

(14.53)

It follows from the Cauchy–Schwarz inequality and the bound $|j|\leqslant\omega_j$ that

$$\left|\sum_{\|k\|\leqslant K}\sideset{}{'}\sum_{|j|\leqslant K}ijz_{-j}^{-k}d_j^k\right|\leqslant\left(\sum_{\|k\|\leqslant K}\sideset{}{'}\sum_{|j|\leqslant K}\omega_j^2\left|z_j^k\right|^2\right)^{1/2}\left(\sum_{\|k\|\leqslant K}\sideset{}{'}\sum_{|j|\leqslant K}\left|d_j^k\right|^2\right)^{1/2}$$

$$\leqslant C\varepsilon\left(\sum_{\|k\|\leqslant K}\sideset{}{'}\sum_{|j|\leqslant K}\frac{\omega_j^2}{\boldsymbol{\omega}^{|k|}}\frac{\varepsilon^{[[k]]}}{\varepsilon^2}\frac{\boldsymbol{\omega}^{|k|}}{\varepsilon^{[[k]]}}\left|z_j^k\right|^2\right)^{1/2}\left(\sum_{\|k\|\leqslant K}\sideset{}{'}\sum_{|j|\leqslant K}\left|d_j^k\right|^2\right)^{1/2}\leqslant C\varepsilon^{N+2}.$$

Furthermore, we note that

$$\sum_{\|k\|\leqslant K}\sideset{}{'}\sum_{|j|\leqslant M}ijz_{-j}^{-k}\nabla_{-j}^{-k}\mathscr{U}_l(\xi(t,\sigma))$$

$$=\sum_{m=2}^{N}\frac{U^{(m+1)}(0)}{m!}\sum_{k^1+\cdots+k^{m+1}=k}\sideset{}{'}\sum_{j_1+\cdots+j_{m+1}=2Ml}z_{j_1}^{k^1}\cdots z_{j_m}^{k^m}\cdot ij_{m+1}z_{j_{m+1}}^{k^{m+1}},$$

is the $2Ml$-th Fourier coefficient of the function (see [3])

$$w(x):=\sum_{m=2}^{N}\frac{U^{(m+1)}(0)}{m!}\sum_{k^1+\cdots+k^{m+1}=k}\mathscr{P}z^{k^1}(x)\cdots\mathscr{P}z^{k^m}(x)\cdot\frac{\mathrm{d}}{\mathrm{d}x}\mathscr{P}z^{k^{m+1}}(x).$$

We then can deduce that $\|w\|_{s-1}\leqslant C\varepsilon^3$, and the $2Ml$-th Fourier coefficient of w is bounded by $C\varepsilon^3\omega_{2Ml}^{-s+1}\leqslant C\varepsilon^3(2Ml)^{-s+1}$, as shown in the proof of Theorem 5.2 of [3]. In such a way, the second statement of (14.48) is confirmed by (14.53).

- **Proof of** (14.49).

We will prove only the second statement of (14.49) since the first one can be dealt with in a similar way.

It follows from the AAVF formula that

$$2h\mathrm{sinc}(h\Omega)\tilde{p}(t)=\tilde{q}(t+h)-\tilde{q}(t-h)+\mathcal{O}(h^2).$$

This shows that

$$\tilde{p}_j(t) = i\omega_j\left(\eta_j^{\langle j\rangle}(\varepsilon t)e^{i\omega_j t} - \eta_j^{-\langle j\rangle}(\varepsilon t)e^{-i\omega_j t}\right) + \mathcal{O}(h\varepsilon^2) + \mathcal{O}(h^3\varepsilon^2).$$

We then have

$$\zeta_j^{\langle j\rangle} = \frac{1}{2}\left(\tilde{q}_j + \frac{1}{i\omega_j}\tilde{p}_j\right) + \mathcal{O}(\varepsilon^2)$$

and

$$\zeta_j^{-\langle j\rangle} = \frac{1}{2}\left(\tilde{q}_j - \frac{1}{i\omega_j}\tilde{p}_j\right) + \mathcal{O}(\varepsilon^2).$$

On the basis of these results, an analysis of $\mathcal{K}$ is presented below:

$$\mathcal{K}[\zeta,\eta](\tau) = \sum_{|j|\leqslant M}{}' j\frac{1}{2}\frac{4\varepsilon h\sin\left(\frac{1}{2}h\omega_j\right)\cos\left(\frac{1}{2}h\omega_j\right)}{2h^2\phi_2(h^2\omega_j^2)}\left(|\zeta_j^{\langle j\rangle}|^2 - |\zeta_j^{-\langle j\rangle}|^2\right) + \mathcal{O}(\varepsilon^3)$$

$$= \sum_{|j|\leqslant M}{}' j\omega_j\frac{\cos\left(\frac{1}{2}h\omega_j\right)}{\mathrm{sinc}\left(\frac{1}{2}h\omega_j\right)}\left(|\zeta_j^{\langle j\rangle}|^2 - |\zeta_j^{-\langle j\rangle}|^2\right) + \mathcal{O}(\varepsilon^3)$$

$$= \sum_{|j|\leqslant M}{}' \frac{j\omega_j}{4}\frac{\cos\left(\frac{1}{2}h\omega_j\right)}{\mathrm{sinc}\left(\frac{1}{2}h\omega_j\right)}\left(|\tilde{q}_j + \frac{1}{i\omega_j}\tilde{p}_j|^2 - |\tilde{q}_j - \frac{1}{i\omega_j}\tilde{p}_j|^2\right) + \mathcal{O}(\varepsilon^3)$$

$$= \sum_{|j|\leqslant M}{}' \frac{\cos\left(\frac{1}{2}h\omega_j\right)}{\mathrm{sinc}\left(\frac{1}{2}h\omega_j\right)}\frac{j\omega_j}{4}4\frac{1}{i\omega_j}\tilde{q}_{-j}\tilde{p}_j + \mathcal{O}(\varepsilon^3)$$

$$= \hat{K}(\tilde{p},\tilde{q}) + \mathcal{O}(\varepsilon^3) + \mathcal{O}(\varepsilon^2 M^{-s}) = \hat{K}(p_n,q_n) + \mathcal{O}(\varepsilon^3) + \mathcal{O}(\varepsilon^2 M^{-s}),$$

where the results (14.37) and (14.45) are used.   □

14.4.10 From Short to Long-Time Intervals

According to the analysis stated above in this chapter, the statement of Theorem 14.3 can be confirmed by patching together many intervals of length ε^{-1} in the same way as that used in [1, 2].

14.5   Analysis for the AAVF Method with a Quadrature Rule

The previous analysis was made for the AAVF method with the integral appearing in (14.5), which usually cannot be solved exactly. Normally a quadrature rule is required. For this reason, we will show that the main result for the AAVF method with the integral is still true for the AAVF method with a quadrature approximation instead of the integral.

As an example, we consider the following AAVF method with the midpoint rule

$$\begin{cases} q_{n+1} = \phi_0(V)q_n + h\phi_1(V)p_n + h^2\phi_2(V)f((q_n + q_{n+1})/2), \\ p_{n+1} = -h\Omega^2\phi_1(V)q_n + \phi_0(V)p_n + h\phi_1(V)f((q_n + q_{n+1})/2). \end{cases} \tag{14.54}$$

The main result presented in Theorem 14.3 can be adapted for this method with the following modifications for the operator and the nonlinearity. We next present only the main differences and omit the details for brevity.

- Modifications for Sect. 14.4.2.

  Since the term $\int_0^1 f((1-\sigma)q_n + \sigma q_{n+1})d\sigma$ is replaced by $f((q_n + q_{n+1})/2)$,

  the function $\xi^k\left(\varepsilon\left(t + \dfrac{h}{2}\right), \sigma\right)$ should be changed to $\xi^k\left(\varepsilon\left(t + \dfrac{h}{2}\right), 1/2\right)$ and

  the operator $L_4^k(\sigma)$ is replaced by $L_4^k(1/2)$. Then all the analyses and results in Sect. 14.4.2 still hold for (14.54).

- Modifications for Sect. 14.4.3.

  For this part, we only need to change $\int_0^1 \left[\left(\xi_{j_1}^{k^1} \cdots \cdots \xi_{j_m}^{k^m}\right)(t\varepsilon, \sigma)\right]d\sigma$ to $\left(\xi_{j_1}^{k^1} \cdots \cdots \xi_{j_m}^{k^m}\right)(t\varepsilon, 1/2)$.

- Modifications for Sect. 14.4.4.

  One part of the function $f_j^k\left(c\xi(\tau)\right)$ here is $\left(c\xi_{j_1}^{k^1} \cdots \cdots c\xi_{j_m}^{k^m}\right)(\tau, 1/2)$ instead of $\int_0^1 \left(c\xi_{j_1}^{k^1} \cdots \cdots c\xi_{j_m}^{k^m}\right)(\tau, \sigma)d\sigma$ and then the property of $f_j^k\left(c\xi(\tau)\right)$ stated in Proposition 14.2 is still true.

- Modifications for Sect. 14.4.7.

  Since the defect expressed by (14.41) needs to be modified according to the

  scheme (14.54), the term $\int_0^1 \left[\left(\xi_{j_1}^{k^1} \cdots \cdots \xi_{j_m}^{k^m}\right)(t\varepsilon, \sigma)\right]d\sigma$ appearing in (14.42)

  should be replaced by $\left(\xi_{j_1}^{k^1} \cdots \cdots \xi_{j_m}^{k^m}\right)(t\varepsilon, 1/2)$. In this situation, we still obtain the same bounds of the defects as those stated previously.

- Modifications for Sect. 14.4.8.

  Here only the expression of Δf should be modified in the light of (14.54).

- Modifications for Sect. 14.4.9.

A new function

$$
\mathscr{U}_l(\xi) = \sum_{m=2}^{N} \frac{U^{(m+1)}(0)}{(m+1)!} \sum_{k^1+\cdots+k^{m+1}=0} \sideset{}{'}\sum_{j_1+\cdots+j_{m+1}=2Ml} \left(\xi_{j_1}^{k^1} \cdot \cdots \cdot \xi_{j_{m+1}}^{k^{m+1}} \right) (t, 1/2)
$$

will be used here instead of the previous one.

At the end of this section, we remark that since the AAVF method with the integral is of only order two, the long-time momentum and actions behaviour does not change for (14.54). For the AAVF method with other higher-order quadrature rules, the main result can also be obtained by following the same approach.

14.6   Conclusions and Discussions

It is known that the preservation of geometric or physical properties of the numerical flow can assist in long-time integration and produce improved qualitative behaviour in comparison with a general-purpose numerical method. In this chapter, we have investigated in detail the long-time behaviour of the AAVF method when applied to semilinear wave equations via spatial spectral semidiscretisations. With the semidiscretisation, the AAVF method exactly preserves the energy and nearly conserves modified actions and modified momentum over long times. The main result has been presented by developing a modulated Fourier expansion of the AAVF method and showing two almost-invariants of the modulated system.

The main result of this chapter explains rigorously the good long-time behaviour of EP methods for the numerical solution of semilinear wave equations. The analysis for multi-dimensional wave equations deserves further investigation. It is also noted that the long-term analysis of many different methods other than EP methods has been given recently for Schrödinger equations and the reader is referred to [18, 38–40]. The Schrödinger equation has become one of the most studied PDEs. It is hoped to obtain near-conservation of actions, momentum and density as well as exact-conservation of energy for some EP schemes when applied to the Schrödinger equation.

The material in this chapter is based on the work by Wang and Wu [41].

References

1. Cohen, D., Hairer, E., Lubich, C.: Conservation of energy, momentum and actions in numerical discretizations of nonlinear wave equations. Numer. Math. **110**, 113–143 (2008)
2. Cohen, D., Hairer, E., Lubich, C.: Long-time analysis of nonlinearly perturbed wave equations via modulated Fourier expansions. Arch. Ration. Mech. Anal. 187, 341–368 (2008)
3. Hairer, E., Lubich, C.: Spectral semi-discretisations of weakly nonlinear wave equations over long times. Found. Comput. Math. **8**, 319–334 (2008)

4. Bambusi, D.: Birkhoff normal form for some nonlinear PDEs. Commun. Math. Phys. **234**, 253–285 (2003)
5. Cano, B.: Conservation of invariants by symmetric multistep cosine methods for second-order partial differential equations. BIT Numer. Math. **53**, 29–56 (2013)
6. Cano, B.: Conserved quantities of some Hamiltonian wave equations after full discretization. Numer. Math. **103**, 197–223 (2006)
7. Cano, B., Moreta, M.J.: Multistep cosine methods for second-order partial differential systems. IMA J. Numer. Anal. **30**, 431–461 (2010)
8. Gauckler, L.: Error analysis of trigonometric integrators for semilinear wave equations. SIAM J. Numer. Anal. **53**, 1082–1106 (2015)
9. Gauckler, L., Weiss, D.: Metastable energy strata in numerical discretizations of weakly nonlinear wave equations. Disc. Contin. Dyn. Syst. **37**, 3721–3747 (2017)
10. Gauckler, L., Lu, J., Marzuola, J., et al.: Trigonometric integrators for quasilinear wave equations. Math. Comput. **88**, 717–749 (2019)
11. Grimm, V.: On the Use of the Gautschi-Type Exponential Integrator for Wave Equations Numerical Mathematics and Advanced Applications. Springer, Berlin (2006), pp. 557–563
12. Liu, C., Iserles, A., Wu, X.: Symmetric and arbitrarily high-order Birkhoff-Hermite time integrators and their long-time behaviour for solving nonlinear Klein-Gordon equations. J. Comput. Phys. **356**, 1–30 (2018)
13. Wang, B., Iserles, A., Wu, X.: Arbitrary-order trigonometric Fourier collocation methods for multi-frequency oscillatory systems. Found. Comput. Math. **16**, 151–181 (2016)
14. Wu, X., Wang, B.: Recent Developments in Structure-Preserving Algorithms for Oscillatory Differential Equations. Springer, Nature Singapore Pte Ltd., Singapore (2018)
15. Gauckler, L., Hairer, E., Lubich, C.: Long-term analysis of semilinear wave equations with slowly varying wave speed. Commun. Partial. Differ. Equ. **41**, 1934–1959 (2016)
16. Gauckler, L., Hairer, E., Lubich, C., et al.: Metastable energy strata in weakly nonlinear wave equations. Commun. Partial. Differ. Equ. **37**, 1391–1413 (2012)
17. Hairer, E., Lubich, C.: Long-time energy conservation of numerical methods for oscillatory differential equations. SIAM J. Numer. Anal. **38**, 414–441 (2000)
18. Cohen, D., Gauckler, L.: One-stage exponential integrators for nonlinear Schrödinger equations over long times. BIT Numer. Math. **52**, 877–903 (2012)
19. Hairer, E., Lubich, C.: Long-term analysis of the Störmer-Verlet method for Hamiltonian systems with a solution-dependent high frequency. Numer. Math. **134**, 119–138 (2016)
20. Hairer, E., Lubich, C., Wanner, G.: Geometric Numerical Integration: Structure-Preserving Algorithms for Ordinary Differential Equations, 2nd edn. Springer, Berlin (2006)
21. McLachlan, R.I., Stern, A.: Modified trigonometric integrators. SIAM J. Numer. Anal. **52**, 1378–1397 (2014)
22. Sanz-Serna, J.M.: Modulated Fourier expansions and heterogeneous multiscale methods. IMA J. Numer. Anal. **29**, 595–605 (2009)
23. Brugnano, L., Frasca Caccia, G., Iavernaro, F.: Energy conservation issues in the numerical solution of the semilinear wave equation. Appl. Math. Comput. **270**, 842–870 (2015)
24. Celledoni, E., Grimm, V., McLachlan, R.I., et al.: Preserving energy resp. dissipation in numerical PDEs using the "Average Vector Field" method. J. Comput. Phys. **231**, 6770–6789 (2012)
25. Li, Y.W., Wu, X.: General local energy-preserving integrators for solving multi-symplectic Hamiltonian PDEs. J. Comput. Phys. **301**, 141–166 (2015)
26. Liu, C., Wu, X.: An energy-preserving and symmetric scheme for nonlinear Hamiltonian wave equations. J. Math. Anal. Appl. **440**, 167–182 (2016)
27. Liu K, Wu X, Shi W. A linearly-fitted conservative (dissipative) scheme for efficiently solving conservative (dissipative) nonlinear wave PDEs. J. Comput. Math. **35**, 780–800 (2017)
28. Mei, L., Liu, C., Wu, X.: An essential extension of the finite-energy condition for extended Runge-Kutta-Nyström integrators when applied to nonlinear wave equations. Commun. Comput. Phys. **22**, 742–764 (2017)

29. Wang, B., Wu, X.: The formulation and analysis of energy-preserving schemes for solving high-dimensional nonlinear Klein-Gordon equations. IMA. J. Numer. Anal. **39**, 2016–2044 (2019)
30. Hairer, E.: Energy-preserving variant of collocation methods. J. Numer. Anal. Ind. Appl. Math. **5**, 73–84 (2010)
31. Li, Y.W., Wu, X.: Exponential integrators preserving first integrals or Lyapunov functions for conservative or dissipative systems. SIAM J. Sci. Comput. **38**, 1876–1895 (2016)
32. Quispel, G.R.W., McLaren, D.I.: A new class of energy-preserving numerical integration methods. J. Phys. A **41**, 045206 (2008)
33. Wang, B., Wu, X.: A new high precision energy preserving integrator for system of oscillatory second-order differential equations. Phys. Lett. A **376**, 1185–1190 (2012)
34. Wu, X., Wang, B., Shi, W.: Efficient energy preserving integrators for oscillatory Hamiltonian systems. J. Comput. Phys. **235**, 587–605 (2013)
35. Hairer, E., Lubich, C.: Long-term analysis of a variational integrator for charged-particle dynamics in a strong magnetic field. Numeri. Math. **144**, 699–728 (2020)
36. Wang, B., Wu, X.: Global error bounds of one-stage extended RKN integrators for semilinear wave equations. Numer. Algor. **81**, 1203–1218 (2019)
37. Feng, K., Qin, M.: The Symplectic Methods for the Computation of Hamiltonian Equations, Numerical Methods for Partial Differential Equations. Springer, Berlin (2006), pp. 1–37
38. Gauckler, L.: Numerical long-time energy conservation for the nonlinear Schrödinger equation. IMA J. Numer. Anal. **37**, 2067–2090 (2017)
39. Gauckler, L., Lubich, C.: Nonlinear Schrödinger equations and their spectral semi-discretizations over long times. Found. Comput. Math. **10**, 141–169 (2010)
40. Gauckler, L., Lubich, C.: Splitting integrators for nonlinear Schrödinger equations over long times. Found. Comput. Math. **10**, 275–302 (2010)
41. Wang, B., Wu, X.: Long-time momentum and actions behaviour of energy-preserving methods for semi-linear wave equations via spatial spectral semi-discretisations. Adv. Comput. Math. **45**, 2921–2952 (2019)

Index

© The Author(s), under exclusive license to Springer Nature Singapore Pte Ltd. 2021
X. Wu, B. Wang, *Geometric Integrators for Differential Equations with Highly
Oscillatory Solutions*, https://doi.org/10.1007/978-981-16-0147-7

Printed in the United States
by Baker & Taylor Publisher Services